WALTER F. LAREDO

AEROSPACE DESIGN ENGINEER
& INVENTOR

ENGINEERING PROJECTS

FOR THE 21st CENTURY

AND FOR PRESENT TIMES

ENGINEERING PROJECTS FOR THE 21ST CENTURY
First Edition

Order this book online at www.trafford.com/07-1585
or email orders@trafford.com

Most Trafford titles are also available at major online book retailers.

Note for Librarians: A cataloguing record for this book is available from Library and Archives Canada at www.collectionscanada.ca/amicus/index-e.html

ISBN: 978-1-4251-3926-1

Laredo Publications
ISBN 0-9629148-2-7

We at Trafford believe that it is the responsibility of us all, as both individuals and corporations, to make choices that are environmentally and socially sound. You, in turn, are supporting this responsible conduct each time you purchase a Trafford book, or make use of our publishing services. To find out how you are helping, please visit www.trafford.com/responsiblepublishing.html

Our mission is to efficiently provide the world's finest, most comprehensive book publishing service, enabling every author to experience success. To find out how to publish your book, your way, and have it available worldwide, visit us online at www.trafford.com/10510

www.trafford.com

North America & international
toll-free: 1 888 232 4444 (USA & Canada)
phone: 250 383 6864 ♦ fax: 250 383 6804
email: info@trafford.com

The United Kingdom & Europe
phone: +44 (0)1865 722 113 ♦ local rate: 0845 230 9601
facsimile: +44 (0)1865 722 868 ♦ email: info.uk@trafford.com

10 9 8 7 6 5 4 3

Contents

ERRATUM

Captions of the following illustrations,
3, 4, 6, 8, 11, 16, 17, 18, 19, 21, 22 and 23
ends with the mistaken number "17,"
instead it should be "33"

1. INTRODUCTION

From the beginning, man has been an alert observer of his surroundings, always trying to understand the mysteries of nature. He became a toolmaker and created primitive weapons necessary for his survival in an unforgiving world.

Man kept evolving and developing his skills through the millennia, all the way to these modern times. Contemporary civilized man began creating great works of engineering. In the last half of the 20th century aerospace engineering was developed and complex projects and programs put man on the moon, and sent robots to Mars.

Today most of those involved in the development of Aerospace Engineering projects, are interested in either of two major objectives: a pragmatic goal, such as to put communications satellites in orbit or a second goal of satisfying human curiosity in exploring the universe.

Depicted in this book are actual engineering projects, whose development was inspired by the assumptions of what advanced civilizations would able to create, and perhaps to a lesser degree by the indirect stimulation of fantasy, such as novels by Jules Verne and movies like *Star Wars*.

Some pages and some Data for this book was compiled from the section of advanced engineering projects and factual futuristic information, extracted from the appendix section of my book *Atlantis Inspiration for the Future (Laredo Publications, 2005)*.

Should be noted that prior to starting the design and the development of each of the projects displayed in this book, extensive technical and scientific research was performed by the author, who also drew on design experience gained during years of engineering work. The result was the development of the real engineering projects shown here.

This is a factual book which strictly focuses on preliminary designs of authentic advanced engineering projects whose conceptualization and development conform to the natural laws of physics and latest state-of-the-art technologies.

If necessary, the highly detailed drawings from this book could be used as a reference to draw the basic layouts for building the real testing prototypes.

2.   FROM STONE AGE TOOLS TO BRONZE AGE MONUMENTS, TO PRESENT-DAY CONSTRUCTION WORKS

Works of simple man

A set of eight pictures of various works of man follows this page.

Figures 1, 2 and 3 of the same set show rough mountain roads made from dirt and built by using simple hand tools.

Ancient methods of construction using extensive manual labor are still used today in developing countries such as Bolivia.

Figure 1. Former road to Chapare, Bolivia; constructed by manual labor.

Figure 2. Highly sloped rough roads built by manual labor and in some places by the use of dynamite; Andes Mountains, Bolivia.

Figure 3. Dangerous road without reinforced edges at Chapare, Bolivia.

Figure 4. Primitive fabrication of mud bricks, still in practice today

Figure 5. A primitive plow, as used at the dawn of agriculture

TEMPLE OF POSEIDON

HORSES

INTERNAL COLUMNS

STATUES FROM OTHER GODS

MONUMENT OF GOD POSEIDON RIDING HIS CHARIOT PULLED BY SIX STEEDS

MAIN ENTRANCES

NEREIDS RIDING ON DOLPHINS, ALL THEM SCULPED ? FROM MARBLE

FOUNTAIN REPRESENTING THE SEAS

PLAN VIEW

W. LAREDO

Figure 6. Ancient structure, assumed to be one of the temples dedicated to Poseidon, as depict in above picture.

Figure 7. Ancient structures, interior view of the legendary temple of Poseidon, as described by Plato.

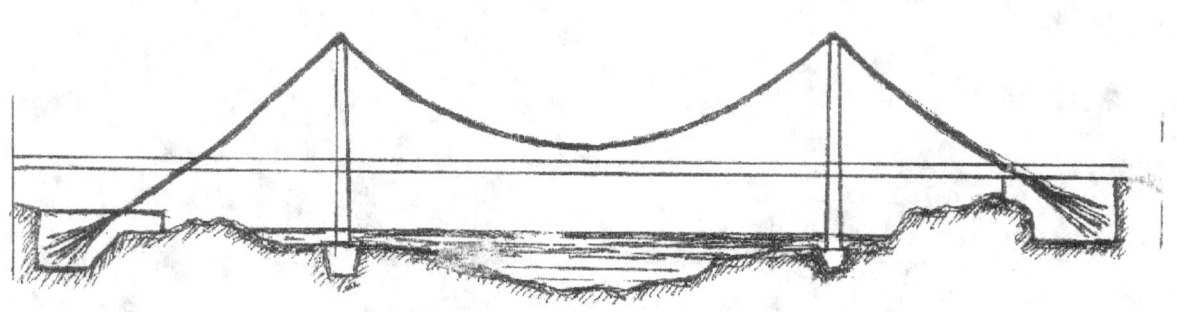

Figure 8. Contemporary engineering work, Golden Gate Bridge,
San Francisco, California

3. GADGETS, SPECIAL OBJECTS AND GREAT WONDERS OF ENGINEERING THAT MAY BECOME OF FREQUENT USE IN THE 21st CENTURY, ARE SHOWN IN FIGURES AND PICTURES OF THE FOLLOWING PAGES

Figures 9, 10, 11, 12, 13, and 14 follows this page

Figure 9. View of a futuristic city, designed and build as a unit

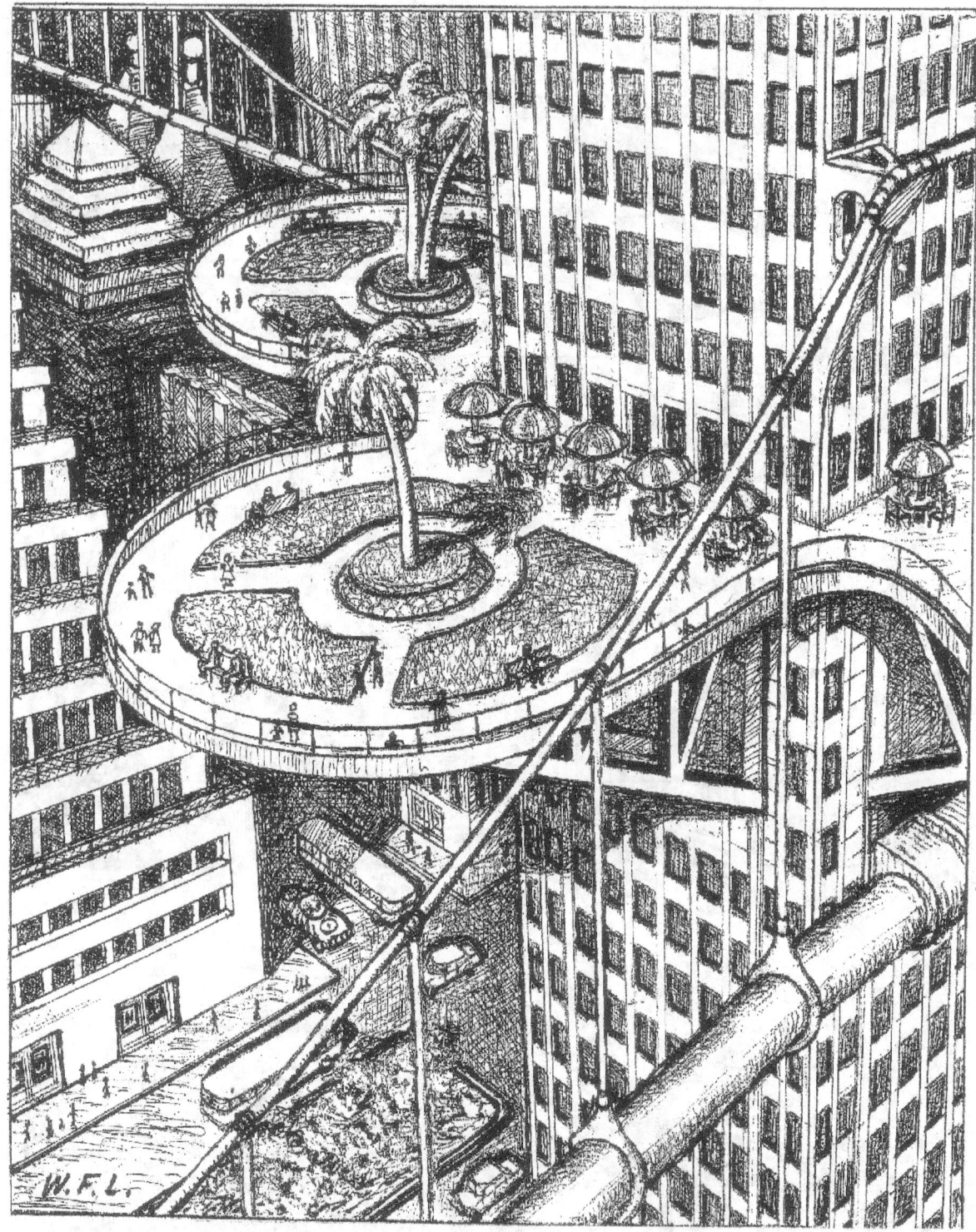

Figure 10. A 21ˢᵗ Century restaurant with hanging gardens, at lower right of picture is shown the tube of a hanging urban train.

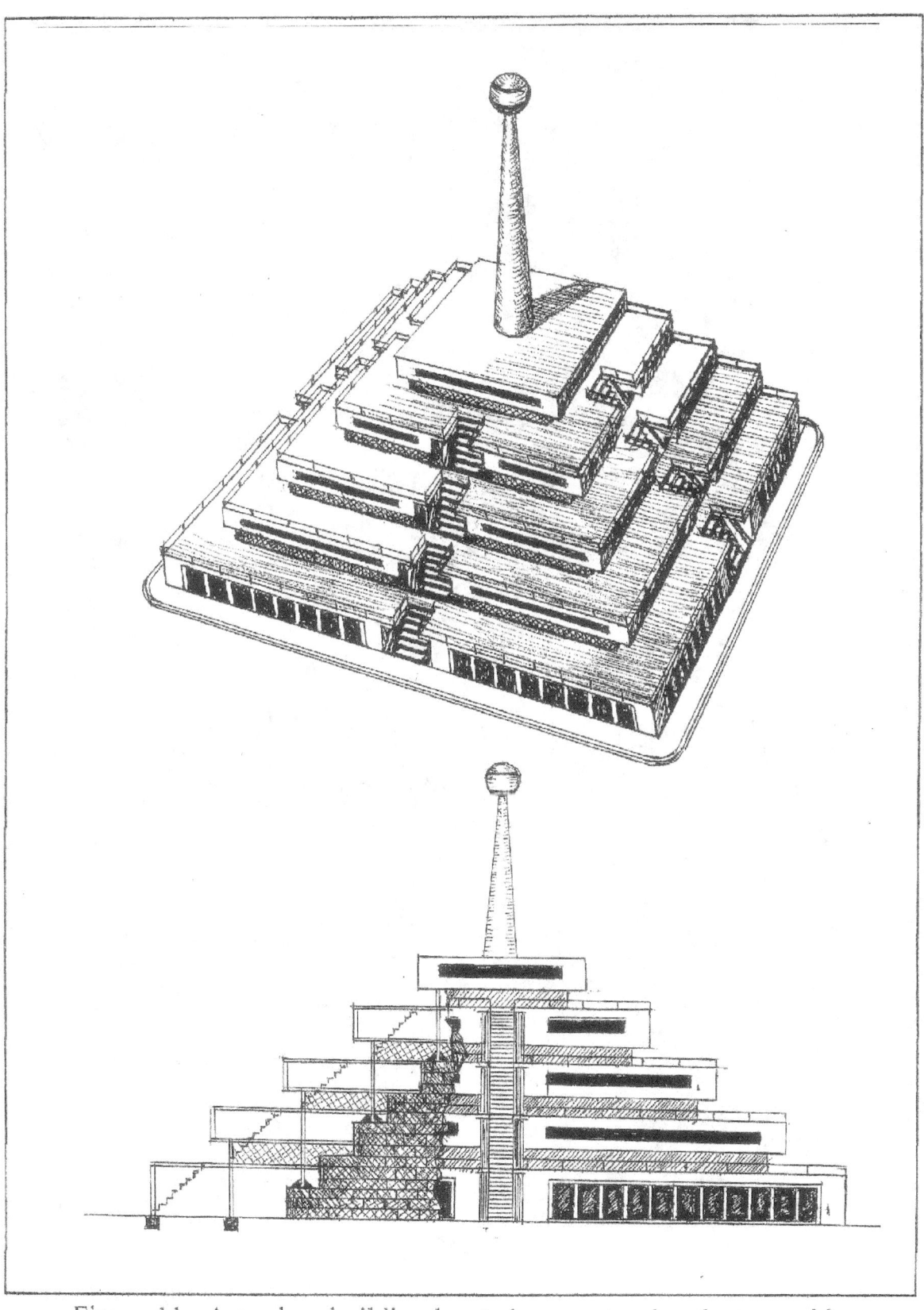

Figure 11. A modern building located on a natural rocky pyramid

Figure 12. A futuristic space base located on a high altitude plateau.

Figure 13. Toys for the rich sportsmen of the future. In the picture, a boy helps his girlfriend to climb onto his mechanical flying horse.

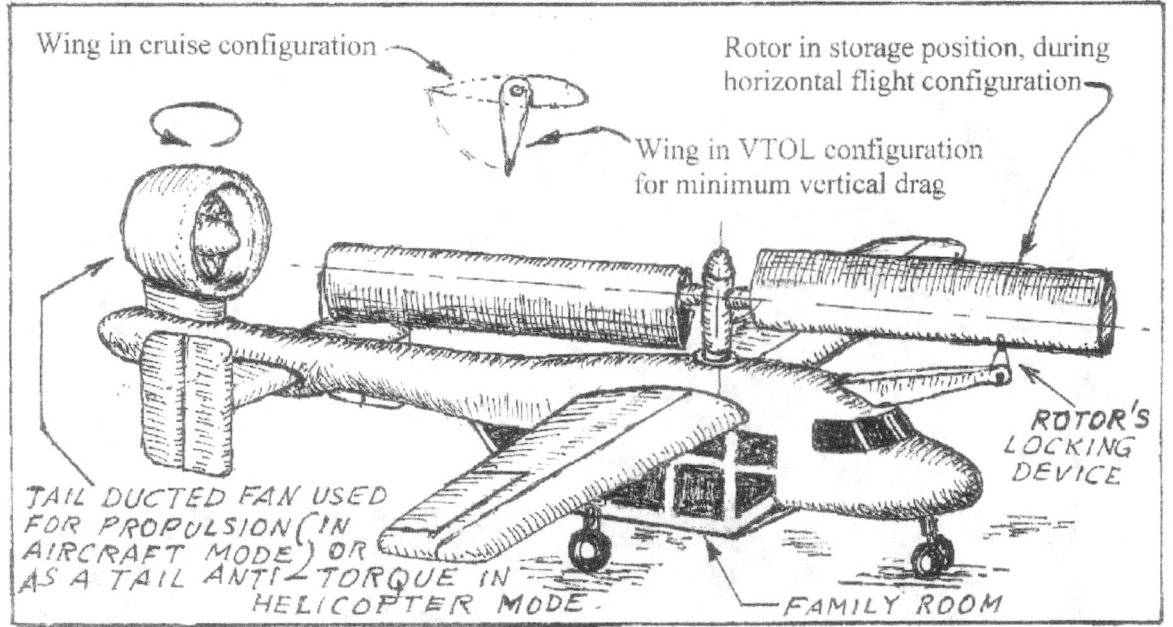

Wing in cruise configuration

Rotor in storage position, during horizontal flight configuration

Wing in VTOL configuration for minimum vertical drag

ROTOR'S LOCKING DEVICE

TAIL DUCTED FAN USED FOR PROPULSION (IN AIRCRAFT MODE) OR AS A TAIL ANTI-TORQUE IN HELICOPTER MODE.

FAMILY ROOM

W.F. LAREDO

Figure 14. Futuristic Flying Camping Vehicle (VTOL). In the picture a toddler feels safe behind the unbreakable glass window

1. BIOENGINEERING CHRONICLES, AS MAY BE PROGNOSTICATED FROM THE NEAR FUTURE, MOST LIKELY FROM THE SECOND DECADE OF THE 21st CENTURY

Design Engineer: W. F. Laredo

Chronicle No. 1: Design of an artificial heart, for its permanent implantation in the
human body

Chronicle No. 2: Medical students practicing surgery

Chronicle No. 3:   Bionic eyes for the blind

List of bioengineering plates :

Mechanical Heart, Plates 1 through 11

Bionic Eye, Plates 12 & 13

CHRONICLE No. 1   (Revised)

BIOENGINEERING FOR THE 21st CENTURY

DESIGN OF ARTIFICIAL HEART FOR PERMANENT IMPLANTATION IN THE HUMAN BODY

(Project presented in April 17, 1998, by Walter F. Laredo to NASA for its review)

A medical documentary was displayed in the screen of a university auditorium, In the documentary, a professor explained his students the different steps to follow in order to develop the artificial heart, a heart designed for permanent implantation in a patient's thorax without being rejected.

On the film the professor advised his students that before starting with the design of an artificial heart, they should get some previous training by designing a simple double pump to pump water, which consisted of a simple box divided into two compartments by a central partition. To each face of this central partition was bonded an inflatable rubber diaphragm membrane, which every time it got inflated, pushed out the surrounding water through a hole in the wall of the same box. Outside this hole was connected to a pipe, mimicking the aorta - the main artery in the human body-. The hydraulic fluid inflated the diaphragms not simultaneously but sequentially.

Hydraulic fluid from a remote hydraulic power system was send by means of long hoses and injected inside the back space of the diaphragms.

By building this crude water pump, the students acquired enough experience to start designing the actual mechanical heart. As our visit to the school finished, our host gave me the following brochure of medical chronicles, which explained with great detail the design and the development of this artificial heart.

BROCHURE CONCERNING THE DESIGN OF AN ARTIFICIAL HEART (Revised)

This heart is just a pump, actuated by a remote hydraulic power system. The heart is light enough to be held in place in a similar way as the living heart is, supported from the top by the same group of main veins and arteries that supported the old discarded heart. The pericardial membrane supports the heart outer surface.

The power package is a separate unit that supplies hydraulic power to the mechanical heart and is located at another place of the body. It is constituted by four components: as the hydraulic pump, the electric motor, the internal battery, and the electronic control. The power package is not installed in the thorax as the artificial heart is, but inside the lower abdomen, tied to bone extensions of the hip bone, which were artificially stimulated to grow after a pre-operation was performed, months before the main open heart operation. This power system provides hydraulic power to the artificial heart through a long, one centimeter in diameter; flexible plastic tubular case. Inside this case or cover, run another two flexible hydraulic lines of smaller diameter, each of which actuates its corresponding flexible membrane in the artificial heart. The tubular case goes through a special fitting and grommet in the person's muscular diaphragm.

During its process of development, this artificial heart was redesigned four times, mainly because it's complex shape and complicated cross-sections. It went through four

design metamorphosis, each time becoming smaller and more efficient, more compact and more refined than the time before, until finally became small enough to fit inside the pericardial cavity. The heart surfaces were treated with a soft coat of anti rejection material. Further refinements were performed without affecting its blood pumping efficiency.

At that moment, Dr. Quetzalco entered the auditorium that was full of people, greeted us and said, "The human body is so compact, from the top of the head to the toes, that it is difficult to implant large artificial organs inside. The same concept applies to the pericardial cavity where there is no room for a large, artificial heart."

The documentary kept displaying the necessary sequential steps in order for the students to follow the development of this heart.

The professor from the film also acted as the Chief Project Engineer for the artificial heart development program. He went through an extended presentation. A transcript of the documentary was made available to me, and it is as follows:

Transcript
DESIGN AND DEVELOPMENT OF THE ARTIFICIAL HEART
This heart was designed for an approximate blood pumping capacity of five liters per minute, and an approximate rate of 70 beats per minute.

DESIGN OF THE HEART COMPONENTS
UPPER STRUCTURE. The upper structure of the mechanical heart is a shell made from titanium, designed with the same complicated features of the living heart. Short pieces of tubes also made from titanium protruded out from the top surface of this mechanical heart, to be coupled with the patient's major blood vessels, it is after the patient's sick heart was removed and disposed. Safe connectors was developed for quick connection between the protruding short tubes of the artificial heart and the major blood vessels, which also were coated with a special substance in order to avoid body rejection. Two atriums constitute the upper part of the titanium heart. In the living heart, the atriums are flexible and act as small pumps. In contrast, for this mechanical heart they are rigid and are used only as small blood reservoirs.

HEART VALVES. The valves of the mechanical heart are designed for a life span of 30 years, equivalent to a billion beats of continuous operation without failure, a real challenge. In designing the valves for an artificial heart are encountered similar engineering problems as in the valve design system for most ordinary machines and equipment as compressors, pumps, and engines. For example, one way to increase the power of a gasoline engine, without increasing its size, is by increasing the size of the valves, or by increasing the number of valves in order to let pass more air for combustion.

In the development of a mechanical heart, if the number of valves is doubled without increasing the size of the heart, the volume of blood flowing through the heart would increase. But this kind of design would turn the laminar blood flow into a turbulent blood flow, also is unfavorable the increased number of internal surfaces and internal cavities. When blood flows over irregular surfaces, it creates internal hydrodynamics problems, such as turbulence and at some tight locations a reversed blood flow, causing oxygen deprivation and mechanical damage to the blood cells and inducing the formation of blood clots.

DESIGN OF AN INDIVIDUAL HEART VALVE.   The simplest, most reliable and efficient valve is the familiar one-way ball valve, used for many years in patients requiring a single valve replacement. A natural heart, using four, of these mechanical valves would become so massive to be practical. Unfortunately the same idea applies to an artificial heart with four ball-type valves which will occupy more space, beyond the size of the patient's pericardial cavity, or will require the removal of part of a lung to make room to install this massive heart.

A flexible flap type of design would be ideal for an artificial valve, similar to the valves from a living heart, but artificial valves made from nonliving material are not self-repairing like the valves of a living heart. Hence its flexible material would be subject to bending millions of times; it will not last and eventually will fail by fatigue. However, this design would be acceptable for a heart that will last only few weeks or months, ideal for patients waiting for a living heart transplant. The illustrations of the mechanical heart show here are for a heart where all four valves are of the flap type, but of special design, for a heart that should last for thirty years. In the illustrations could be seeing the gradual development of these valves, from concept to final.

INFERIOR STRUCTURE
The inferior structure is made from titanium and includes two rigid ventricles.

PUMPING DIAPHRAGMS.   The central partition that separates the cavities of the two ventricles also supports on each of its faces an inflatable flexible membrane called diaphragm, which is supported by a circular frame, attached to the partition. Also on each side of the partition and behind the membranes there is a recessed space, where hydraulic fluid is injected.

Hydraulic fluid, under pressure, coming from a remote electric pump, is injected inside those shallow recessed spaces forcing the membranes to expand outwards as inflated balloons, compressing and pushing out the blood contained in between the outside face of the diaphragm and the inside of the rigid ventricle. The blood pressure will open both valves and blood will flow into the aorta and the pulmonary arteries. Pumping will not happen simultaneously on both ventricles, but sequentially, controlled by a hydraulic sequence valve, located inside the remote power unit. The whole process is controlled electronically.

Temporary artificial hearts for short life spans don't need to have a sophisticated valve system as previously explained but just a much simpler design, where the pumping membranes are made of a special rubber. For this simplest heart, during each cycle of expansion, the rubber membranes stretch radial and tangentially simultaneously. After few months, because of the material limited life it would fail as expected by fatigue.

In the more sophisticated heart for permanent implantation, the pumping membranes are of special long lasting design, where material fatigue is minimized by decreasing the tensile deflection of the membrane in all directions; the secret is to use only minute bending deflections from each minute element of the membrane. These membranes are built into plies, each from a different material; the inner ply is considered as the main structure of the system and never should touch blood and it is built from a special, durable, metallic alloy, and is called tireless devil, because of its great resistance to fatigue. The outer ply with smooth surface that touches blood is coated with a special material that doesn't allow the formation of blood clots and doesn't allow blood to adhere it. This smooth surface of the pumping diaphragms doesn't mechanically damage blood cells.

POWER SUPPLY FOR THE ARTIFICIAL HEART

The internal power system permanently implanted inside the lower abdominal cavity of the patient; includes batteries, an electric hydraulic pump, a sequence valve, and electronic controls. All these components are enclosed inside a container the size of a grapefruit. As mentioned before the power pack and the artificial heart are connected by a coated flexible plastic hose, two feet long which interior holds two hoses of smaller diameter. This hose's system is routed in between organs like a snake, passing through a special grommet implanted on the diaphragm, at a place where it doesn't interfere with the movement of the muscle. Each of the two tubes located inside the larger hose or case supplies hydraulic power to each of the inflating mechanical diaphragms located inside the ventricles of the mechanical heart.

The internal batteries from the power unit are constantly recharged by the external batteries carried in a portable pack, which is carried externally by being strapped to the patient's body with a harness. The portable pack weighs about three kilograms, which battery has a short cord with a connector at its end, which can be plugged into an electric bony socket permanently implanted into the skin of the patient's abdomen. The electric bony socket slightly protrudes from the abdomen skin, similar as when horns protruded from an animal head skin.

When the patient wants to take a shower or go swimming, he could unplug his pack from the electric socket in his abdomen skin, which never should remain unplugged for more than four hours, the maximum time that the internal batteries inside the body can remain charged and still keep the patient alive.

Four months before the principal operation, for the implantation of the artificial heart was performed, the patient was subjected to a previous operation, but it was in the lower abdomen, where four small scaffolds were installed on the iliac bones which function was to stimulate the growing of bony supports directly from the pelvis iliac bones. Protrusion looking as rings or small donuts, through which holes will pass the artificial ligaments supporting the weight of the power supply.

Growing those bony protrusions was a slow process, which began by installing first a scaffold for each of them. The scaffolds established the shape of what will become the bony protrusion. The scaffolds were made from fibers of a biodegradable polymer implanted on the edge of the hipbone, with time the scaffolds will be reabsorbed leaving in its place new bone as extension from the hipbone itself.

Four months later took place the principal operation, the implantation of the heart. Two different teams of surgeons operated simultaneously on the patient. One team was in charge of the artificial heart implant inside the thorax, and the other team was in charge of the implantation of the internal power system inside the lower abdomen, where it was secured to those artificially grown ringed bony protrusions by means of artificial ligaments. The ligaments were implanted in such a way that neither, the ligaments nor the power system interfered with the other organs. The surgeons inserted a specially designed seal, a grommet at the location where the long hose system went through the diaphragm muscle. This is the hose system that provides hydraulic power to the heart. Finally, the abdomen was closed and sutured.

ANOTHER ALTERNATIVE, THE HYBRID HEART

The last part of this chronicle describes another type of heart design, a hybrid, half-living and half-artificial. The lower part of this heart with its ventricles corresponded to the living portion. Either from extracted or from modified living heart cells living ventricles were stimulated to grow cells that months before were temporarily implanted

in the patient lower abdomen. There those grew into fully developed ventricles. Months later, at the time of the principal operation the ventricles were fully growth and ready to be removed from there to be implanted at the proper place, inside the patient's thorax.

The two ventricles including its respective natural coronary blood vessels constituted the living portion of this heart. Because it was made from actual heart tissue, its muscles have the same three functions as in any living heart which are 1) An integral contracting motor with a pump, 2) an electric power source which adjust by itself. 3) Self-repairing as any living organ.

A pacemaker implanted in the patient also contributed to control the functioning of the ventricles. This completes the description of the living portion of the heart.

Next will be described, the upper part of the heart which is artificial, it includes the valves, two small atriums and the connection nipples. The nipples are used to be coupled with the large living vessels. In addition there were small-protruded nipples to be coupled with the coronary arteries in order to bypass blood to the living ventricles. The material of the artificial part of this heart is either a special plastic or titanium, which surfaces should be treated with a special coating in order not to be rejected by the body. The internal surfaces also should be covered with a lining of a special inert coating. On this lining, endothelial cells are stimulated to grow, cells extracted from the same patient which would multiply and attaching by themselves there. This inner lining made from protein deposits covers all the internal surfaces of the artificial part of the heart to avoid direct blood contact, hence avoiding blood clothing.

A natural heart, could be repair by transplanting in it, some parts made by using tissue engineering. Here, undifferentiated cells after turned into heart cells, are growth in laboratories, cells that are used to create heart ventricles and heart valves. Since these organs were produced with cells extracted from the same patient; there was no risk of rejection.

When scientists tried to clone whole hearts, but because those always grew with defective valves, it was common practice to discard the upper portion of this living heart, to be replaced by an artificial part, as described before.

CHRONICLE No. 2   (Revised)

MEDICAL STUDENTS OPERATING ON PLASTIC DUMMIES
By Dr. Nubis, surgeon from the future

Long ago, our students practiced surgery by dissecting and performing mock operations on cadavers, a nauseating process because of foul smells. Today they practice the same operations on full-size dummy patients. A mannequin manufacturer got the idea of mass-producing those full-size dolls, made from various soft plastic materials, each mimicking different human tissue. The anatomical complexity of the dummies simulated closely to a real human. The dummies have body systems with similar anatomy to the real human body, as the skeletal, the muscular, the circulatory, the digestive, the nervous, etc. The internal organs of the dummies were made from different kinds of materials, and were perfect copies from those in the human body.

Red liquid circulated inside the mannequin blood vessels, mimicking the same features and same ability to coagulate as the real blood. It circulated by means of a small electric pump. During mock surgical operations, the dummies bled like a real person

when its veins were cut then cauterized. These dummies even could get a transfusion of fake blood.

Some models resembled humans afflicted by different kinds of diseases and different health problems. For medical students was less intimidating to practice simulated surgery on dummies, than performing real operations on real people. Working with dummies was much cleaner than with cadavers, after so much use dummies were discarded.

Concurrently with the courses of mock surgery was a much less scientific course, however a very important one, it concerned with suturing different kinds of organs, a pragmatic technical course to improve surgeon's manual dexterity.

Do you know that in suturing, some surgeons were more skilled than tailors could be? believe or not a surgeon could repair a shirt or a trouser better than a tailor. The most advanced part of this course was related with microsurgery suturing performed with the help of a microscope and by using remote controlled micro-instruments. For mock surgical operations for some particular part of the body partial anatomical models were available, as the head, the limbs, the trunk and so on. Models of the thorax are very popular; our students often practice open-heart operations on them. Usually after models got too damaged after being operated several times, were thrown in trash boxes to be recycled. Young doctors before performing their first real surgical operation on real people, already had great experience after performing hundreds of operations on those dummies.

CHRONICLE No. 3   (Revised)

BIONIC EYES FOR THE BLIND

This patient used to be totally blind," said the doctor. "He suffered from dead, detached retinas on both eyes. We operated on his left eye several times to provide him with a bionic eye. Although the patient only sees in black and white, this eye is a marvel of microelectronic engineering.

"Inside the outer structure after the choroid of the living eyeball was emptied was implanted a small, spherical camera with a lens in the front. Back inside this bionic eyeball or camera there is a miniature cap, which is the artificial retina made up of 10,000 special micro-photoreceptors. The artificial retina is divided into two semicircles, left and right. Each corresponds to half of the retina and is made up of 5,000 special micro-photoreceptors; most of them concentrated in the artificial fovea, where more acute vision is required. Each photoreceptor is connected to a long conductor, a gold wire, much thinner than a human hair, which function as an artificial nerve fiber. Wires coming out from the bionic eyeball are arranged into two bundles of 5,000 each. The forward end of each bundle is connected to the corresponding half of the artificial retina, and the two bundles together constitutes the artificial optic nerve."

The doctor said that the patient's living eyeball, including its cornea, was not removed, only its contents. The sclera and the choroid were modified to support inside that bionic camera, a permanently implanted bionic eyeball. The aqueous humor, the crystalline lens, and the vitreous body were removed to make room for the spherical bionic eye. The choroid was maintained a life by its highly vascular blood vessels, it remained attach externally to the existing muscles that provided eye movements.

Took years to develop special materials to be used in the construction of the bionic eye, including its special external coating, so that the bionic eyeball would not to be rejected by the living choroid. The two flexible bundles of wires began behind the eyeball, deep inside the eye socket. The bundles or bionic optic nerves were flexible enough to bend allowing eye movements. The bionic optic nerve did not follow the usual internal routing as the living optic nerve, passing by the optic chiasma. The bionic optic nerve made by a bundle of thin gold fibers was first routed through a hole drilled in the lower left frontal bone along the left side of the eye socket, then outside the skull under the scalp. To avoid infection by external exposure, the external bionic optic nerve, was covered along its entire length with a long bony ridge that were, artificially stimulated to grow as an extension of the skull bone where it grew as a ridge under the scalp. To do this, the patient's scalp was opened and the bionic optic nerve laid touching the left side of the skull and extended all the way back to the occipital area of the head.

He continued explaining that each of the two bundles of the bionic optic nerve, ended connected into a contoured, flexible plate installed under the skull's occipital bone, each plate supported five thousand minute probes, to which were connected the 5000 gold fibers. The material from those flexible plastic plates was able to breathe, because it was made from a material full of tiny breathing holes, allowing air and gas molecules to pass through the plastic plates. The plastic plates looked like miniature brushes with tiny metal bristles. Plates connected to the corresponding bionic optic nerve. To bring these plates inside the occipital area of the skull, and in direct contact with the brain, a rectangular portion of the occipital bone, with part of the arachnoid, the dura matter and the pia matter membranes was removed and saved to be restored in its proper place near the end of the operation. Each plastic plate was flexible enough to be extended and laid like a miniature carpet over a mapped, visual sensory area of the brain's cerebral cortex.

Rectangular plastic plates with rounded corners specially molded for each patient to match exactly the convolution of the brain surfaces for that particular patient. The plates were molded quickly in the laboratory located in the next room; it took no more than few minutes for the technician, to prepare them. The plates were laid by pressing smoothly against the brain in order that the microscopic pin probes would pierce the corresponding surfaces of the brain convolutions, penetrating the outer layer of the primary visual cortex. All this was done without damaging tiny veins and arteries. The illustration drawings shows that the plastic plates have a pin guidance device that doesn't allow pins to buckle at the time of their insertion on the brain. The microscopic probes were so small that their penetration produced negligible damage to the brain cells. For narrow areas of the brain with small arteries and veins, the plate inner surface was locally smooth and had no pins.

The patient was able to move his bionic eye as a living eye, because it was implanted inside the hollow living eyeball, the same muscles for eye motion remained attach to the original living eye structure. After a series of operations, this patient went through many months of rehabilitation. A lengthy process that allowed his brain to reprogram itself, learning to decode and interpret what it saw with the new eye, similar to what happens with two months old babies that are trying to see the world. This kind of operations was performed only on people 25 years old or younger, because their brains was still able to reprogram by itself. The bionic eye rigid lens was designed to focus mostly on objects that are one-meter distance. After healing, those former patients carried two monocles in their pocket, one used for reading and the other for long distance. The former patients were able to read only large print. The quality of the restored vision was not perfect, but acceptable.

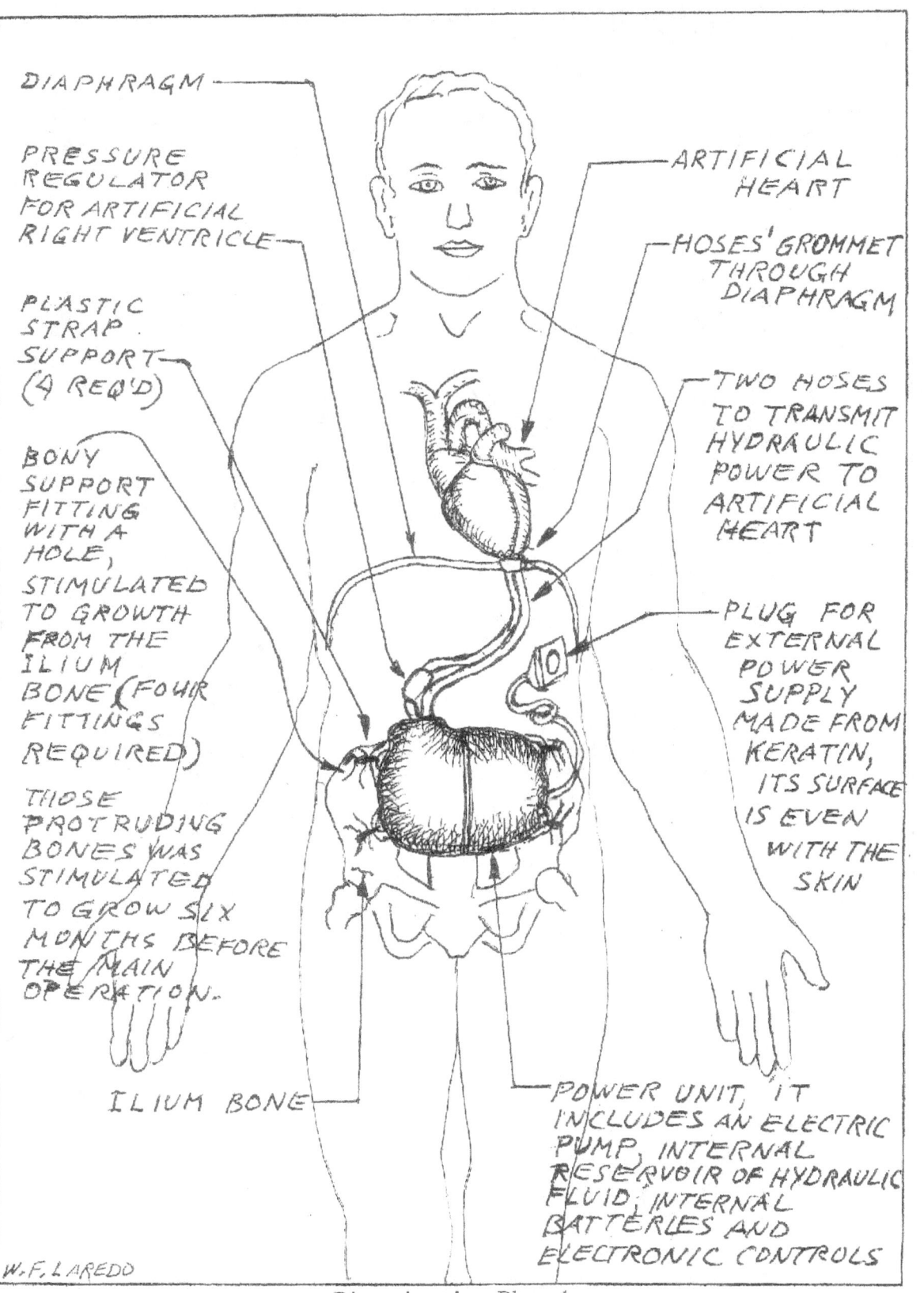

DIAPHRAGM

PRESSURE
REGULATOR
FOR ARTIFICIAL
RIGHT VENTRICLE

PLASTIC
STRAP
SUPPORT
(4 REQ'D)

BONY
SUPPORT
FITTING
WITH A
HOLE,
STIMULATED
TO GROWTH
FROM THE
ILIUM
BONE (FOUR
FITTINGS
REQUIRED)

THOSE
PROTRUDING
BONES WAS
STIMULATED
TO GROW SIX
MONTHS BEFORE
THE MAIN
OPERATION.

ILIUM BONE

ARTIFICIAL
HEART

HOSES' GROMMET
THROUGH
DIAPHRAGM

TWO HOSES
TO TRANSMIT
HYDRAULIC
POWER TO
ARTIFICIAL
HEART

PLUG FOR
EXTERNAL
POWER
SUPPLY
MADE FROM
KERATIN,
ITS SURFACE
IS EVEN
WITH THE
SKIN

POWER UNIT, IT
INCLUDES AN ELECTRIC
PUMP, INTERNAL
RESERVOIR OF HYDRAULIC
FLUID, INTERNAL
BATTERIES AND
ELECTRONIC CONTROLS

W. F. LAREDO

Bioengineering Plate 1

NATURAL HEARTH
FOUR-PHASES PUMPING CYCLE

DIASTOLIC PHASE, THE EARTH RELAX	ATRIAL SYSTOLE WITH VENTRICLE STILL IN DIASTOLE	VENTRICLES CONTRACT, VENTRICULAR SYSTOLE	ATRIA EXPANDS AND FILLS WITH BLOOD (CYCLE REPEATS)

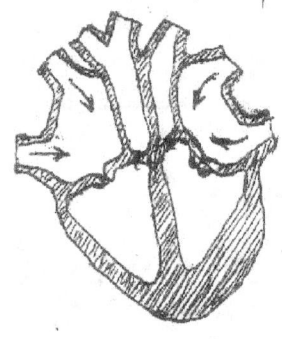 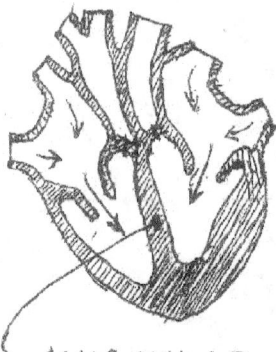 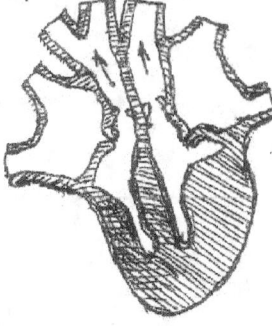 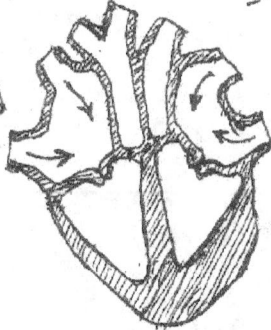

MUSCULAR INTERVENTRICULAR SEPTUM

MECHANICAL HEART
TWO-PHASE PUMPING CYCLE

INFLATABLE MEMBRANES

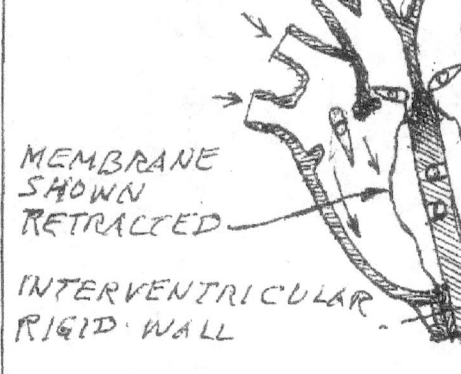

MEMBRANE SHOWN RETRACTED

INTERVENTRICULAR RIGID WALL

MEMBRANE SHOWN EXPANDED

DIASTOLIC PHASE THE CHAMBER INSIDE THE RIGID VENTRICLES EXPANDS AND FILLS WITH BLOOD.

VENTRICULAR SYSTOLE, THE BLOOD IN THE CHAMBERS IS COMPRESSED BY THE INFLATABLE MEMBRANES AND ENTERS THE AORTA AND THE PULMONARY ARTERY.

PHASES OF OPERATION
COMPARISON BETWEEN THE NATURAL HEART AND THE MECHANICAL HEART
W.F. LAREDO

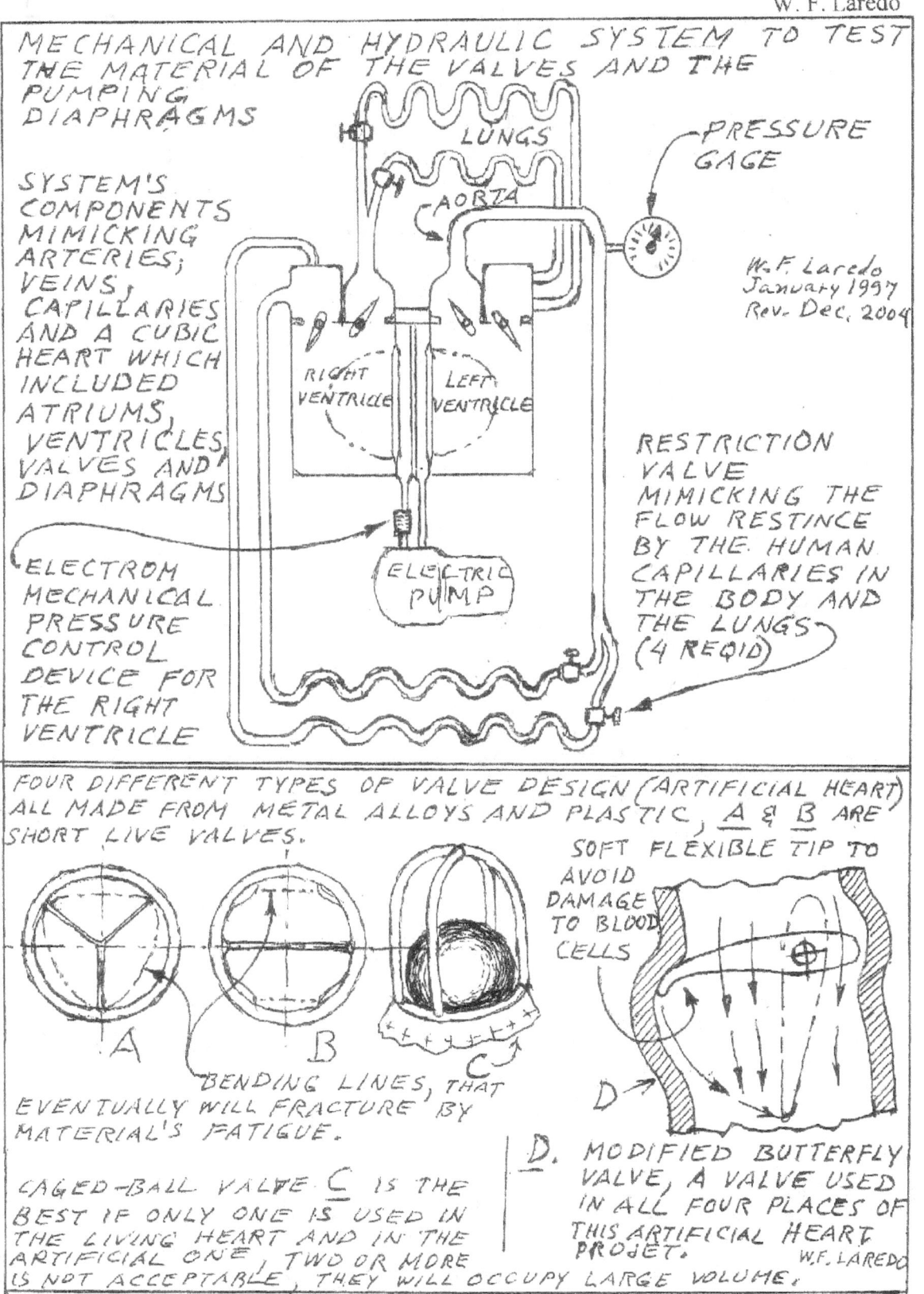

MECHANICAL AND HYDRAULIC SYSTEM TO TEST THE MATERIAL OF THE VALVES AND THE PUMPING DIAPHRAGMS

SYSTEM'S COMPONENTS MIMICKING ARTERIES; VEINS, CAPILLARIES AND A CUBIC HEART WHICH INCLUDED ATRIUMS, VENTRICLES, VALVES AND DIAPHRAGMS

LUNGS

AORTA

PRESSURE GAGE

W. F. Laredo
January 1997
Rev. Dec. 2004

RIGHT VENTRICLE

LEFT VENTRICLE

RESTRICTION VALVE MIMICKING THE FLOW RESTINCE BY THE HUMAN CAPILLARIES IN THE BODY AND THE LUNGS (4 REQID)

ELECTROM MECHANICAL PRESSURE CONTROL DEVICE FOR THE RIGHT VENTRICLE

ELECTRIC PUMP

FOUR DIFFERENT TYPES OF VALVE DESIGN (ARTIFICIAL HEART) ALL MADE FROM METAL ALLOYS AND PLASTIC, A & B ARE SHORT LIVE VALVES.

SOFT FLEXIBLE TIP TO AVOID DAMAGE TO BLOOD CELLS

A

B

C

D

BENDING LINES, THAT EVENTUALLY WILL FRACTURE BY MATERIAL'S FATIGUE.

CAGED-BALL VALVE C IS THE BEST IF ONLY ONE IS USED IN THE LIVING HEART AND IN THE ARTIFICIAL ONE, TWO OR MORE IS NOT ACCEPTABLE, THEY WILL OCCUPY LARGE VOLUME.

D. MODIFIED BUTTERFLY VALVE, A VALVE USED IN ALL FOUR PLACES OF THIS ARTIFICIAL HEART PROJET.
W.F. LAREDO

Bioengineering Plate 3

ARTIFICIAL HEART MADE FROM TITANIUM WITH
RIGID EXTERNAL WALLS, IT IS IN CONTRAST TO THE
LIVING HEART WHICH HAVE FLEXIBLE WALLS.
THIS MECHANICAL HEART USES ITS ATRIUMS AS
BLOOD RESERVOIRS AND THESE ATRIUMS ARE
SMALLER THAN IN THE NATURAL HEART.
ALL EXTERNAL SURFACES ARE COATED WITH AN
ANTI-REJECTION SUBSTANCE WHILE THE INTERNAL
SURFACES IN CONTACT WITH THE BLOOD ARE
COATED WITH AN ANTI-CLOTHING SUBSTANCE.

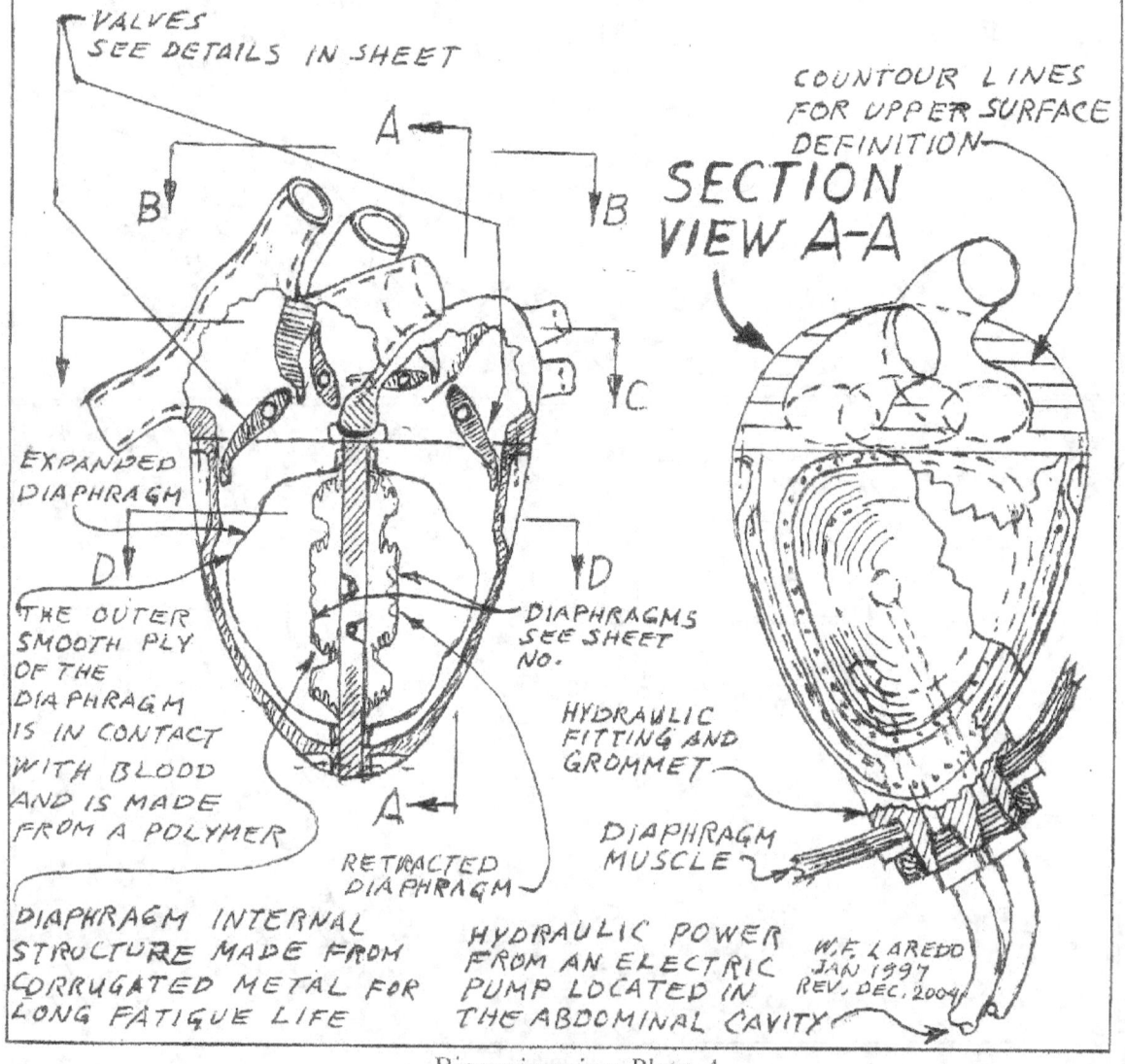

VALVES
SEE DETAILS IN SHEET

COUNTOUR LINES
FOR UPPER SURFACE
DEFINITION

SECTION
VIEW A-A

EXPANDED
DIAPHRAGM

THE OUTER
SMOOTH PLY
OF THE
DIAPHRAGM
IS IN CONTACT
WITH BLOOD
AND IS MADE
FROM A POLYMER

DIAPHRAGMS
SEE SHEET
NO.

HYDRAULIC
FITTING AND
GROMMET

RETRACTED
DIAPHRAGM

DIAPHRAGM
MUSCLE

DIAPHRAGM INTERNAL
STRUCTURE MADE FROM
CORRUGATED METAL FOR
LONG FATIGUE LIFE

HYDRAULIC POWER
FROM AN ELECTRIC
PUMP LOCATED IN
THE ABDOMINAL CAVITY

W.F. LAREDO
JAN 1997
REV. DEC. 2004

Bioengineering Plate 4

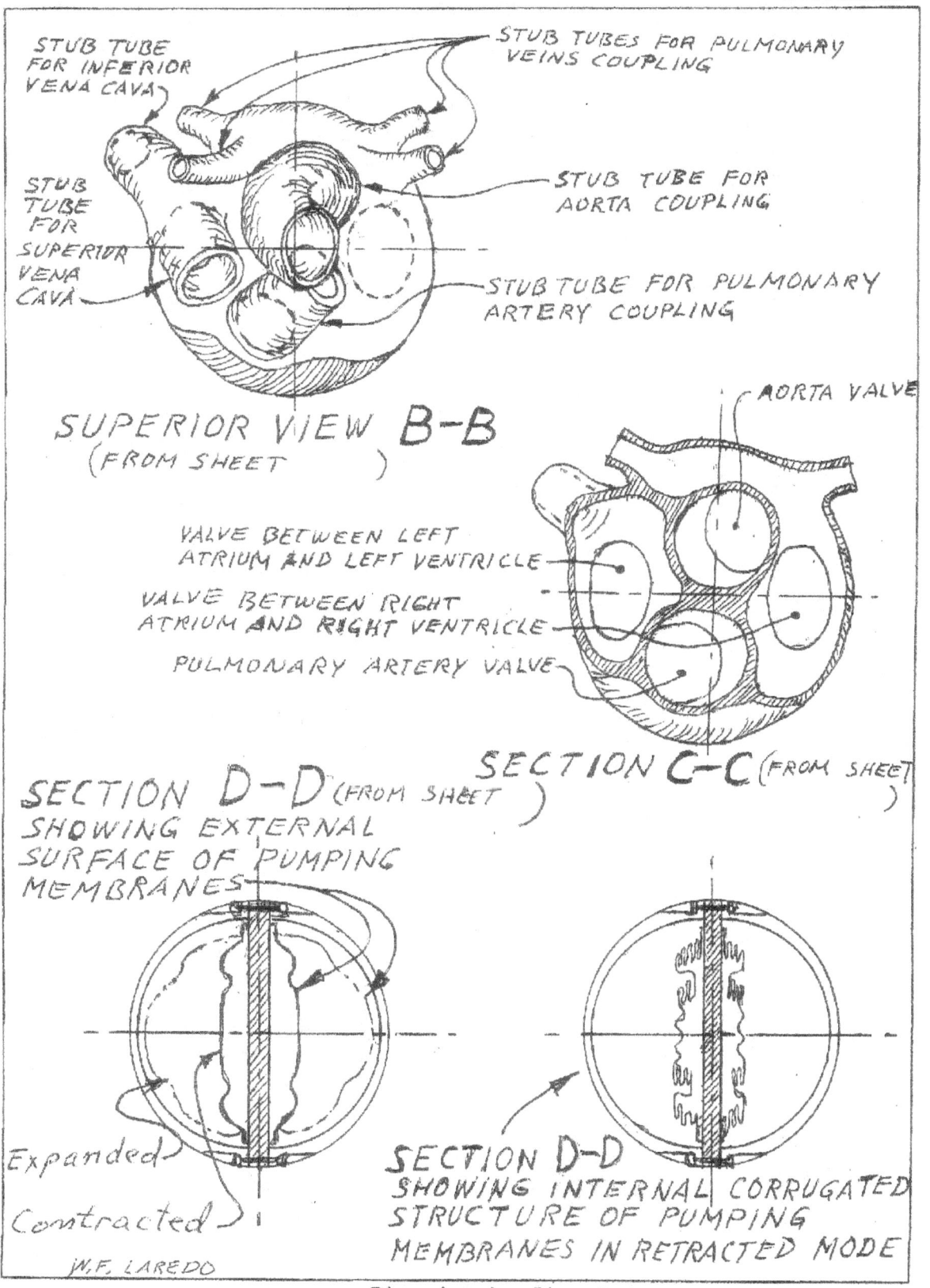

STUB TUBE FOR INFERIOR VENA CAVA

STUB TUBES FOR PULMONARY VEINS COUPLING

STUB TUBE FOR SUPERIOR VENA CAVA

STUB TUBE FOR AORTA COUPLING

STUB TUBE FOR PULMONARY ARTERY COUPLING

SUPERIOR VIEW B-B
(FROM SHEET)

AORTA VALVE

VALVE BETWEEN LEFT ATRIUM AND LEFT VENTRICLE

VALVE BETWEEN RIGHT ATRIUM AND RIGHT VENTRICLE

PULMONARY ARTERY VALVE

SECTION C-C (FROM SHEET)

SECTION D-D (FROM SHEET)
SHOWING EXTERNAL SURFACE OF PUMPING MEMBRANES

Expanded

Contracted

W.F. LAREDO

SECTION D-D
SHOWING INTERNAL CORRUGATED STRUCTURE OF PUMPING MEMBRANES IN RETRACTED MODE

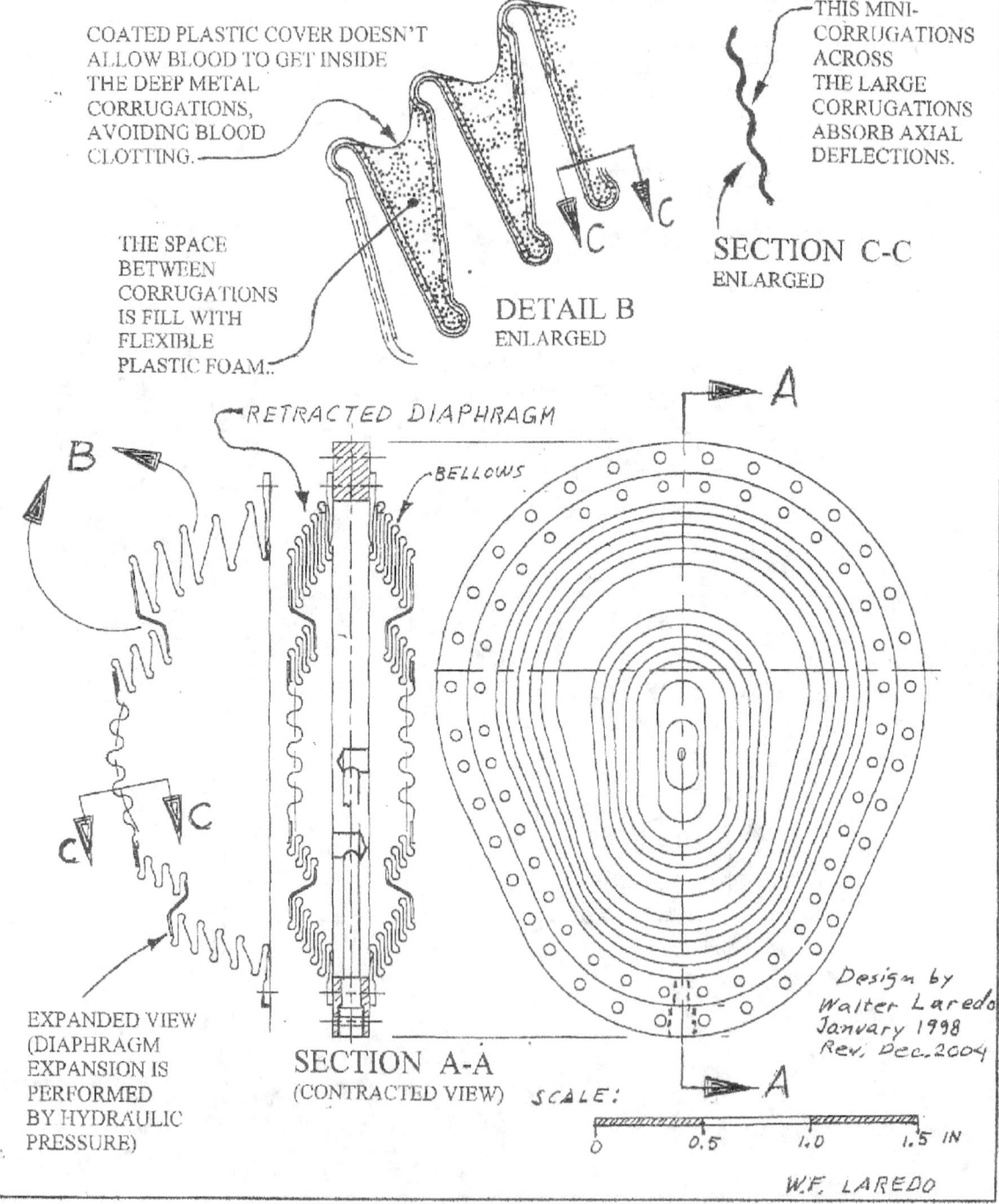

INTERNAL STRUCTURE OF THE BLOOD PUMPING DIAPHRAGM

STRUCTURE IN CONTACT WITH THE PUMPING HYDRAULIC FLUID; IT IS MADE FROM SPECIAL ALLOY TO WITHSTAND ONE BILLION CYCLES. THIS STRUCTURE AND ITS MATERIAL ARE DESIGNED TO WITHSTAND FATIGUE, DESIGNED IN SUCH WAY THAT MOST DEFLECTION IS BY BENDING AND NOT BY STRETCHING.

COATED PLASTIC COVER DOESN'T ALLOW BLOOD TO GET INSIDE THE DEEP METAL CORRUGATIONS, AVOIDING BLOOD CLOTTING.

THIS MINI-CORRUGATIONS ACROSS THE LARGE CORRUGATIONS ABSORB AXIAL DEFLECTIONS.

THE SPACE BETWEEN CORRUGATIONS IS FILL WITH FLEXIBLE PLASTIC FOAM.

DETAIL B
ENLARGED

SECTION C-C
ENLARGED

RETRACTED DIAPHRAGM

BELLOWS

EXPANDED VIEW (DIAPHRAGM EXPANSION IS PERFORMED BY HYDRAULIC PRESSURE)

SECTION A-A
(CONTRACTED VIEW)

SCALE:

0   0.5   1.0   1.5 IN

Design by
Walter Laredo
January 1998
Rev. Dec. 2004

W.F. LAREDO

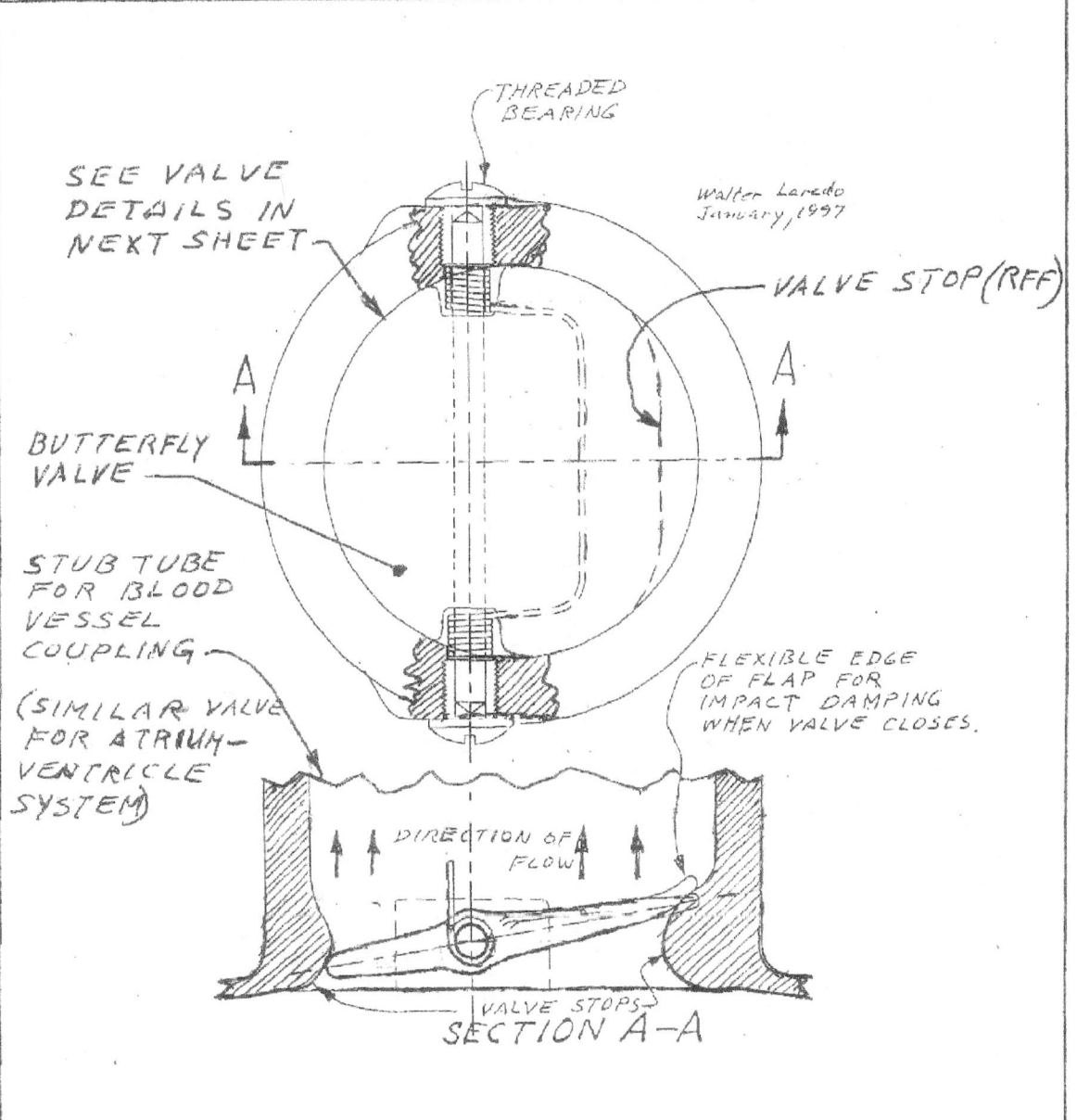

SEE VALVE DETAILS IN NEXT SHEET

THREADED BEARING

Walter Laredo
January, 1997

VALVE STOP (REF)

BUTTERFLY VALVE

STUB TUBE FOR BLOOD VESSEL COUPLING.

(SIMILAR VALVE FOR ATRIUM-VENTRICLE SYSTEM)

FLEXIBLE EDGE OF FLAP FOR IMPACT DAMPING WHEN VALVE CLOSES.

DIRECTION OF FLOW

VALVE STOPS

SECTION A-A

VALVE ASSEMBLY SYSTEM
FOR ARTIFICIAL HEART

INTERNAL DUCT AND VALVE SURFACES ARE COATED WITH A SUBSTANCE THAT DOESN'T ALLOW BLOOD TO CLOTH.

W.F. LAREDO

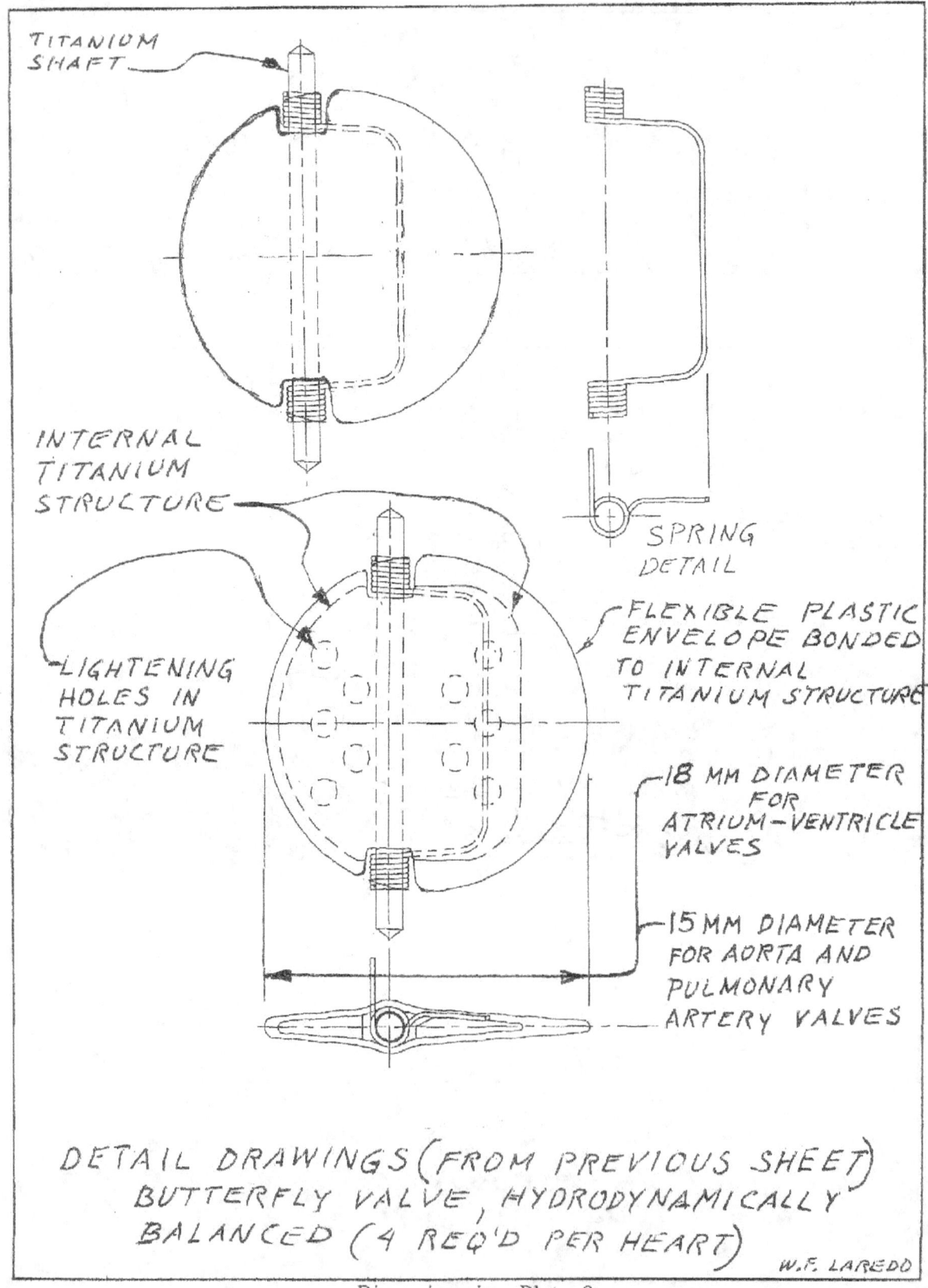

TITANIUM SHAFT

INTERNAL TITANIUM STRUCTURE

SPRING DETAIL

LIGHTENING HOLES IN TITANIUM STRUCTURE

FLEXIBLE PLASTIC ENVELOPE BONDED TO INTERNAL TITANIUM STRUCTURE

18 MM DIAMETER FOR ATRIUM-VENTRICLE VALVES

15 MM DIAMETER FOR AORTA AND PULMONARY ARTERY VALVES

DETAIL DRAWINGS (FROM PREVIOUS SHEET) BUTTERFLY VALVE, HYDRODYNAMICALLY BALANCED (4 REQ'D PER HEART)

W.F. LAREDO

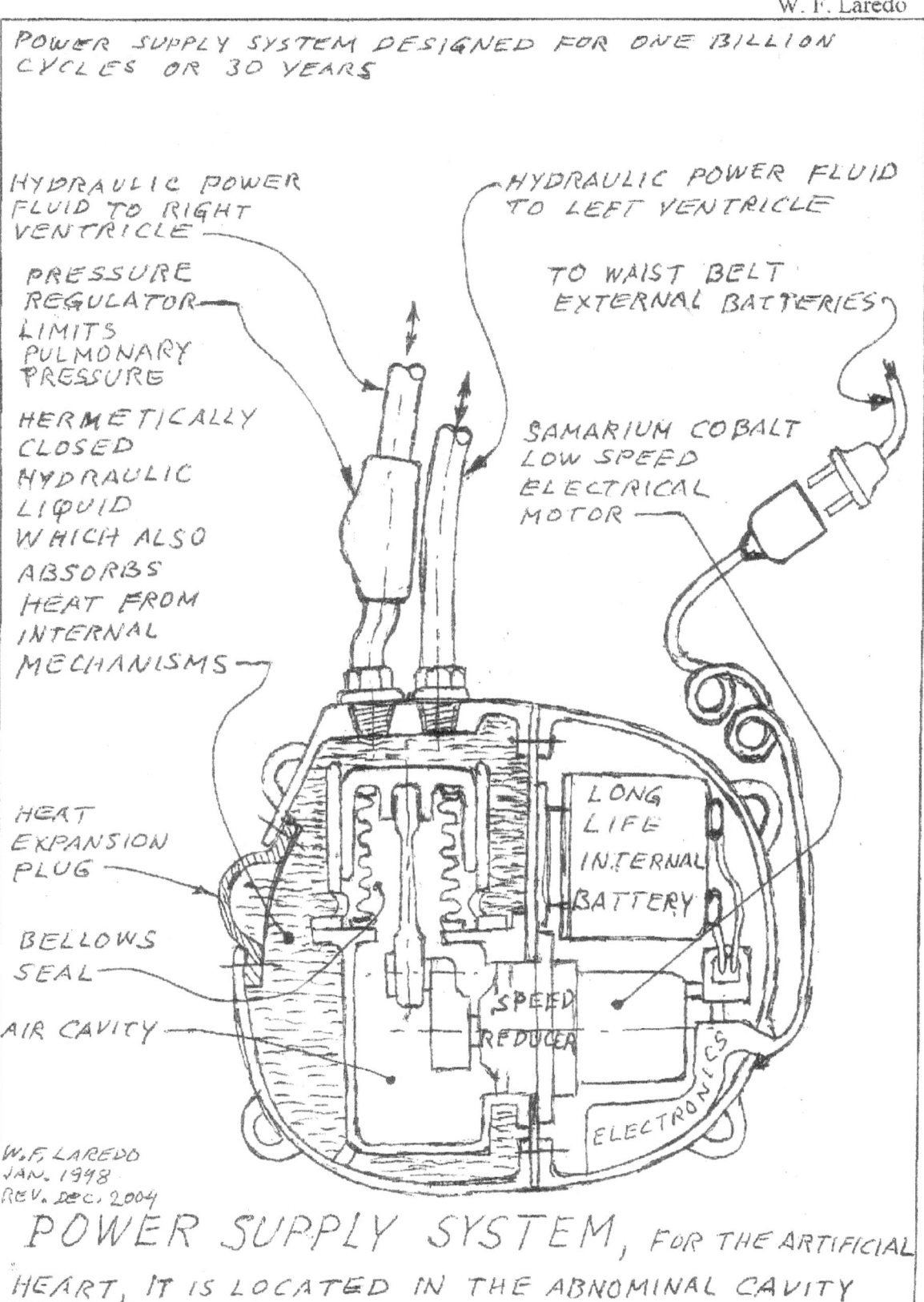

POWER SUPPLY SYSTEM DESIGNED FOR ONE BILLION CYCLES OR 30 YEARS

HYDRAULIC POWER FLUID TO RIGHT VENTRICLE

HYDRAULIC POWER FLUID TO LEFT VENTRICLE

PRESSURE REGULATOR LIMITS PULMONARY PRESSURE

TO WAIST BELT EXTERNAL BATTERIES

HERMETICALLY CLOSED HYDRAULIC LIQUID WHICH ALSO ABSORBS HEAT FROM INTERNAL MECHANISMS

SAMARIUM COBALT LOW SPEED ELECTRICAL MOTOR

HEAT EXPANSION PLUG

LONG LIFE INTERNAL BATTERY

BELLOWS SEAL

AIR CAVITY

SPEED REDUCER

ELECTRONICS

W. F. LAREDO
JAN. 1998
REV. DEC. 2004

POWER SUPPLY SYSTEM, FOR THE ARTIFICIAL HEART, IT IS LOCATED IN THE ABNOMINAL CAVITY

MECHANICAL HEART, PUMPING POWER CALCULATION

70 pulses a minute with cycles of 0.86 seconds.

To simplify calculations assume that blood density and its relative density is equal to that of water, 62.4 lb/cu ft and 1 respectively.

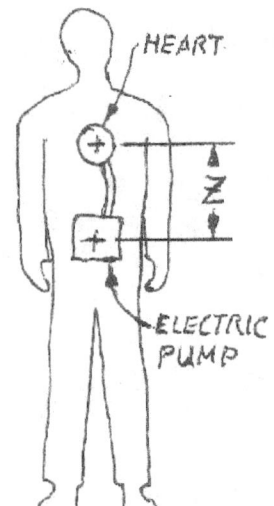

$Z = 15$ in. $= 1.25$ ft

$h_1 =$ Systolic blood pressure of 120 mm of mercury.

$h_2 =$ Occasional blood pressure rise of 50.28 mm of mercury, when artificial sensors detects an state of emergency or emotional state.

$F =$ Force to overcome the stiffness of the deflecting steel diaphragm.

$h_1 = 120$ mm Hg $= 1636.8$ mm water $= 5.37$ ft
$h_2 = 50.28$ mmHg $= 685.8$ mm water $= 2.25$ ft

Head $\quad H = h_1 + h_2 + Z = 8.87$ ft

Pressure at point A $=$ Head $\times$ Density
$(8.87$ ft$) \times (62.4$ lb/cu ft$) = 553.5$ lb/sq ft

Ventricle volume $= 78.33$ cu cm
Blood pumped volume (both ventricles)
$2 (78.33) = 156.66$ cu cm $= .005232$ cu ft
Per min. $0.005232 \times 70 = 0.36624$ cu ft/min
Per sec. $0.36624 / 60 = .0061$ cu ft / sec.

Power $=$ (Pressure $\times$ Volume)$/$t
553.5 lb/sq ft $\times .0061$ cu ft/ sec
$= 3.38$ lb. ft/sec

Hydraulic HP
$\quad$ WHP $= 3.38/550 = .0061$
Break HP required for motor,
with a motor-pump efficiency of 0.7
$\quad$ bHP $= .0061/ 0.7 = .0087$

ADDITIONAL OPERATING POWER TO DEFLECT THE HEART PUMPING DIAPHRAGMS

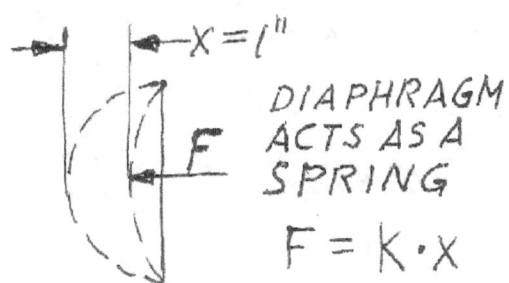

$X = 1''$

DIAPHRAGM ACTS AS A SPRING

$F = K \cdot X$

70 pulses per minute, with 0.86 sec cycles
$K = 2$ lb/in $= 24$ lb/ft,
average deflection $= 1$ inch

Energy for a cycle of 0.86 sec
$F = k \cdot x = 24$ lb/ft $\times x$

$$\text{Energy} = \int_0^1 F \cdot dx = \int_0^1 k \cdot x \cdot dx$$
$$= 24 \left[\frac{x^2}{2} \right]_0^1 = 12 \text{ ft. lb}$$

Energy per minute, $\quad 12 \times 70 = 840$ ft. lb
Energy per second, $\quad 840/60 = 14$ ft. lb
$\quad$ For 2 springs $= 28$ ft.lb
The diaphragms' springs return most of this energy to the system, assuming the system absorbs only 20 per cent of it.
$\quad$ Energy $= 0.20 \times 28 = 5.60$ ft. lb/sec
$\quad$ HP spring $= 5.60/550 = 0.010$ HP

Total pump HP $=$ breakHP $+$ HP spring
$\quad 0.87 + 0.010 = 0.0187$ HP
Better use a,

0.03 HP or 17.16 watts

electric motor.

W.F. LAREDO
JAN 1997

National Aeronautics and
Space Administration

Lewis Research Center
Cleveland, OH 44135-3191

APR 1 7 1998

6000

Reply to Attn of:

Mr. Walter F. Laredo
780 Oak Grove Rd
Apt. C-216
Concord, CA 94518-2708

Dear Mr. Laredo:

I have had several members of my staff review your unsolicited proposal for design of bionic heart pump and for a reusable Earth-to-Orbit launch vehicle. Obviously, your proposals in both areas are very detailed, and contain some thoughtful engineering ideas. I thank you very much for your interest. The assessment and suggestions follow.

Bionic Heart:

NASA Reviewer: Joseph P. Veres
Org Code: 2900
Telephone: (216) 433-2436
E-mail: joseph.veres@lerc.nasa.gov

The concept has numerous interesting features that appear to address some fundamental design criteria for artificial heart pumps. However, there appears to be unresolved mechanical, as well as biological, issues that need to be addressed by technology development programs. A summary of the challenges to the proposed bionic heart concept include the following:

1. There are many mechanical moving parts in the proposed "bionic heart" pumping device. The numerous moving parts in this concept may be prone to increased risk of failure over time. (Note: other artificial heart pumps that are currently under development are simpler in concept, with significantly fewer moving parts, and reduced risk of failure.)

2. Long-term lubrication of the motor bearing is not addressed in the bionic heart concept. If the motor were to use conventional rolling element bearings, they would be prone to failure over long periods of time.

3. The speed reduction gear is also a complex mechanical part. The proposal states that the pump pulsations would simulate the human heart at 70 beats per minute.

2

If the electric motor were to rotate at 3000RPM, the corresponding speed reduction gear ratio would be about 50:1. A high gear ratio such as this may also have correspondingly high losses. The gear and its bearings would need lubrication and a cooling scheme for heat removal. Similarly, the linkage driving the reciprocating piston pump has the same lubrication and cooling requirements for its bearings or bushings.

4. The special durable metallic alloy of the flexible pumping bellows, and the expandable pumping membrane, are both materials that need to be developed. The proposal refers to this yet to be discovered material as "tireless devil" because of its resistance to fatigue.

5. The system operation of the bionic heart does not completely simulate the human heart. The two halves of the human heart pump in series; that is, the right side pulses while the left side is at rest, and the left side pulses while the right side is at rest. The bionic heart proposes to pressurize both the right and left ventricles simultaneously, and this difference from the operation of the human heart may be biologically significant. In addition, the left side of the human heart does most of the work, creating a pressure rise of approximately 100mm of mercury, while the right side creates a 25mm pressure rise.

6. The design of long life mechanical valves that are acceptable for blood flow is an area that is currently the subject of research in the medical community. Movement of the valve may cause damage to the blood cells.

7. Uneven flow swirling and vortices within the large cavities of the bionic heart can create areas of stagnant flow. The residence time of blood within the pump is a known key parameter that is responsible for the onset of coagulation. With partially stagnant flow in the cavities and bellows, the time the blood is in contact with the foreign material increases and triggers the blood to coagulate, and in time collect on the internal walls.

8. The proposed "grommet seal" in the human diaphragm muscle may be a challenge.

9. The bio compatibility of all of the materials that are in contact with blood is an issue on all blood development projects. This proposal has a large variety of materials that contact the blood, and for each of them the bio compatibility issue has to be addressed.

3

In closing, Mr. Laredo's proposal has numerous novel concepts for blood flow. Some of these concepts appear to be low risk and can be pursued in a routine design and development program. However, some of the issues inherent to this pumping system, such as long life flexible materials and high potential for blood clotting, need to be addressed by long term technology development programs. My suggestion is for Mr. Laredo to also contact a medical research institution, such as the Cleveland Clinic or University Hospitals, that can provide him more insight into the typical rheological issues associated with blood.

Earth-to-Orbit Launch Vehicle:

Mr. Laredo's recent paper on the World Aerospace Center and Space Plane concepts contains several very impressive drawings and clearly indicates that he is aware of many of the primary engineering issues and concerns associated with reusable launch vehicles and their supporting infrastructure. Advancing the longer term state of the art in space transportation has always been part of the National Aeronautics and Space Administration's plan for the future development of space. His concept for enabling single and two stage to orbit vehicles is generally consistent with related concepts which NASA Headquarters and Marshall Space Flight Center(MSFC) are studying under the Reusable Launch Vehicle (RLV) program.

In particular, the set of nine detailed design drawings was most impressive. I suggest that Mr. Laredo send a copy of the proposal to the MSFC. MSFC is NASA's lead center for Earth-to-Orbit and space propulsion vehicle concepts. NASA Lewis has been largely devoted to propulsion systems (the engine), not necessarily the entire vehicle or its supporting infrastructure. I also recommend sending a copy to NASA Headquarters, where much of what he is proposing has been in the evaluation process over the last few years. Both organizations are better positioned to evaluate and perhaps act on your concept than the staff at NASA Lewis. Please contact:

Garry Lyles
NASA Marshall Space Flight Center
Huntsville, AL
Telephone: (205)544-9203
E-mail   garry.lyles@msfc.nasa.gov

John Mankins
NASA Headquarters
Washington, DC
Telephone: (202)358-4659
E-mail: john.mankins@hq.nasa.gov

I would like to make a suggestion to improve Mr. Laredo's concept. The primary goal of the current major NASA effort is to explore Earth-to-Orbit concepts that offer the potential to significantly reduce, by one to two orders of magnitude, the cost per pound of payload to low Earth orbit. A comprehensive, total life cycle cost analysis would be a powerful and essential addition to his work. This should also include the cost of ground infrastructure, labor-intensive maintenance, and cost of failure, i.e. replacement cost. Such an analysis is needed to truly assess the merits of any reusable infrastructure.

I hope our comments will be taken as constructive criticism to improve your concepts in both areas. Again, we appreciate your interest and the effort you expended in these proposals, and wish you the best in your future endeavors.

Sincerely,

Donald J. Campbell
Director

NASA'S REVIEW OF THE ARTIFICIAL HEART DESIGNED BY W. F. LAREDO

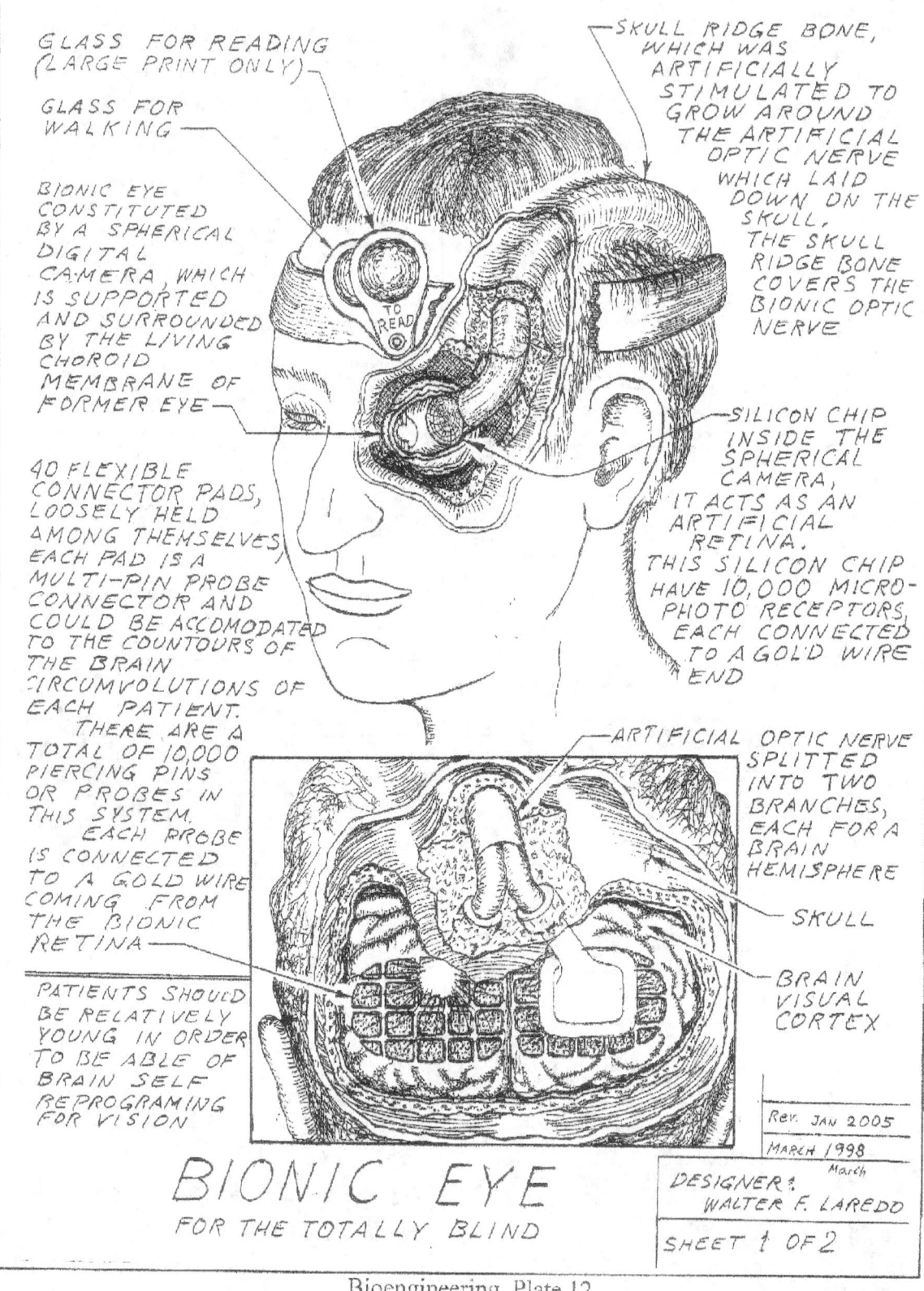

GLASS FOR READING (LARGE PRINT ONLY)

GLASS FOR WALKING

BIONIC EYE CONSTITUTED BY A SPHERICAL DIGITAL CAMERA, WHICH IS SUPPORTED AND SURROUNDED BY THE LIVING CHOROID MEMBRANE OF FORMER EYE

40 FLEXIBLE CONNECTOR PADS, LOOSELY HELD AMONG THEMSELVES, EACH PAD IS A MULTI-PIN PROBE CONNECTOR AND COULD BE ACCOMODATED TO THE COUNTOURS OF THE BRAIN CIRCUMVOLUTIONS OF EACH PATIENT.
   THERE ARE A TOTAL OF 10,000 PIERCING PINS OR PROBES IN THIS SYSTEM.
   EACH PROBE IS CONNECTED TO A GOLD WIRE COMING FROM THE BIONIC RETINA

PATIENTS SHOULD BE RELATIVELY YOUNG IN ORDER TO BE ABLE OF BRAIN SELF REPROGRAMING FOR VISION

TO READ

SKULL RIDGE BONE, WHICH WAS ARTIFICIALLY STIMULATED TO GROW AROUND THE ARTIFICIAL OPTIC NERVE WHICH LAID DOWN ON THE SKULL.
THE SKULL RIDGE BONE COVERS THE BIONIC OPTIC NERVE

SILICON CHIP INSIDE THE SPHERICAL CAMERA, IT ACTS AS AN ARTIFICIAL RETINA.
THIS SILICON CHIP HAVE 10,000 MICRO-PHOTO RECEPTORS, EACH CONNECTED TO A GOLD WIRE END

ARTIFICIAL OPTIC NERVE SPLITTED INTO TWO BRANCHES, EACH FOR A BRAIN HEMISPHERE

SKULL

BRAIN VISUAL CORTEX

BIONIC EYE
FOR THE TOTALLY BLIND

Rev. JAN 2005
MARCH 1998
DESIGNER:
   WALTER F. LAREDO
SHEET 1 OF 2

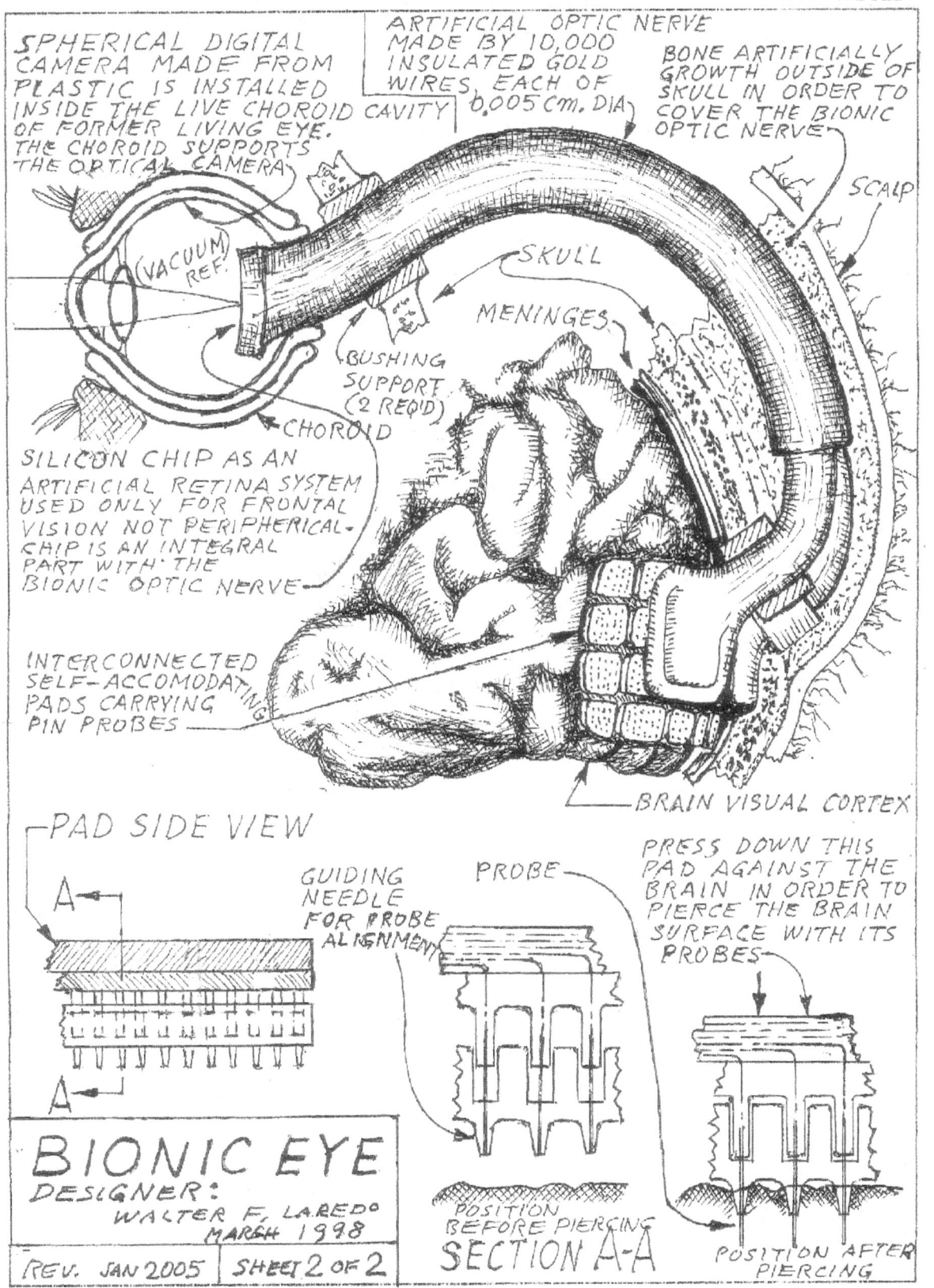

SPHERICAL DIGITAL CAMERA MADE FROM PLASTIC IS INSTALLED INSIDE THE LIVE CHOROID OF FORMER LIVING EYE. THE CHOROID SUPPORTS THE OPTICAL CAMERA

ARTIFICIAL OPTIC NERVE MADE BY 10,000 INSULATED GOLD WIRES, EACH OF 0.005 Cm. DIA

CAVITY

BONE ARTIFICIALLY GROWTH OUTSIDE OF SKULL IN ORDER TO COVER THE BIONIC OPTIC NERVE

SCALP

(VACUUM) REF.

SKULL

MENINGES

BUSHING SUPPORT (2 REQ'D)

CHOROID

SILICON CHIP AS AN ARTIFICIAL RETINA SYSTEM USED ONLY FOR FRONTAL VISION NOT PERIPHERICAL. CHIP IS AN INTEGRAL PART WITH THE BIONIC OPTIC NERVE

INTERCONNECTED SELF-ACCOMODATING PADS CARRYING PIN PROBES

BRAIN VISUAL CORTEX

PAD SIDE VIEW

A

A

GUIDING NEEDLE FOR PROBE ALIGNMENT

PROBE

PRESS DOWN THIS PAD AGAINST THE BRAIN IN ORDER TO PIERCE THE BRAIN SURFACE WITH ITS PROBES

BIONIC EYE
DESIGNER:
WALTER F. LAREDO
MARCH 1998

REV. JAN 2005 | SHEET 2 OF 2

POSITION BEFORE PIERCING
SECTION A-A

POSITION AFTER PIERCING

Bioengineering Plate 13

5. ENGINEERING CHRONICLES AND PROJECTS FOR THE 21ˢᵗ CENTURY

CONTENTS:

a.   Engineering Chronicle No. 1
Technical description of the "Mars Spaceship,"
including its propulsion system.

See illustrations 13 through 15, in the list of
Engineering Projects for the 21ˢᵗ Century

b.   Engineering Chronicle No. 2
Commuter twinjet for 30 passengers

See illustrations 4 to 5, in the list of
Engineering Projects for the 21ˢᵗ Century

c.   Engineering Chronicle No. 3
Flying boat aircraft with nuclear propulsion system

See illustration 16, in the list of Engineering
Projects for the 21ˢᵗ Century

**d.   List of Engineering Projects
for the 21ˢᵗ Century**

e.   Highly detailed Preliminary Design Drawings

ENGINEERING CHRONICLE No. 1

"MARS " SPACESHIP

TECHNICAL DESCRIPTION FOR THIS FUTURE PROJECT

The main structural support of the spaceship is a long heavy hollow tube, it is as a shaft around which everything else is mounted, and at its mid-length is a group of three huge rotating fat rings looking like huge donuts. The biggest one located at the center counter-rotates with respect to the other two, a rotation that produces enough centrifugal force that acts as an artificial gravity, with an intensity equivalent to that in-between Mars and the moon, which is enough to make people feel comfortable during the long voyages.

The interior of the hollow shaft also was used as an aisle, allowing passengers to visit other ship compartments. A thick disk plate was welded as a plug at each end of the hollow shaft. Those disk plates were strong enough to support other attached structures, including the head and the neck of the vehicle at one end and the long tail at the other. The tail boom known as the scorpion tail, is a light structure consisting mainly of a 250-meter long trust made of slender bars? The forward end of the main shaft has an internal door and at this end is coupled the neck with the spherical head of the vehicle.

The three floors full of double-pane windows constituted the spherical head of the ship. The head's upper floor constituted the observation deck where there was a telescope. The next lower floor of the same head was the bridge from where Captain Cronus gave his orders, and the third floor below was the small dormitory assigned only for part of the crew, only for the ones on duty. The head of the ship didn't rotate; hence it didn't have artificial gravity and was fixed with respect to the stars for navigational purposes. Captain Cronus and the crew on-duty spent most of their time on the bridge and in the living quarters inside the head, where only trained astronauts could work in an environment of zero gravity.

The group of donut-shaped rotating structures located near the center of the ship, contained most of the major components and systems of the vehicle, including the living quarters for passengers and for part of the crew, the entertainment and sports rooms, and a theater that doubled as a meeting place. The huge rotating structures also contained the ship's life-support equipment, including air and water recycling systems, the hydroponics gardens, repair shops, laboratories, and a one-room hospital. For protection against space radiation, inside each of the fat rings was enclosed another less fat ring, similar to the inner tube of a giant tire. This inner tube contained the living quarters with bedrooms. The space in between the internal and external tubes was not empty, but was surrounded by most of the equipment and accessories, providing enough material thickness which acted as a shield protecting humans and the hydroponics gardens from cosmic radiation and occasional sun flares. This enclosed design that allowed all bedrooms to be kept away from the ship's external walls, was another reason not to have real windows but artificial ones (flat TV screens bonded to the walls).

Because the Command Bridge didn't rotate and was fixed with respect to the stars, cameras installed in it transmitted the TV images from the universe to the faked windows of all bedrooms.

A magnetic field protected the whole ship from sun flares.

Now, in order to protect crew and passengers from another radiation, the one coming from the rear of the ship, from where the nuclear reactors and the nuclear power group are installed, it is about 250 meters away from the center of the ship, at the tip of that long scorpion tail.

A propulsion system just located behind a semi-spherical radiation shield, which is a lead shield in the form of an umbrella that covers the front part of the propulsion system.

There is another nuclear power group that produces electrical energy for internal use of the ship, and is attached at several points of the tail truss, it is a group constituted by small nuclear fission reactors and the plutonium radio-isotope devices, both used to supplied internal energy for the ship's needs. However most of this internal energy was used to energize the laser system that triggered the pellets in the fusion parabolic reactors, which are the main components of the propulsion system.

PROPULSION SYSTEM

In this adventure from the future, an hour or so after landing on an icy moonlet at the edge of the outer Saturn ring, teams of space workers descended to its surface and started to work. Among them was a team that operated special equipment with a sign that read HEAVY WATER SEPARATOR. Complicated equipment used for the separation of heavy water from a previously melted block of ice, which was extracted from the moonlet's surface. Once the heavy water was separated it was pumped to special tanks inside the ship.

Captain Cronus said, "After we depart, an electrolysis equipment will operate continuously separating the heavy water into its basic elements, deuterium (hydrogen's heavy isotope) and oxygen."

"The oxygen will be stored in aluminum bottles to use later to produce a more breathable atmosphere inside the ship. The deuterium gas will flow from its storage tank into a special machine inside the ship, where with pulverized lithium, will be continuously synthesized into lithium deuteride, a white powder. This new form will be sent to a miniature factory inside the ship to be encapsulated inside hollow microspheres, a continuous process of producing millions of miniature, lithium deuteride pellets."

"These pellets are the actual nuclear fuel for the spaceship's fusion engines. Millions of these pellets will be stored in a tank, enough fuel to last for the rest of the trip."

The outer shell from each microsphere or pellet was covered with several coats, each from a different substance with a special characteristic. The pellet's most outer coat was made from a special alloy to get suspended electromagnetically, at the exact focus point in the interior of the parabolic chamber, where several laser beams will fire simultaneously. Each beam leaves from the end of a crystal rod. The ends of the crystal rods protruded inside the parabolic chamber. All of them were arranged as in a ring. Each pellet is a miniature hydrogen bomb. The series of explosions produced a continuous series of thrust pulses which reactions propelled the ship. See corresponding engineering illustrations in the preliminary design section of this book, where is shown an older design that used a mechanical method to suspend pellets at the focus point, where each individual support for each individual pellet also vaporized with each explosion. Later designs used electromagnetic suspension.

Pellets were sent, one by one, into the engine's open parabolic chambers for controlled fusion, (nuclear mini explosion), which were not actually chambers; chambers would melt. Instead they were as open gigantic rocket nozzles made from beryllium used as parabolic reflectors for electromagnetic energy. The inside surfaces of the chamber and the nozzle walls were coated. Among them there was a special coat that reflected neutrons. On the ship's tail, the fusion engines were mounted side-by-side. In this engine design, the explosion chamber and the nozzle were integrated into a simple parabolic mirror with walls continuously cooled by serpentine systems with circulating liquid. The fusion reaction from each pellet took place at the focus of the parabolic reflector, and its energy reflected away from the rear of the ship as a powerful beam whose momentum effect was the thrust that propelled the ship.

Preliminary design drawings for this interplanetary spaceship are shown in the preliminary design section of this book. In the drawing zone where the nuclear fusion engine should be located, are shown blank spaces where many details and section views are missing because I never got the required information for their complete mental visualization.

I am not aware if today some laboratories and aerospace corporations already generated the necessary data to design a very small and very compact multi-rod laser beam system, able to ignite pellets made from lithium deuteride. However, there are still some components of the engine that could be seen in the drawing, including some mechanical fuel feeding system and part of the parabolic nozzles.

Some day in the near future when the construction of this modern spaceship will be accomplished and will be ready to received its engines an if by that time no Fusion propulsion system would be still available, then an alternates, a more practical and less sophisticated propulsion system, will be mounted on this ship, as any of the electrical propulsion systems, including the ion. Although inadequate for round trips through the solar system, those are good enough for round trips to Mars, and perhaps to Jupiter and Saturn.

The preliminary designs for this modern spaceship are shown in the preliminary design section of this book. The drawings include versatile engine mounting provisions and an adapter plate that in the future will allow the mounting of any type of propulsion systems that will be available at that time, from the most sophisticated fusion type, to the more realistic and simpler electric ones. Electric ones may get developed sooner, satisfactory enough for the round trips to Mars.

ENGINEERING CHRONICLE No. 2

COMMUTER TWIN JET AIRLINER FOR 30 PASSENGERS

PROJECT PERFORMED BY A GROUP OF TALENTED HIGH SCHOOL STUDENTS FROM THE NEAR FUTURE   (21ˢᵗ CENTURY)

The professor of the class was also the Chief Project Engineer for this project; seven teenage students assisted him, and each of them working in his own computer and in the specialty that was assigned to that person. In order to stimulate the youth's self-confidence, each got a temporary title as a design engineer.

Their desks were numbered 1 through 7.

Month's prior to the initiation of the project, an airliner executive gave the professor a list with all desired characteristics and specifications the wanted aircraft should have. With this information in hand, the professor did some preliminary calculations for the performance and weight chart for this aircraft. On the first day of work, the professor delivered this information to a student, starting with the one at desk number 1, a student with the title of configuration engineer, who with some help of the professor became the architect for the aircraft. He was responsible to create the initial concept, the design of preliminary layouts, which included the general arrangement drawings for the aircraft, representing the entire aircraft but without too much detail at this time.

The preliminary design drawings he did, shown the top, the side, the front and cross sections views of the airplane. In another set of drawings he shown the aircraft interior including the plan view of the passengers' seating arrangement, the cockpit arrangement, the cargo compartments, and also the locations of doors, windows, toilets, galleys, engines, fuel tanks, and landing gear.

After this student finished to create all the original drawings several blueprint copies from his original set of plans were printed and distributed to his six colleagues. Each of them would use a copy to revise and correct it with red. The same copies also were used as guidance for further development. Each youth did some corrections on those blueprints and added more details in the field of their own expertise.

The student at desk number two, was specialized in aerodynamics and aerodynamic stability and control. He analyzed many parameters including lift and drag for several flight conditions and speeds. With these calculations, he corrected the layouts representing the external basic shape of the aircraft for the proper aerodynamic shape.

The student at desk number three acted as the structural engineer designing the whole internal structure including the structure of the wings, body and tail. With the help of his computer he created some mechanical systems, he was also responsible to perform the total structural analysis of the vehicle.

The student at desk number four, working as the mechanical engineer, designed the landing gear, flap mechanisms, door locking systems, and various other mechanical systems that goes inside the aircraft structure.

The student at desk five acted as the mechanical and also as the propulsion engineer and was in charge to choose the type of engines that should be used on this aircraft. He also did the drawings for the fuel system and the installation of the propulsion system including the design of the engine mountings, the engine cowlings and the installation of the oil coolers.

The student at desk number six was the chemist and the material's property specialist, providing technical support, consultation, and the necessary data concerning materials, including metal alloys, plastics and composites to his six colleagues.

At desk number seven was a student with two specialties: Electrical and electronics engineering.

The first design cycle of the aircraft was completed but it was not perfect yet. By reviewing and correcting each previous cycle, it started a new cycle for further development and refinement of the project. Better copies or blueprints were re-circulated again in between the same students, until after two or more cycles of refinements, the project became perfect and was released. At this point the project was completed the students had nothing else to do and left. However the student at desk number one, the architect of the project, remained and was re-assigned to a new task by transforming himself into an artist. He was re-assigned to create a comfortable and aesthetic interior design for the passenger's cabin, which would be pleasant to the eye.

The project was completed. The construction drawings for this new commuter twinjet were shipped to the manufacturer, who built two aircraft prototypes, which were tested for hundreds of flying hours. After passing all flying tests, the aircraft was certified to be mass-produced for the airline. Two months later these seven students graduated from high school and for their extra work in that aircraft project each received an aeronautical certificate and additional school credits. They got a useful practical experience, and a grant to pay in advance their entire career as full time students at the 21ˢᵗ century University.

ENGINEERING CHRONICLE No. 3
A CHRONICLE FROM THE NEAR FUTURE

FLYING BOAT AIRCRAFT WITH NUCLEAR PROPULSION

AIR CARGO TRANSPORT, FILLED WITH ENOUGH NUCLEAR FUEL FOR TEN YEARS OF CONTINUOS OPERATION BEFORE IT GETS REFUELED AGAIN.

Flying monsters, weighing hundreds of tons, were actually flying cargo boats and were able to fly with its nuclear propulsion for more than ten years without getting refueled. Aircraft was "goose" because of their canard shape. The vehicle had a long, thick hollow neck, wide enough to be used as a compartment for cargo and passengers.

The cockpit was the head of the bird and its nuclear engines were located inside its massive rear.

The main problem in developing this aircraft was the extra heavy shielding from lead, a protection against radiation. It was the cover of the nuclear reactor that protected the environment, the crew and the passengers from radiation. This shielding wrapped around the nuclear reactor was a spherical container with walls made by two layers. The outer layer was a thick lead clad and the inner layer was made from a thin reflective material. This kind of shielding, would be so heavy that the airplane would not be able to fly.

The only solution to reduce weight was by using a partial shielding design, where the forward semispherical half of this container would have full wall thickness, while the rear half, would be only one-fourth the thickness of what the front half would have. The forward semispherical part of the shielded container was like a thick umbrella, blocking most of the radiation projected in the forward direction, where the pilot's cockpit, the passenger cabin, and most systems were located.

When this seaplane was parked in a pier, it was forbidden to approach its rear, where the radioactive nuclear reactor was located. Every time after arrival, in order to dock, tugboats pushed the aircraft in reverse until its protruding rear entered inside a deep pocket or hole on the 20-foot thick concrete wall, a wall raising up into the air from below the waters, similar to a harbor water breaker. For safety the face of that concrete wall was covered with lead in order to protect people and the environment from the reactor radiation. For the advancement of technology in the 21ˢᵗ century only few people objected to the construction of nuclear projects including nuclear aircraft.

For long trips, this aircraft flew no higher than 300 feet above the ocean, due to its frequent descents close to the ocean's surface, barely touching it, for just few seconds, in order to scoop enough seawater to fill its depleted seawater tanks. During flight, seawater circulated through heat exchangers in order to condense the clean steam exhaust of the turbines' also the aircraft had additional aircraft condensers exposed to the free air stream, using additional free air-stream condensers is increased the rate of steam condensation. Injectors sprayed seawater from a tank into the air stream going into the condenser's inlet. This condenser was made from a special alloy, and the cooled air went through the condensers, and then was exhausted into the atmosphere. There was no reason to recycle the seawater; it would require a system of excessive weight. Instead the lost water was recuperated by descending the aircraft periodically to the sea surface to scoop up more water.

After the clean steam of the secondary closed circuit passed by the turbines, it was condensed in heat exchangers and radiators. Then in liquid form it was brought back and re-circulated again through the hot and pressurized helium heat exchanger belonging to the primary circuit of the nuclear reactor where this clean water was converted again into the hot steam that moved the turbines.

The nuclear reactor was of the pressurized helium type with uranium fuel elements in the form of ceramic multicoated pellets, designed for operation at high temperatures.

The high-pressure primary circuit with helium included the high temperature nuclear reactor, the heat exchanger, the helium circulating pumps and the tube system that connected all of them into the primary circuit. The reactor thermal power was 67,000 kilowatts and the turbine-input power was 22,371 kilowatts (30,000 SHP).

d.   LIST OF ENGINEERING PROJECTS FOR THE 21ˢᵀ CENTURY

All designs made by W. F. Laredo

NOTE: For additional information and more
detailed description concerning the
development and the design of each of
the above projects, see page 33

1. Architecture. Futuristic mansion

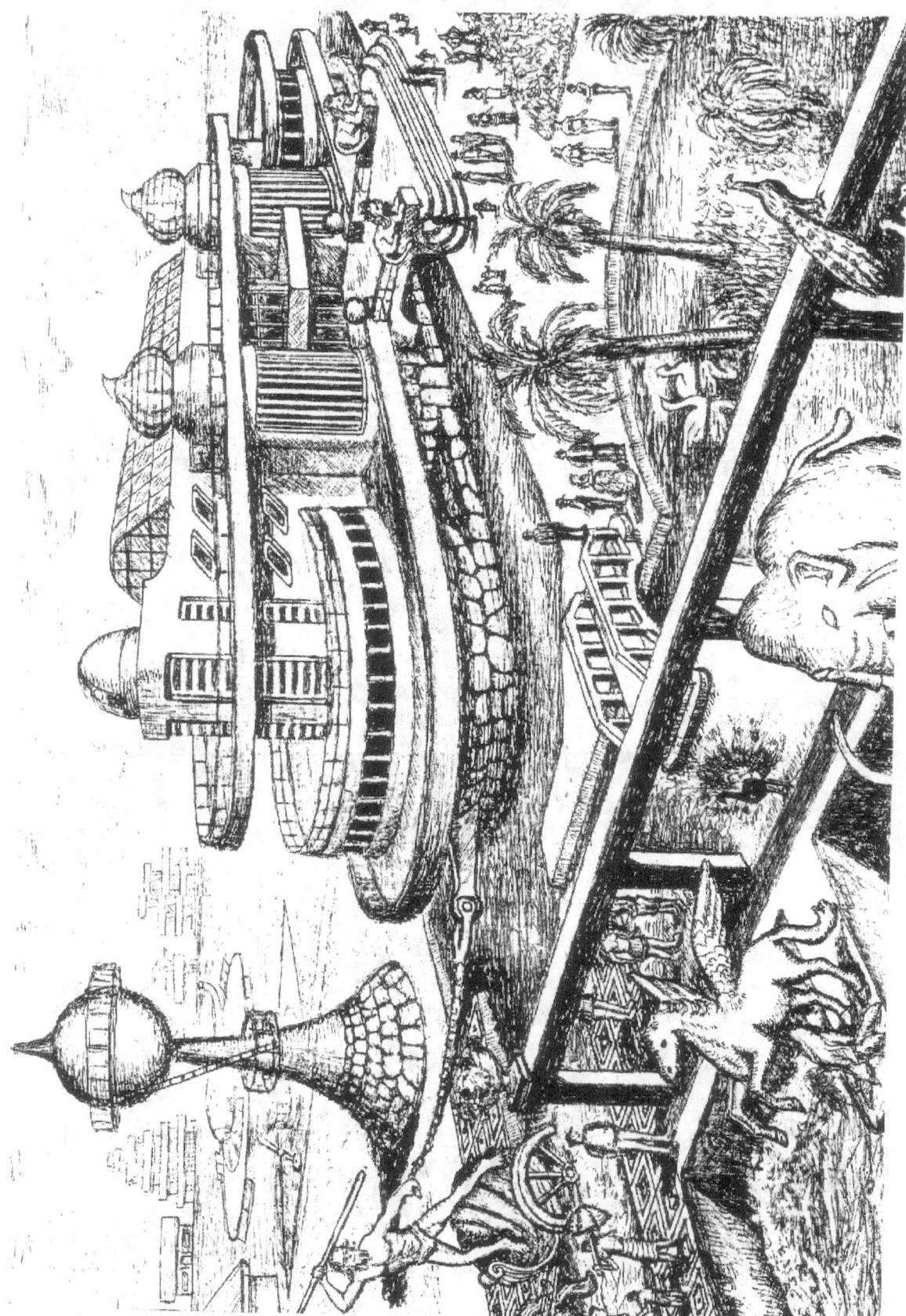

2. Detail of illustration 1. Futuristic mansion

Illustration 2bis. STATUE OF NEPTUNE. Detail of illustration 1

"ULTRA MODERN" PERSONAL JET

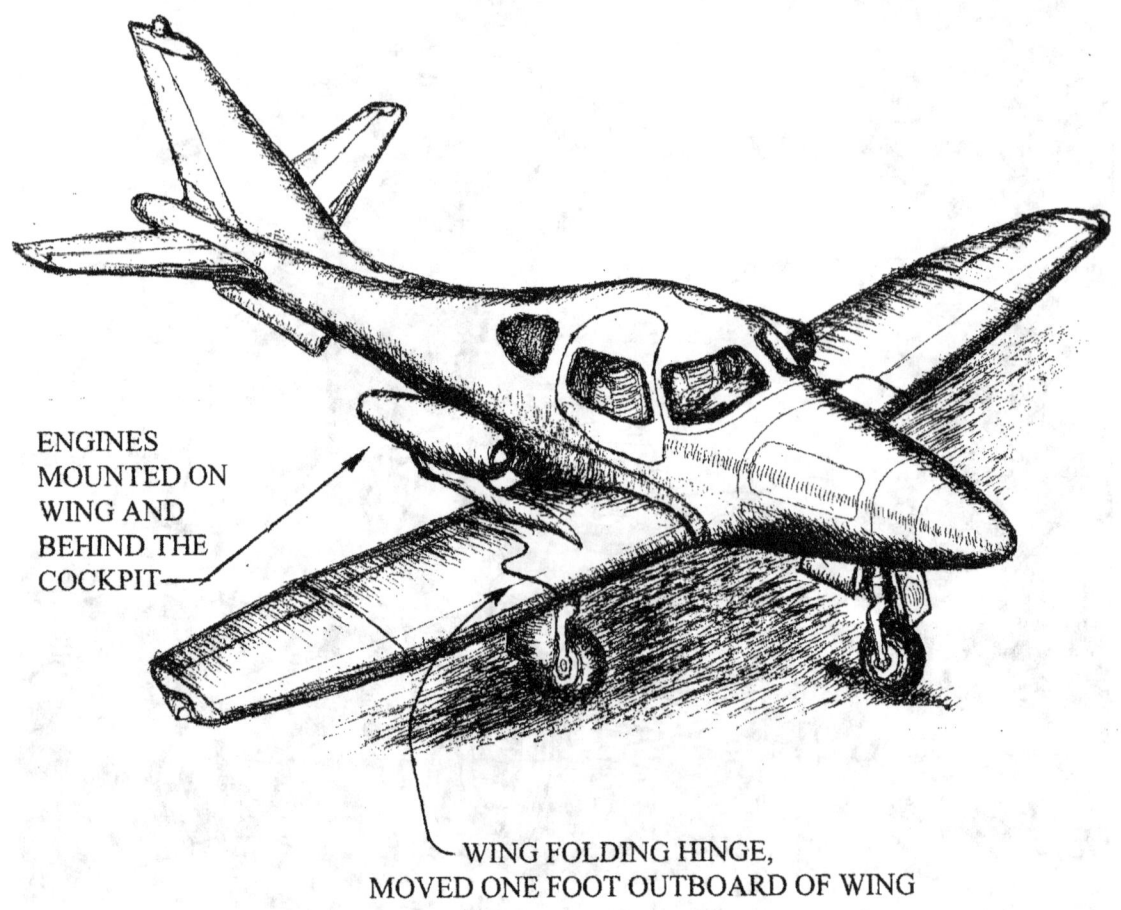

ENGINES
MOUNTED ON
WING AND
BEHIND THE
COCKPIT

WING FOLDING HINGE,
MOVED ONE FOOT OUTBOARD OF WING

NOTE:
THIS CONFIGURATION WAS SLIGHTLY IMPROVED BY MOVING THE
ENGINES FROM THE NOSE SECTION TO ABOVE THE WING.

BECAUSE THE WINGS ARE NOT BUILT YET, ITS STRUCTURAL DESIGN
COULD BE MODIFIED IN ORDER TO INSTALL ON ITS UPPER SURFACE
THE PYLONS THAT SUPPORT THE ENGINES.

Illustration 3 bis.   Latest revised Configuration
"Ultra Modern" personal Jet
For details see project 800 in page 33

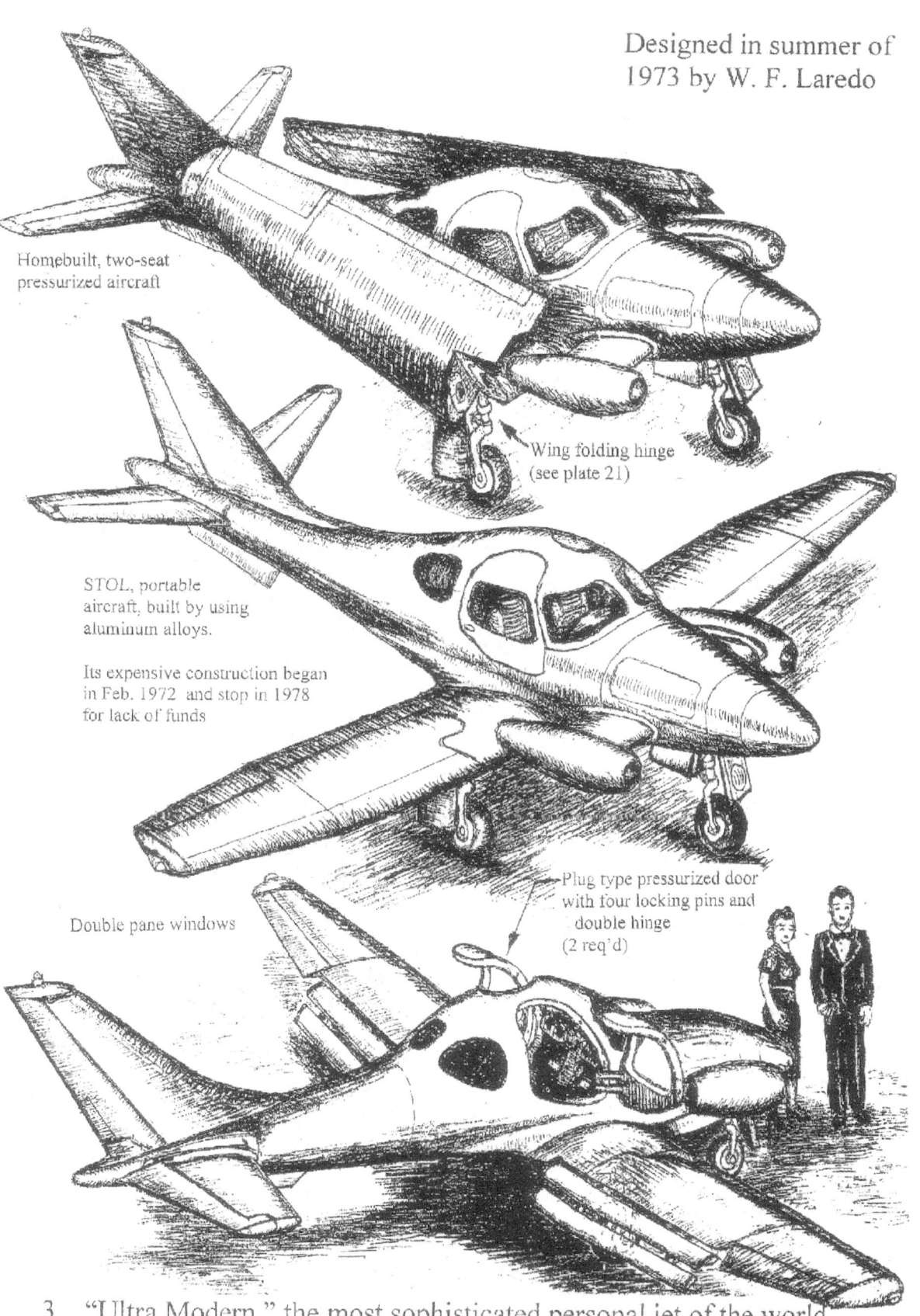

Designed in summer of
1973 by W. F. Laredo

Homebuilt, two-seat
pressurized aircraft

Wing folding hinge
(see plate 21)

STOL, portable
aircraft, built by using
aluminum alloys.

Its expensive construction began
in Feb. 1972 and stop in 1978
for lack of funds

Plug type pressurized door
with four locking pins and
double hinge
(2 req'd)

Double pane windows

3. "Ultra Modern," the most sophisticated personal jet of the world.
For details see project 800 in page 17

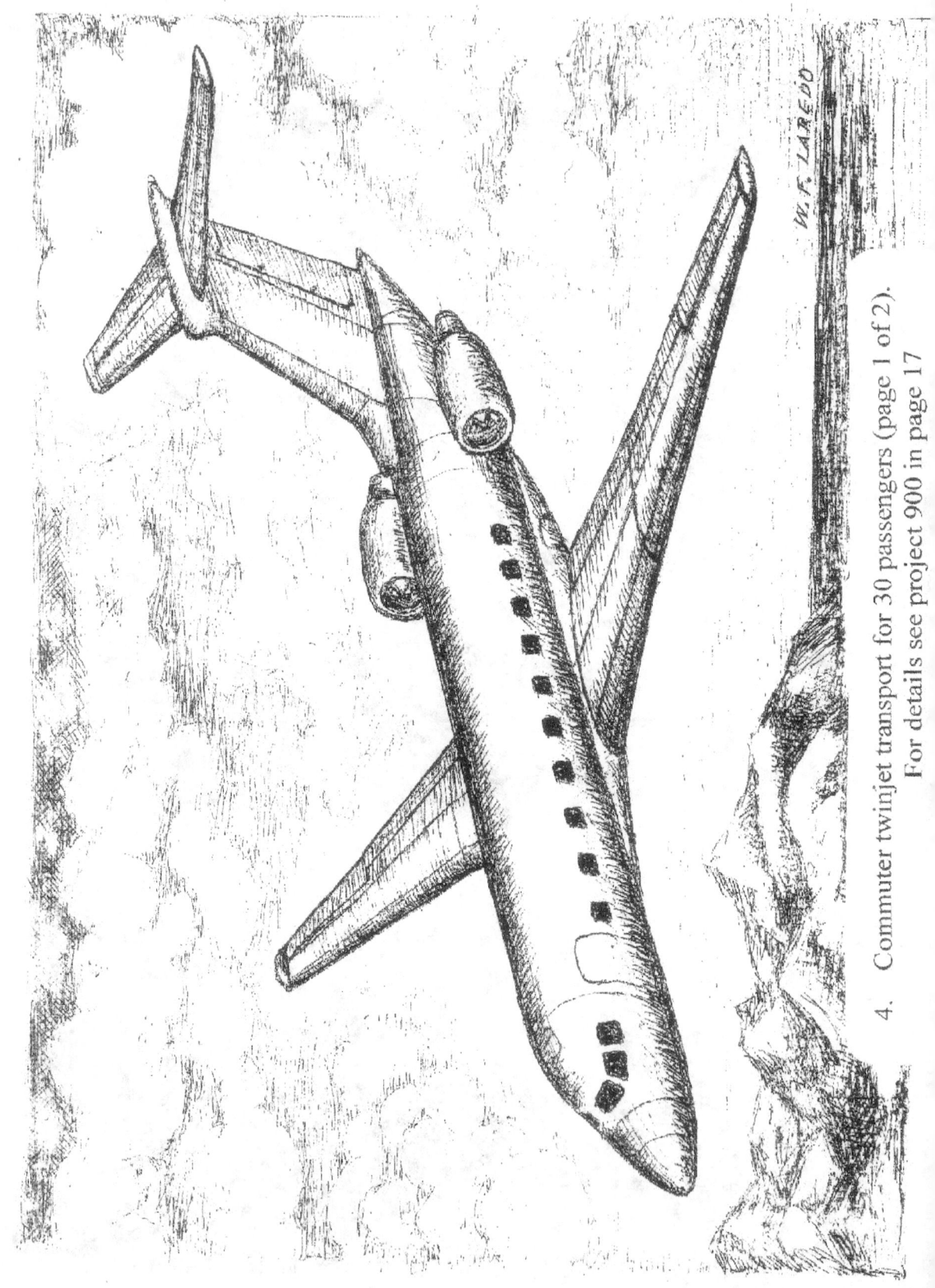

W.F. LAREDO

4.   Commuter twinjet transport for 30 passengers (page 1 of 2).
For details see project 900 in page 17

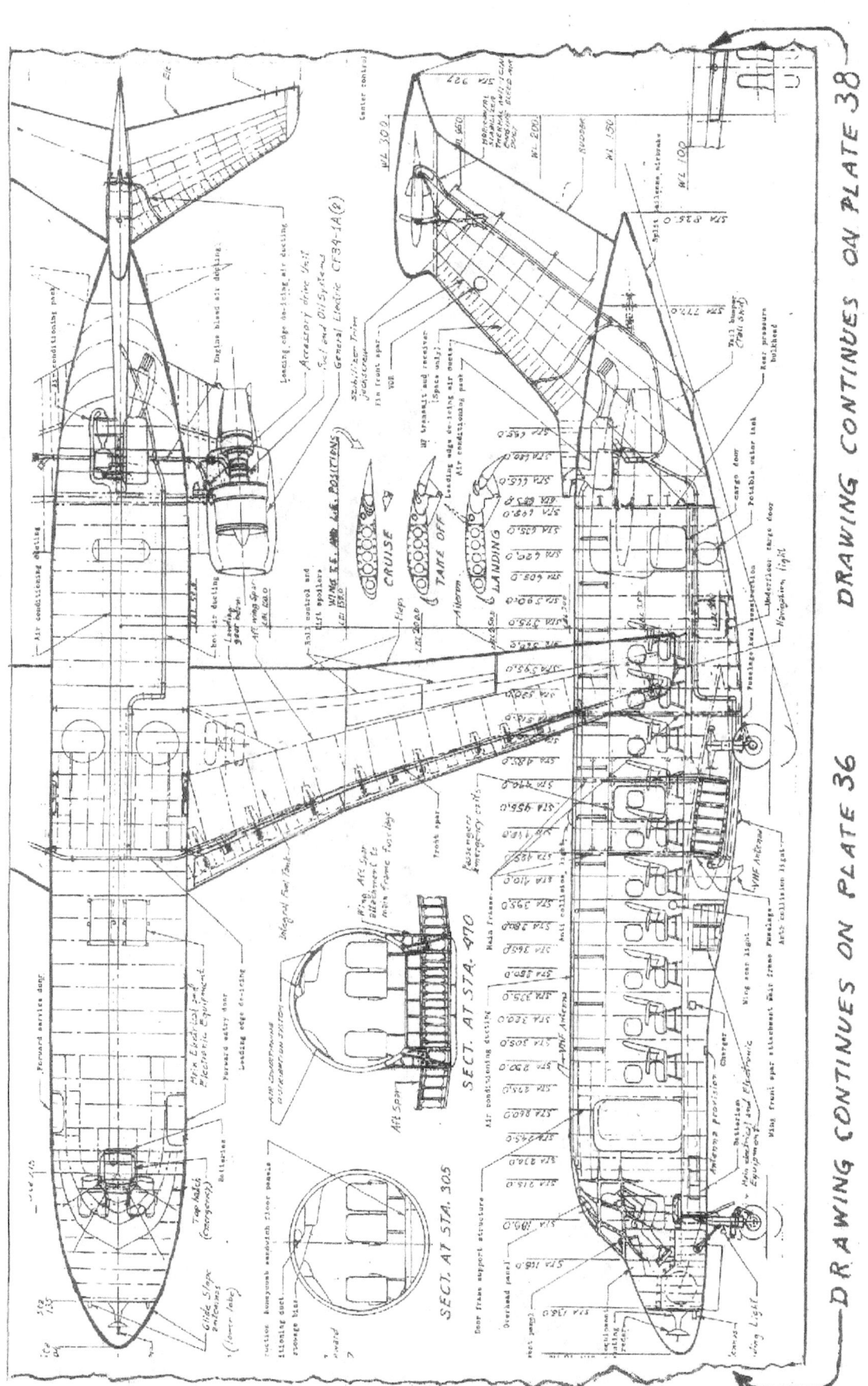

5.   Commuter twinjet transport, general arrangement (page 2 of 2)

Designed in spring of 1987 by W. F. Laredo, and later proposed separately to Brazil, Argentina, Spain and Italy for the construction of the first prototype

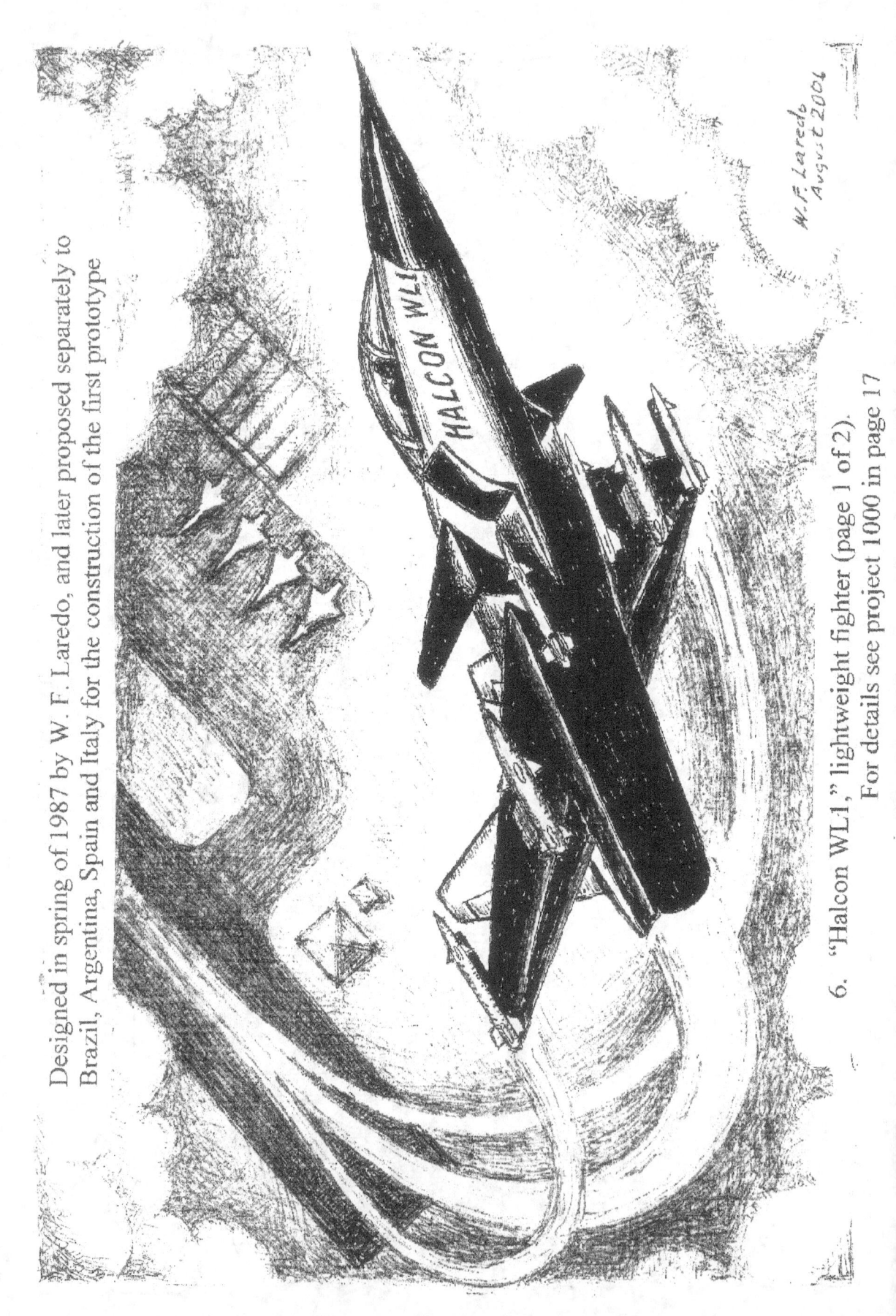

W.F. Laredo
August 2006

6. "Halcon WL1," lightweight fighter (page 1 of 2).
For details see project 1000 in page 17

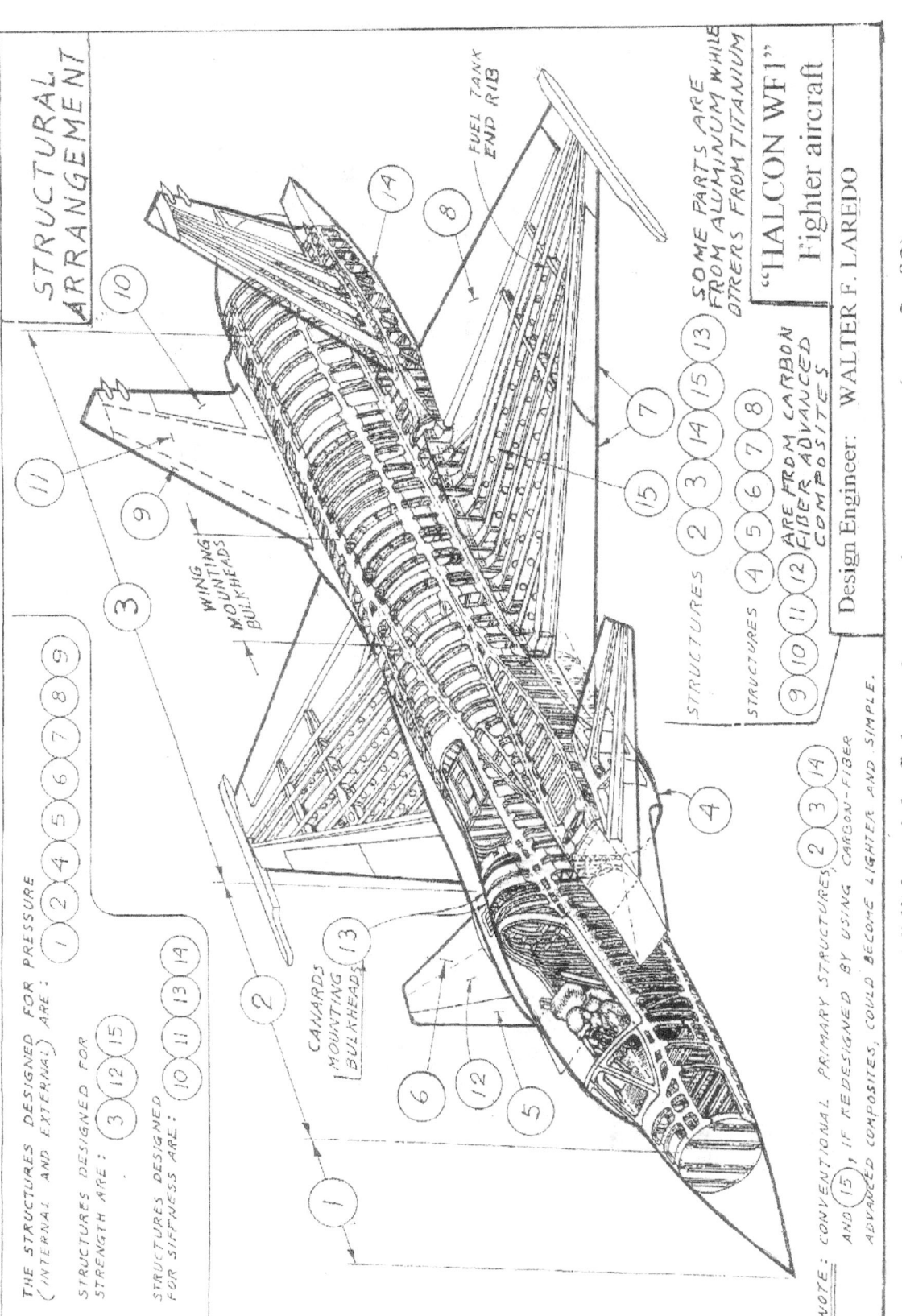

7. "Halcon WL1," lightweight fighter. Structural arrangement (page 2 of 2)

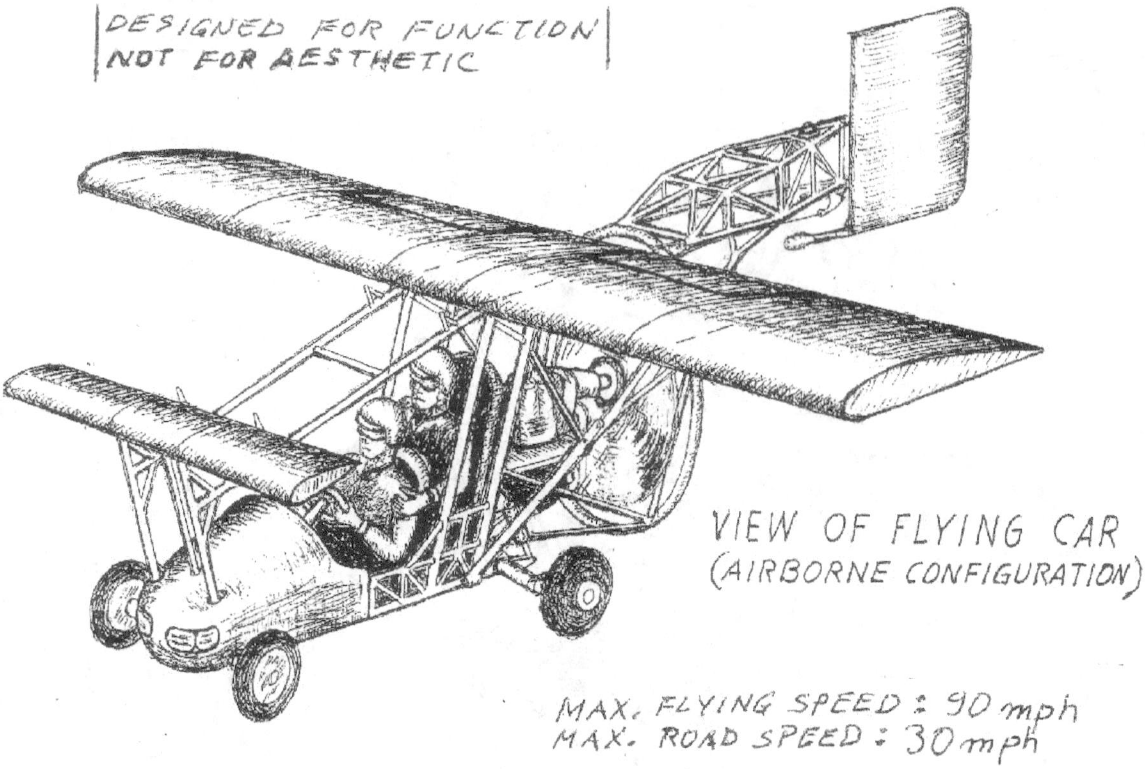

DESIGNED FOR FUNCTION
NOT FOR AESTHETIC

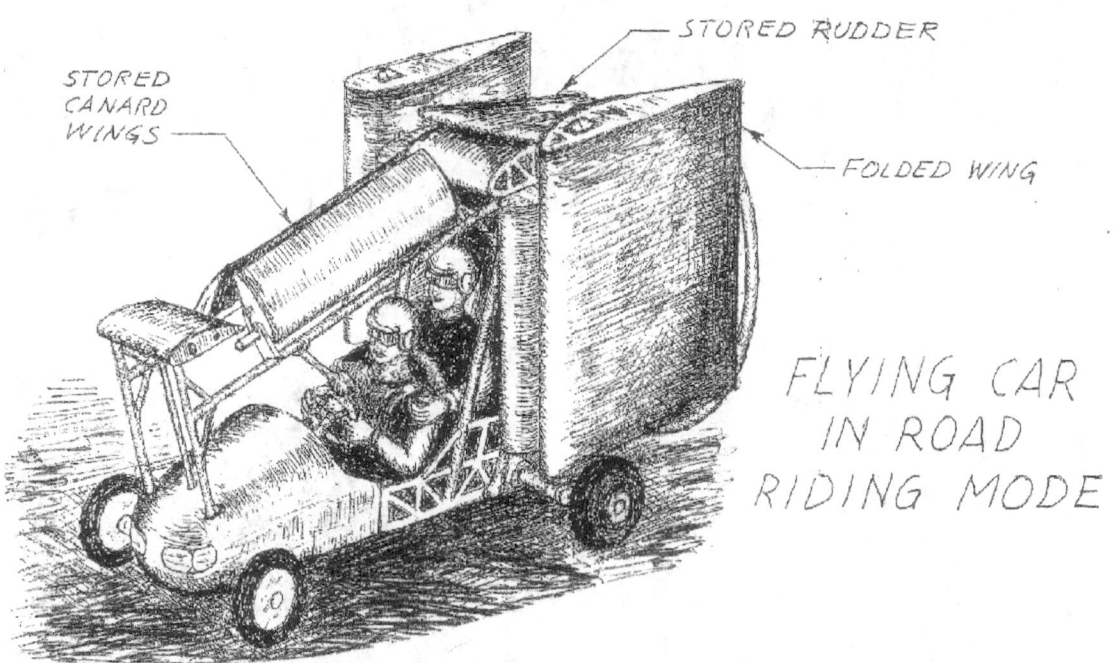

VIEW OF FLYING CAR
(AIRBORNE CONFIGURATION)

MAX. FLYING SPEED : 90 mph
MAX. ROAD SPEED : 30 mph

STORED CANARD WINGS

STORED RUDDER

FOLDED WING

FLYING CAR
IN ROAD
RIDING MODE

8.   Aerocar, Flying Car (page 1 of 3).   For details see project 1200 in page 17

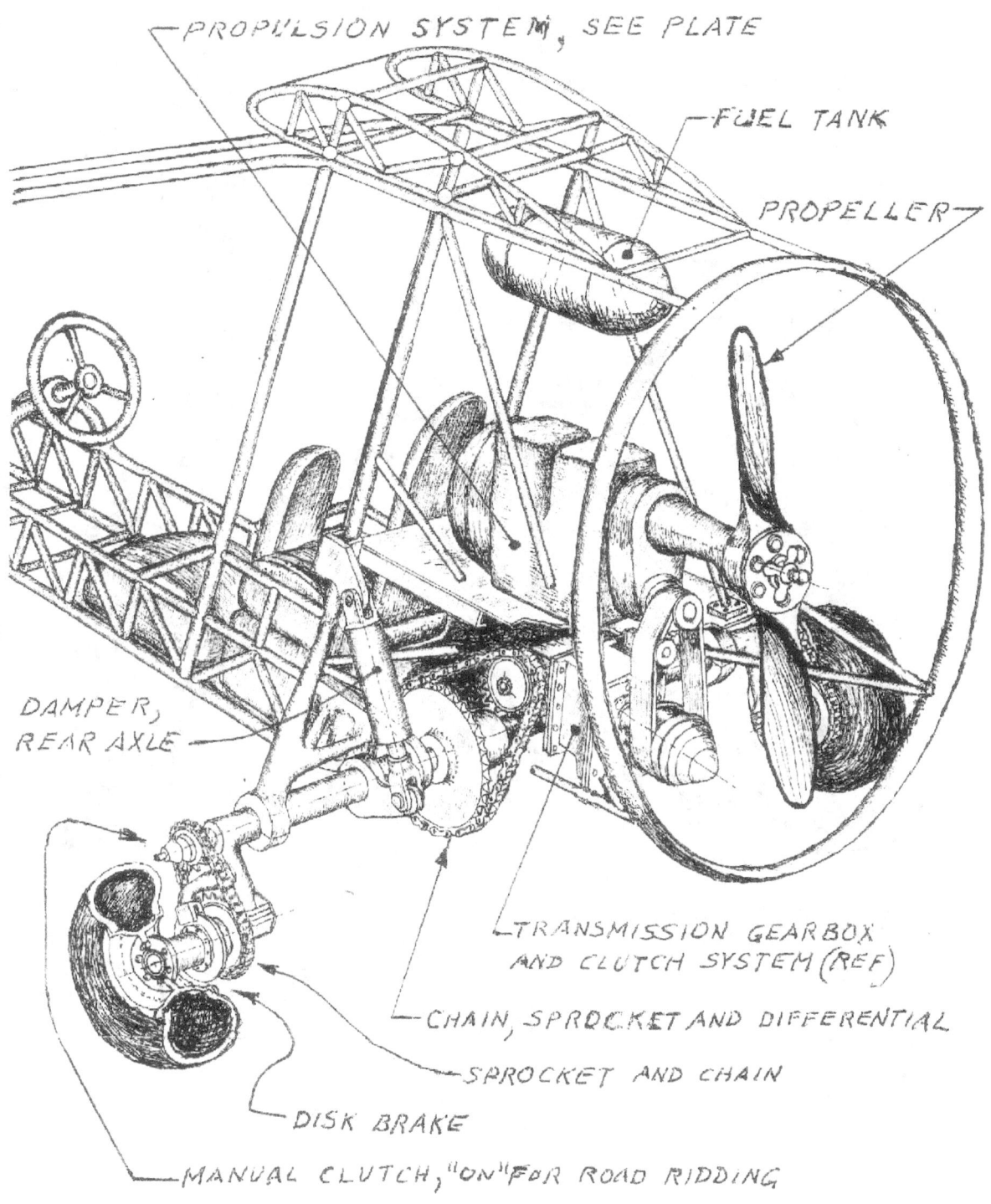

PROPULSION SYSTEM, SEE PLATE

FUEL TANK

PROPELLER

DAMPER,
REAR AXLE

TRANSMISSION GEARBOX
AND CLUTCH SYSTEM (REF)

CHAIN, SPROCKET AND DIFFERENTIAL

SPROCKET AND CHAIN

DISK BRAKE

MANUAL CLUTCH, "ON" FOR ROAD RIDDING

9.   Aerocar's rear view, showing main components (page 2 of 3)

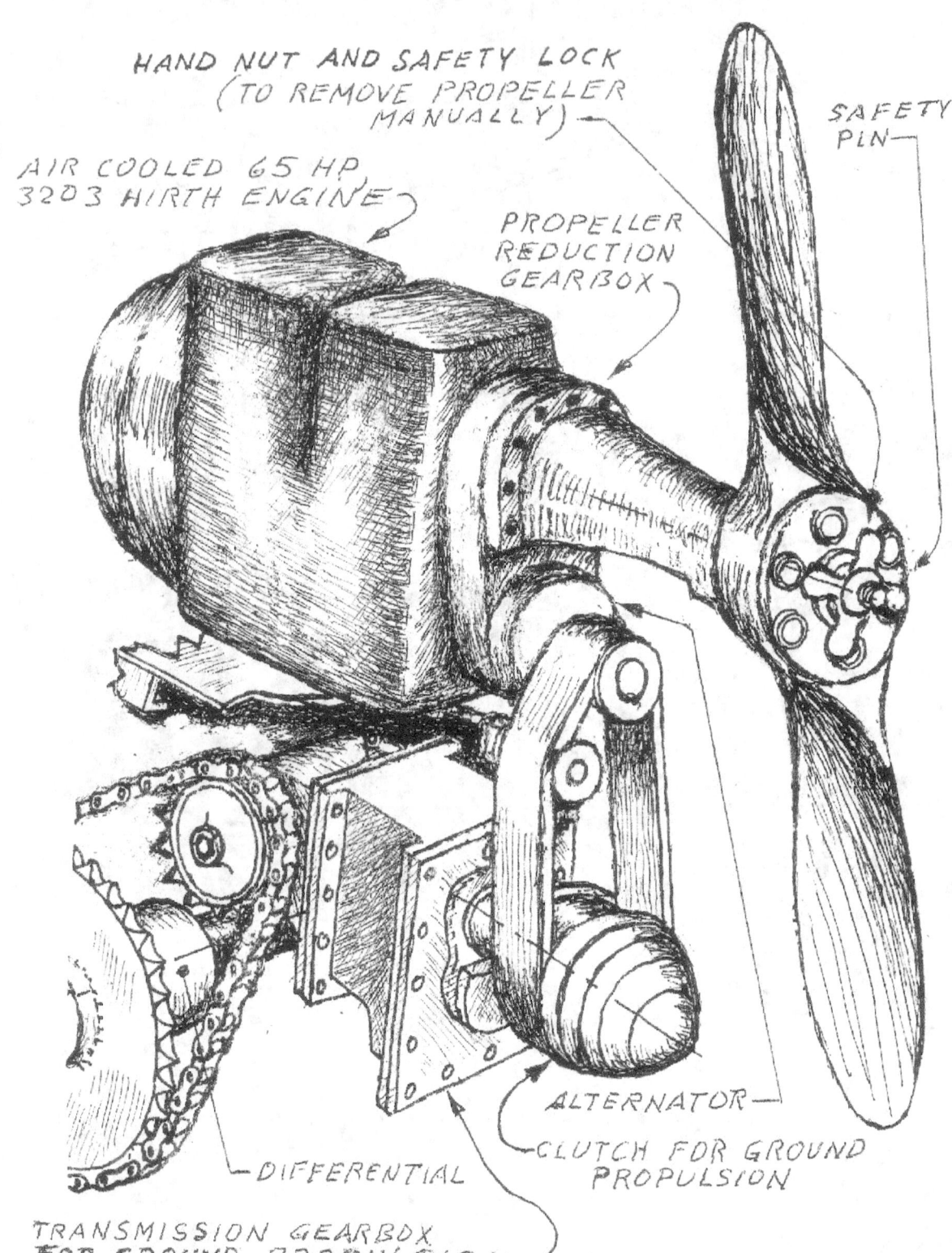

HAND NUT AND SAFETY LOCK
(TO REMOVE PROPELLER
MANUALLY)

AIR COOLED 65 HP
3203 HIRTH ENGINE

PROPELLER
REDUCTION
GEARBOX

SAFETY
PIN

ALTERNATOR

CLUTCH FOR GROUND
PROPULSION

DIFFERENTIAL

TRANSMISSION GEARBOX
FOR GROUND PROPULSION

10. Aerocar, detail of illustration 9. Propulsion system

Engineering projects for the 21st century

GIANT ROBOT

Can transport heavy structures above mountains with no roads

Design Engineer: Walter F. Laredo	Date: August 8, 2002

11. Super giant walking robot (page 1 of 2),
For details see project 1300 in page 17

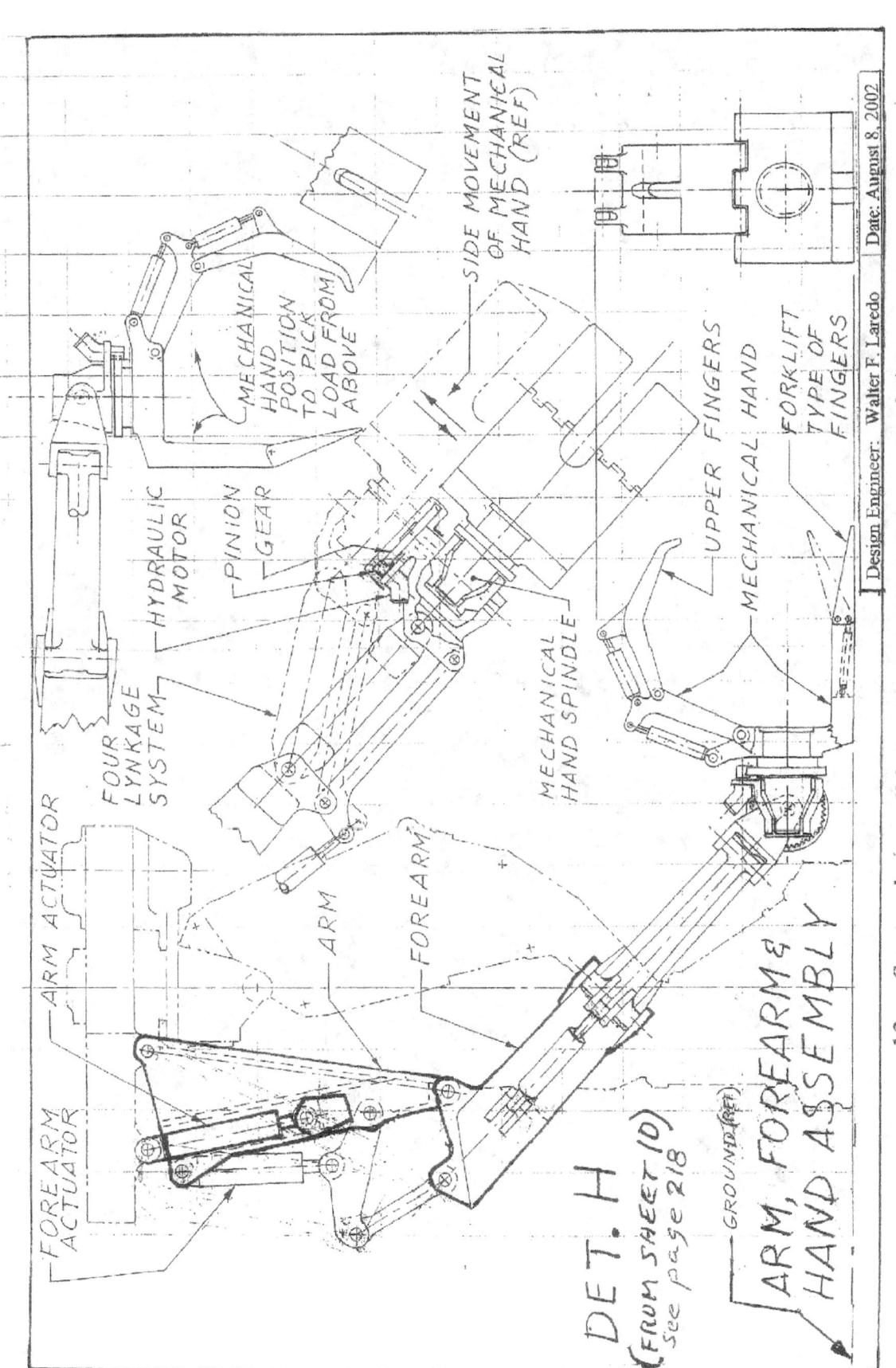

Design Engineer: Walter F. Laredo — Date: August 8, 2002

12. Super giant walking robot, arm detail (page 2 of 2)

DET. H
(FROM SHEET (D)
See page 218)

ARM, FOREARM &
HAND ASSEMBLY

FOREARM
ACTUATOR

ARM ACTUATOR

FOUR
LYNKAGE
SYSTEM

HYDRAULIC
MOTOR

PINION
GEAR

ARM

FOREARM

GROUND (REF)

MECHANICAL
HAND SPINDLE

MECHANICAL HAND
POSITION
TO PICK
LOAD FROM
ABOVE

SIDE MOVEMENT
OF MECHANICAL
HAND (REF)

UPPER FINGERS

MECHANICAL HAND

FORKLIFT
TYPE OF
FINGERS

13. Interplanetary "Mars Spaceship" (page 1 of 3).
For details see project 1400 in page 17

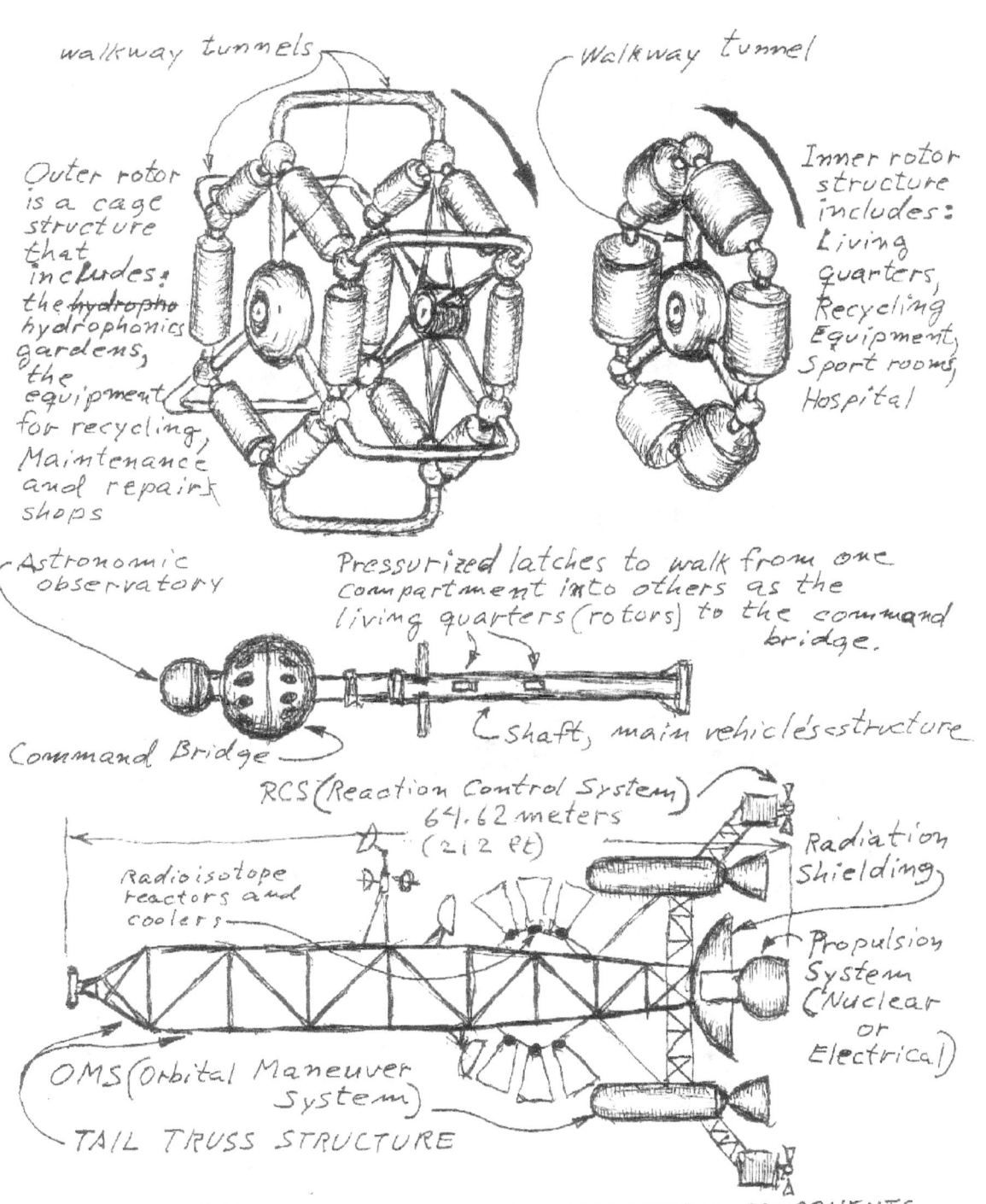

walkway tunnels

Walkway tunnel

Outer rotor is a cage structure that includes: the ~~hydropho~~ hydrophonics gardens, the equipment for recycling, Maintenance and repairs shops

Inner rotor structure includes: Living quarters, Recycling Equipment, Sport rooms, Hospital

Astronomic observatory

Pressurized latches to walk from one compartment into others as the living quarters (rotors) to the command bridge.

Command Bridge

shaft, main vehicle's structure

RCS (Reaction Control System)
64.62 meters
(212 ft)

Radiation Shielding

Radioisotope reactors and coolers

Propulsion System (Nuclear or Electrical)

OMS (Orbital Maneuver System)

TAIL TRUSS STRUCTURE

ABOVE ARE SHOWN THE FOUR MAIN STRUCTURAL COMPONENTS OF THE SPACESHIP

14. "Mars Spaceship" Main Structural Components (page 2 of 3)

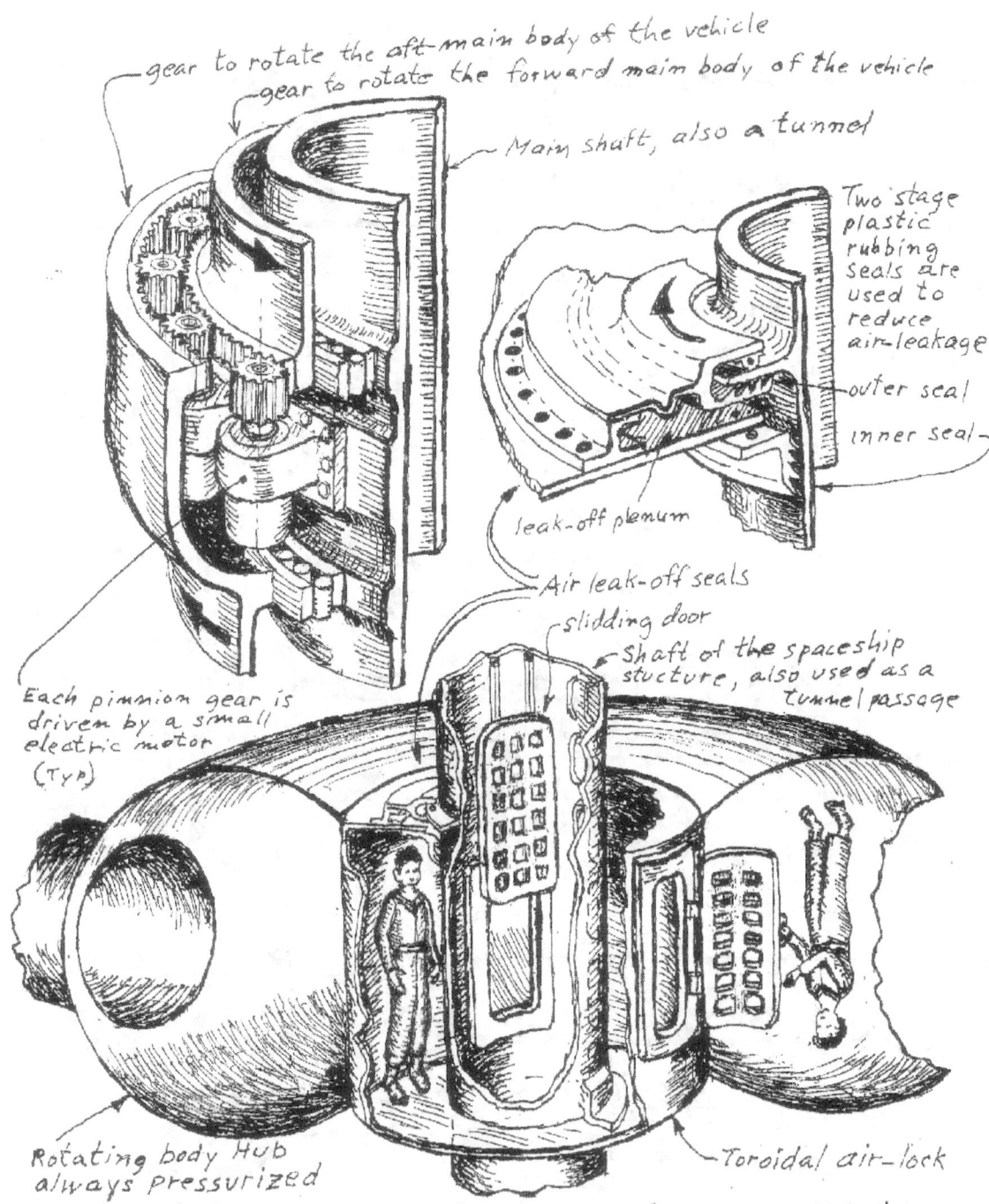

gear to rotate the aft-main body of the vehicle

gear to rotate the forward main body of the vehicle

Main shaft, also a tunnel

Two stage plastic rubbing seals are used to reduce air-leakage

outer seal

inner seal

leak-off plenum

Air leak-off seals

slidding door

Shaft of the spaceship stucture, also used as a tunnel passage

Each pinnion gear is driven by a small electric motor (Typ)

Rotating body Hub always pressurized

Toroidal air-lock

15. (above left) Planetary-gear actuation system, used to counter-rotate the round main bodies of the Mars Spaceship. The airlock shown in picture (below) is used for the transfer of objects and people in between the different sections of the spaceship (page 3 of 3).

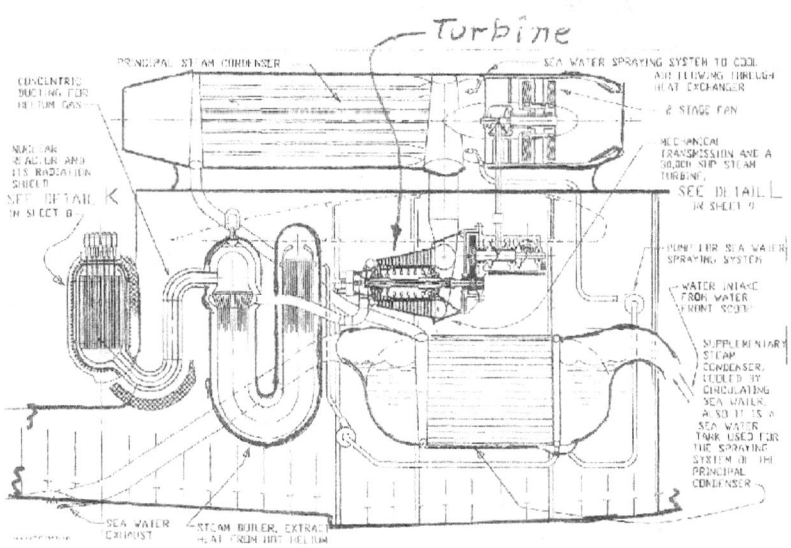

Nuclear Reactor and Shielding

Turbine

16. Cargo aircraft with nuclear propulsion system.
For details see project 1500 in page 17

17. Blended wing-body airliner. For details see project 1600 in page 17

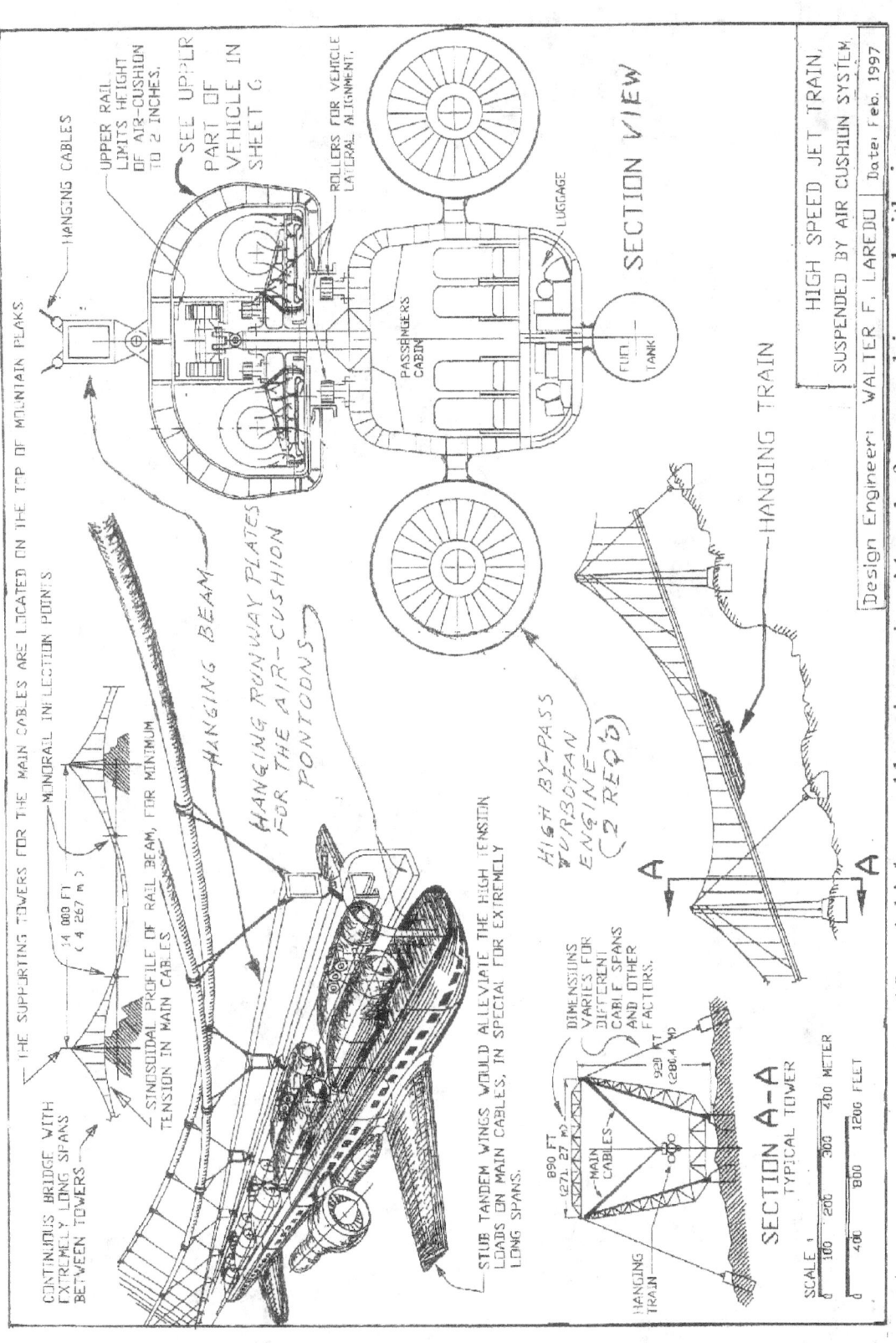

SECTION VIEW

HANGING CABLES

UPPER RAIL LIMITS HEIGHT OF AIR-CUSHION TO 2 INCHES.

SEE UPPER PART OF VEHICLE IN SHEET 6

ROLLERS FOR VEHICLE LATERAL ALIGNMENT.

LUGGAGE

PASSENGERS CABIN

FUEL TANK

HANGING TRAIN

THE SUPPORTING TOWERS FOR THE MAIN CABLES ARE LOCATED ON THE TOP OF MOUNTAIN PEAKS

MINIMAL INFLECTION POINTS

HANGING BEAM

HANGING RUNWAY PLATES FOR THE AIR-CUSHION PONTOONS

CONTINUOUS BRIDGE WITH EXTREMELY LONG SPANS BETWEEN TOWERS

14 000 FT (4 267 m)

SINUSOIDAL PROFILE OF RAIL BEAM, FOR MINIMUM TENSION IN MAIN CABLES

HIGH BY-PASS TURBOFAN ENGINE (2 REQD)

STUB TANDEM WINGS WOULD ALLEVIATE THE HIGH TENSION LOADS ON MAIN CABLES, IN SPECIAL FOR EXTREMELY LONG SPANS.

A
A

HANGING TRAIN

SECTION A-A
TYPICAL TOWER

DIMENSIONS VARIES FOR DIFFERENT CABLE SPANS AND OTHER FACTORS.

920 FT (280.4 m)

890 FT (271.27 m)

MAIN CABLES

HANGING TRAIN

SCALE :
0 100 200 300 400 METER
0 400 800 1200 FEET

HIGH SPEED JET TRAIN.
SUSPENDED BY AIR CUSHION SYSTEM.

Design Engineer: WALTER F. LAREDO | Date: Feb. 1997

18. A high speed hanging train with turbofan propulsion and with air cushion suspension. For details see project 1700 in page 17

19. Hypersonic aircraft with scramjet propulsion, uses LH2 fuel, (page 1 of 2). For details see project 1800 in page 17

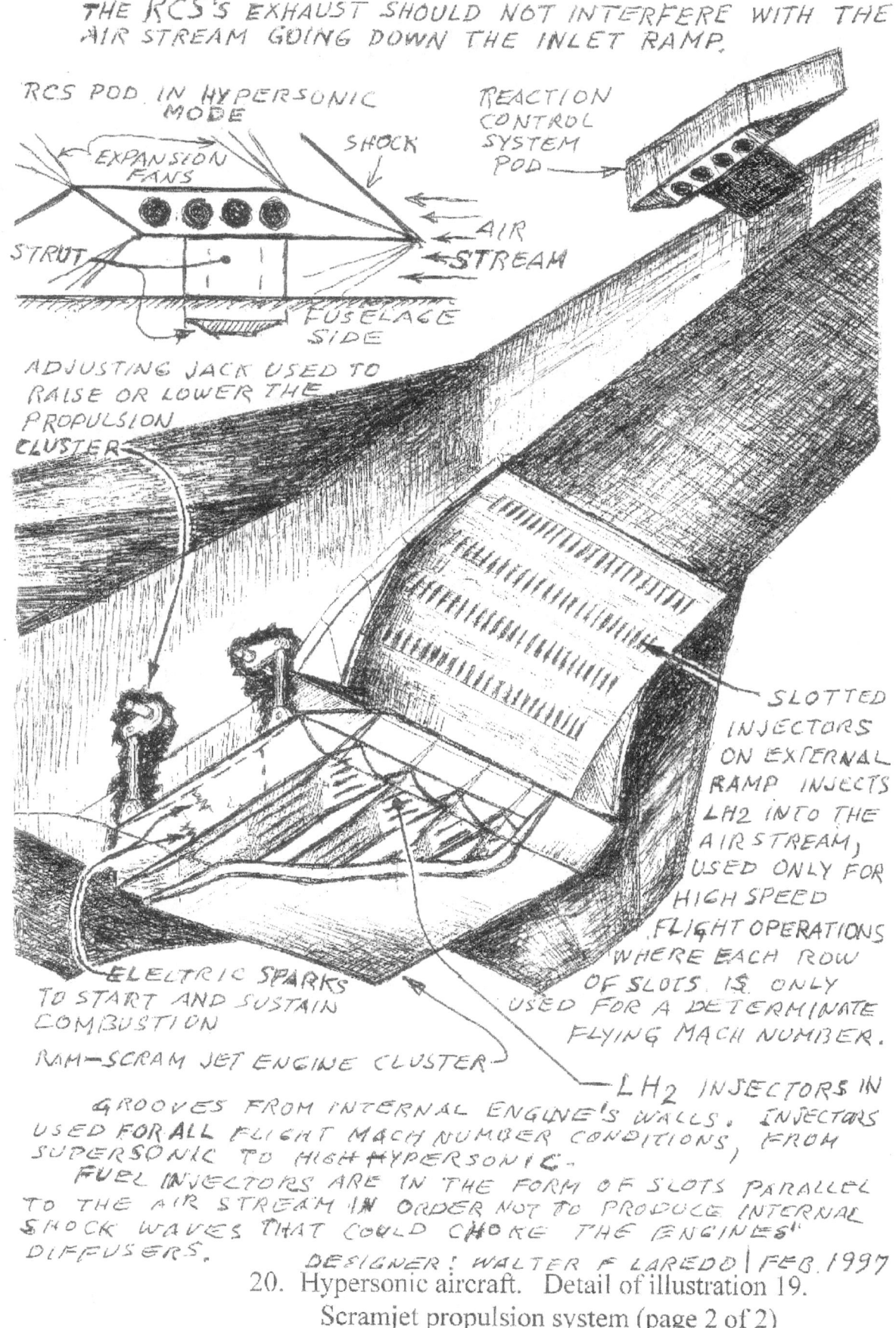

THE RCS'S EXHAUST SHOULD NOT INTERFERE WITH THE AIR STREAM GOING DOWN THE INLET RAMP.

RCS POD IN HYPERSONIC MODE

EXPANSION FANS

SHOCK

REACTION CONTROL SYSTEM POD

STRUT

AIR STREAM

FUSELAGE SIDE

ADJUSTING JACK USED TO RAISE OR LOWER THE PROPULSION CLUSTER

SLOTTED INJECTORS ON EXTERNAL RAMP INJECTS LH2 INTO THE AIR STREAM, USED ONLY FOR HIGH SPEED FLIGHT OPERATIONS WHERE EACH ROW OF SLOTS IS ONLY USED FOR A DETERMINATE FLYING MACH NUMBER.

ELECTRIC SPARKS TO START AND SUSTAIN COMBUSTION

RAM-SCRAM JET ENGINE CLUSTER

LH2 INJECTORS IN GROOVES FROM INTERNAL ENGINE'S WALLS. INJECTORS USED FOR ALL FLIGHT MACH NUMBER CONDITIONS, FROM SUPERSONIC TO HIGH HYPERSONIC.
FUEL INJECTORS ARE IN THE FORM OF SLOTS PARALLEL TO THE AIR STREAM IN ORDER NOT TO PRODUCE INTERNAL SHOCK WAVES THAT COULD CHOKE THE ENGINES' DIFFUSERS.

DESIGNER: WALTER F LAREDO | FEB. 1997

20. Hypersonic aircraft. Detail of illustration 19.
Scramjet propulsion system (page 2 of 2)

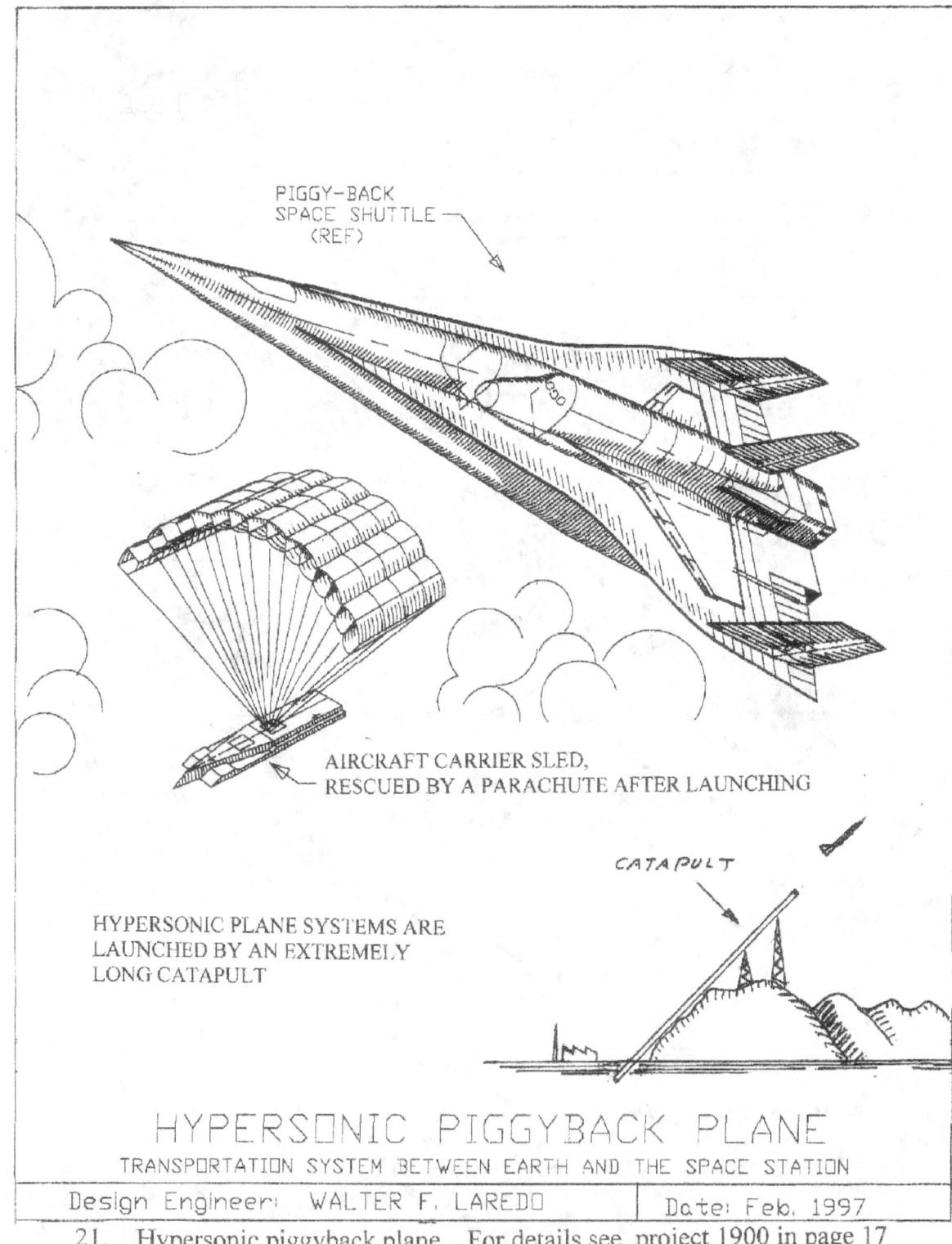

PIGGY-BACK
SPACE SHUTTLE
(REF)

AIRCRAFT CARRIER SLED,
RESCUED BY A PARACHUTE AFTER LAUNCHING

CATAPULT

HYPERSONIC PLANE SYSTEMS ARE
LAUNCHED BY AN EXTREMELY
LONG CATAPULT

HYPERSONIC PIGGYBACK PLANE
TRANSPORTATION SYSTEM BETWEEN EARTH AND THE SPACE STATION

Design Engineer: WALTER F. LAREDO   Date: Feb. 1997

21.   Hypersonic piggyback plane. For details see project 1900 in page 17

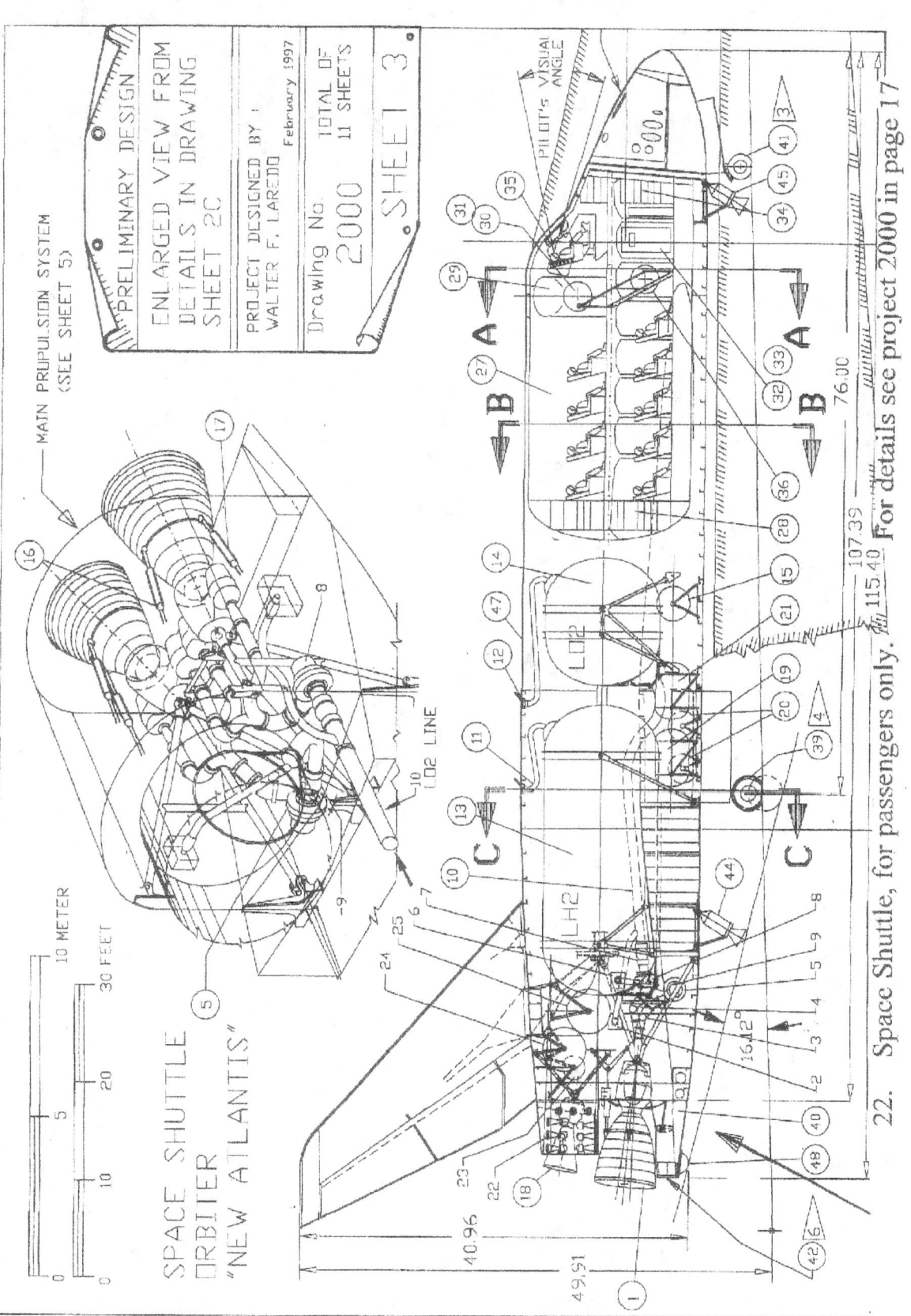

22. Space Shuttle, for passengers only. For details see project 2000 in page 17

WARNING:
Vibrations and shaking caused by wing flapping could harm the rider's back and brain.
Would be better to ride a fix-wing mechanical horse.

23. Toys for the rich sportsmen of the future, a mechanical flying Horse (page 1 of 4). For details see project 1100 in page 17

FLYING MOTORCYCLE

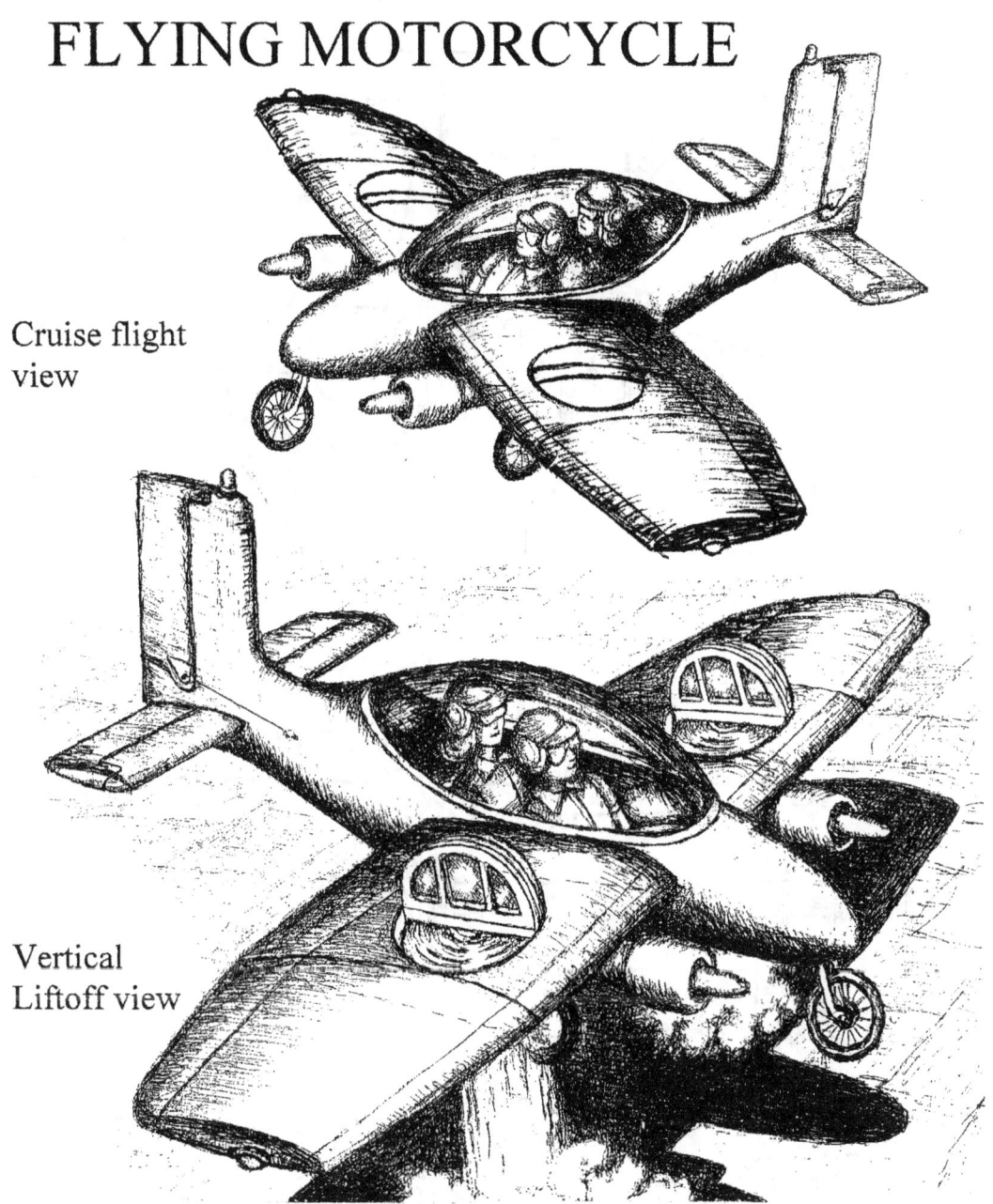

Cruise flight
view

Vertical
Liftoff view

The commonality of components between the Flying Motorcycle and the Flying
Horse are the following:
The fuselage tubular structure, the wing including its structural and mechanical
systems, Landing gear, the propulsion system (jet engines, liftoff fans,
gear boxes, cross shafted transmission, upper and lower liftoff fan doors).
The location of the exhaust of the Reaction Control System (RCS) for both
vehicles, for roll control are the same, but for pitch control are different.
The nose and tail assemblies for both vehicles are different.

Illustration 23 bis, Flying Motorcycle

FLYING
MOTORCYCLE

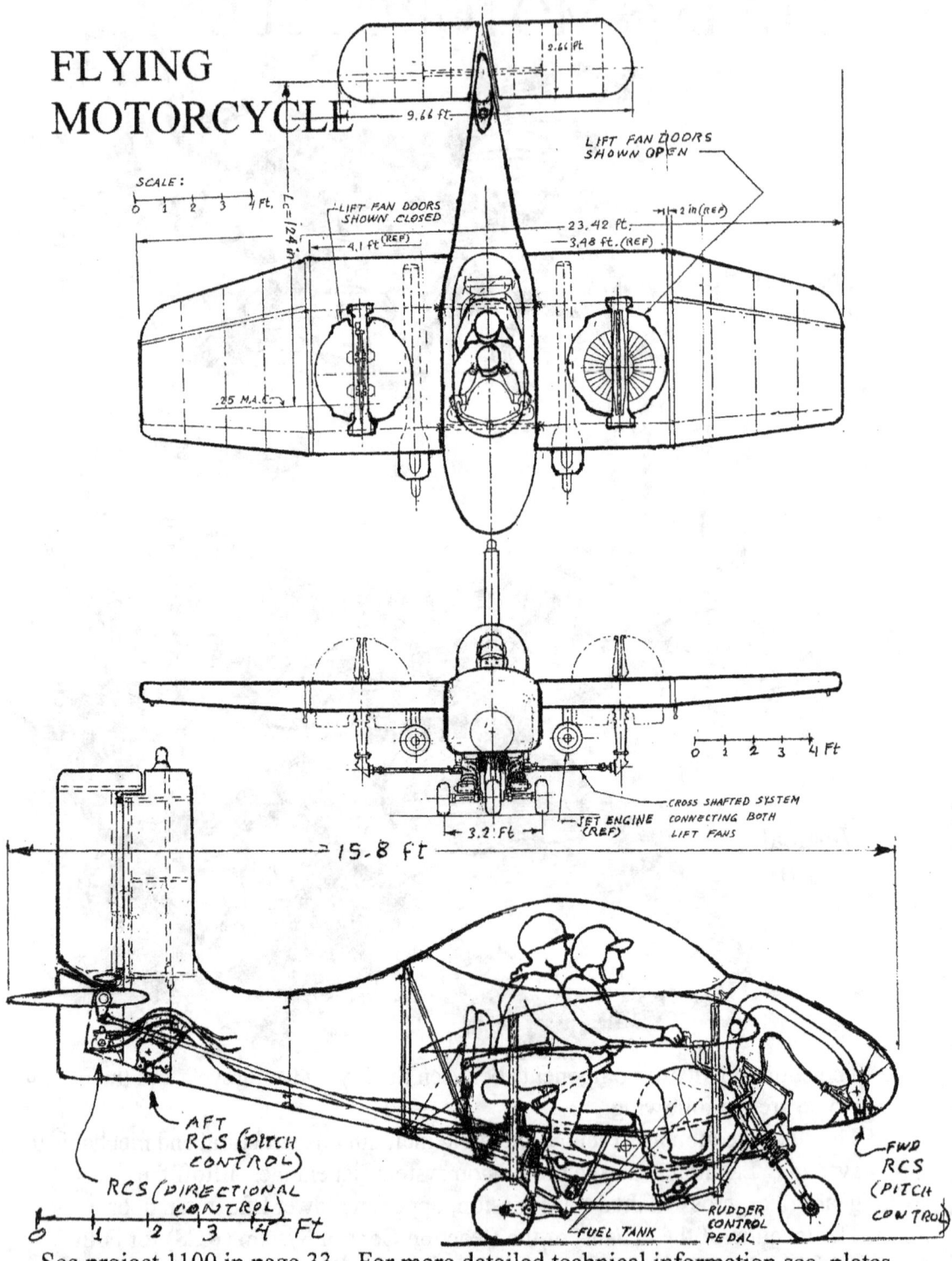

SCALE:
0 1 2 3 4 Ft.

LIFT FAN DOORS
SHOWN CLOSED

LIFT FAN DOORS
SHOWN OPEN

2.66 ft.

9.66 ft.

Lc= 124 in

4.1 ft (REF)

23.42 ft.

3.48 ft. (REF)

2 in (REF)

.25 M.A.C.

0 1 2 3 4 Ft

CROSS SHAFTED SYSTEM
CONNECTING BOTH
LIFT FANS

JET ENGINE
(REF)

3.2 Ft

15.8 ft

AFT
RCS (PITCH
CONTROL)

RCS (DIRECTIONAL
CONTROL)

FWD
RCS
(PITCH
CONTROL)

FUEL TANK

RUDDER
CONTROL
PEDAL

0 1 2 3 4 Ft

See project 1100 in page 33, For more detailed technical information see plates
81, 85 and plates 87 through 98.
    Additional information could be extracted from plates 73.74, 83, 84 and 86

Continuation of Illustration 23 bis, Flying Motorcycle

24. Mechanical Flying Horse, in flying mode (above),
and in lift-off mode (below), page 2 of 4

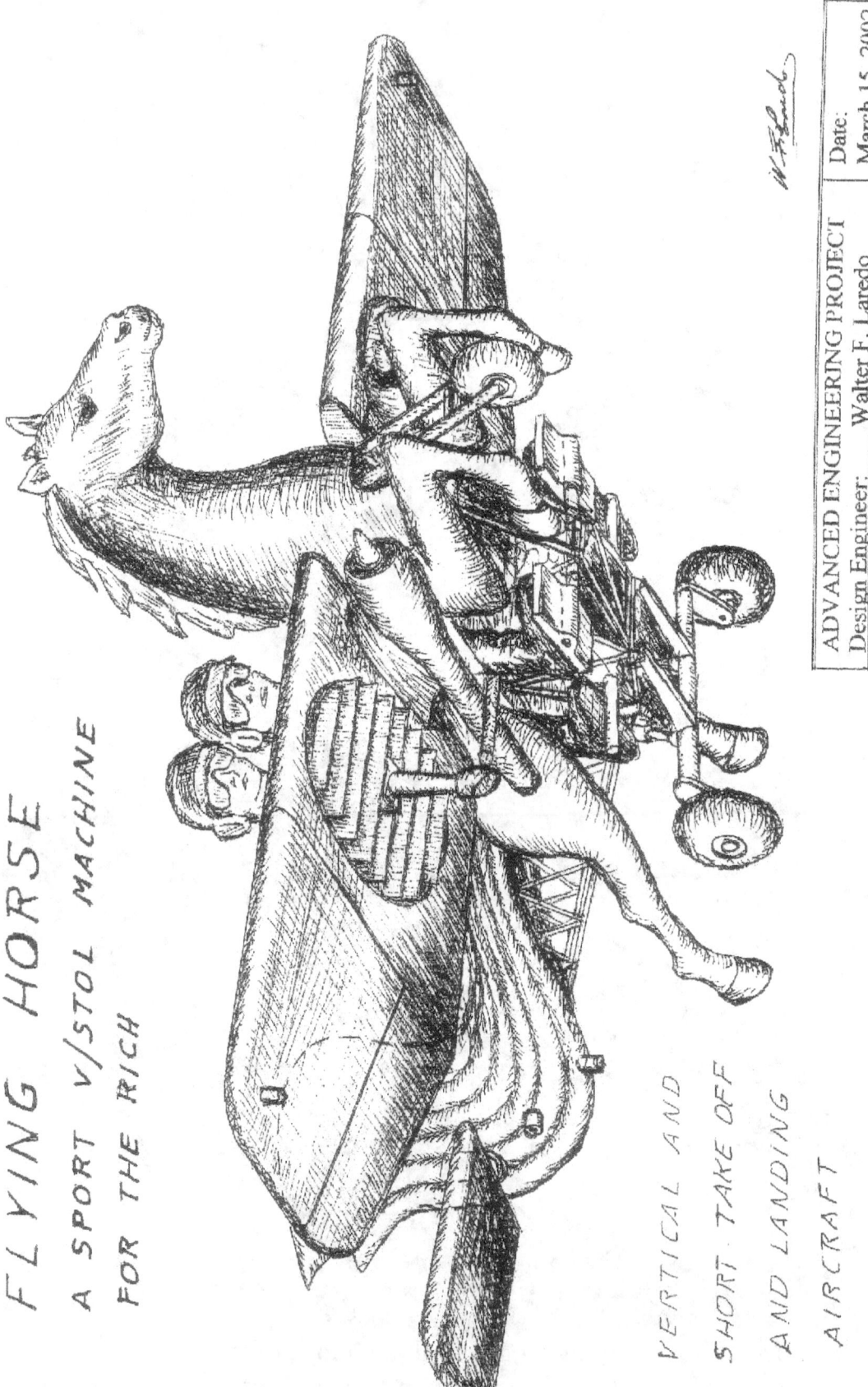

FLYING HORSE

A SPORT V/STOL MACHINE

FOR THE RICH

VERTICAL AND

SHORT TAKE OFF

AND LANDING

AIRCRAFT

| ADVANCED ENGINEERING PROJECT | Date: |
| Design Engineer: Walter F. Laredo | March 15, 2002 |

25. Mechanical Flying Horse. Vertical lift-off view (page 3 of 4)

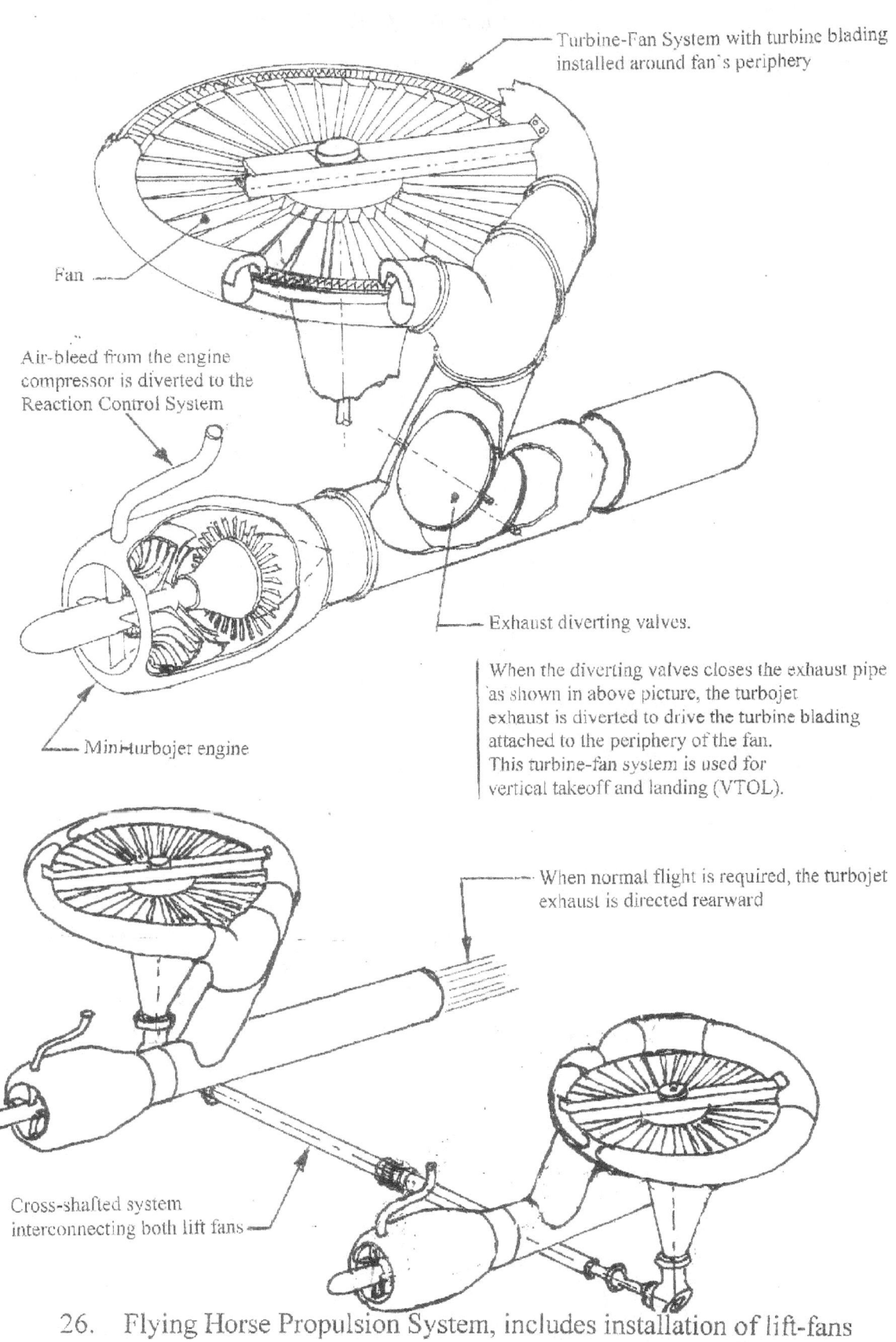

Turbine-Fan System with turbine blading installed around fan's periphery

Fan

Air-bleed from the engine compressor is diverted to the Reaction Control System

Exhaust diverting valves.

When the diverting valves closes the exhaust pipe as shown in above picture, the turbojet exhaust is diverted to drive the turbine blading attached to the periphery of the fan. This turbine-fan system is used for vertical takeoff and landing (VTOL).

Mini-turbojet engine

When normal flight is required, the turbojet exhaust is directed rearward

Cross-shafted system interconnecting both lift fans

26.   Flying Horse Propulsion System, includes installation of lift-fans
(page 4 of 4)

Engineering projects for the 21st century

Mountain climbing robot, Designed to find water on planet Mars

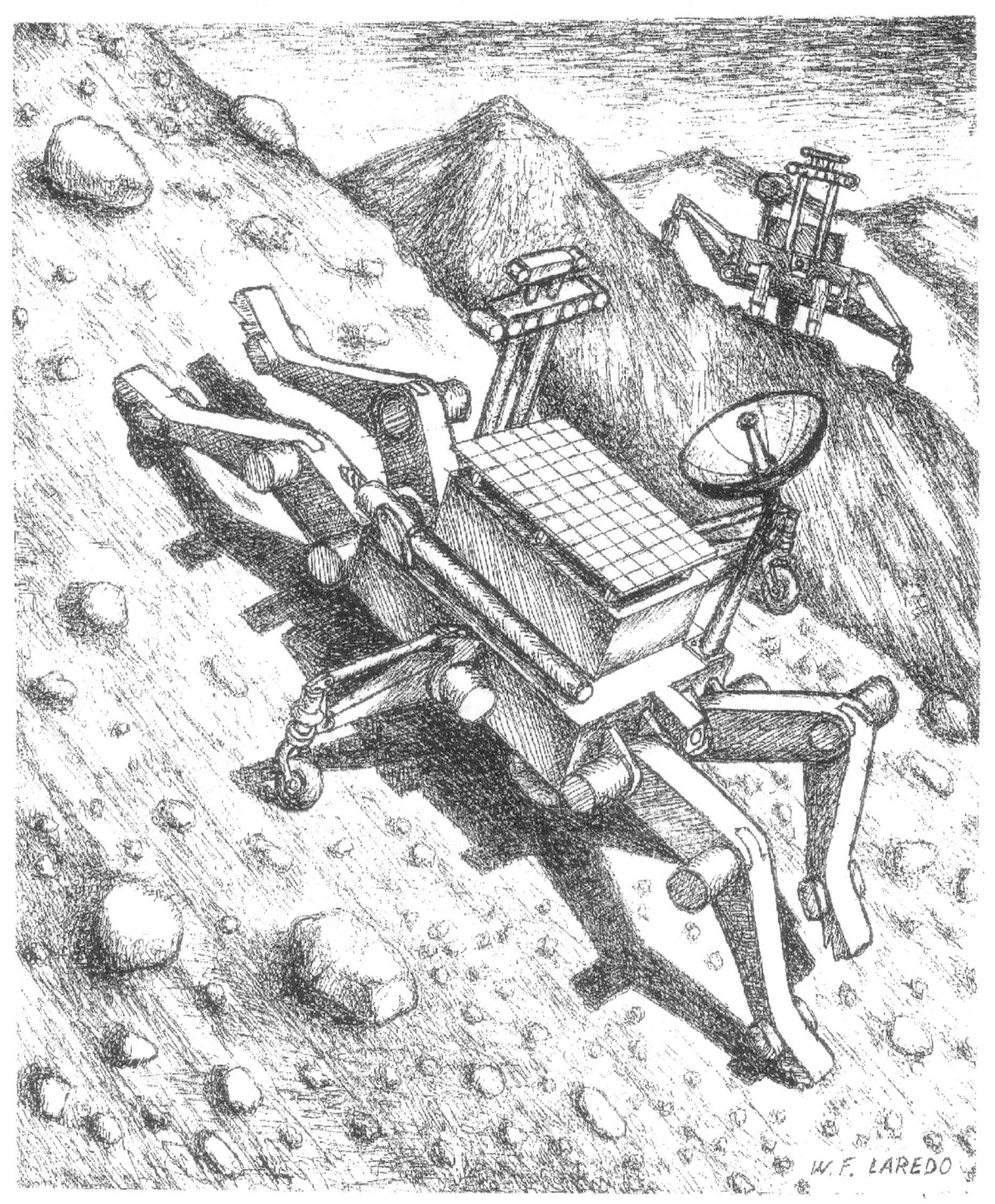

27.　Mountain climbing robot, designed to find water on planet Mars
(page 1 of 12)

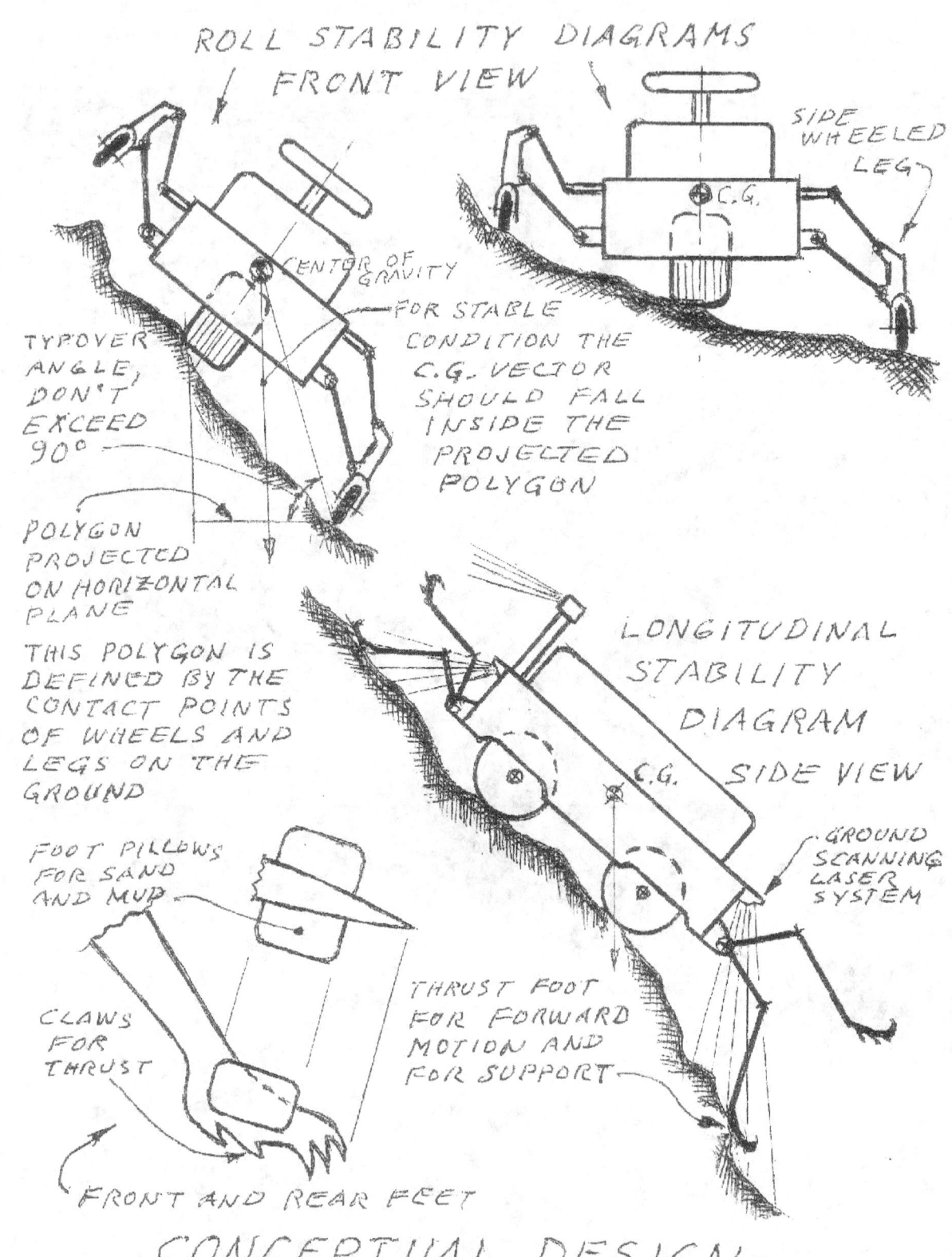

ROLL STABILITY DIAGRAMS
FRONT VIEW

SIDE WHEELED LEG

CENTER OF GRAVITY

C.G.

TYPOVER ANGLE, DON'T EXCEED 90°

FOR STABLE CONDITION THE C.G. VECTOR SHOULD FALL INSIDE THE PROJECTED POLYGON

POLYGON PROJECTED ON HORIZONTAL PLANE

THIS POLYGON IS DEFINED BY THE CONTACT POINTS OF WHEELS AND LEGS ON THE GROUND

LONGITUDINAL STABILITY DIAGRAM SIDE VIEW

C.G.

GROUND SCANNING LASER SYSTEM

FOOT PILLOWS FOR SAND AND MUD

CLAWS FOR THRUST

THRUST FOOT FOR FORWARD MOTION AND FOR SUPPORT

FRONT AND REAR FEET

CONCEPTUAL DESIGN

28. Conceptual design and stability diagrams.
Mountain climbing robot (page 2 of 12)

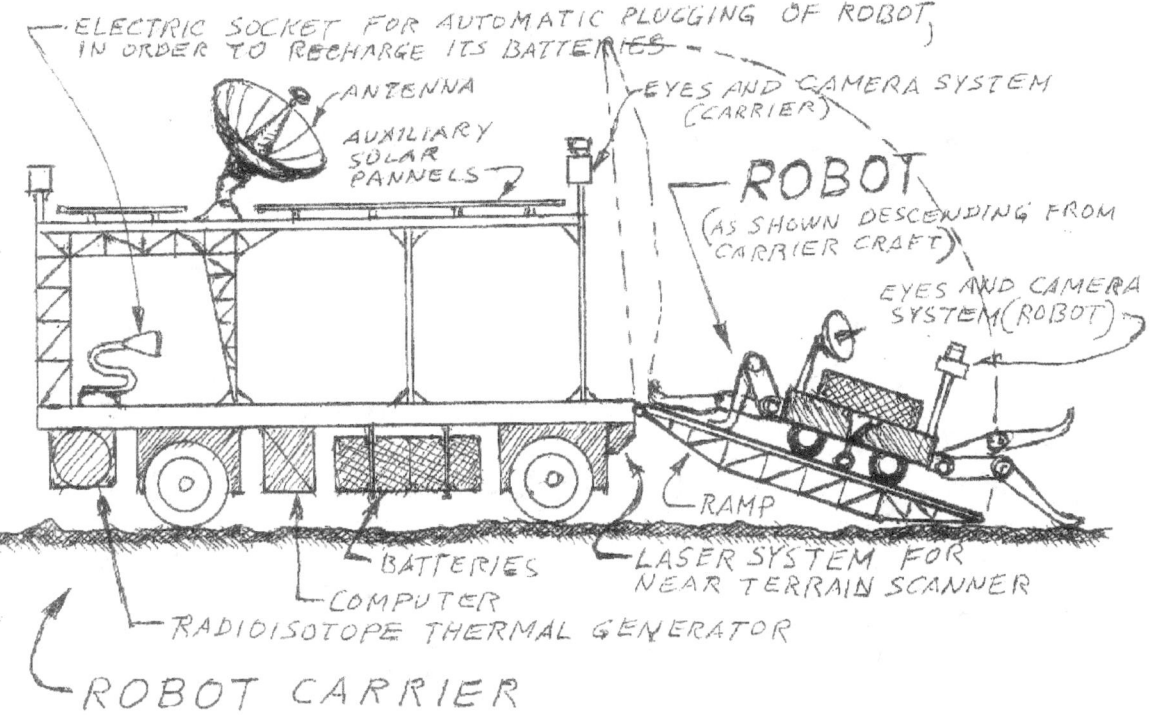

ELECTRIC SOCKET FOR AUTOMATIC PLUGGING OF ROBOT, IN ORDER TO RECHARGE ITS BATTERIES

ANTENNA

AUXILIARY SOLAR PANNELS

EYES AND CAMERA SYSTEM (CARRIER)

ROBOT
(AS SHOWN DESCENDING FROM CARRIER CRAFT)

EYES AND CAMERA SYSTEM (ROBOT)

RAMP

LASER SYSTEM FOR NEAR TERRAIN SCANNER

BATTERIES

COMPUTER

RADIOISOTOPE THERMAL GENERATOR

ROBOT CARRIER

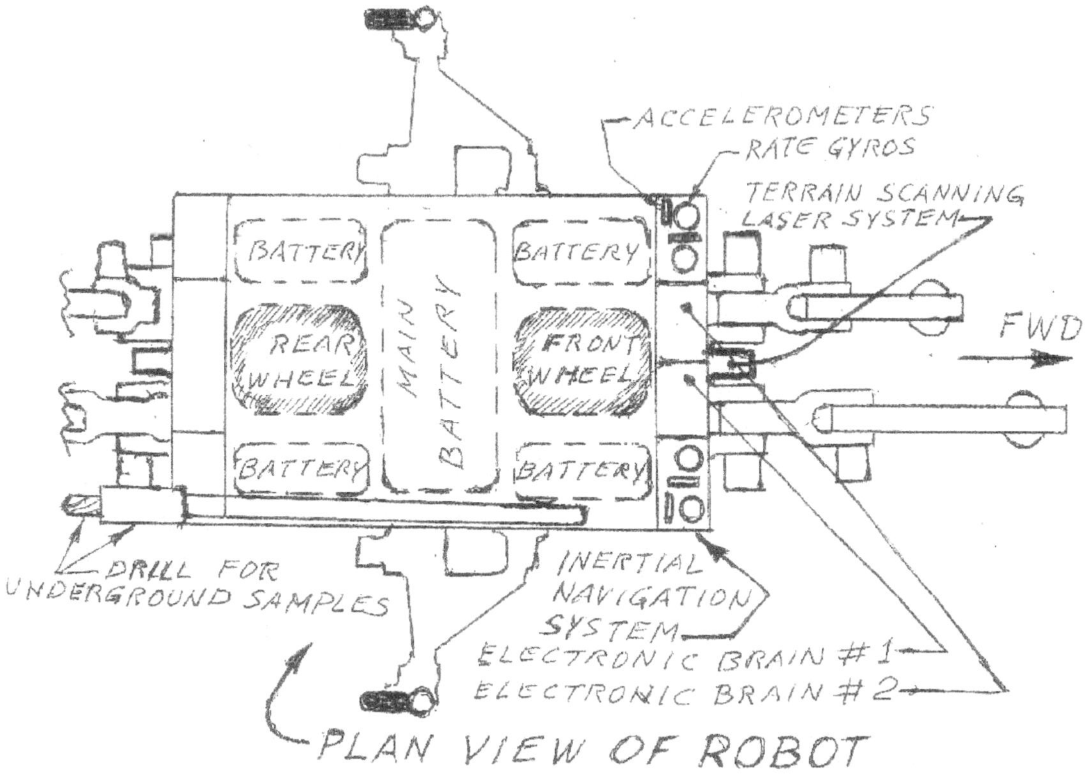

ACCELEROMETERS
RATE GYROS

TERRAIN SCANNING LASER SYSTEM

BATTERY

BATTERY

MAIN BATTERY

REAR WHEEL

FRONT WHEEL

FWD

BATTERY

BATTERY

DRILL FOR UNDERGROUND SAMPLES

INERTIAL NAVIGATION SYSTEM

ELECTRONIC BRAIN #1
ELECTRONIC BRAIN #2

PLAN VIEW OF ROBOT

29. Robot's carrier vehicle and robot's plan view (page 3 of 12)

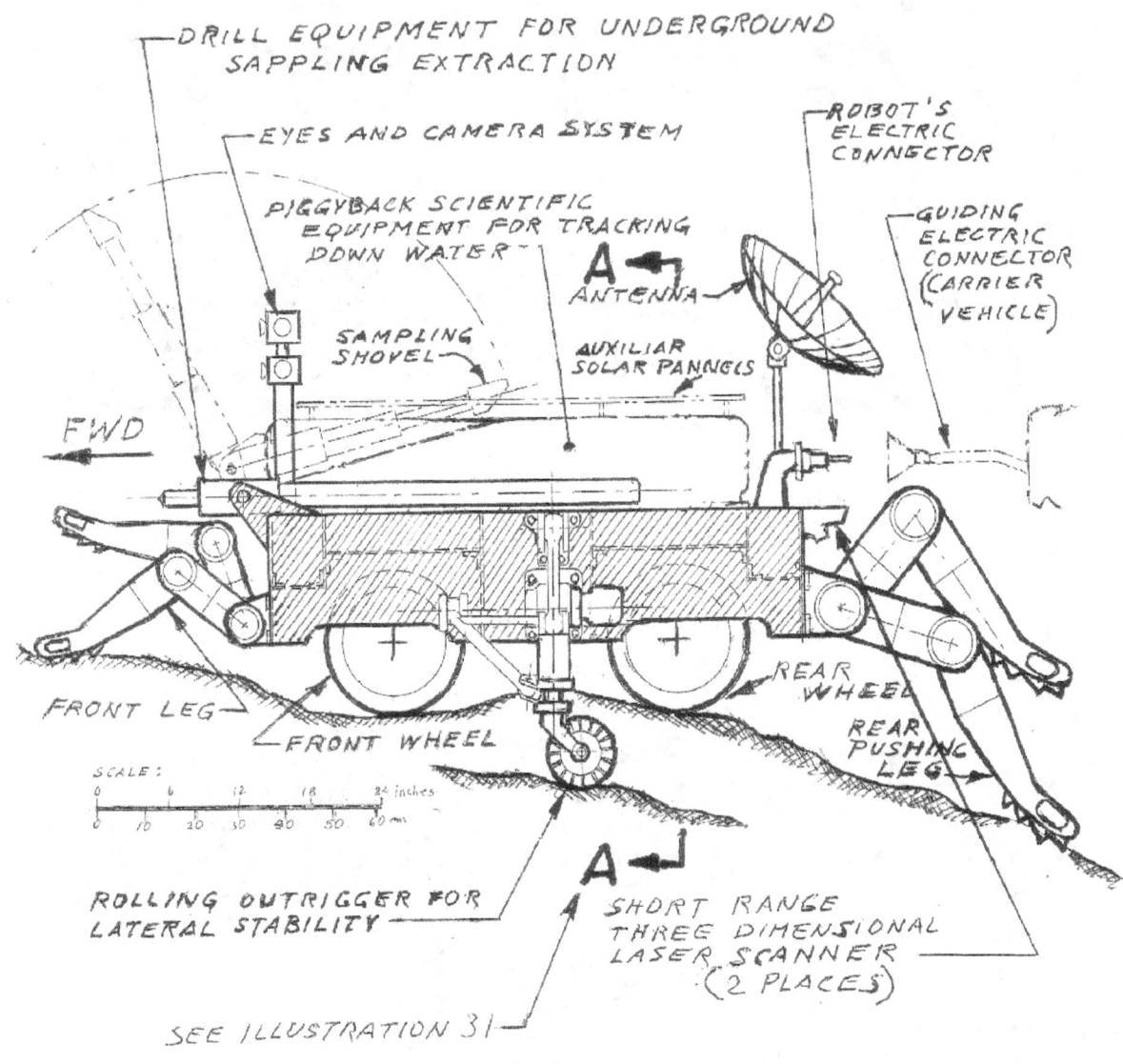

DRILL EQUIPMENT FOR UNDERGROUND
SAPPLING EXTRACTION

EYES AND CAMERA SYSTEM

ROBOT'S
ELECTRIC
CONNECTOR

PIGGYBACK SCIENTIFIC
EQUIPMENT FOR TRACKING
DOWN WATER

GUIDING
ELECTRIC
CONNECTOR
(CARRIER
VEHICLE)

A

ANTENNA

SAMPLING
SHOVEL

AUXILIAR
SOLAR PANNELS

FWD

FRONT LEG

FRONT WHEEL

REAR
WHEEL

REAR
PUSHING
LEG

SCALE:
0   6   12   18   24 inches
0   10   20   30   40   50   60 mm

ROLLING OUTRIGGER FOR
LATERAL STABILITY

A

SHORT RANGE
THREE DIMENSIONAL
LASER SCANNER
(2 PLACES)

SEE ILLUSTRATION 31

30.   Side view. Mountain climbing robot (page 4 of 12)

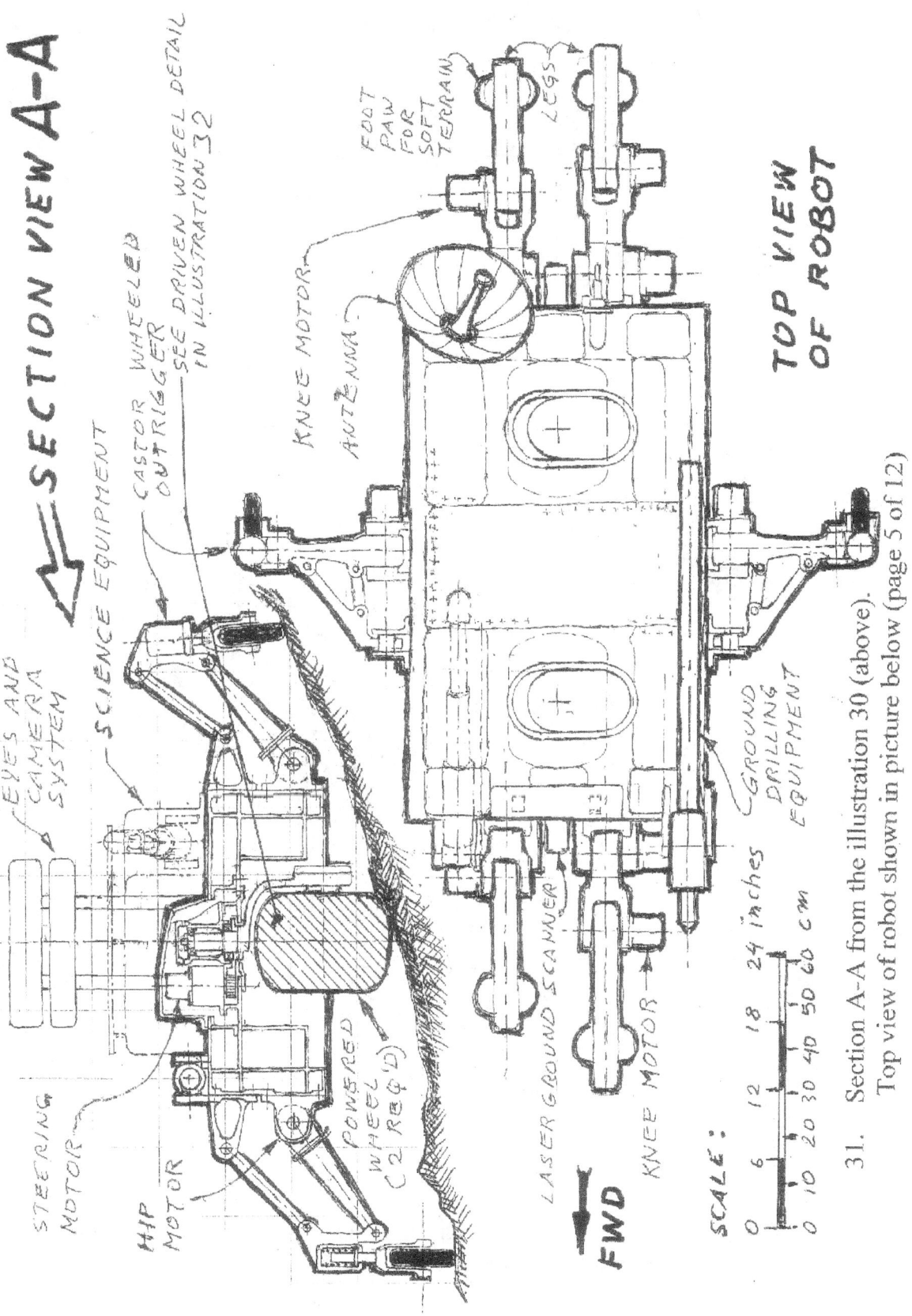

SECTION VIEW A-A

EYES AND CAMERA SYSTEM

SCIENCE EQUIPMENT

CASTOR WHEELED OUTRIGGER

SEE DRIVEN WHEEL DETAIL IN ILLUSTRATION 32

KNEE MOTOR

ANTENNA

FOOT PAW FOR SOFT TERRAIN

LEGS

STEERING MOTOR

HIP MOTOR

POWERED WHEEL (2 REQ'D)

LASER GROUND SCANNER

GROUND DRILLING EQUIPMENT

KNEE MOTOR

FWD

TOP VIEW OF ROBOT

SCALE:

```
0        6      12      18      24 inches
0   10  20  30  40  50  60  cm
```

31.   Section A-A from the illustration 30 (above).
      Top view of robot shown in picture below (page 5 of 12)

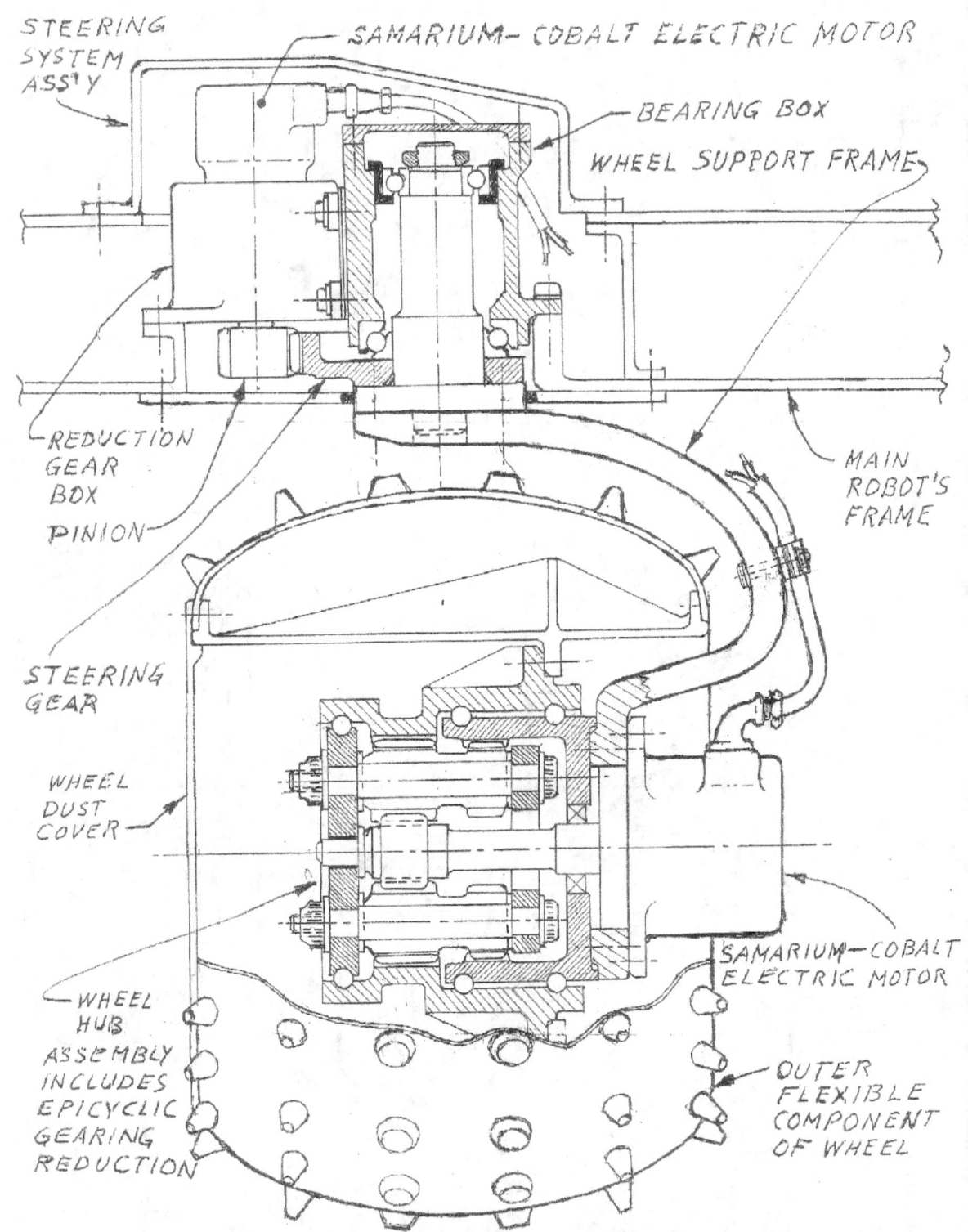

STEERING SYSTEM ASS'Y

SAMARIUM – COBALT ELECTRIC MOTOR

BEARING BOX

WHEEL SUPPORT FRAME

MAIN ROBOT'S FRAME

REDUCTION GEAR BOX

PINION

STEERING GEAR

WHEEL DUST COVER

SAMARIUM – COBALT ELECTRIC MOTOR

WHEEL HUB ASSEMBLY INCLUDES EPICYCLIC GEARING REDUCTION

OUTER FLEXIBLE COMPONENT OF WHEEL

32. Detail from illustration 31. Driving wheel system, includes motors, reduction gears and steering system (page 6 of 12)

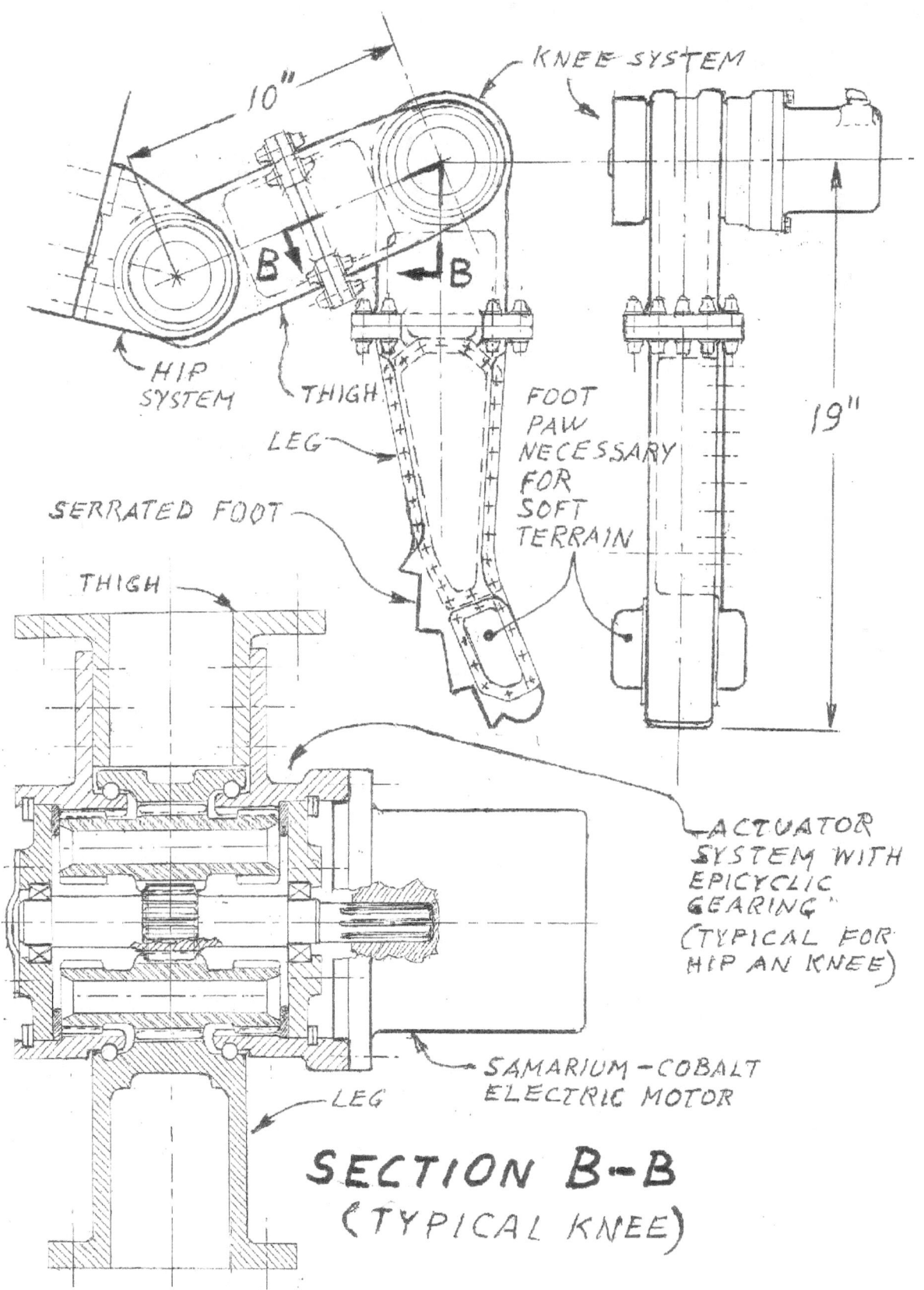

10"

KNEE SYSTEM

B

B

19"

HIP
SYSTEM

THIGH

LEG

FOOT
PAW
NECESSARY
FOR
SOFT
TERRAIN

SERRATED FOOT

THIGH

ACTUATOR
SYSTEM WITH
EPICYCLIC
GEARING
(TYPICAL FOR
HIP AN KNEE)

SAMARIUM - COBALT
ELECTRIC MOTOR

LEG

SECTION B-B
(TYPICAL KNEE)

33.   Leg actuation system, typical for all legs.
Mountain climbing robot (page 7 of 12)

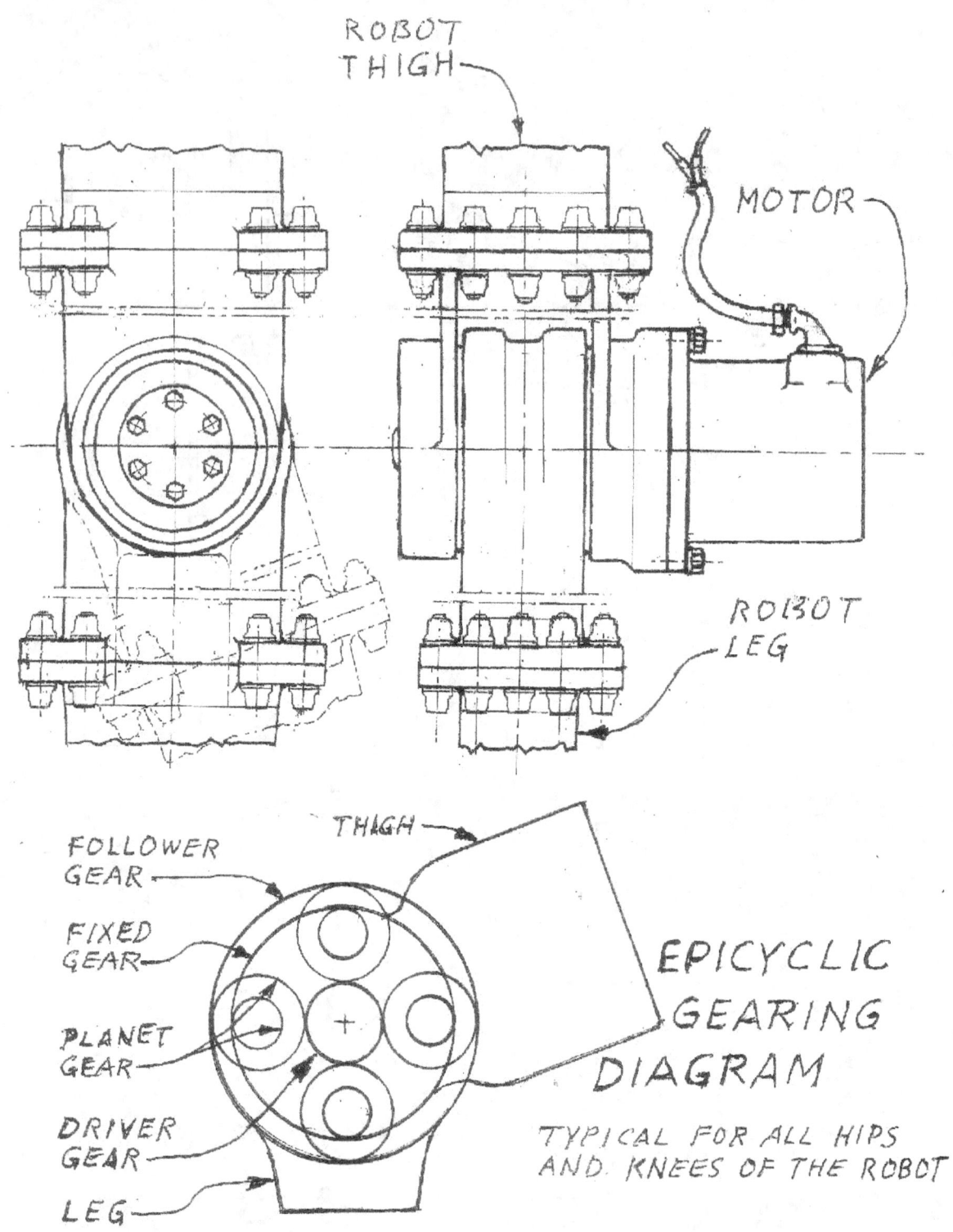

ROBOT THIGH

MOTOR

ROBOT LEG

FOLLOWER GEAR

FIXED GEAR

PLANET GEAR

DRIVER GEAR

LEG

THIGH

EPICYCLIC GEARING DIAGRAM

TYPICAL FOR ALL HIPS AND KNEES OF THE ROBOT

34.   Typical leg's knee, includes motors and reduction gears.
Mountain climbing robot (page 8 of 12)

MOUNTAIN CLIMBING ROBOT

FOR WATER EXPLORATION ON MARS,

By Walter F. Laredo

The robot is designed to verify if there is water at some presumed locations of Mars, an important step prior to the first trip to Mars with human beings.

This design is a modification of a previous robot design created by the author as an unsolicited proposal (June 2, 1983).

This robot is an all-electrical vehicle platform powered by batteries. The first three or four robots will be made from aluminum alloy and assigned for testing here on Earth. Later robots to be sent to Mars will be made from beryllium, a good material for lightness and strength. The robot will be able to walk on highly sloped mountain terrain at a speed of five to eight inches per second.

Once on Mars, the robot would be transported in a four-wheeled carrier vehicle to the place where exploration is to start. After arrival, the robot will descend from its carrier on a tilt ramp and will explore the assigned area.

After the robot returns from its mission, it will climb back into the carrier's platform and get automatically plugged to an electric socket in the carrier vehicle to recharge its depleted batteries for the next mission.

In an emergency situation, for example if the robot wants to return to its carrier but the batteries are dead, some relatively small solar cells on the robot's back can provide it with enough power for its slow return.

The large carrier craft also runs with batteries, which are constantly charged by a radioisotope thermal generator located on the rear of the vehicle. Large solar cell panels located on its roof also contribute to charging the carrier batteries.

Inputs will provide for some of the autonomous reflexes that contributes to the 8robot's motion coordination, such as walking or climbing over an object. They also allow the robot to avoid obstacles and to find openings to pass through. It has sensors on its head, body and legs. A microprocessor in each leg dictates when it should be lifted up or set down by means of its own electric motors. The whole leg's system is synchronized to move in a sequence that lets the robot walk. It is an autonomous robot that can roam through mountains and fields, if it has enough information about the terrain.

Electronic brains in general operate in a series of steps. However, it would be faster if some of their perceptions were arranged in parallel. The robot head is outfitted with a special camera, roll and pith inclinometers for balance, and a range finder, which sends out a scanning laser beam to judge how far away are obstacles which will be in the robot's path scheduled for the robot's path, and whether those rocks are too big to climb over. Gyroscopes give the robot the concept of horizontality and directionality. With its gyroscopes the robot would follow a constant heading.

A drill mounted on the robot's frame will be used to extract samples from below the Martian surface. A manipulatable sampler scoop, a soil-water sensor and on-board sample analysis system are part of the piggyback mini-laboratory. This piggyback equipment also includes a facsimile camera system and has the capability for direct robot-Earth communications. The robot has two brains. one of the robot's two brains is able to interpret the mapping image of the ground surface, and also is connected to hazard sensors and to an automatic steering system for collision avoidance. This robot has a range radius of two-miles.

Among the tasks of the brain number one is to connect it directly with Earth for remote control and for the continuous transmission of views from the robot's surroundings, including the views of the ground in front of the robot's path. This brain also controls the transmission of the results of the research done by the piggyback mini-laboratory in search of water.

The second brain of the robot will control the movement of the wheels, the legs and the wheeled outriggers, synchronizing the movements of the entire leg system for different walking conditions. This second brain is also connected to the three-dimensional laser scanning and the disaster avoidance system.

A HYPOTHETICAL PROGRAM FOR THE DEVELOPMENT, DESIGN AND TESTING OF SYSTEMS FOR THE ROBOT, INCLUDING THE ELECTRONICS, COMPUTERS AND THE CHEMICAL MICRO-PROCESSOR, ALL WHICH WILL BE DEVELOPED BY SOMEONE OTHER THAN THE AUTHOR:

The author will create only the hardware and the frame of the robot, and in it will provide the necessary space provisions for the avionics' developers. Probably it will be necessary to build three or four identical robots with frames from aluminum alloy. Each of them would be delivered to a different government agency or to private contractors. This is in order to allow competition for the best design of the robot's systems, such as the piggy-back laboratory, electronic brains, navigation system and the stability control with gyros and accelerometers.

36. Mountain climbing robot, for planet Mars (page 10 of 12)

After the robot, including its systems, is completed, it will be subjected to rigorous testing of its terrain walking ability, including volcano explorations. The systems designed by the bid winner will be incorporated later into the robots that will go to Mars.

A warning for the contractors and the developers of the piggyback scientific package and the systems that go on top of the robot: it should be designed with a center of gravity as low as possible for stability.

The preliminary designs for the robot shown in this book, which includes the structural and the mechanical part of the transport platform (robot), could be used directly as basic layouts for the drafting of the construction detail drawings for the robot.

If requested, the author could develop a complete set of construction plans for the structural-mechanical part and the hardware for the construction of the aluminum testing robots. The design of these testing robots will later be revised and modified for weight savings and be applied for the construction of the robots made from beryllium to go to Mars.

The existing preliminary designs already include spatial volume provisions, which envelope dimensions would be use by the other developers for their systems that includes electronic brains, sensors, laser scanners, inertial navigation system and the stability control system, with gyros and accelerometers.

PROPOSED STABILITY AND CONTROL SYSTEM
The robot operates automatically by its own electronic brain; however, if necessary it could be remotely operated in slow motion from Earth.
The robot has three laser scanners. Each acts as a special eye. The one on top of the robot's head is for distance scanning, 25 feet and beyond, and up to 320 degrees field of view. The robot laser range finder warns the robot's brain, when the robot is approaching to dangerous objects such as rocks, rocky walls or places as precipice edges. The other two are short-range laser scanners, one located in the front and the other at the rear of the robot. Both face down to scan a ground terrain not larger than a half square meter, mainly at the prospected spot, where the robot's foot will step next.

Depending on whether the robot is going forward or backwards, the laser scanners will transfer 3D terrain mapping information to the brain, which will order one of the insect-like legs to move forward or back. The leg will move first up, then down to the spot establishes in advance by the robot's brain, then by pushing back against the ground, it thrust the robot forward. As the robot steps on the terrain thick hairy sensors sticking out from its legs would sense the terrain characteristics, such as softness, hardness, humidity, dryness or muddiness.

For any static or dynamic attitude of the robot plus the equipment system, in order to remain stable, the vertical vector of the CG (Center of Gravity), should always fall inside a polygon projected on the ground. The corners of the polygon are formed by the ground contact of the main wheels and the wheeler outriggers. A computer, a guidance system and the load pressure sensors located under the feet, also contributes to the robot's stability and control.

Each step of each of the robot's insect legs represents a complete dynamics cycle of a set of operations for that particular leg. After this cycle is completed, that leg is ready for a new step and a new cycle of operations, a process that repeats over and over.

The dynamics of the robot's walking and its stability are controlled by the stability control and guidance system, which consist of computers, load distribution, terrain pressure sensors, laser scanner range finders, and a guidance system with gyros and accelerometers. The guidance system considers the robot's platform as the theoretical plane of reference. Systems that will be developed and designed by somebody other according to their own criteria. The author's design include only the structure, the mechanisms and the hardware for the robot.

FUTURISTIC FIGHTER CONCEPT

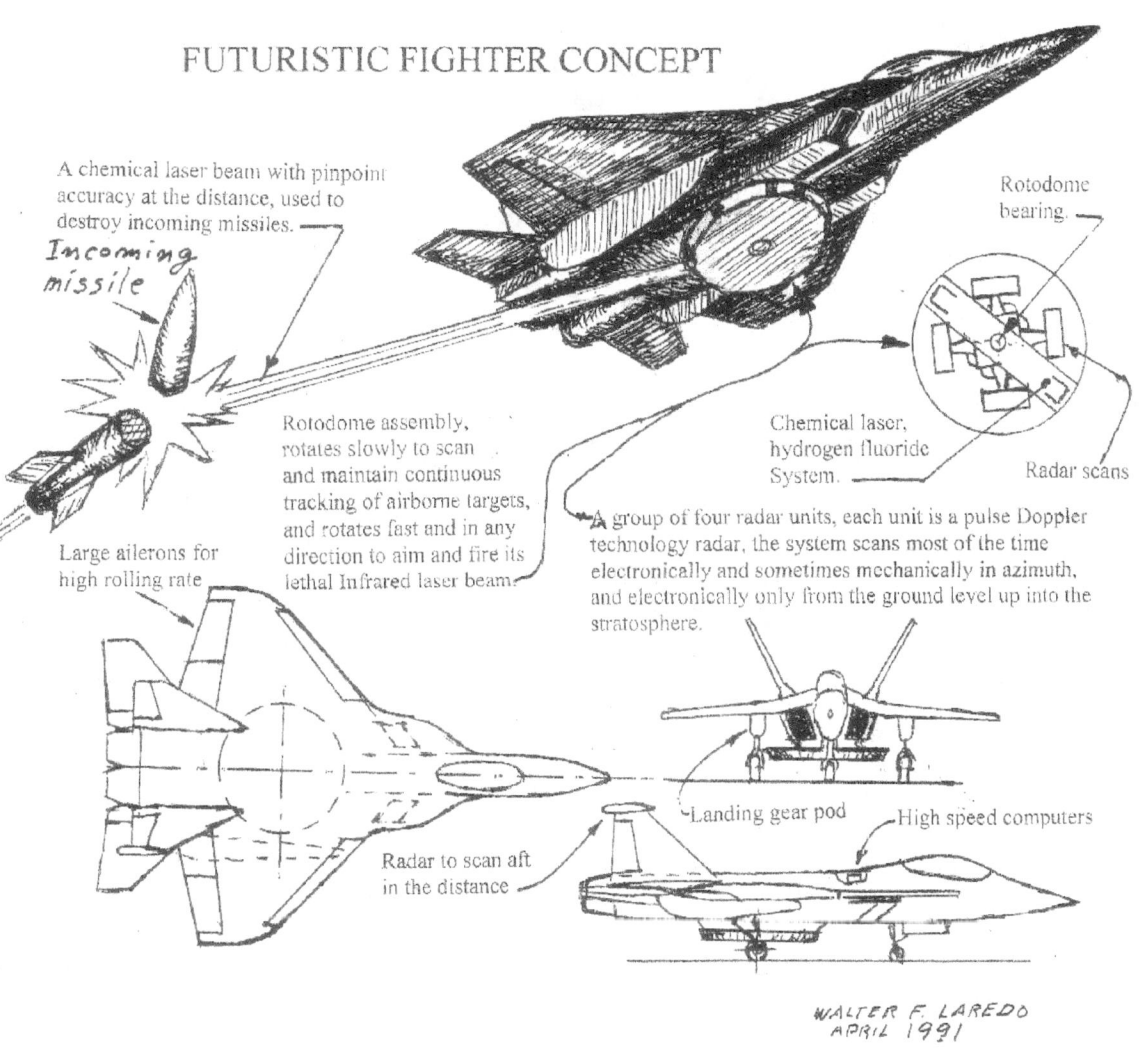

A chemical laser beam with pinpoint accuracy at the distance, used to destroy incoming missiles.

Incoming missile

Rotodome bearing.

Rotodome assembly, rotates slowly to scan and maintain continuous tracking of airborne targets, and rotates fast and in any direction to aim and fire its lethal Infrared laser beam.

Chemical laser, hydrogen fluoride System.

Radar scans

A group of four radar units, each unit is a pulse Doppler technology radar, the system scans most of the time electronically and sometimes mechanically in azimuth, and electronically only from the ground level up into the stratosphere.

Large ailerons for high rolling rate

Landing gear pod

High speed computers

Radar to scan aft in the distance

WALTER F. LAREDO
APRIL 1991

The aircraft carry a hydrogen fluorine chemical laser used as a weapon for defense as well as for offense.

39. Futuristic fighter concept with laser weapons

Engineering projects for the 21ˢᵗ century

SECTION OF PRELIMINARY DESIGN DRAWINGS

e. NOTE:

The pictures shown in this section are not just simple illustrations, but real Preliminary Design Drawings draw to scale in order to be use to develop real engineering projects.

Highly detailed designs, corresponding to each of the aerospace engineering projects from the previous section, as shown in list of page 29

Real Preliminary Design Drawings, draw to scale. Designs that may be use as guidance for further development of real aerospace projects.

227 plates follow this page.

The list below shows plates subdivided by project.

PLATE 1

Designed in summer of 1973 by W. F. Laredo

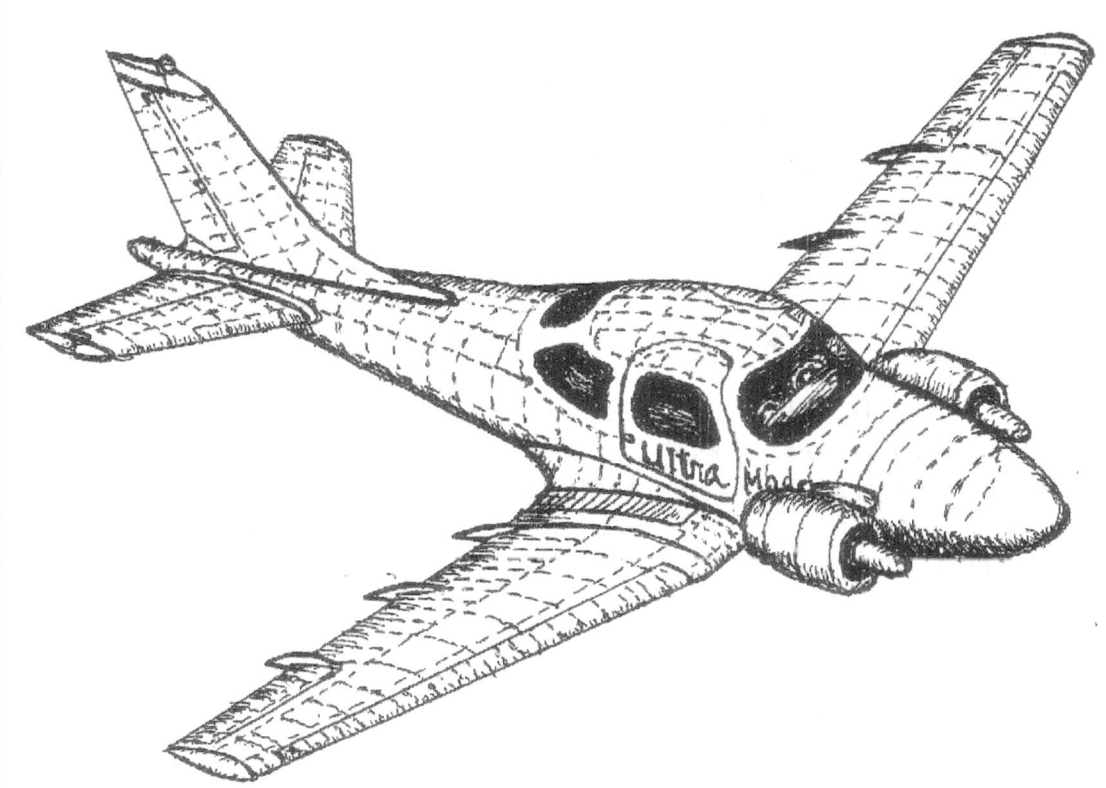

THE MOST SOPHISTICATED SMALL
PRESSURIZED JET OF THE WORLD

ADVANCED ENGINEERING PROJECTS

Name of project:

"ULTRA MODERN" AIRCRAFT

Designer and builder:	WALTER F. LAREDO	Date:
Drawing Number: 800 SERIES		Sheet 1 of 24

PLATE 2

While working in the Structural Engineering Department of McDonnell Douglas Corp., where the Douglas DC-10 transport was developed, learned to design long life airframe structures with good fatigue resistance and applied to big jets. Then one day decided to design and build his own miniature executive aircraft named "ULTRA MODERN", although the size of a small Cessna its structure is much more sophisticated.

"ULTRA MODERN " AIRCRAFT, A DREAM FOR THE SPORTLY RICH PILOT

An expensive leisure private plane, which cockpit surrounded with windows all around and with modern instruments is for 2 people, it resembles the cabin of a small sport car.

It is also a STOL aircraft for short takeoff and short landing as most big birds do. For landings or takeoffs could use grass fields or short dirt runways. Ideal private plane for picnics and camping, and because of its long range, ideal for long trips, safaris, and vacations to different places of the world, from the hot deserts to the polar caps regions.

It cruises at high altitude of around 36,000 feet, the cabin's structure and its pressure bulkheads are designed for pressurization, including the double pane windows and the two doors, each door with a double hinge and four sliding safe locks.

A special feature of this aircraft is the foldable wings for towing and easy storage. The wing box structure is also an integral fuel tank of great capacity. The span of the horizontal stabilizer was established in order its tips should clear by one inch a single car garage door for storage.

For short takeoffs and landings the wings are provided with double slotted flaps and with wing leading edge slats, all them similar to the Boeing 747 but in miniature.

The structure of this aircraft is so sophisticated that to build it will take the same number of man hours than to build an average size executive aircraft.

The aluminum structure is designed for long life span (30 years), made by a large number of components, each extremely light, highly elaborated and good against fatigue.

CHARACTERISTICS AND PERFORMANCE
GENERAL DESCRIPTION:

High performance STOL (Short Takeoff, and Landing) aircraft. Takeoff under 560 ft over 50 ft obstacle, pressurized for a service ceiling of 37,000 ft., and a speed of Mach 0.5 (332 mph) at 36,000 ft. with a range near to 800 miles.

Sophisticated structure, Very light but super strong at flight, designed for 50, 000 cycles (flights).

Wings are easy to fold, whenever fuel contained in each is less than 11 gallons.

The inboard constant section of the wing have a Supercritical NASA airfoil GA (W)-1, while the outboard trapezoidal section have a constant transition airfoil

"ULTRA MODERN" AIRCRAFT		
Designer and builder: WALTER F. LAREDO		
Drawing Number:	800 series	Sheet 2 of 24

PLATE 3

between NASA GA (W)-1 and NACA 4412.

The NASA GA (W)-1 is a laminar flow and low drag airfoil. Because this airfoil is thicker than ordinary ones, the wing contains more fuel, also for a thicker wing the structural weight would be lighter than for a thin one, subject to the same loads.

POWERPLANT

Turbojet engine, 2 required
Designer: WALTER F. LAREDO
Power output per engine, sea level, static, standard day

Takeoff Thrust (lb.)	326
Max. Continuous Thrust (lb.)	310
Max. Cruise Thrust (lb.)	290
75 % Max. Continuous (lb.)	233
Max. Cruise Thrust (lb.)	
at 36.089 ft. Mach 0.45	83

Fuel consumption per engine:

Max Cruise, standard day, Sea level, static (lb./hr)	184
Max Cruise. at 35,000 ft., Standard day (lb./hr)	98

Mechanical Speed limits:

Spool speed (RPM)	50,000
Exhaust Gas temperature limit at Takeoff	1.090 deg. F
Weight	88 lb.
Power/ Weight Ratio	3.70 : 1
Fuels:	JP-4, JP-5, Jet A
Oils:	Mil-L-7808, Mil-L-23699
Maximum oil consumption	0.10 pt/hr

AIRCRAFT DIMENSIONS

Overall length, ft.	20' 8"
Height, ft.	7' 1"
Seating Capacity	2
Cabin Door, h x w, in.	38 x 29
Headroom, sidewise front, in.	35
Legroom, front, in	48
Hiproom, front, in.	40.5
Elbow to elbow room front, in.	41
Baggage Capacity, lb	100
Size, in.	24x24x30
If air conditioning is removed, in.	36x30x30
Wheel base, in.	79
Thread, in. C/L Tires	72

"ULTRA MODERN" AIRCRAFT

Designer and builder: WALTER F. LAREDO

Drawing Number:	800 series	Sheet 3 of 24

PLATE 4

Main tire size, in.	6.00 x 6
Nose tire size, in	5.00 x 5

DESIGN DATA AND WEIGHTS

Empty weight, lb.	1150
Useful load, lb.	
(payload plus fuel)	900
Gross weight, lb.	2050
G Limit Load, pos.	7 (semi-aerobatic)
Neg.	5
Datum station	Perimeter of Fwd. Pressure Bulkhead
C.G. Limits, fore, in.	35.50
aft, in.	39.00

CONTROL SURFACES

Aileron area, sq. ft.	6.8
Up deflection, degrees.	25
Down deflection, degrees.	20
Elevator area, sq.ft.	6
Up deflection, degrees.	30
Down deflection, degrees.	25
Horizontal stabilizer, sq. ft.	9.3
Up incidence travel, degrees	5
Down incidence travel, degrees	8
Rudder area, sq. ft.	3.4
Deflection, L & R, degrees	30
Slat, wing leading edge	most of the span
Double slotted flap, area, sq. ft.	17
No of positions	variable in between the two following positions
Takeoff setting, deg.	30
Landing setting, deg.	45

SYSTEMS OPERATION

Double slotted flaps and leading edge slat	Electric
Oleopneumatic landing gear, retractable	Electric & manual (backup)
Brakes	Hydraulic
Air conditioning	Electric
Pressurization	Bleed air from engine compressor
PSI differential	7.0

FUEL SYSTEM

Capacity, main tanks (standard) gal.	75
With auxiliary tanks, (over-gross-weight aircraft, requires long runway for takeoff, no landing is permitted for this weight condition) gal.	104

"ULTRA MODERN" AIRCRAFT

Designer and builder: WALTER F. LAREDO

Drawing Number:	800 series	sheet 4 of 24

PLATE 5

WING

Wing loading, lb/sq ft	21.23
Wing Airfoil, section of constant chord	Supercritical NASA GA(W)-1
Wing Airfoil, outboard trapezoidal section	Constant transition Airfoil between
	Between NASA GAW-1 and NACA 4412
Airfoil, wing tip	NACA 4412
Incidence angle, degrees	2.5
Dihedral, degrees	5
Wingspan, ft.	30
Wing Area, sq. ft.	96.57
Wing Chord, root, in.	42
Aspect Ratio	9.32
MAC, in.	38.6

PERFORMANCE

Service ceiling, ft	37000
Glide ratio (flaps retracted)	10.2
Takeoff roll, ft.	330
Over 50 ft.	650
Landing roll, ft.	310
Over 50 ft.	603

SPEEDS

Normal operating at 36,000 ft, mph	332
Economy cruising at 36,000 ft, mph	301
Max. Speed to extend flaps, IAS, mph	120
Max. Speed to extend gear, IAS, mph	130
Slow flight, IAS, mph (flaps and slat extended)	35
Best Rate of Climb, IAS, at sea level, ft/min	1610
Liftoff, IAS, mph	40
Touchdown, IAS, mph	35
Stall, flaps down, slat extended, IAS, mph	30

RANGE (with no reserve)

With only main tanks, at 36,089 ft. and Mach 0.45, miles	794
With main and auxiliary tanks, at 36,089 ft. and Mach 0.45	1100

"ULTRA MODERN" AIRCRAFT

Designer and builder: WALTER F. LAREDO

Drawing Number:	800 series	Sheet 5 of 24

PLATE 6

TOWING VIEW OF THE "ULTRA MODERN"

FAA Inspections:
During the various steps of the structural construction for the "ULTRA MODERN"
aircraft and previous to the closure of each aerodynamic surface, it was inspected and
approved by the inspectors of the Federal Aviation Administration of The United States,
Long Beach branch, CA.

The construction of this aircraft stopped, after the builder injured his left arm in a traffic
accident. In the aerospace industry, usually two specialists are required to do the
riveting, but for the rare case that there is only one, this person must use both of his hands
for a precise riveting with conventional rivets. Grabbing simultaneously the pneumatic
riveter with the right hand and the bucking bar with the other.

"ULTRA MODERN" AIRCRAFT

Designer and builder: WALTER F. LAREDO

| Drawing Number: | 800 series | Sheet 6 of 24 |

PLATE 7

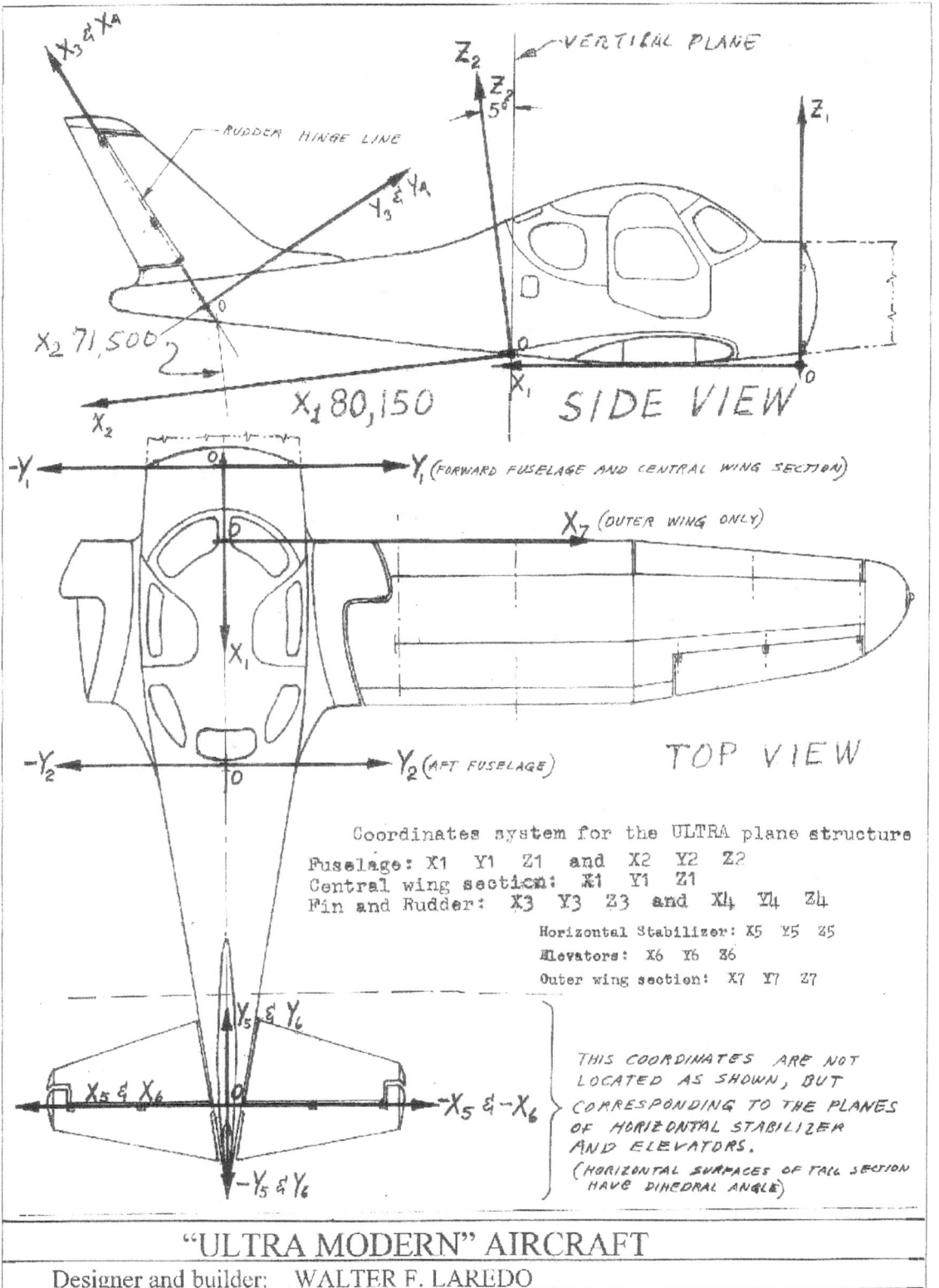

VERTICAL PLANE

RUDDER HINGE LINE

SIDE VIEW

X_2 71,500

X_1 80,150

$-Y_1$

Y_1 (FORWARD FUSELAGE AND CENTRAL WING SECTION)

X_7 (OUTER WING ONLY)

X_1

$-Y_2$

Y_2 (AFT FUSELAGE)

TOP VIEW

Coordinates system for the ULTRA plane structure
Fuselage: X1 Y1 Z1 and X2 Y2 Z2
Central wing section: X1 Y1 Z1
Fin and Rudder: X3 Y3 Z3 and X4 Y4 Z4

Horizontal Stabilizer: X5 Y5 Z5

Elevators: X6 Y6 Z6

Outer wing section: X7 Y7 Z7

Y_5 & Y_6

X_5 & X_6

$-X_5$ & $-X_6$

$-Y_5$ & Y_6

THIS COORDINATES ARE NOT
LOCATED AS SHOWN, BUT
CORRESPONDING TO THE PLANES
OF HORIZONTAL STABILIZER
AND ELEVATORS.
(HORIZONTAL SURFACES OF TAIL SECTION
HAVE DIHEDRAL ANGLE)

"ULTRA MODERN" AIRCRAFT

Designer and builder: WALTER F. LAREDO

Drawing Number: 800 series

Sheet 7 of 24

PLATE 8

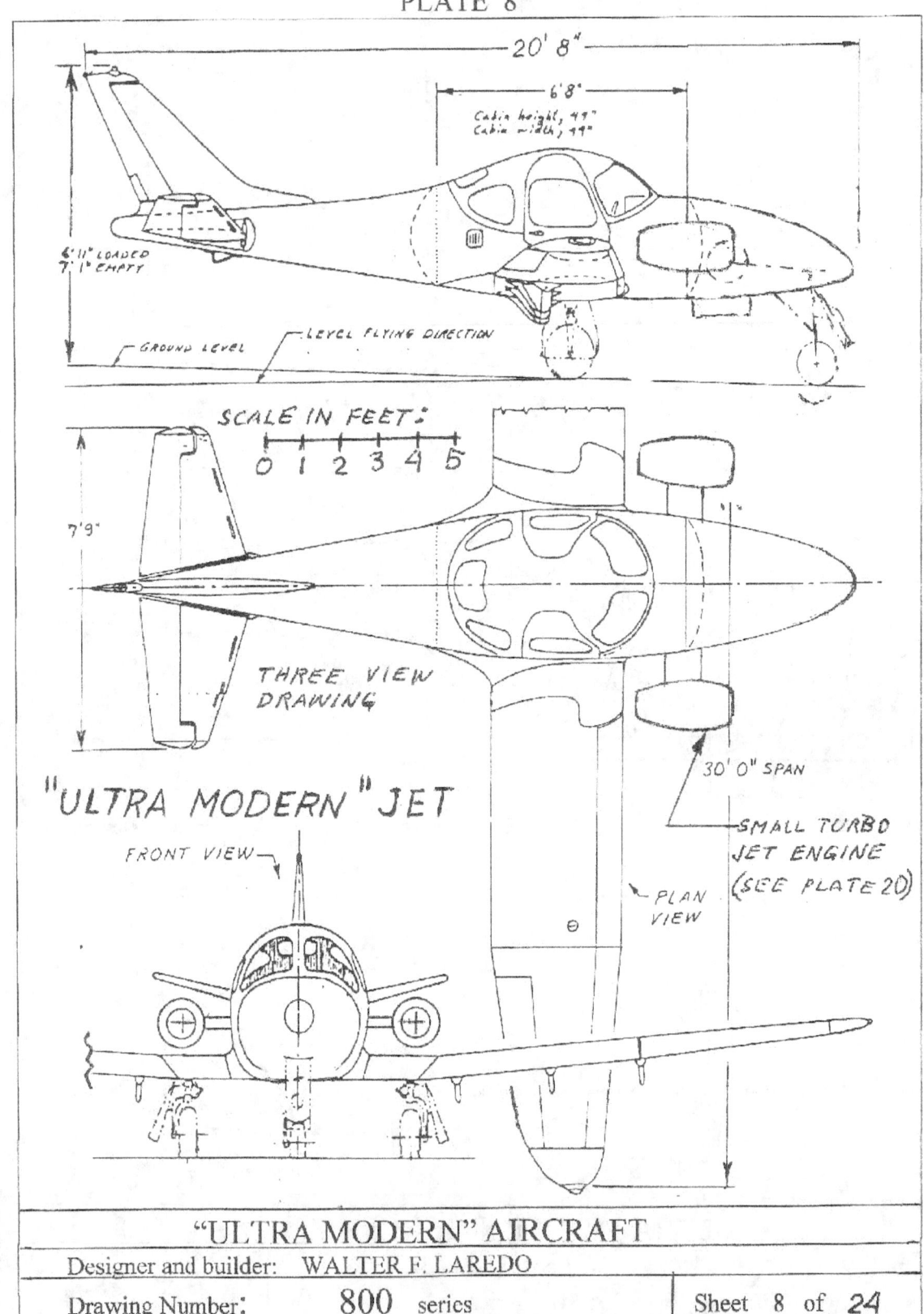

20' 8"

6'8"

Cabin height, 44"
Cabin width, 44"

6'11" LOADED
7'1" EMPTY

GROUND LEVEL   LEVEL FLYING DIRECTION

SCALE IN FEET:

0 1 2 3 4 5

7'9"

THREE VIEW
DRAWING

"ULTRA MODERN" JET

FRONT VIEW

30' 0" SPAN

SMALL TURBO
JET ENGINE
(SEE PLATE 20)

PLAN
VIEW

"ULTRA MODERN" AIRCRAFT

Designer and builder: WALTER F. LAREDO

Drawing Number: 800   series   Sheet 8 of 24

PLATE 9

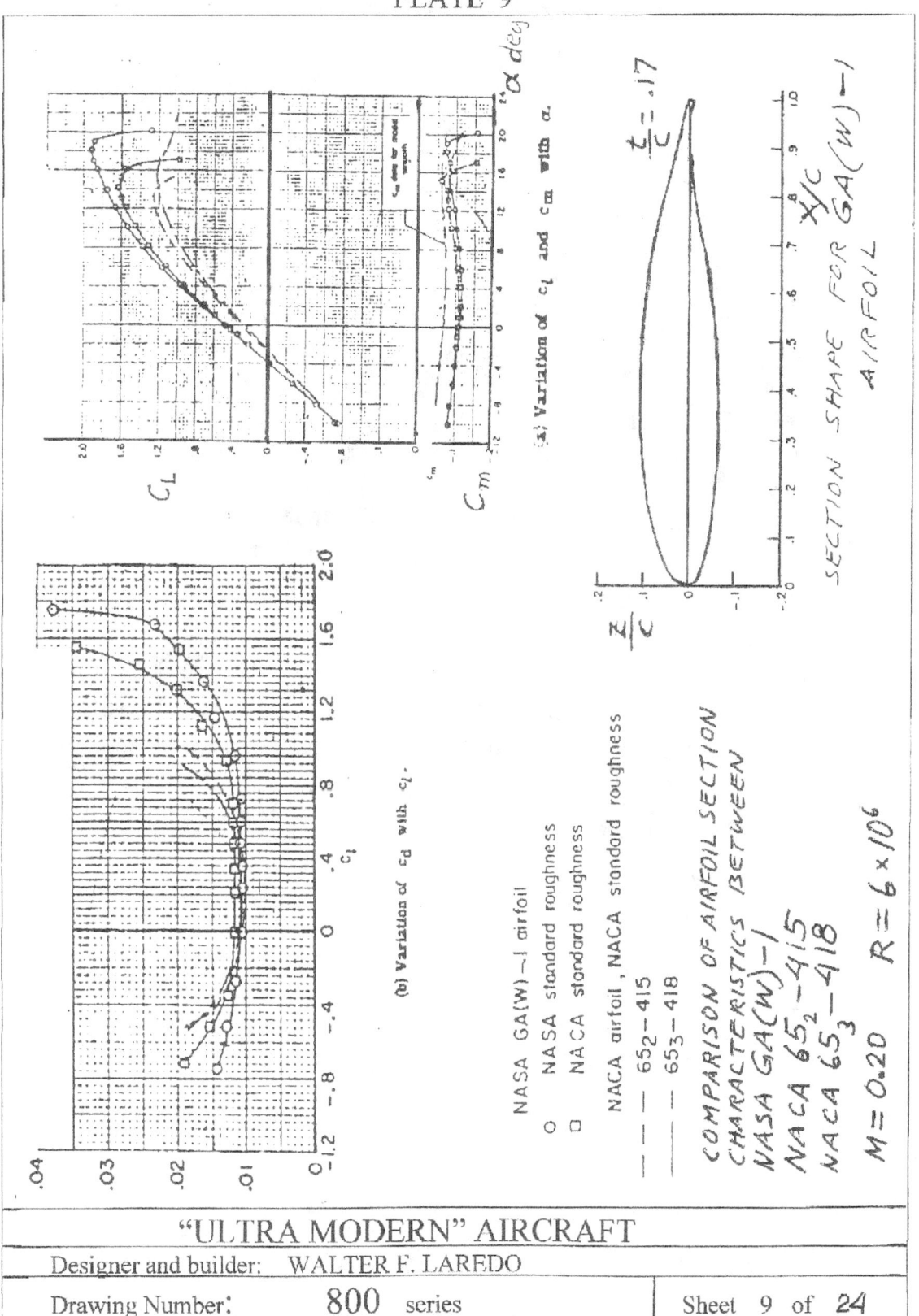

(a) Variation of c_l and c_m with α.

(b) Variation of c_d with c_l.

$\frac{t}{c} = .17$

SECTION SHAPE FOR GA(W)—1 AIRFOIL

NASA GA(W)—1 airfoil

o   NASA standard roughness

□   NACA standard roughness

NACA airfoil, NACA standard roughness

- - 65_2—415

- - 65_3—418

COMPARISON OF AIRFOIL SECTION
CHARACTERISTICS BETWEEN
NASA GA(W)—1
NACA 65_2—415
NACA 65_3—418

$M = 0.20 \qquad R = 6 \times 10^6$

"ULTRA MODERN" AIRCRAFT

Designer and builder: WALTER F. LAREDO

Drawing Number: 800   series

Sheet 9 of 24

PLATE 10

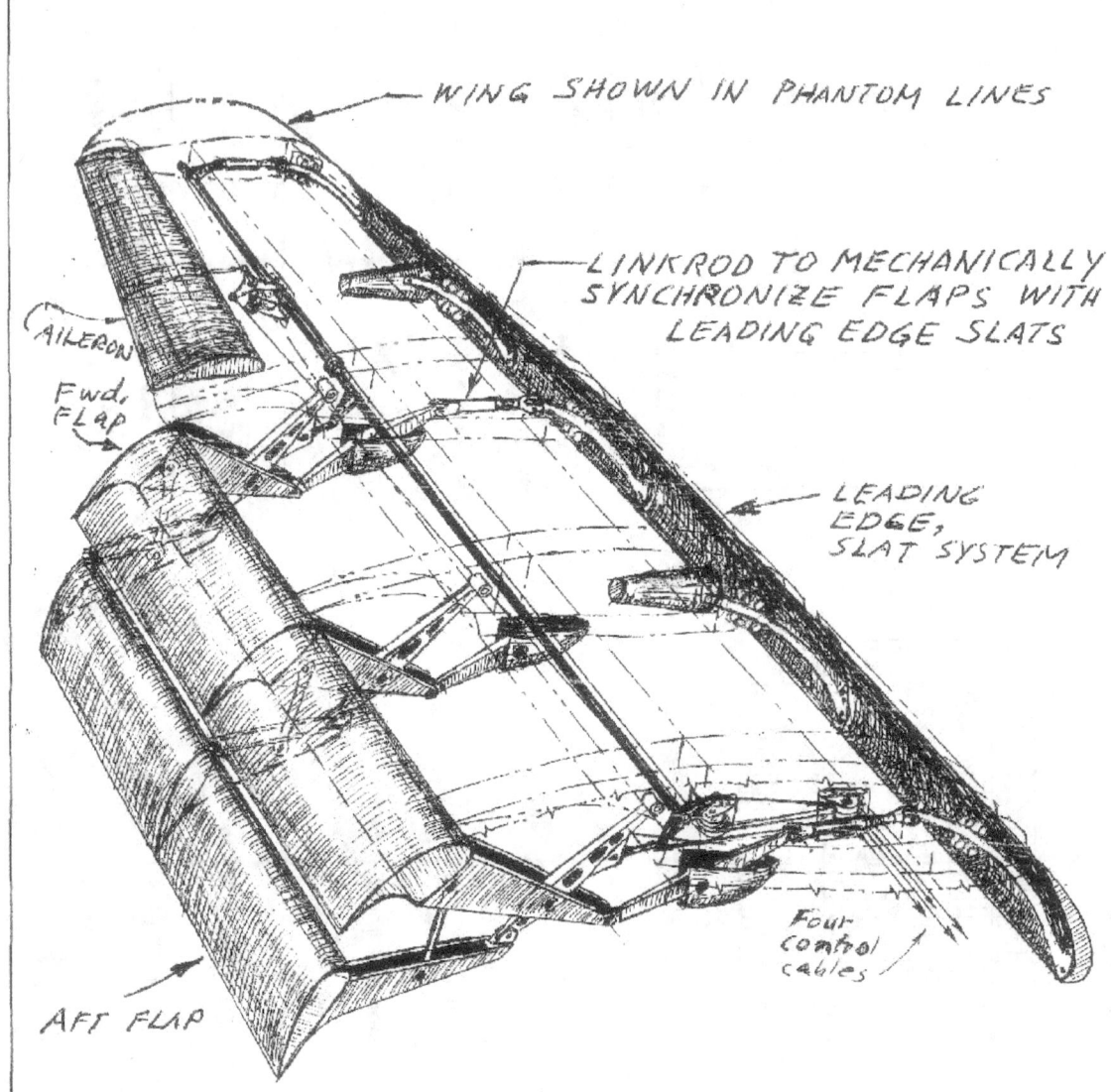

WING SHOWN IN PHANTOM LINES

LINKROD TO MECHANICALLY SYNCHRONIZE FLAPS WITH LEADING EDGE SLATS

AILERON

Fwd. FLAP

LEADING EDGE, SLAT SYSTEM

Four control cables

AFT FLAP

MECHANISM OF FLAPS AND SLATS SHOWN DEPLOYED

"ULTRA MODERN" AIRCRAFT		
Designer and builder: WALTER F. LAREDO		
Drawing Number: 800 series		Sheet 10 of 24

PLATE 11

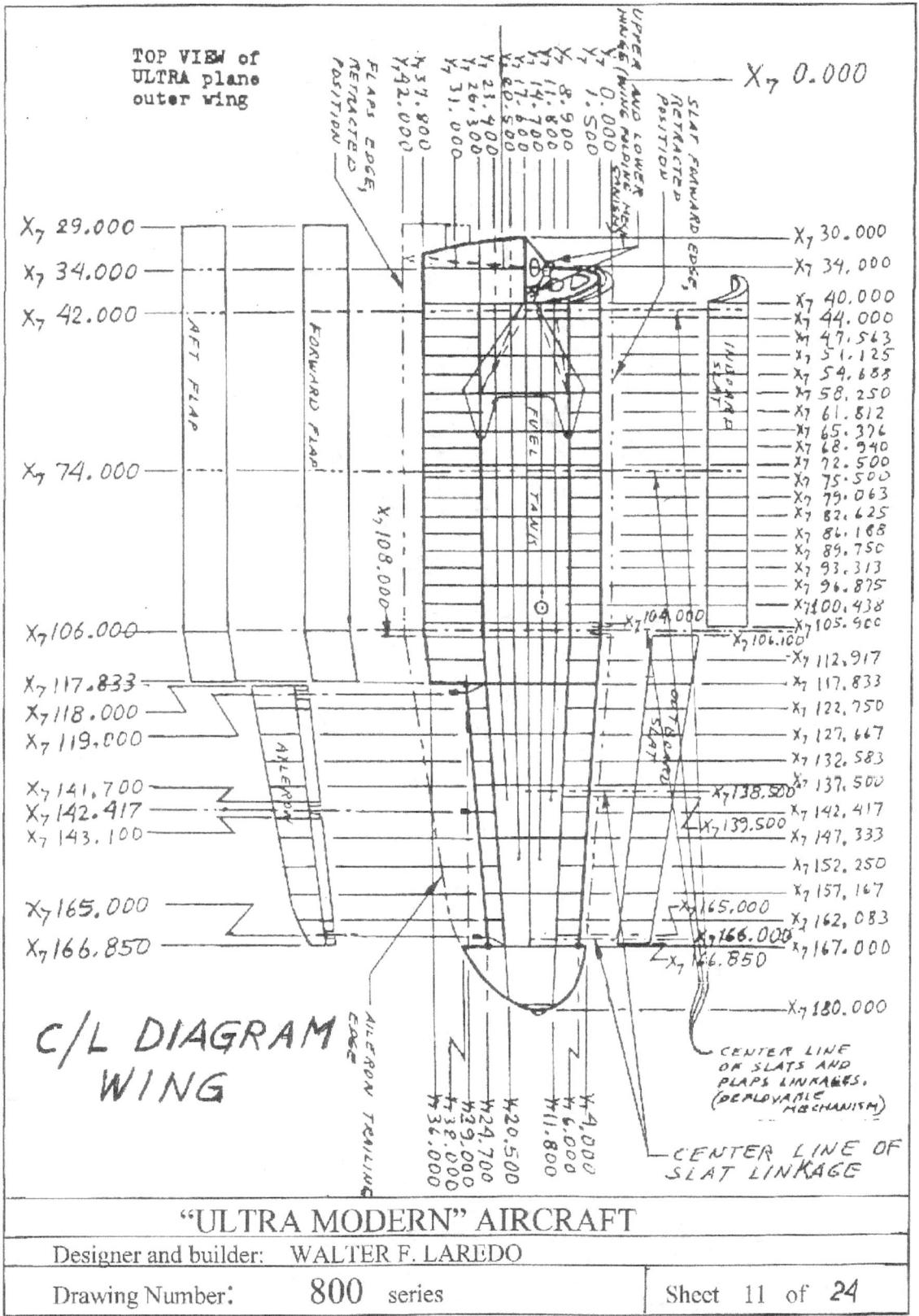

TOP VIEW of
ULTRA plane
outer wing

C/L DIAGRAM
WING

CENTER LINE
OF SLATS AND
FLAPS LINKAGES.
(DEPLOYABLE
MECHANISM)

CENTER LINE OF
SLAT LINKAGE

"ULTRA MODERN" AIRCRAFT

Designer and builder:	WALTER F. LAREDO	
Drawing Number:	800 series	Sheet 11 of 24

PLATE 12

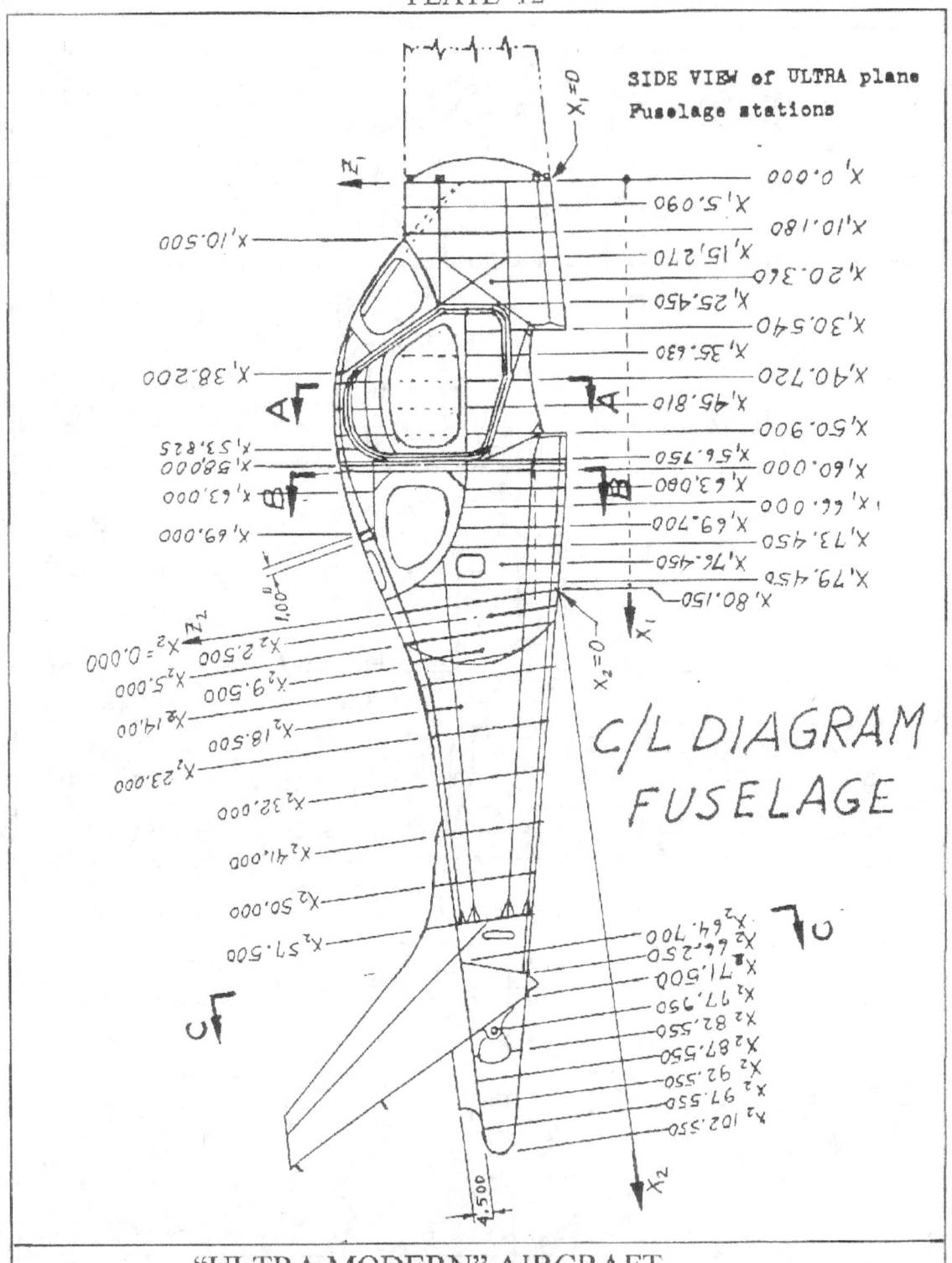

SIDE VIEW of ULTRA plane
Fuselage stations

C/L DIAGRAM
FUSELAGE

PLATE 13

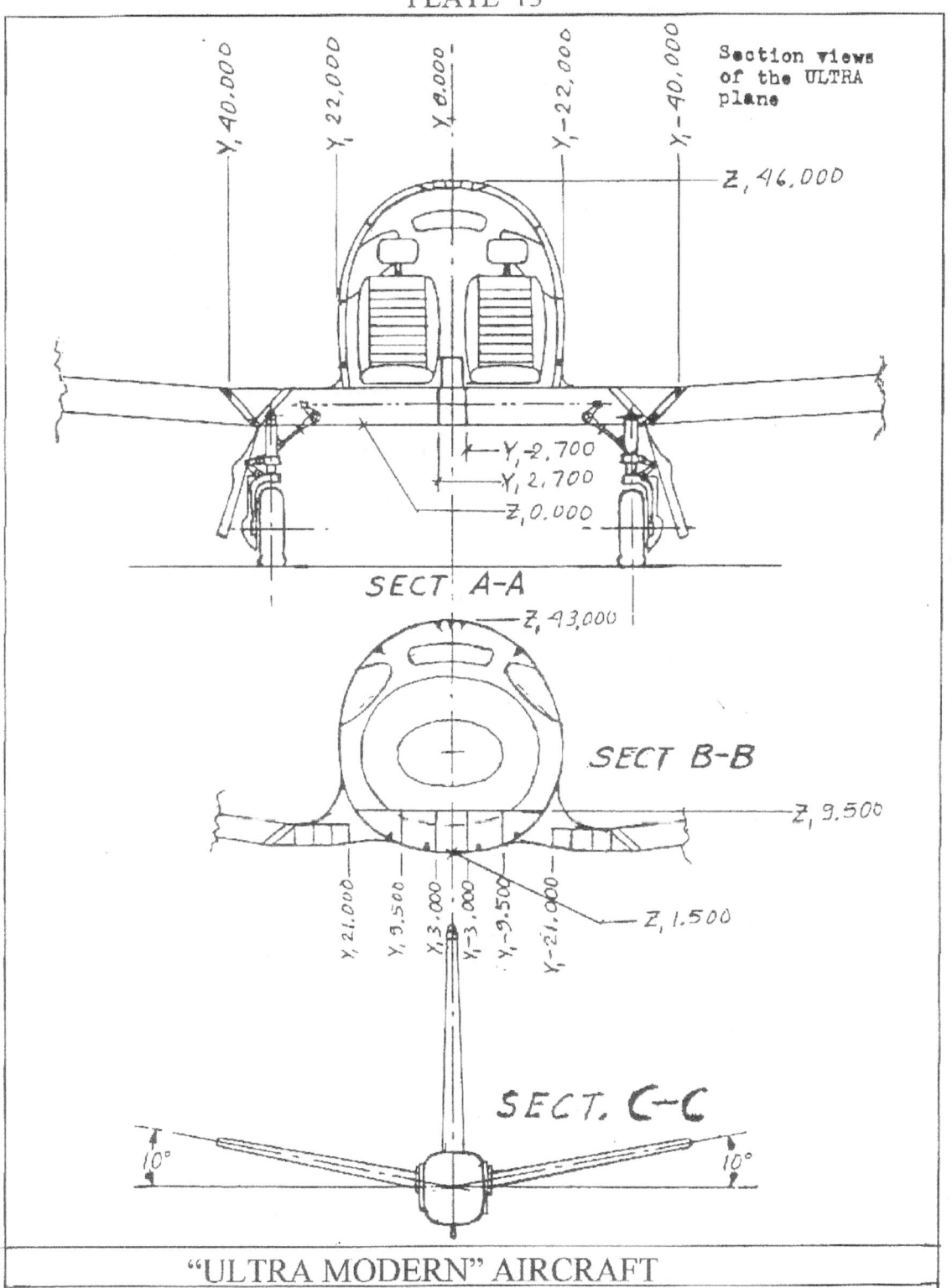

Section views
of the ULTRA
plane

Z, 46.000

Y, 40.000
Y, 22.000
Y, 0.000
Y, -22.000
Y, -40.000

Y, -2.700
Y, 2.700
Z, 0.000

SECT A-A

Z, 43.000

SECT B-B

Z, 9.500

Y, 21.000
Y, 9.500
Y, 3.000
Y, -3.000
Y, -9.500
Y, -21.000

Z, 1.500

SECT. C-C

10°   10°

"ULTRA MODERN" AIRCRAFT		
Designer and builder: WALTER F. LAREDO		
Drawing Number: 800 series		Sheet 13 of 24

PLATE 14

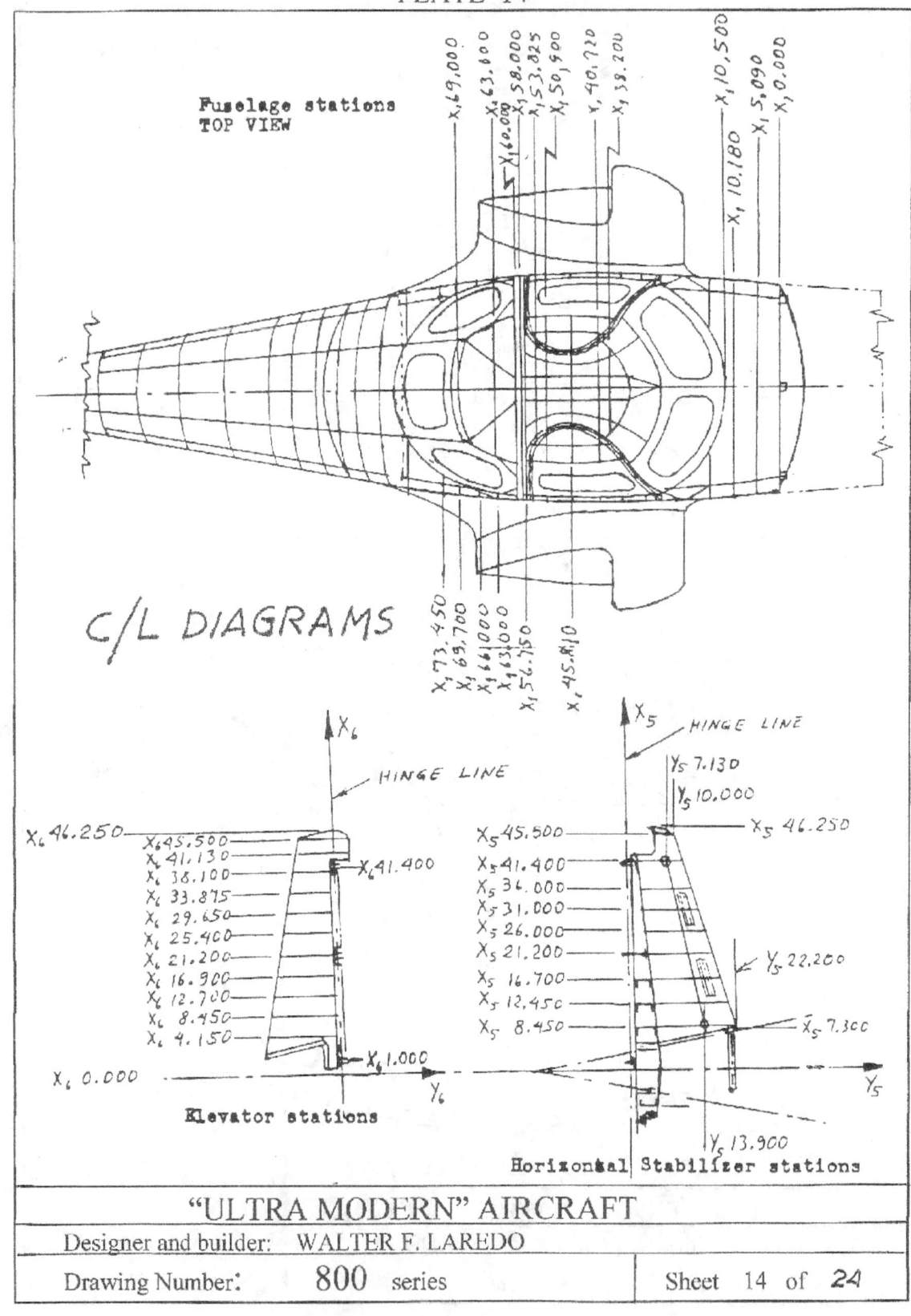

Fuselage stations
TOP VIEW

$X_1 69.000$
$X_1 63.600$
$X_1 60.000$
$X_1 58.000$
$X_1 53.825$
$X_1 50.900$
$X_1 40.720$
$X_1 38.200$
$X_1 10.500$
$X_1 10.180$
$X_1 5.090$
$X_1 0.000$

C/L DIAGRAMS

$X_1 73.450$
$X_1 69.700$
$X_1 64.000$
$X_1 63.000$
$X_1 56.750$
$X_1 45.810$

X_6
X_5
HINGE LINE
$Y_5 7.130$
$Y_5 10.000$
$X_5 46.250$
HINGE LINE

$X_6 46.250$
$X_6 45.500$
$X_6 41.130$
$X_6 38.100$
$X_6 33.875$
$X_6 29.650$
$X_6 25.400$
$X_6 21.200$
$X_6 16.900$
$X_6 12.700$
$X_6 8.450$
$X_6 4.150$
$X_6 41.400$

$X_5 45.500$
$X_5 41.400$
$X_5 36.000$
$X_5 31.000$
$X_5 26.000$
$X_5 21.200$
$X_5 16.700$
$X_5 12.450$
$X_5 8.450$
$Y_5 22.200$
$X_5 7.300$

$X_6 0.000$
$X_6 1.000$
Y_6

Elevator stations

Y_5
$Y_5 13.900$

Horizontal Stabilizer stations

"ULTRA MODERN" AIRCRAFT

Designer and builder: WALTER F. LAREDO

Drawing Number: 800 series

Sheet 14 of 24

PLATE 15

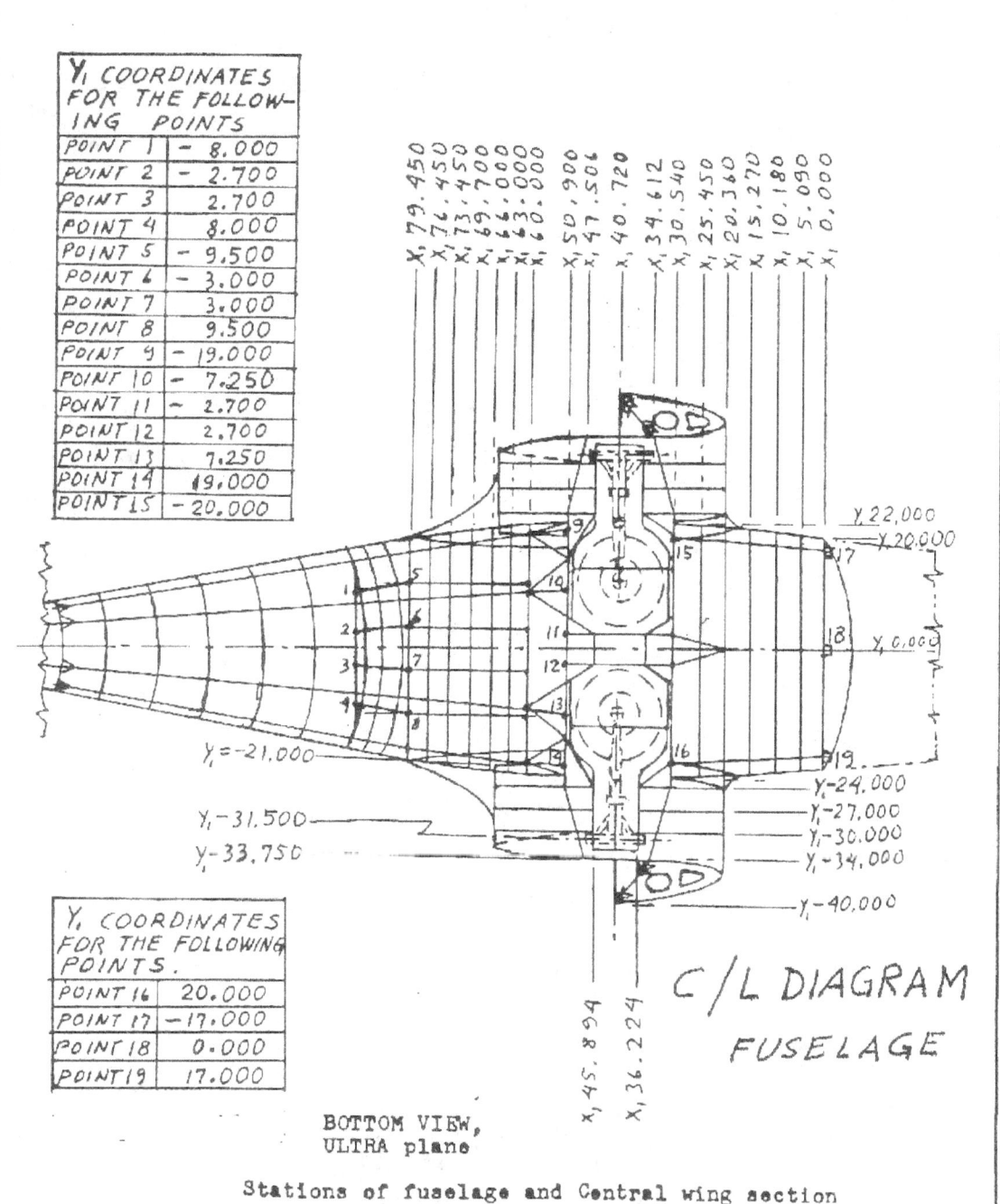

Y₁ COORDINATES FOR THE FOLLOWING POINTS

POINT	
POINT 1	− 8.000
POINT 2	− 2.700
POINT 3	2.700
POINT 4	8.000
POINT 5	− 9.500
POINT 6	− 3.000
POINT 7	3.000
POINT 8	9.500
POINT 9	− 19.000
POINT 10	− 7.250
POINT 11	− 2.700
POINT 12	2.700
POINT 13	7.250
POINT 14	19.000
POINT 15	− 20.000

Y₁ COORDINATES FOR THE FOLLOWING POINTS.

POINT	
POINT 16	20.000
POINT 17	−17.000
POINT 18	0.000
POINT 19	17.000

C/L DIAGRAM

FUSELAGE

BOTTOM VIEW,
ULTRA plane

Stations of fuselage and Central wing section

"ULTRA MODERN" AIRCRAFT

Designer and builder: WALTER F. LAREDO

Drawing Number:	800 series	Sheet 15 of 24

PLATE 16

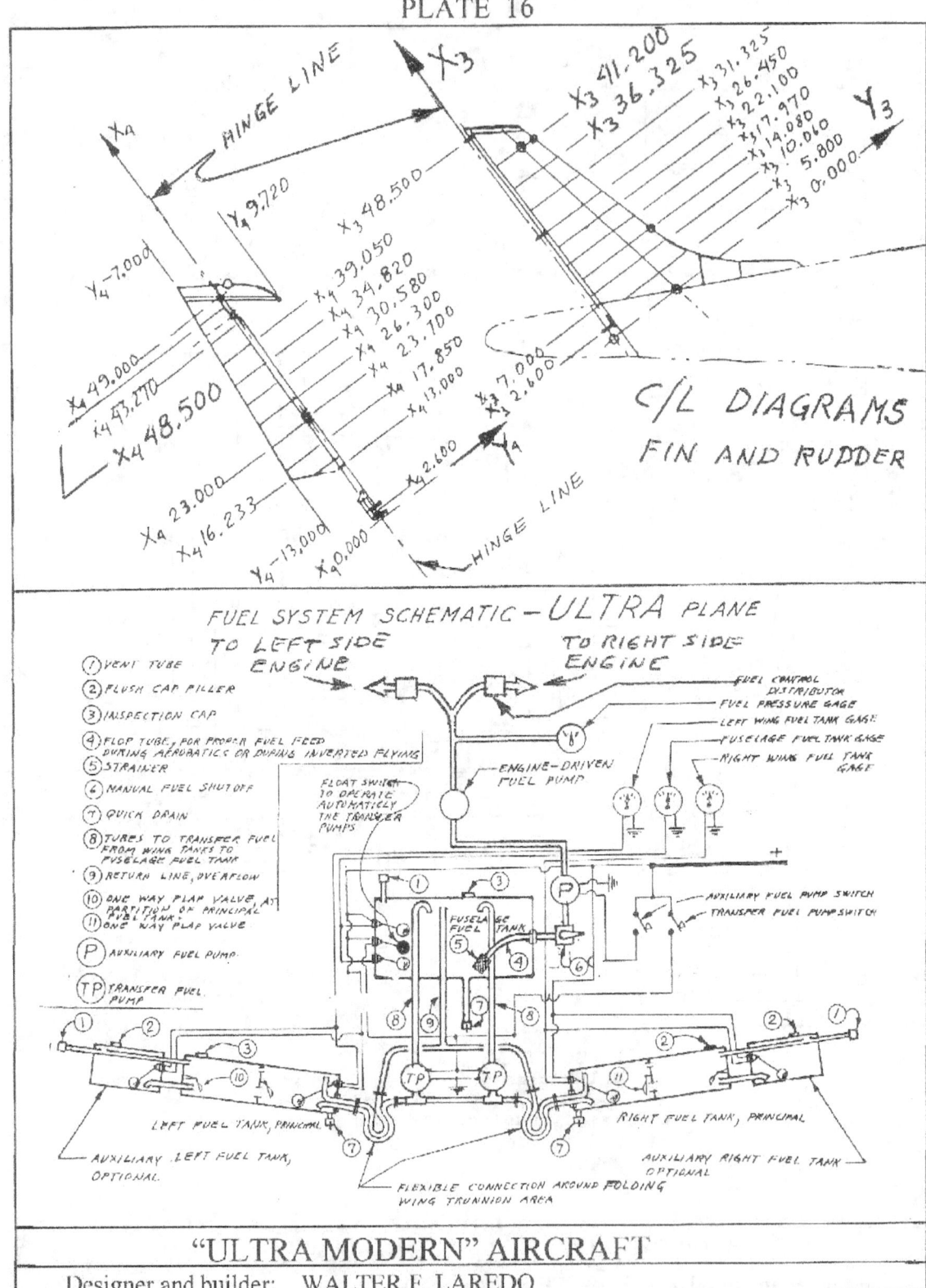

C/L DIAGRAMS
FIN AND RUDDER

FUEL SYSTEM SCHEMATIC — ULTRA PLANE

TO LEFT SIDE ENGINE

TO RIGHT SIDE ENGINE

① VENT TUBE
② FLUSH CAP FILLER
③ INSPECTION CAP
④ FLOP TUBE, FOR PROPER FUEL FEED DURING AEROBATICS OR DURING INVERTED FLYING
⑤ STRAINER
⑥ MANUAL FUEL SHUTOFF
⑦ QUICK DRAIN
⑧ TUBES TO TRANSFER FUEL FROM WING TANKS TO FUSELAGE FUEL TANK
⑨ RETURN LINE, OVERFLOW
⑩ ONE WAY FLAP VALVE, AT PARTITION OF PRINCIPAL FUEL TANK
⑪ ONE WAY FLAP VALVE
Ⓟ AUXILIARY FUEL PUMP.
ⓉⓅ TRANSFER FUEL PUMP

FUEL CONTROL DISTRIBUTOR
FUEL PRESSURE GAGE
LEFT WING FUEL TANK GAGE
FUSELAGE FUEL TANK GAGE
RIGHT WING FUEL TANK GAGE

ENGINE-DRIVEN FUEL PUMP

FLOAT SWITCH TO OPERATE AUTOMATICLY THE TRANSFER PUMPS

AUXILIARY FUEL PUMP SWITCH
TRANSFER FUEL PUMP SWITCH

FUSELAGE FUEL TANK

LEFT FUEL TANK, PRINCIPAL

RIGHT FUEL TANK, PRINCIPAL

AUXILIARY LEFT FUEL TANK, OPTIONAL.

AUXILIARY RIGHT FUEL TANK OPTIONAL

FLEXIBLE CONNECTION AROUND FOLDING WING TRUNNION AREA

"ULTRA MODERN" AIRCRAFT

Designer and builder: WALTER F. LAREDO

| Drawing Number: | 800 series | Sheet 16 of 24 |

PLATE 17

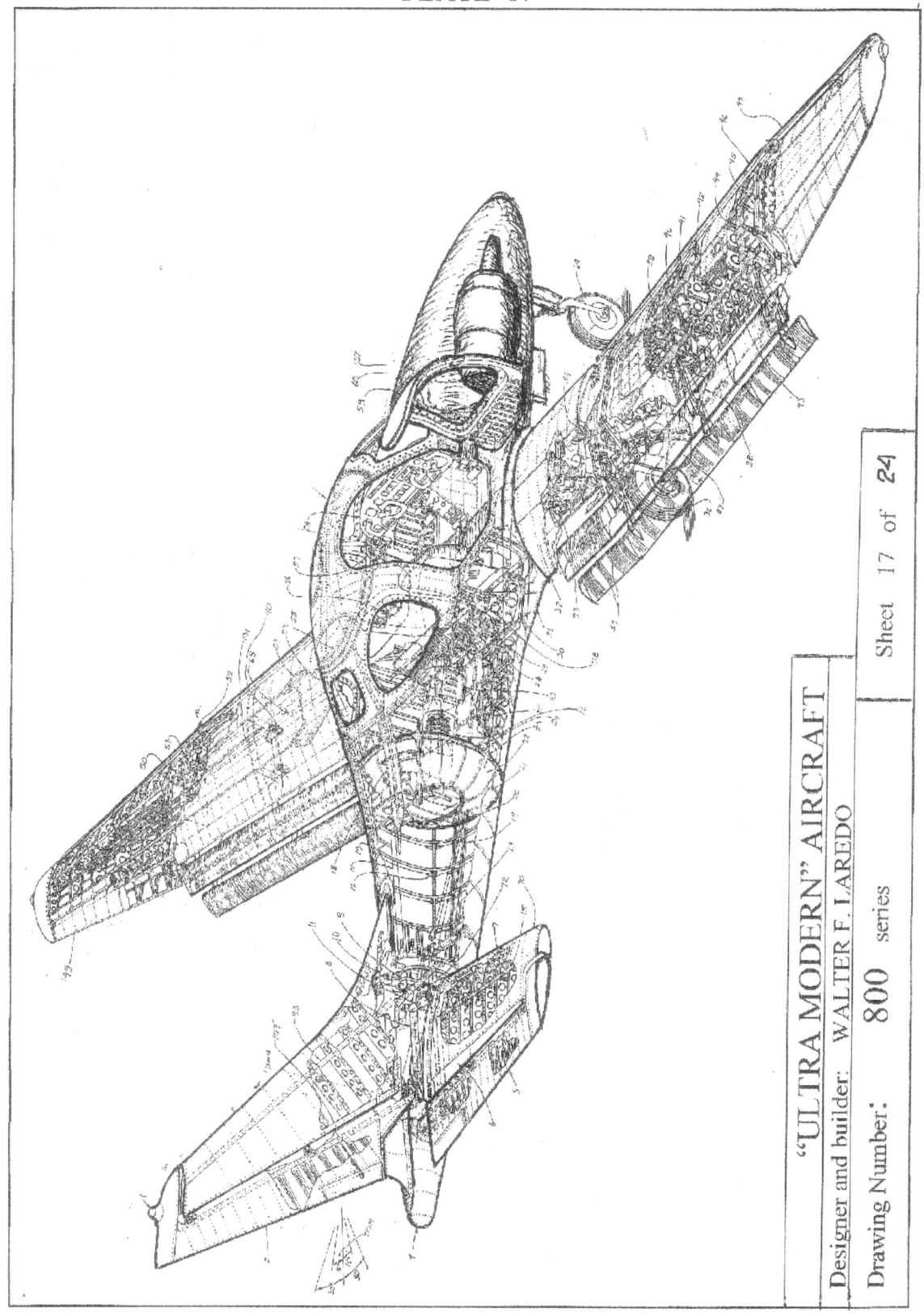

"ULTRA MODERN" AIRCRAFT

Designer and builder: WALTER F. LAREDO

Drawing Number: 800 series

Sheet 17 of 24

PLATE 18

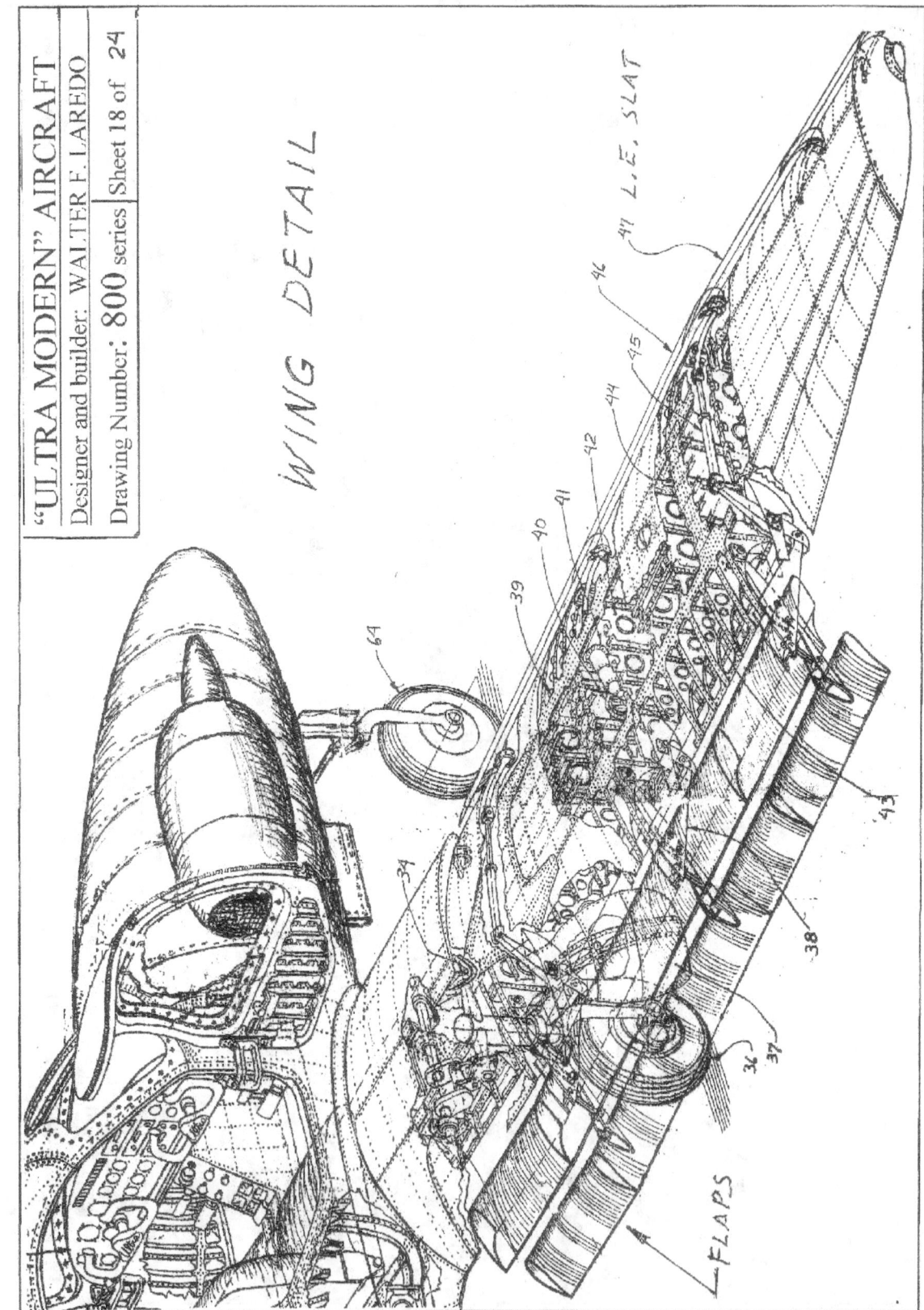

"ULTRA MODERN" AIRCRAFT
Designer and builder: WALTER F. LAREDO
Drawing Number: 800 series | Sheet 18 of 24

WING DETAIL

L.E. SLAT

FLAPS

PLATE 19

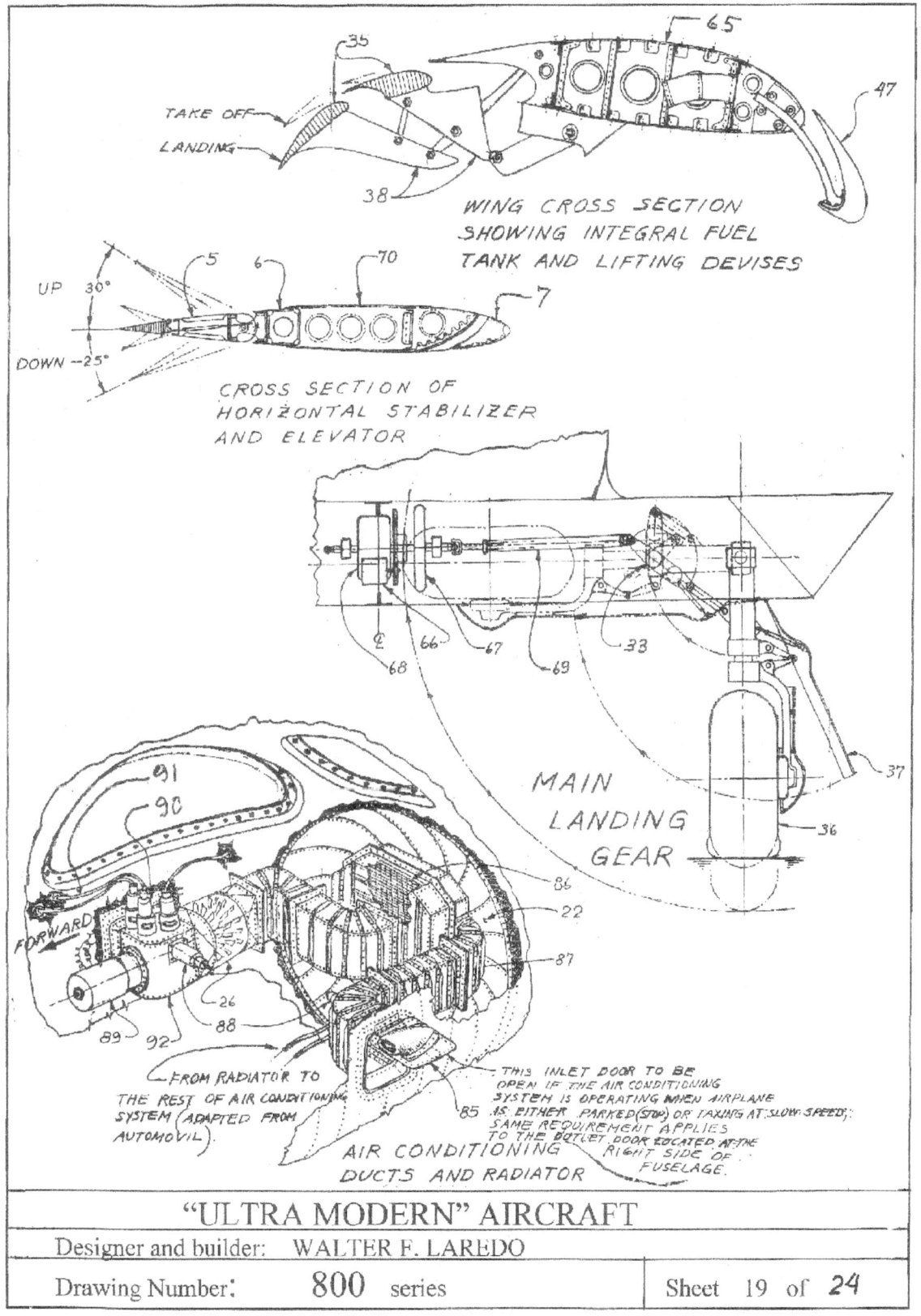

TAKE OFF

LANDING

35

65

47

38

WING CROSS SECTION
SHOWING INTEGRAL FUEL
TANK AND LIFTING DEVISES

UP 30°

DOWN ~25°

5   6   70

7

CROSS SECTION OF
HORIZONTAL STABILIZER
AND ELEVATOR

62   66   67   33   69

68

37

36

MAIN

LANDING

GEAR

91

90

FORWARD

26

89   92   88

86

22

87

FROM RADIATOR TO
THE REST OF AIR CONDITIONING
SYSTEM (ADAPTED FROM
AUTOMOVIL).

85

THIS INLET DOOR TO BE
OPEN IF THE AIR CONDITIONING
SYSTEM IS OPERATING WHEN AIRPLANE
IS EITHER PARKED (STOP) OR TAXING AT SLOW SPEED;
SAME REQUIREMENT APPLIES
TO THE OUTLET DOOR LOCATED AT THE
RIGHT SIDE OF
FUSELAGE.

AIR CONDITIONING
DUCTS AND RADIATOR

"ULTRA MODERN" AIRCRAFT

Designer and builder: WALTER F. LAREDO	
Drawing Number: 800 series	Sheet 19 of 24

PLATE 20

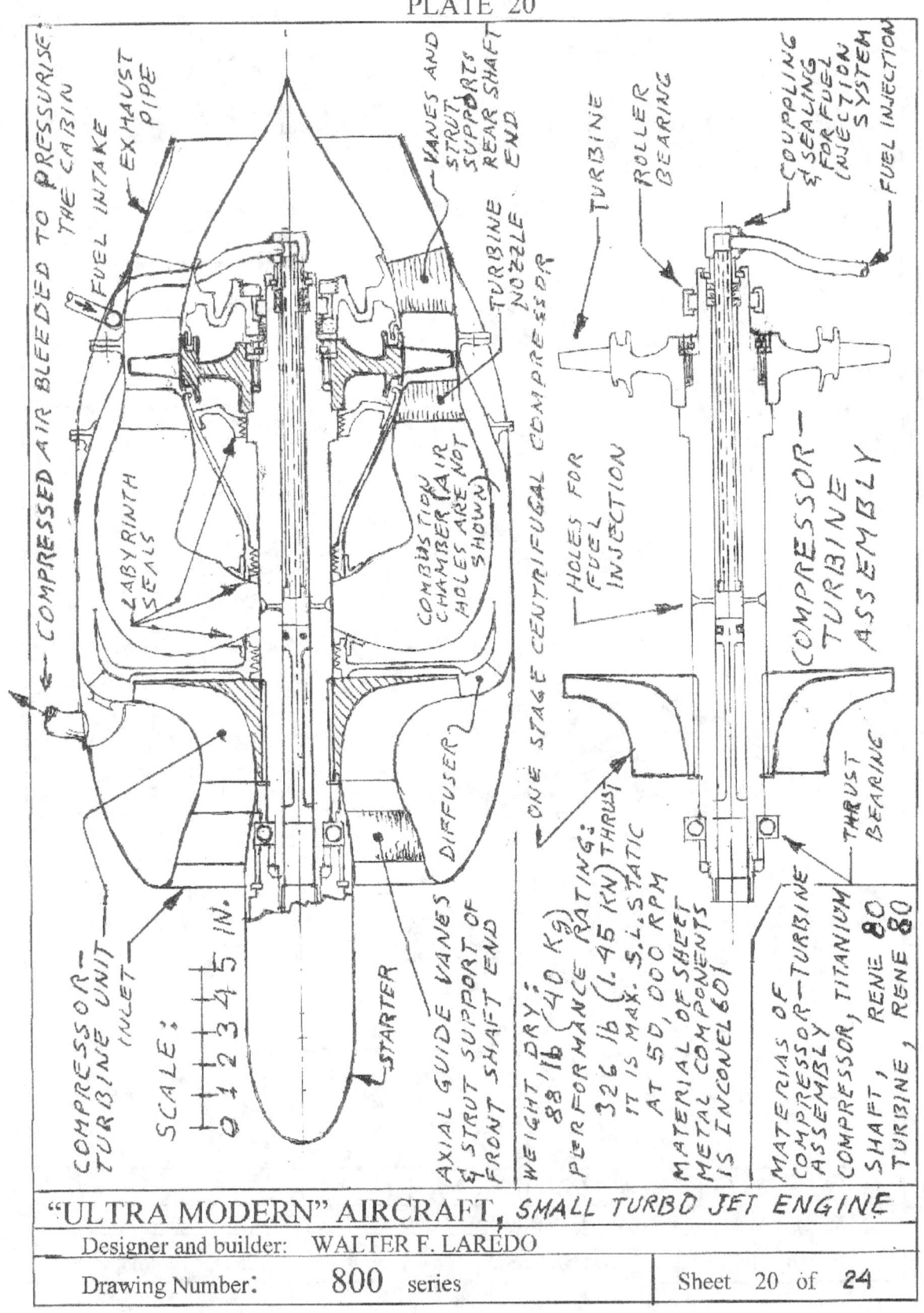

COMPRESSED AIR BLEEDED TO PRESSURISE THE CABIN

FUEL INTAKE

EXHAUST PIPE

VANES AND STRUT SUPPORTS REAR SHAFT END

TURBINE NOZZLE

TURBINE

ROLLER BEARING

COUPPLING & SEALING FOR FUEL INJECTION SYSTEM

FUEL INJECTION

LABYRINTH SEALS

COMBUSTION CHAMBER (AIR HOLES ARE NOT SHOWN)

ONE STAGE CENTRIFUGAL COMPRESSOR

HOLES FOR FUEL INJECTION

COMPRESSOR-TURBINE ASSEMBLY

COMPRESSOR-TURBINE UNIT

INLET

SCALE: 0 1 2 3 4 5 IN.

STARTER

DIFFUSER

AXIAL GUIDE VANES & STRUT SUPPORT OF FRONT SHAFT END

WEIGHT, DRY: 88 lb (40 Kg)

PERFORMANCE RATING: 326 lb (1.45 KN) THRUST IT IS MAX. S.L. STATIC AT 50,000 RPM

MATERIAL OF SHEET METAL COMPONENTS IS INCONEL 601

MATERIALS OF COMPRESSOR-TURBINE ASSEMBLY COMPRESSOR, TITANIUM THRUST BEARING

SHAFT, RENE 80 TURBINE, RENE 80

"ULTRA MODERN" AIRCRAFT, SMALL TURBO JET ENGINE

Designer and builder:	WALTER F. LAREDO	
Drawing Number:	800 series	Sheet 20 of 24

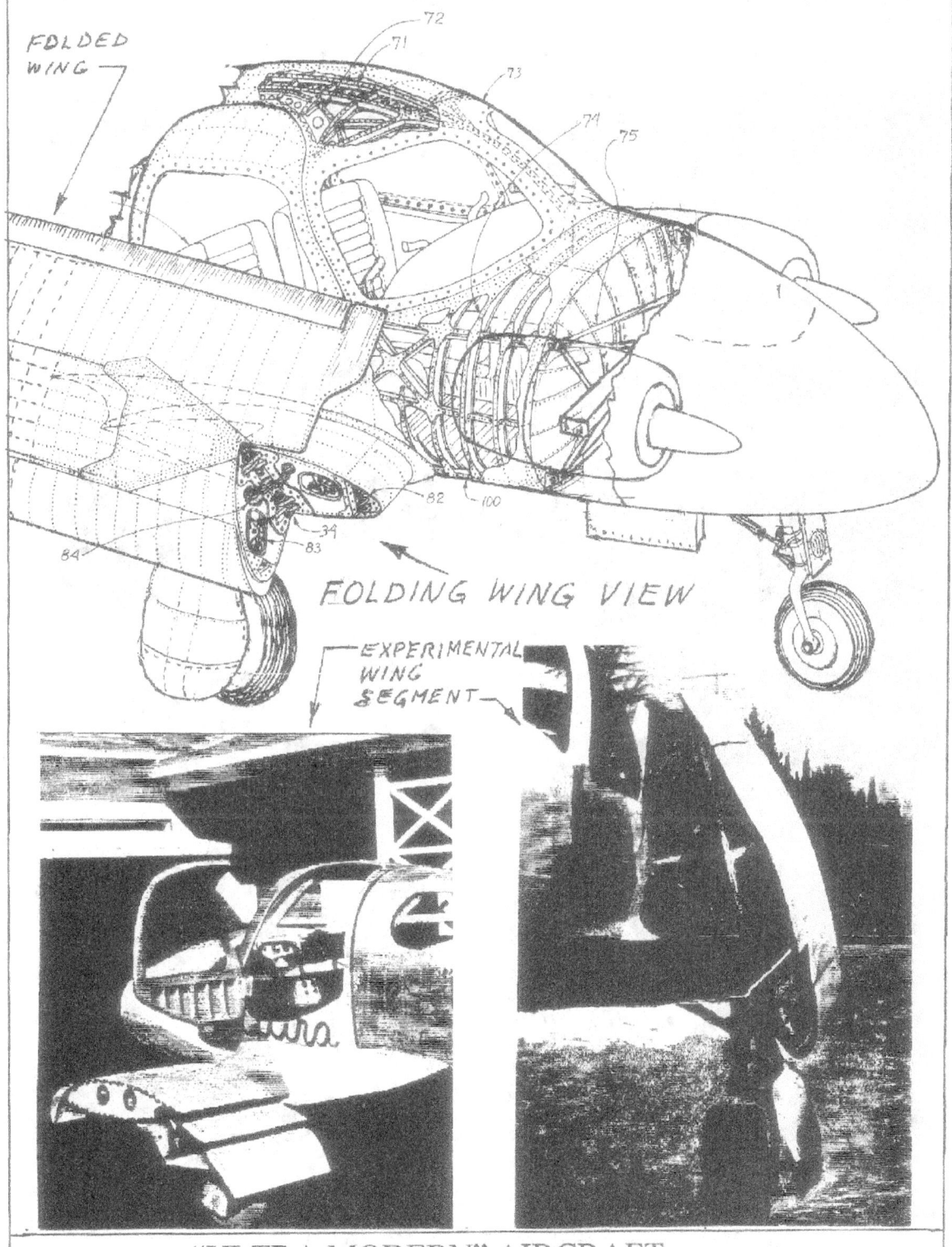

FOLDED WING

FOLDING WING VIEW

EXPERIMENTAL WING SEGMENT

"ULTRA MODERN" AIRCRAFT		
Designer and builder: WALTER F. LAREDO		
Drawing Number: 800 series	Sheet 21 of 24	

PLATE 22

CUSTOM MADE PARTS, FORMED FROM ALUMINUM ALLOY SHEET ON HARD WOOD DIES

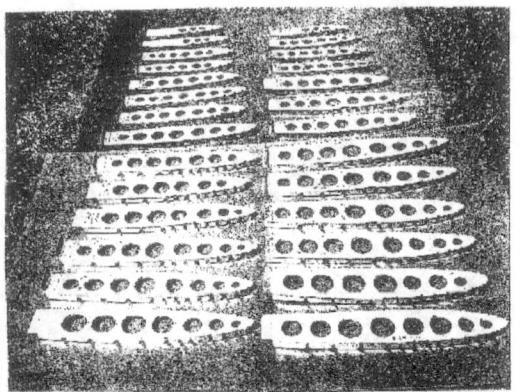

RIBS FOR THE STABILIZERS

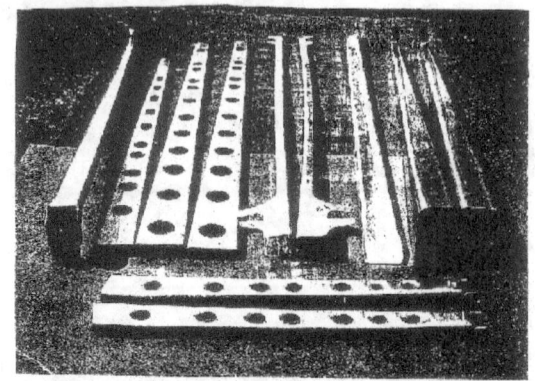

TAIL SECTION SPARS

CONSTRUCTION VIEW OF HORIZONTAL STABILIZER

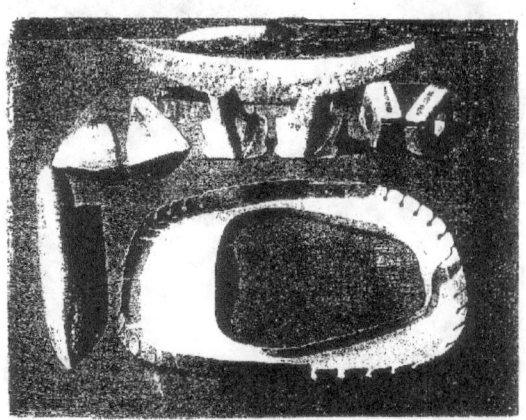

FITTINGS STAMPED FROM ALUMINUM SHEET

HORIZONTAL STABILIZER READY FOR RIVETING (AL. SKIN HOLD BY CLICOS)

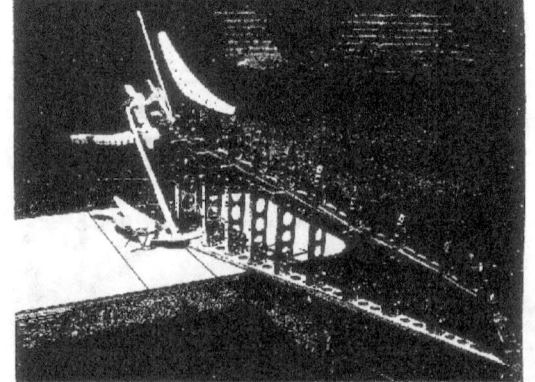

CONSTRUCTION OF VERTICAL STABILIZER

"ULTRA MODERN" AIRCRAFT

Designer and builder: WALTER F. LAREDO

| Drawing Number: 800 series | Sheet 22 of 24 |

AFT
FUSELAGE
AIRFRAME

AIRCRAFT
STRUCTURE
IS HALF
COMPLETED

PRESSURE
BULKHEAD
AFT OF
PRESSURIZED
AREA

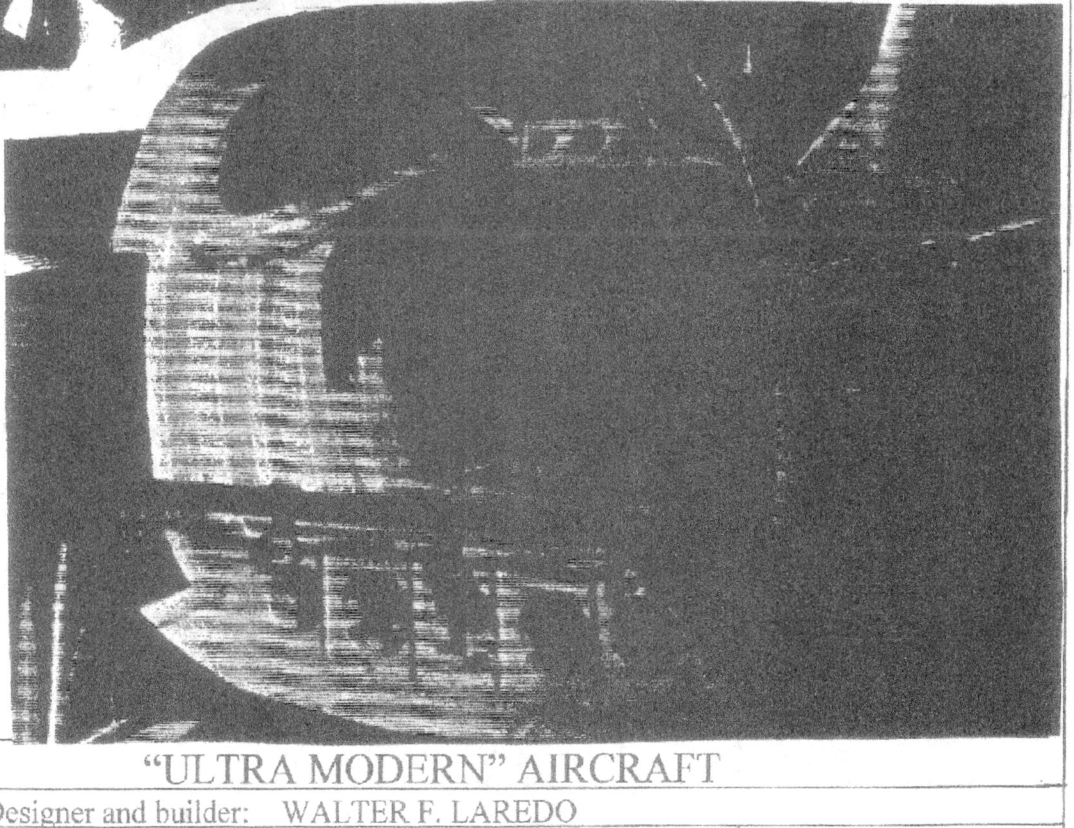

"ULTRA MODERN" AIRCRAFT

Designer and builder: WALTER F. LAREDO	
Drawing Number: 800 series	Sheet 23 of 24

PLATE 24

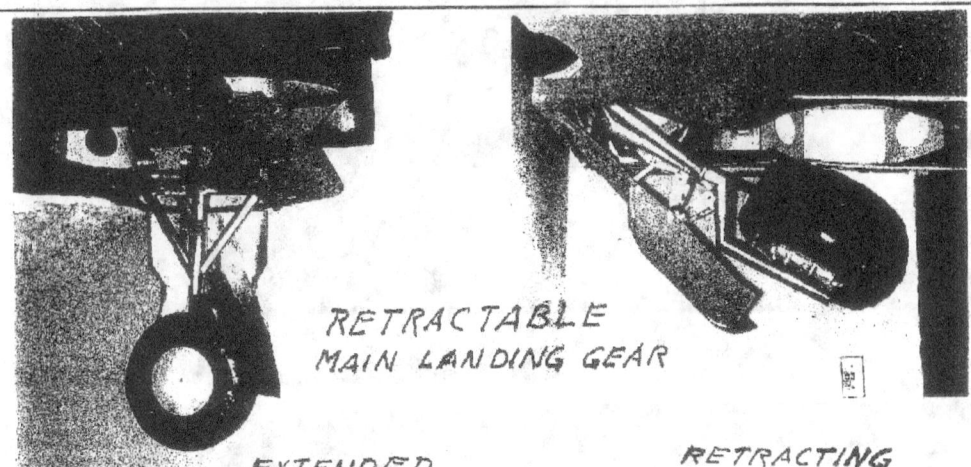

RETRACTABLE
MAIN LANDING GEAR

EXTENDED

RETRACTING

FUSELAGE CONSTRUCTION

L.E. SLAT

MAIN LANDING GEAR

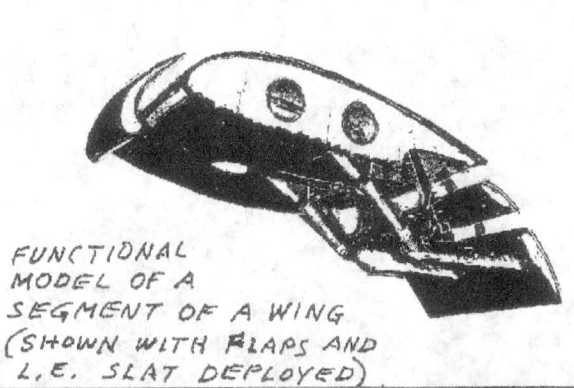

FUNCTIONAL MODEL OF A SEGMENT OF A WING (SHOWN WITH FLAPS AND L.E. SLAT DEPLOYED)

BUILDING A WOOD CRADDLE MAIN TOOL TO ASSEMBLY THE AIRFRAME

"ULTRA MODERN" AIRCRAFT

Designer and builder: WALTER F. LAREDO	
Drawing Number: 800 series	Sheet 24 of 24

PLATE 25

TWIN TURBOFAN TRANSPORT FOR 30 PASSENGERS

ADVANCED ENGINEERING PROJECTS

Name of project:

"ULTRA TWIN JET" AIRCRAFT

Design Engineer:	WALTER F. LAREDO	Date:	
Drawing Number:	900 SERIES	Sheet 1 of 14	

PLATE 26

INTRODUCCION

The "Ultra Twin Jet " is a thirty passenger twin-turbofan aircraft. Though its primary role is as commercial transport, it can be suited to other roles included commuter, business, VIP and express cargo. The aircraft have high by pass ratio engines, advanced airframe, swept wing design, advanced composite structure and digital avionics will enable this new aircraft to achieve better quietness and fuel efficiency. Aircraft sized effectively for profitably entry in variety of markets without the burden of surplus seats.

Designed to meet future demand in high or low-density routes, the Ultra Twin Jet of advanced technology could be implemented to serve routes of average flight time up to six continuous hours. For later production models, structural materials such as aluminum-lithium will replace to conventional aluminum alloys. With those considerations, the Ultra Twin Jet will become one of the most advanced aircraft ever built by any aircraft industries in to-day's and future short-to-medium and long range twin jet standards for durability, reliability, maintainability and profitability.

FLIGHT TEST OF PROTOTYPE AND FLIGHT CERTIFICATION

Fight test of prototype and flight certification for a thirty (30) passenger commercial twin jet aircraft shall meet the normal category requirements of the United States Federal Aviation Administration.

GENERAL DESCRIPTION

The Ultra Twin Jet will be a fuel efficient airplane and will have lower cost per passenger-mile than most passenger jets flying today, efficiency will be the result of its aerodynamics, an advanced wing design and high-bypass-ratio engines will provide excellent performance. In fact the Ultra Twin Jet will operate from most general airfields. The innovative wing will give the plane more lift with less drag and less weight, this last as a result of improved aluminum alloys and lightweight composites.

Ailerons, spoilers, elevators and rudder will be build as a sandwich structure with woven or with tape form graphite skins over a Du Pond "Nomex" honeycomb. Parts exposed to small foreign object impact as landing gear doors, nacelle-cowl components and wing body fairing which will be covered with an additional outer layer of extremely thought woven Du Pond "Kevlar" fibers imbedded in an epoxy matrix.

This medium range aircraft will have integral fuel tanks in the wings and in the medium long configuration, probably will have extra fuel tanks in its body. Carrying additional fuel in external wing tip tanks to extend range as in many executive airplanes was not considered as acceptable alternative, wing tanks present problems as high roll inertia with detrimental effects on lateral and directional handling qualities. Conventional wings could not have the volume to hold the required amount of fuel, but the advanced technology wing is about 20 per cent thicker, increasing in this way its fuel volume.

"ULTRA TWIN JET", aircraft for 30 passengers		
Design Engineer: WALTER F. LAREDO		
Drawing number: 900 series		Sheet 2 of 14

PLATE 27

The advanced-technology-wing with supercritical airfoils will be able to retain the high critical number and reasonable drag-rise characteristics. This wing will be thicker, less swept with advanced high lift devices as double slotted Fowler flaps and full-span leading-edge devices will be incorporated in this airplane to compensate for its relative higher wing loading that offers good ride. Lifting heavy loads from high hot fields, because its leading edge devices the airplane will require about 30 per cent less runway length in addition to that will also reduce its stalling speed. Good wing for higher cruising efficiency and higher cruising altitudes where the thin air will produce less drag, faster climbing, quieter approaches and all that was indicated will result in excellent fuel savings.

A good maximum lift capability will be obtain, by incorporating to this wing full-span leading-edge devices, plus double slotted Fowler flaps, with all this considerations this airplane will offer good a good ride because its high wing loading per square feet at cruise.

STRUCTURES DESIGN PHILOSOPHY
Safety will be the prime consideration in structural design of the Ultra Twin Jet airplane, the primary structure of wing and fuselage are of conventional semi-monocoque construction. Every part, every assembly will be designed to minimize corrosion and fatigue and for fail-safe.

Latest-State-of-the-Art in fatigue-resistance design, airframe designed for fatigue life in excess of 40 000 hours. Damage-tolerant Structure, corrosion resistance and low maintenance aircraft.

WINGS
One piece wing utilizes a two-spar box for strength, full span leading edge slats and two section double-slotted flaps, two spoilers used concurrently with the ailerons. A low swept wing design with supercritical airfoil enable me to design a more efficient and lighter structure than for a wing of conventional design. The wing box have Zee and Jay stringers, fastened to a constant tapered machine skins(*), spar caps are also machined (*), spar webs are chemically milled(*).

Materials for wing components:
Upper Wing Panels: Aluminum alloys 7000 series. (*)
Lower Wing Panels: Aluminum alloys 2000 series. (*)
The upper and lower panels are Machine tapered and shot peen formed. (*)
(*) At the beginning of production, for low budget construction could be use build-up parts by riveting constant thickness stock, build this way instead of machined tapered parts, sacrificing a little payload.
Stringers: Zee and jay extrusion sections, 7000 series aluminum alloys.
Spar caps, spar web, ribs and rib chords are from series 2000 and 7000 to meet special Requirements.
Leading edge slats and its tracks cover or fairing, trailing edge flaps, flaps veins, aileron and spoilers are built from advanced composites, by using some of the following materials:

"ULTRA TWIN JET", aircraft for 30 passengers		
Design Engineer: WALTER F. LAREDO		
Drawing number:	900 series	Sheet 3 of 14

PLATE 28

Graphite, Kevlar, fiberglass and "Nomex" core, all embedded with epoxy resin. By using advanced technology materials, wing will become 20 per cent lighter.

BODY
Damage-tolerant, Fail-safe type of construction, semi-monocoque design. The body is of circular cross section. Skins from 2024-T3 aluminum clad, stringers from 7075-T6 or 2024-T3511 aluminum alloys, the bonded tear straps are of 7075-T6 aluminum alloy. In later production airplanes of more advanced technology the skins could be made by employing mechanically or chemically milled aluminum alloys.

Body pressurization to a maximum differential pressure of 8.5 psi.

EMPENAGE
The T tail configuration design, uses an incident-adjustable horizontal stabilizer with inverted airfoil is made by a two-spar box structure and skins with zee stringers.

The vertical stabilizer box is a multiple-spar structure, with fittings at its upper end which houses the bearings that supports the horizontal stabilizer hinges. The leading edge and the trailing edge of the elevator and rudder structure are made from graphite-epoxy.

The aft end of the fuselage behind the pressure bulkhead is not pressurized and contains the air conditioning pack, bleeding-ducts from the engines, accessories and parts of the fuel system. The forward spars of the engine struts are integral with the pressure bulkhead structure.

FLIGHT CONTROLS
Ailerons, rudder and elevators are driven by hydraulic servo-actuators combined with push-pull rod linkages, backed by a mechanical cable system. A motor jack is attached to the stabilizer for pitch trim. Full span leading edge slats, double slotted flaps and spoilers are actuated by an electric motor or as alternative by a hydraulic motor.

HYDRAULIC SYSTEM
Two independent 3000 psi systems powered by the two pumps mounted on the two engines. System 1 and system 2, both independently powers the flight controls, Leading Edge Slats, Trailing Edge Flaps, Main and Nose Landing Gears, Landing Gear Doors (which could be actuated by hydraulic power or by a direct mechanical linkage connected to the landing gear struts). Hydraulic system number 1, also powers the brakes.

ENVIRONMENT CONTROL SYSTEM
Pressurization in climb and cruise flight conditions are obtained by bleeding air from an intermediate compressor stage. During idle, descend and other low flight conditions, pressurization is obtained from the last compressor stage.
Engine bleed air powers the Air Cycle Machine, then this air passes through the Ram Heat Exchanger to be cool, then to a dehumidifier and other processes. At 36000 ft (10800 m) is maintained an equivalent cabin pressure of 8000 ft (2400 m).

"ULTRA TWIN JET", aircraft for 30 passengers		
Design Engineer: WALTER F. LAREDO		
Drawing number:	900 series	Sheet 4 of 14

PLATE 29

OXYGEN SYSTEM: Conventional Gaseous System
FIRE EXTINGUISHER: Conventional
ELECTRIC SYSTEM: Conventional
ELECTRONICS: Small Radar
ANTI-ICING SYSTEM: Will prevent the formation of ice on the Engine Nose Cone. The engine inlet Anti-Icing System uses hot engine air bleed. Pitot Static and Probes are heated by Electric Heaters.
PRESSURIZATION: Normal cabin pressure differential is 8.6 psi.

EXPECTED PERFORMANCE AND WEIGHTS DATA

PERFORMANCE

SPEEDS
Maximum cruise.............528 mph   (850 km/hr)   (0.80 Mach)
Normal Cruise................501 mph   (806 km/hr)   (0.76 Mach)
Long-Range Cruise...........469 mph   (754 km/hr)   (0.71 Mach)

CEILING
Maximum Operating Altitude........42,000 ft.   (12,802 m)

CLIMB
Time of Climb to Initial Cruise Altitude......23 minutes

AIRFIELD PERFORMANCE
Balance Field Length at Max. Takeoff Weight.......5500 ft. (1676 m.)
Landing Distance at Max. Landing Weight...........3800 ft. (1158 m.)

RANGE
2095 nautical miles (2408 miles) (3874 km) with 30 passengers and a crew of two.

WEIGHTS
Maximum ramp weight.....................39100 lb.   (17735 kg.)
Maximum takeoff weight...................39000 lb.   (17689 kg.)
Maximum landing weight...................36000 lb.   (16329 kg.)
Maximum zero fuel weight.................25500 lb.   (11566 kg.)
Approximate weight empty.................17500 lb.   (7938 kg.)
Typical operating weight empty...........22450 lb.   (10183 kg.)
Maximum fuel load........................... 9771 lb.   (4432 kg.)
Payload with full fuel....................... 6879 lb.   (3120 kg.)
Maximum payload weight.................. 8000 lb.   (3629 kg.)

FWD   C.G. LIMIT (% MAC)............17 inch (0.432 m.)
AFT   C.G. LIMIT (% MAC)............34 inch (0.864 m.)

"ULTRA TWIN JET", aircraft for 30 passengers

Design Engineer: WALTER F. LAREDO

| Drawing number: | 900 series | Sheet 5 of 14 |

PLATE 30

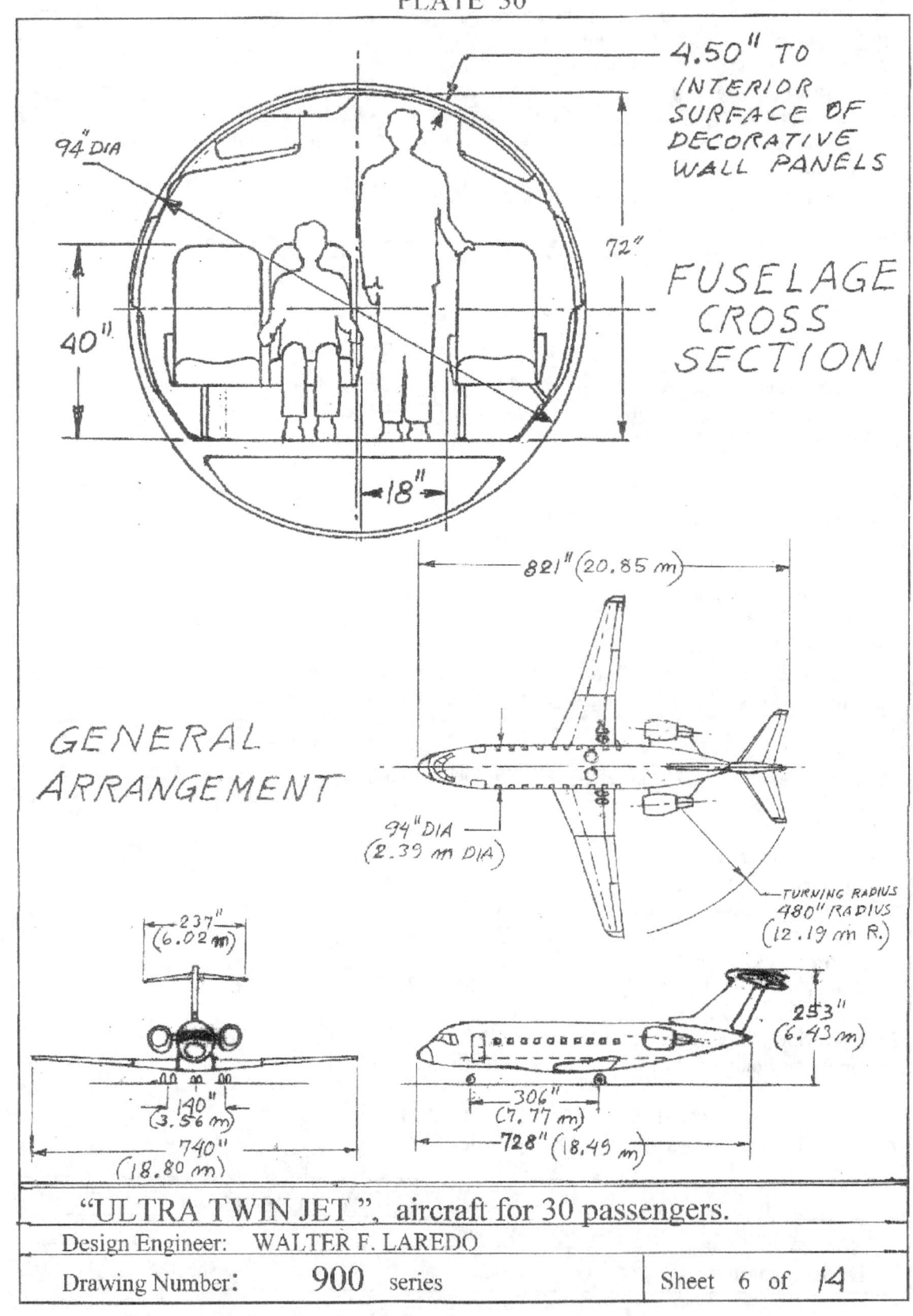

4.50" TO INTERIOR SURFACE OF DECORATIVE WALL PANELS

94" DIA

72"

40"

FUSELAGE CROSS SECTION

18"

821" (20.85 m)

GENERAL ARRANGEMENT

94" DIA
(2.39 m DIA)

TURNING RADIUS
480" RADIUS
(12.19 m R.)

237"
(6.02 m)

253"
(6.43 m)

140"
(3.56 m)

306"
(7.77 m)

728" (18.49 m)

740"
(18.80 m)

"ULTRA TWIN JET", aircraft for 30 passengers.

Design Engineer: WALTER F. LAREDO

Drawing Number: **900**   series

Sheet 6 of 14

PLATE 31

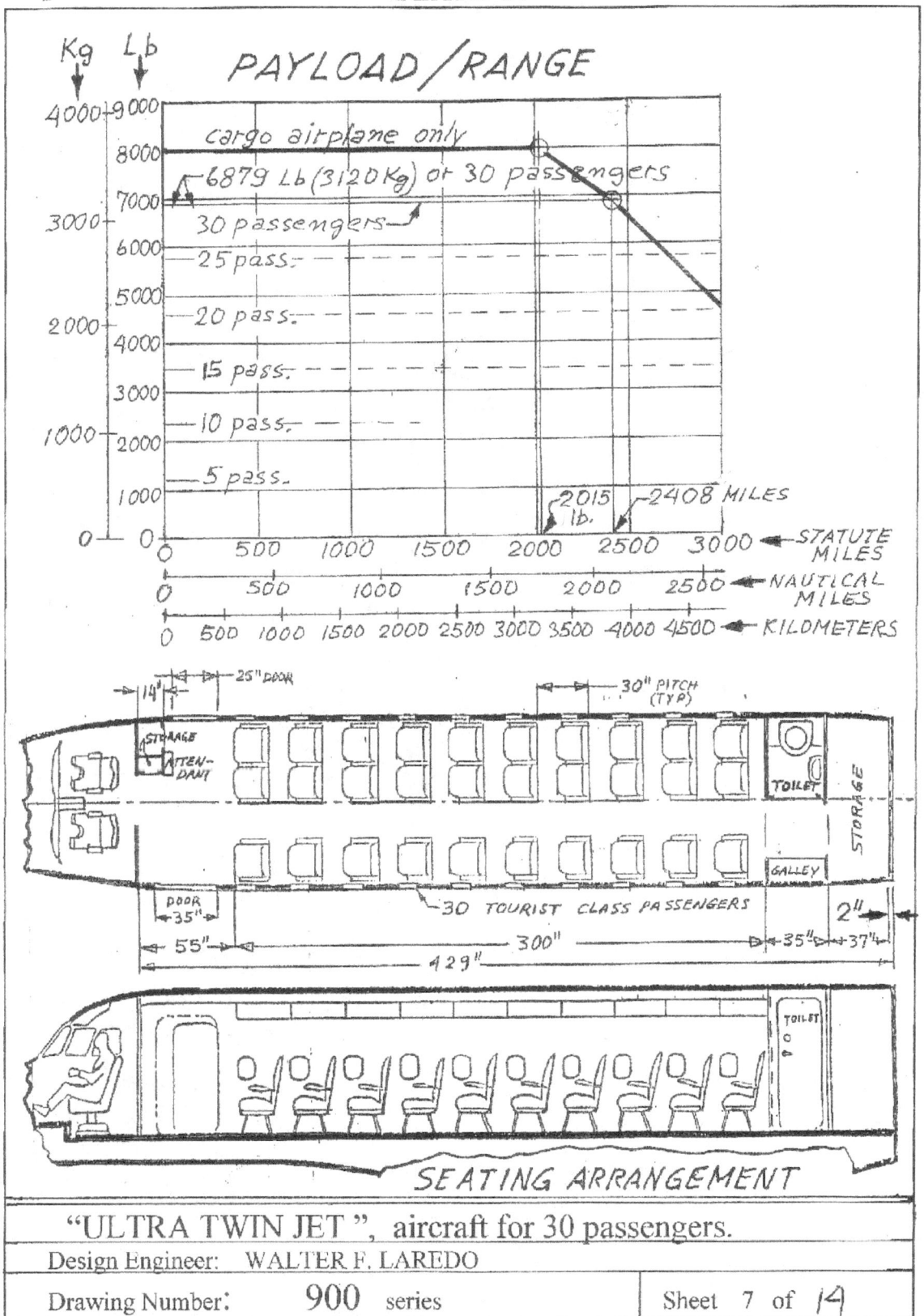

PAYLOAD/RANGE

cargo airplane only

6879 Lb (3120 Kg) or 30 passengers

30 passengers
25 pass.
20 pass.
15 pass.
10 pass.
5 pass.

2015 lb. 2408 MILES

STATUTE MILES
NAUTICAL MILES
KILOMETERS

25" DOOR
14"
STORAGE
ATTEN-DANT
30" PITCH (TYP)
TOILET
STORAGE
GALLEY

DOOR 35"
30 TOURIST CLASS PASSENGERS
2"
55"
300"
35"
37"
429"

TOILET

SEATING ARRANGEMENT

"ULTRA TWIN JET", aircraft for 30 passengers.

Design Engineer: WALTER F. LAREDO

Drawing Number: 900 series

Sheet 7 of 14

PLATE 3 2

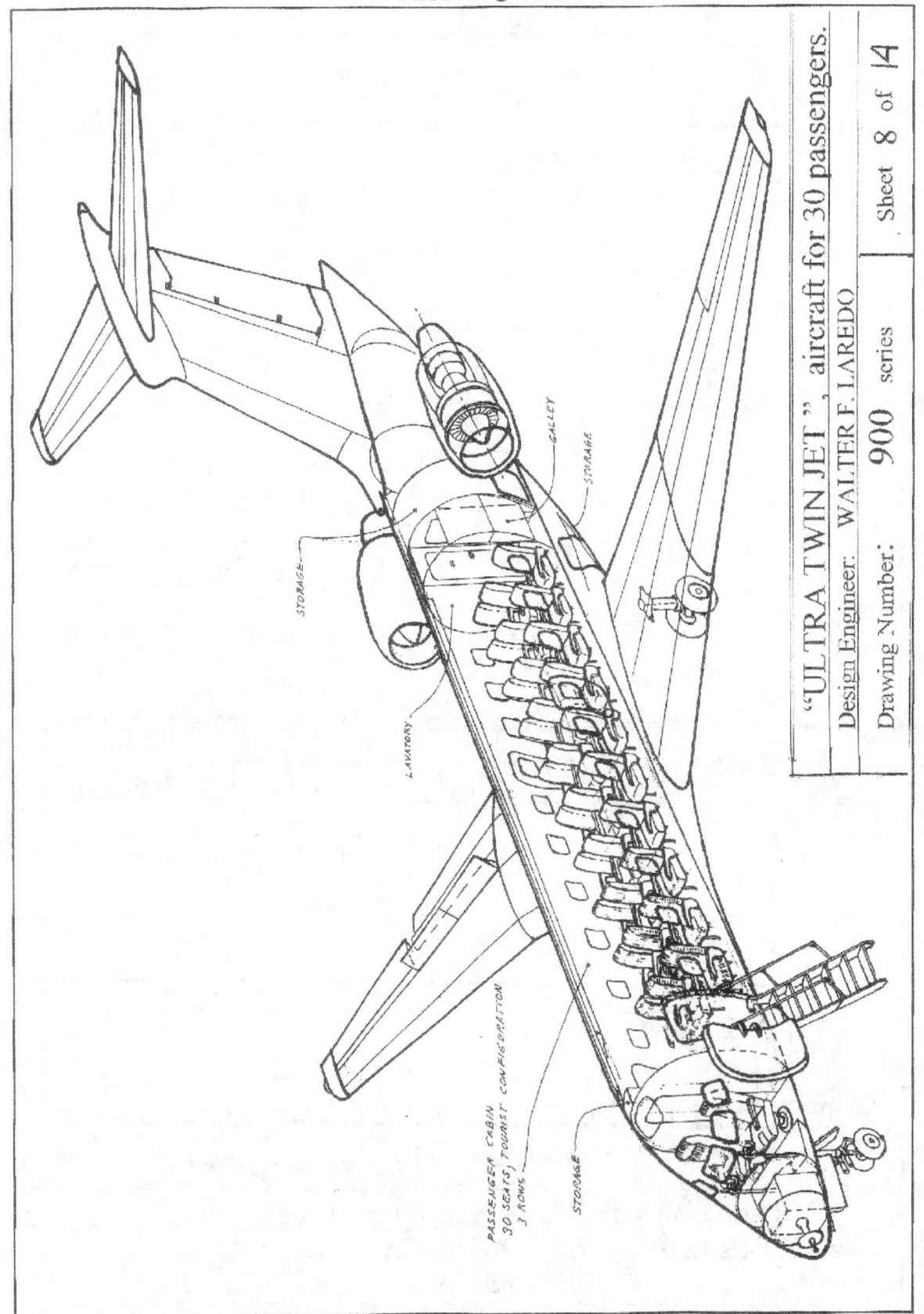

STORAGE

GALLEY

STORAGE

LAVATORY

PASSENGER CABIN
30 SEATS, TOURIST CONFIGURATION
3 ROWS

STORAGE

"ULTRA TWIN JET", aircraft for 30 passengers.

Design Engineer: WALTER F. LAREDO

Drawing Number: 900 series

Sheet 8 of 14

PLATE 33

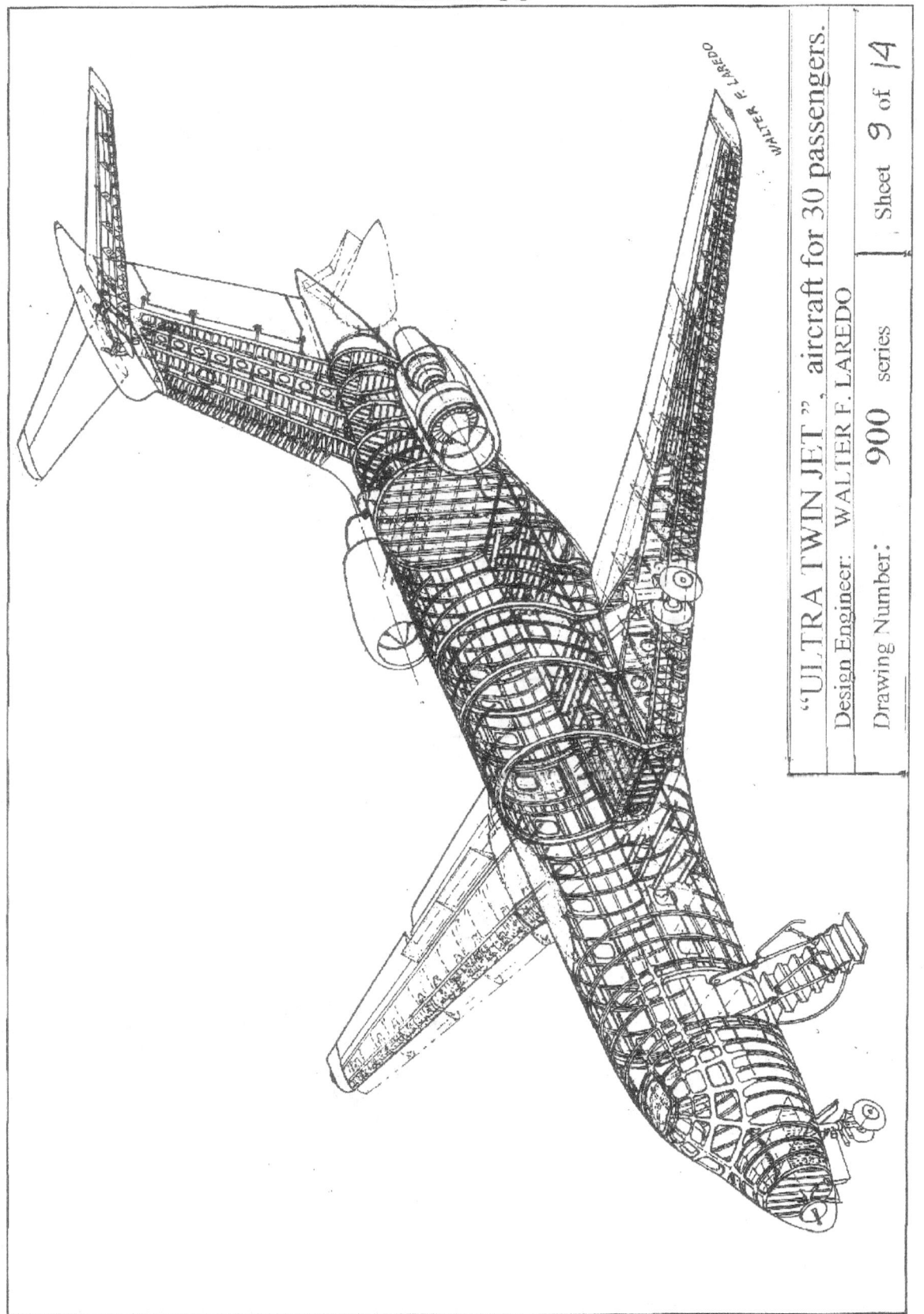

WALTER F. LAREDO

"ULTRA TWIN JET", aircraft for 30 passengers.

Design Engineer: WALTER F. LAREDO

Drawing Number: 900 series

Sheet 9 of 14

PLATE 34

GENERAL ELECTRIC CF 34 HIGH BYPASS TURBOFAN ENGINE TO BE USE IN THE "ULTRA TWINJET" AIRCRAFT

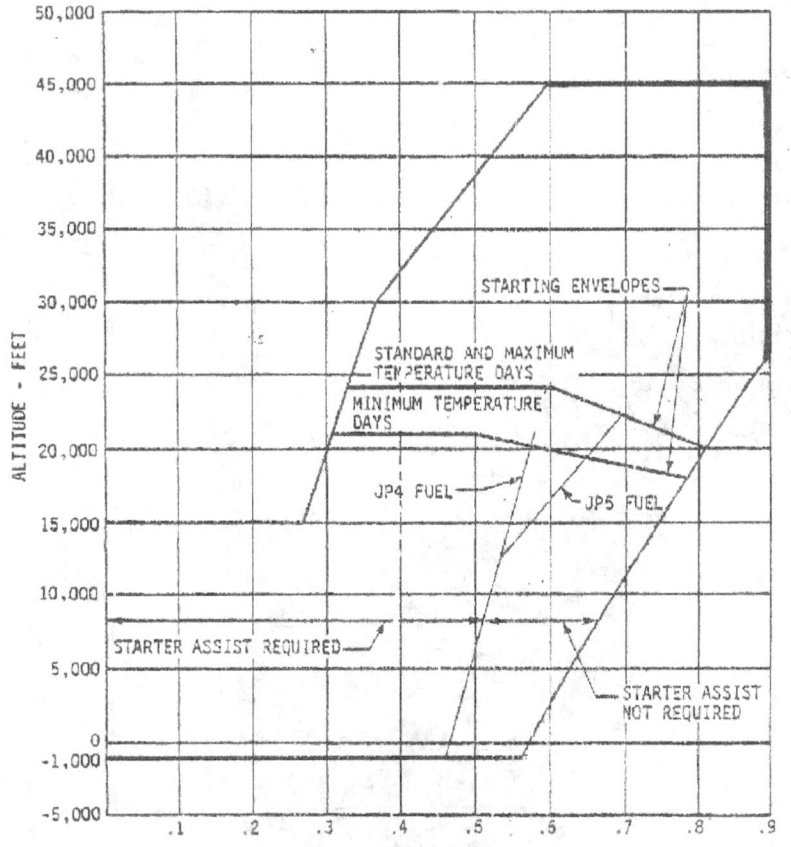

Flight Envelope Altitude/Mach No. Extremes

SPECIFICATIONS

Takeoff Thrust (lb)	7990
7990 (flat rated to 73°F)	
Max Continuous Thrust (.8M/36K) (lb)	1760
Specific Fuel Consumption at Takeoff	.359
Max Continuous Specific Fuel Consumption (.8M/36K)	.687
Weight (lb)	1525
Length (in.)	100
Max Diameter (in.)	49

"ULTRA TWIN JET", aircraft for 30 passengers.

Design Engineer: WALTER F. LAREDO

Drawing Number: 900 series

Sheet 10 of 14

PLATE 35

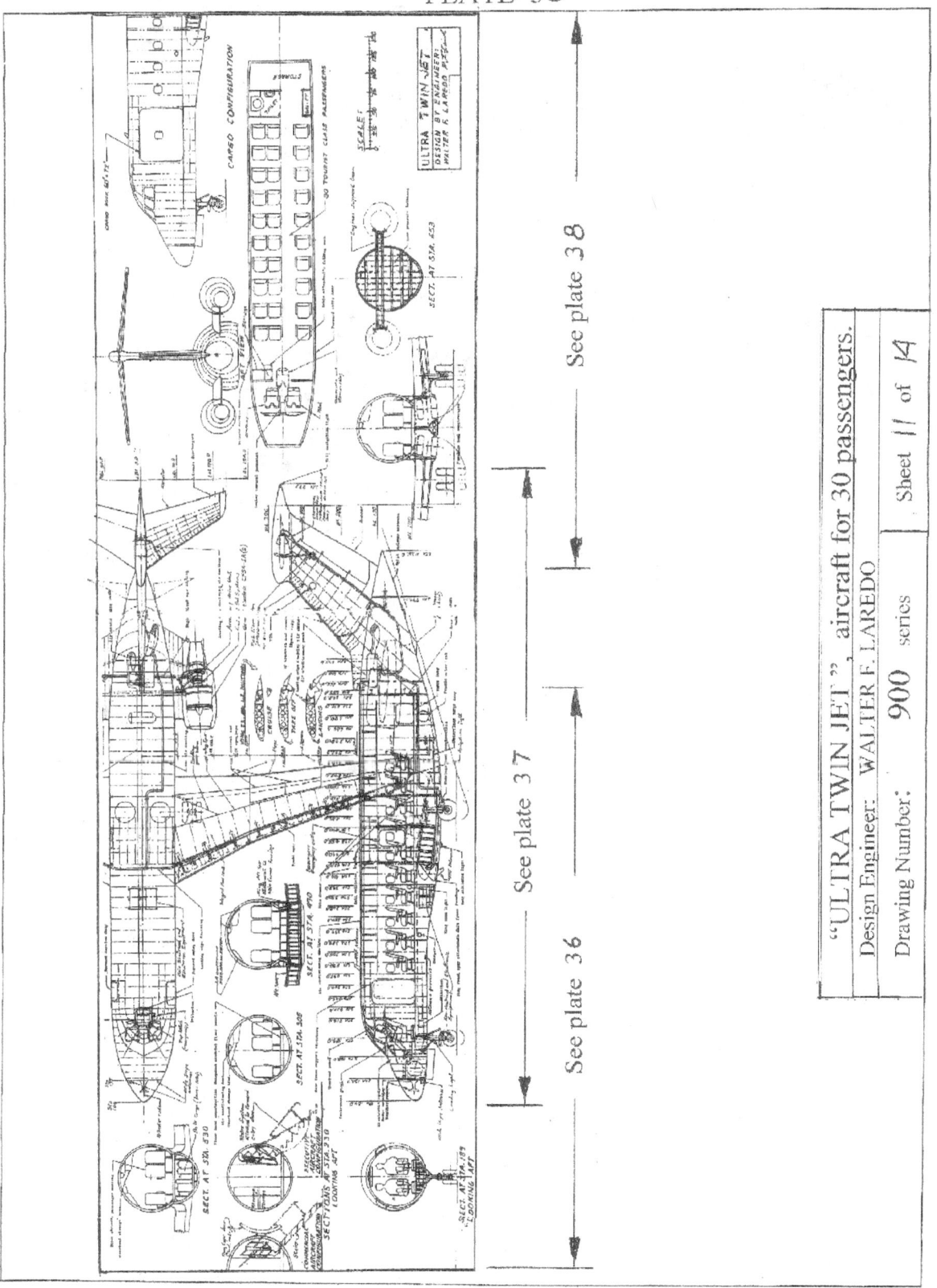

CARGO CONFIGURATION

30 TOURIST CLASS PASSENGERS

SCALE:

ULTRA TWIN JET
DESIGN BY ENGINEER:
WALTER F. LAREDO M.E.

See plate 38

See plate 37

See plate 36

"ULTRA TWIN JET", aircraft for 30 passengers.		
Design Engineer: WALTER F. LAREDO		
Drawing Number:	900 series	Sheet 11 of 14

PLATE 36

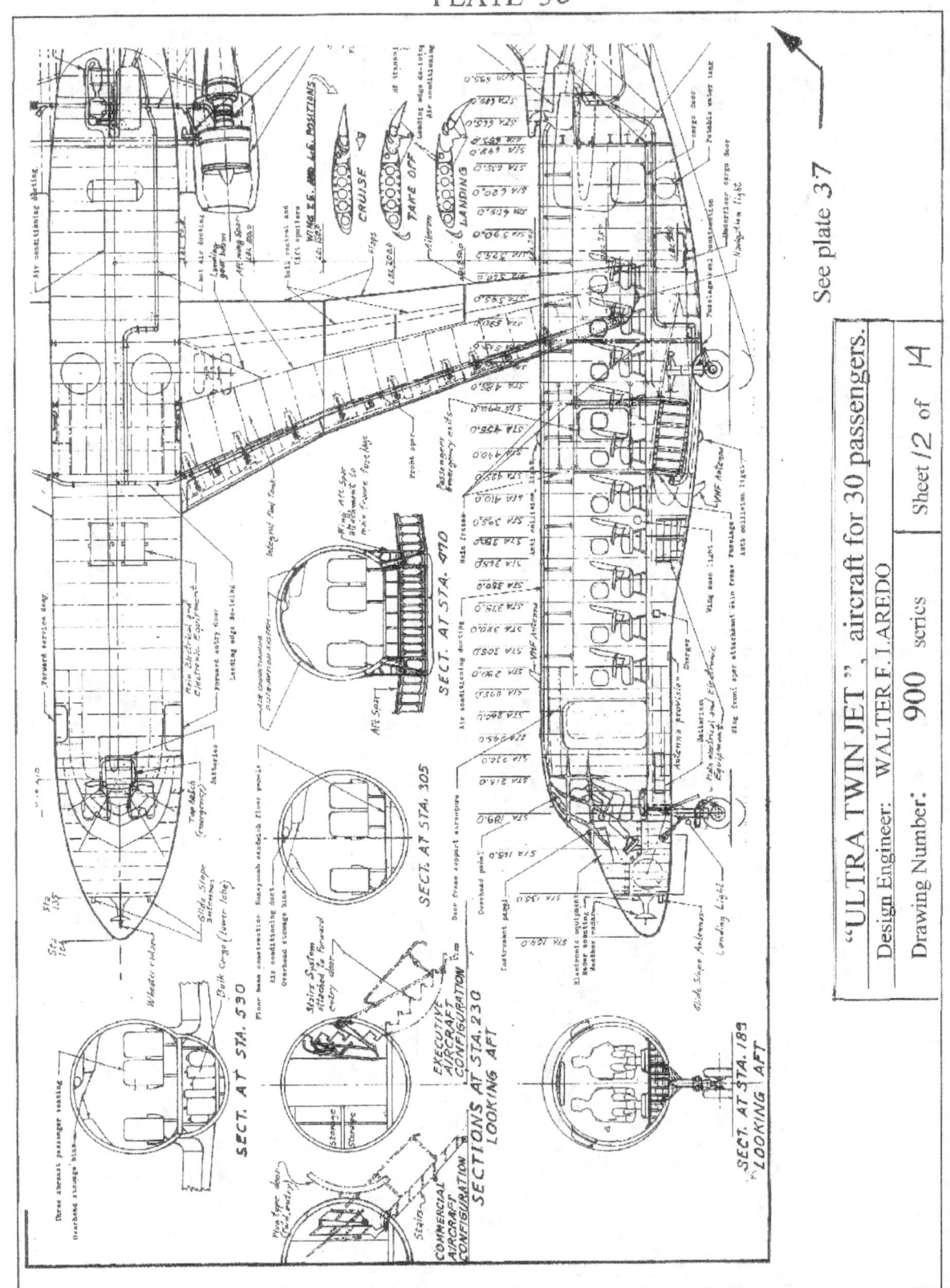

See plate 37

"ULTRA TWIN JET ", aircraft for 30 passengers.

Design Engineer: WALTER F. LAREDO

Drawing Number: 900 series   Sheet 12 of 14

PLATE 37

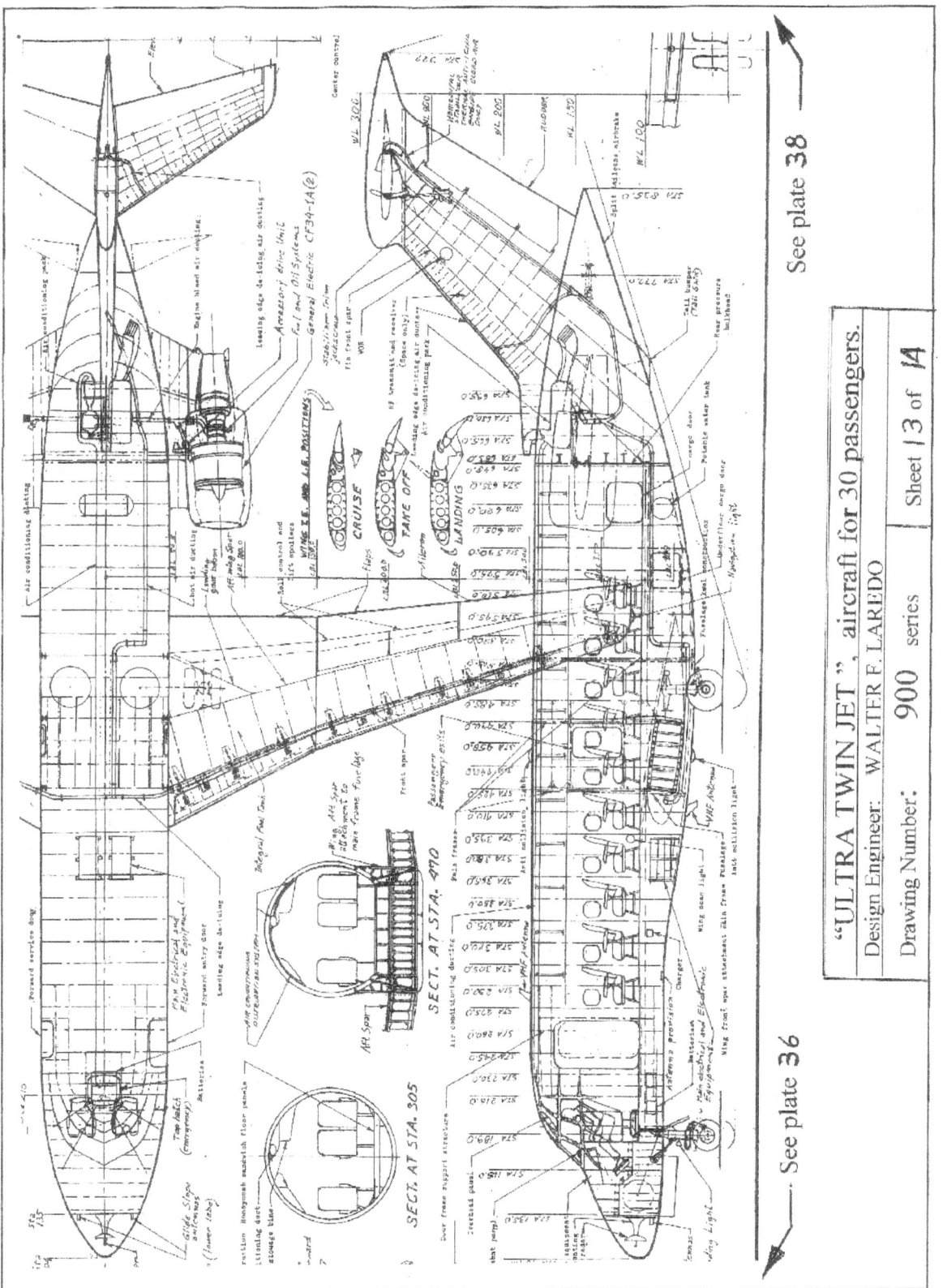

See plate 38

See plate 36

"ULTRA TWIN JET", aircraft for 30 passengers.

Design Engineer: WALTER F. LAREDO	
Drawing Number: 900 series	Sheet 13 of 1A

PLATE 38

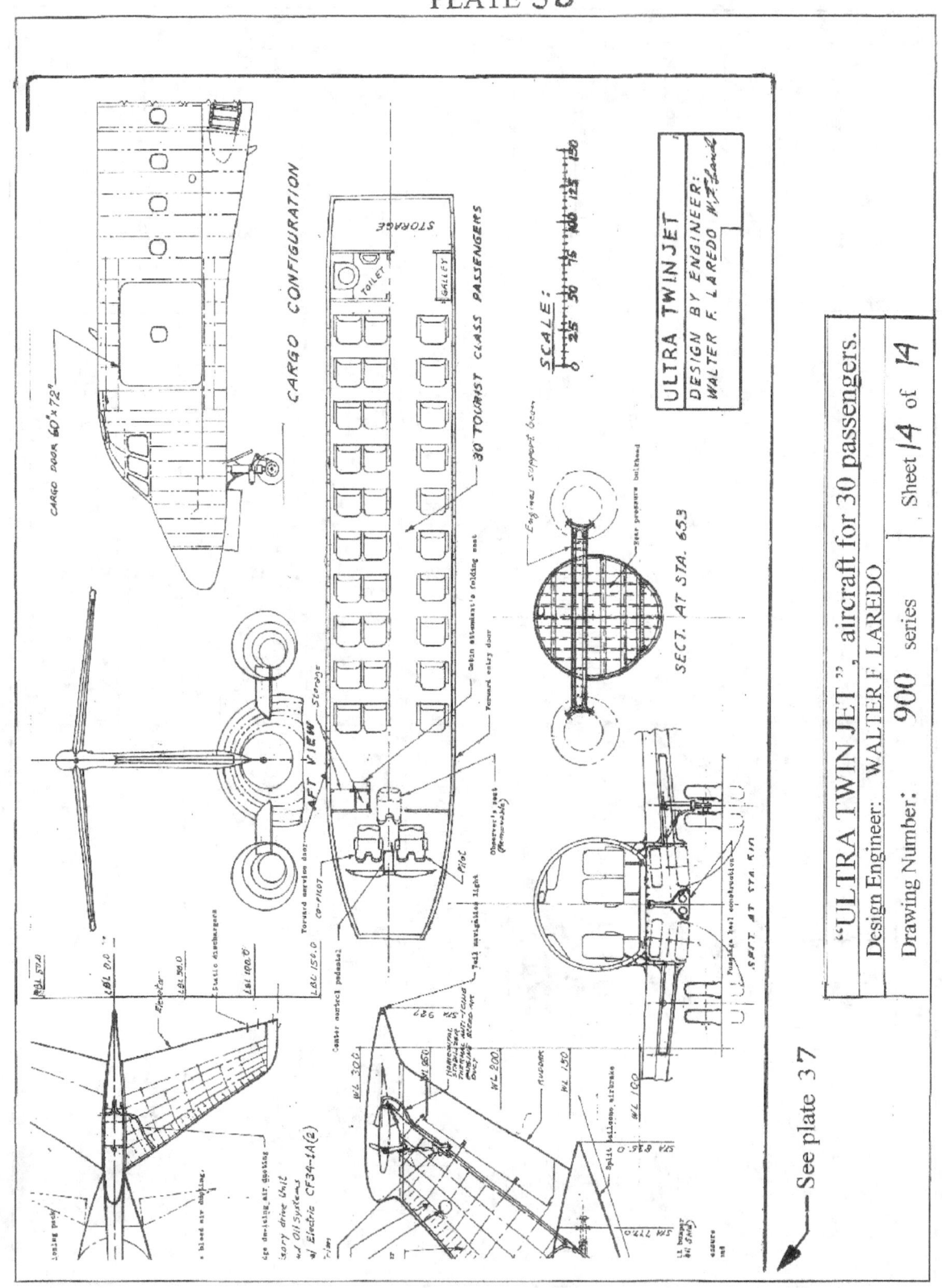

CARGO CONFIGURATION

CARGO DOOR 60" x 72"

STORAGE

TOILET

GALLEY

30 TOURIST CLASS PASSENGERS

Cabin attendant's folding seat

Forward entry door

AFT VIEW

SCALE:
0 25 50 75 100 125 150

Engines support beam

Rear pressure bulkhead

SECT. AT STA. 65.3

SECT. AT STA. 5.0

See plate 37

ULTRA TWIN JET
DESIGN BY ENGINEER: WALTER F. LAREDO

"ULTRA TWIN JET", aircraft for 30 passengers.	
Design Engineer: WALTER F. LAREDO	
Drawing Number: 900 series	Sheet 14 of 14

PLATE 39

Designed in spring of 1987 by W. F. Laredo, and later proposed separately to Brazil, Argentina, Spain and Italy for the construction of the first prototype

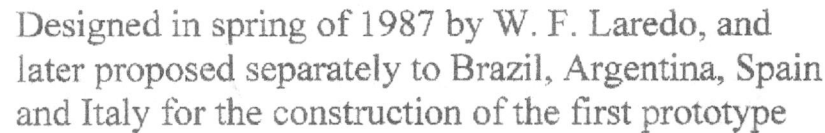

FOR ADDITIONAL INFORMATION SEE
"HALCON WL1" PROPOSAL BOOK
BY W.F. LAREDO

ADVANCED ENGINEERING PROJECTS

Name of project:	PRELIMINARY DESIGN	
"HALCON WF1", Lightweight fighter		
Design Engineer: WALTER F. LAREDO		Date: May 1987
Drawing Number:	1000 SERIES	Sheet 1 of 34

PLATE 40

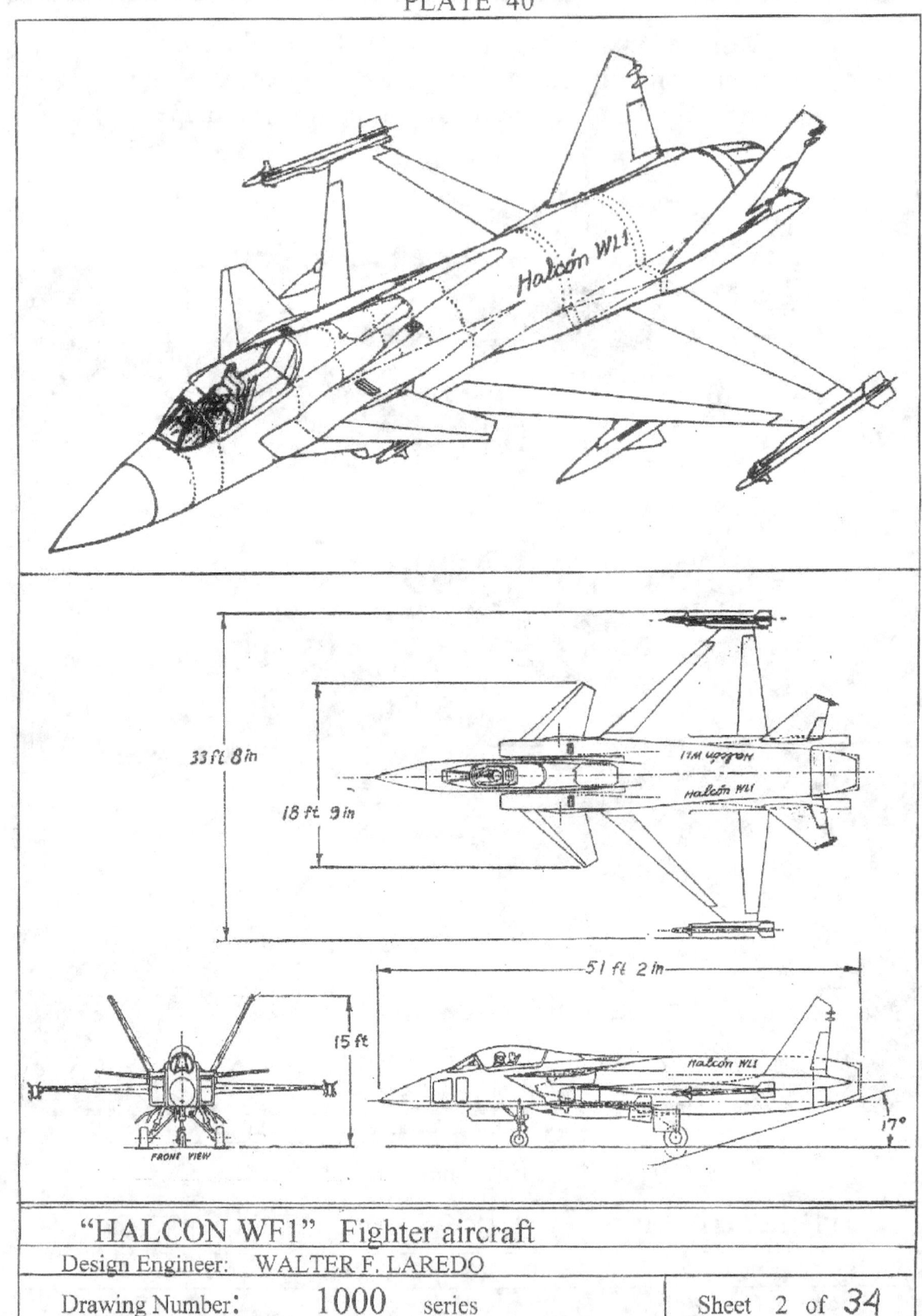

Halcón WL1

33 ft 8 in

18 ft 9 in

51 ft 2 in

15 ft

17°

FRONT VIEW

"HALCON WF1" Fighter aircraft

Design Engineer: WALTER F. LAREDO

Drawing Number: 1000 series

Sheet 2 of 34

PLATE 41

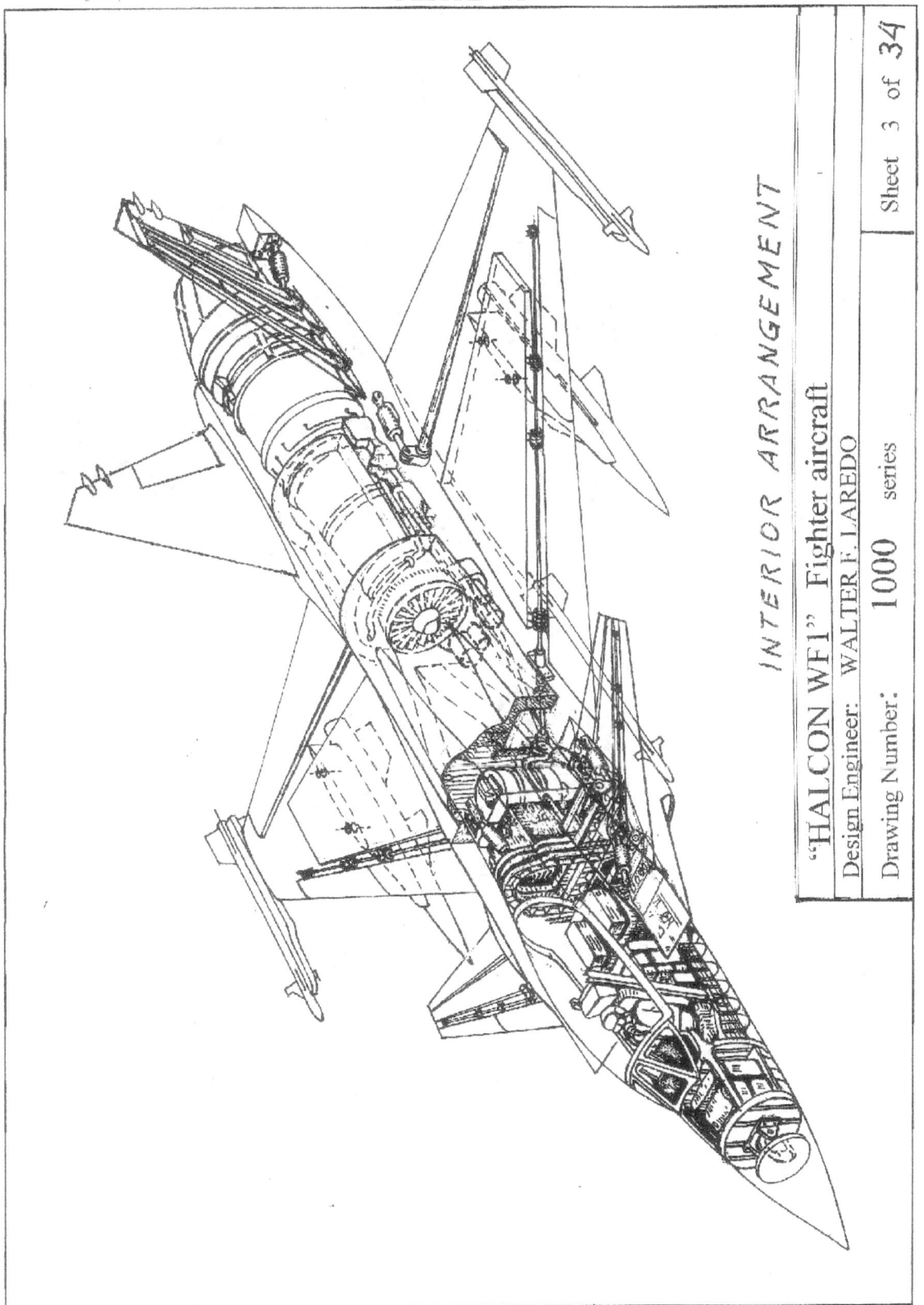

INTERIOR ARRANGEMENT

"HALCON WF1" Fighter aircraft

Design Engineer: WALTER F. LAREDO

Drawing Number: 1000 series

Sheet 3 of 34

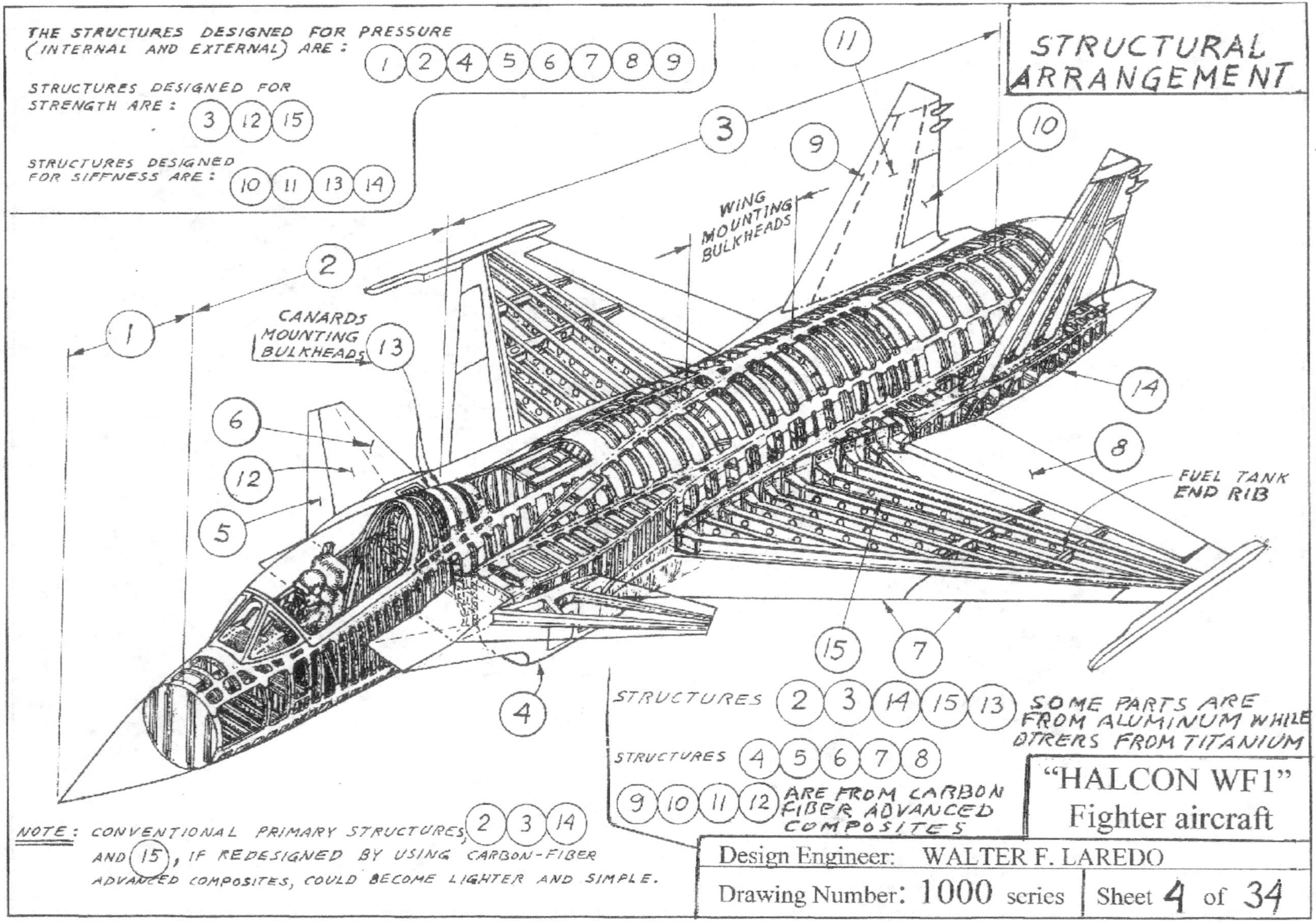

PLATE 42

STRUCTURAL ARRANGEMENT

THE STRUCTURES DESIGNED FOR PRESSURE (INTERNAL AND EXTERNAL) ARE: ① ② ④ ⑤ ⑥ ⑦ ⑧ ⑨

STRUCTURES DESIGNED FOR STRENGTH ARE: ③ ⑫ ⑮

STRUCTURES DESIGNED FOR STIFFNESS ARE: ⑩ ⑪ ⑬ ⑭

CANARDS MOUNTING BULKHEADS

WING MOUNTING BULKHEADS

FUEL TANK END RIB

SOME PARTS ARE FROM ALUMINUM WHILE OTHERS FROM TITANIUM

STRUCTURES ② ③ ⑭ ⑮ ⑬ ARE FROM CARBON FIBER ADVANCED COMPOSITES

STRUCTURES ④ ⑤ ⑥ ⑦ ⑧

⑨ ⑩ ⑪ ⑫

NOTE: CONVENTIONAL PRIMARY STRUCTURES, ② ③ ⑭ AND ⑮, IF REDESIGNED BY USING CARBON-FIBER ADVANCED COMPOSITES, COULD BECOME LIGHTER AND SIMPLE.

"HALCON WF1" Fighter aircraft

Design Engineer: WALTER F. LAREDO

Drawing Number: 1000 series | Sheet 4 of 34

PLATE 43

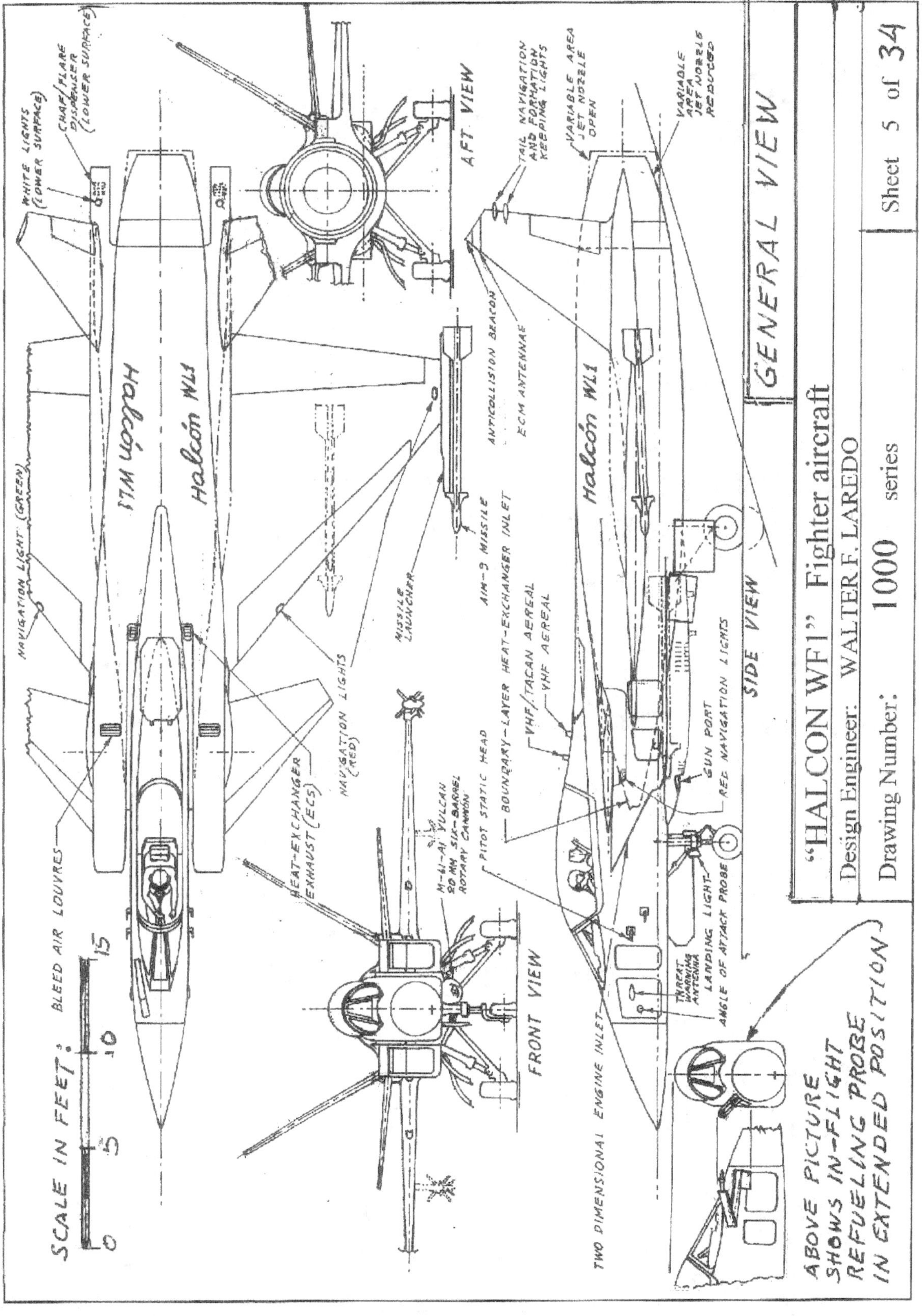

SCALE IN FEET:

BLEED AIR LOUVRES

WHITE LIGHTS (LOWER SURFACE)

CHAF/FLARE DISPENSER (LOWER SURFACE)

NAVIGATION LIGHT (GREEN)

Halcón WL1

Halcón WL1

AFT VIEW

HEAT-EXCHANGER EXHAUST (ECS)

NAVIGATION LIGHTS (RED)

MISSILE LAUNCHER

AIM-9 MISSILE

BOUNDARY-LAYER HEAT-EXCHANGER INLET

VHF/TACAN AERIAL

VHF AERIAL

PITOT STATIC HEAD

ANTICOLLISION BEACON

ECM ANTENNAE

TAIL NAVIGATION AND FORMATION KEEPING LIGHTS

VARIABLE AREA JET NOZZLE OPEN

VARIABLE AREA JET NOZZLE REDUCED

Halcón WL1

M-61-A1 VULCAN 20 MM SIX-BARREL ROTARY CANNON

FRONT VIEW

TWO DIMENSIONAL ENGINE INLET

THREAT WARNING ANTENNA

LANDING LIGHT

ANGLE OF ATTACK PROBE

GUN PORT

REG NAVIGATION LIGHTS

SIDE VIEW

GENERAL VIEW

"HALCON WF1" Fighter aircraft

Design Engineer: WALTER F. LAREDO

Drawing Number: 1000 series

Sheet 5 of 34

ABOVE PICTURE SHOWS IN-FLIGHT REFUELING PROBE IN EXTENDED POSITION

PLATE 44

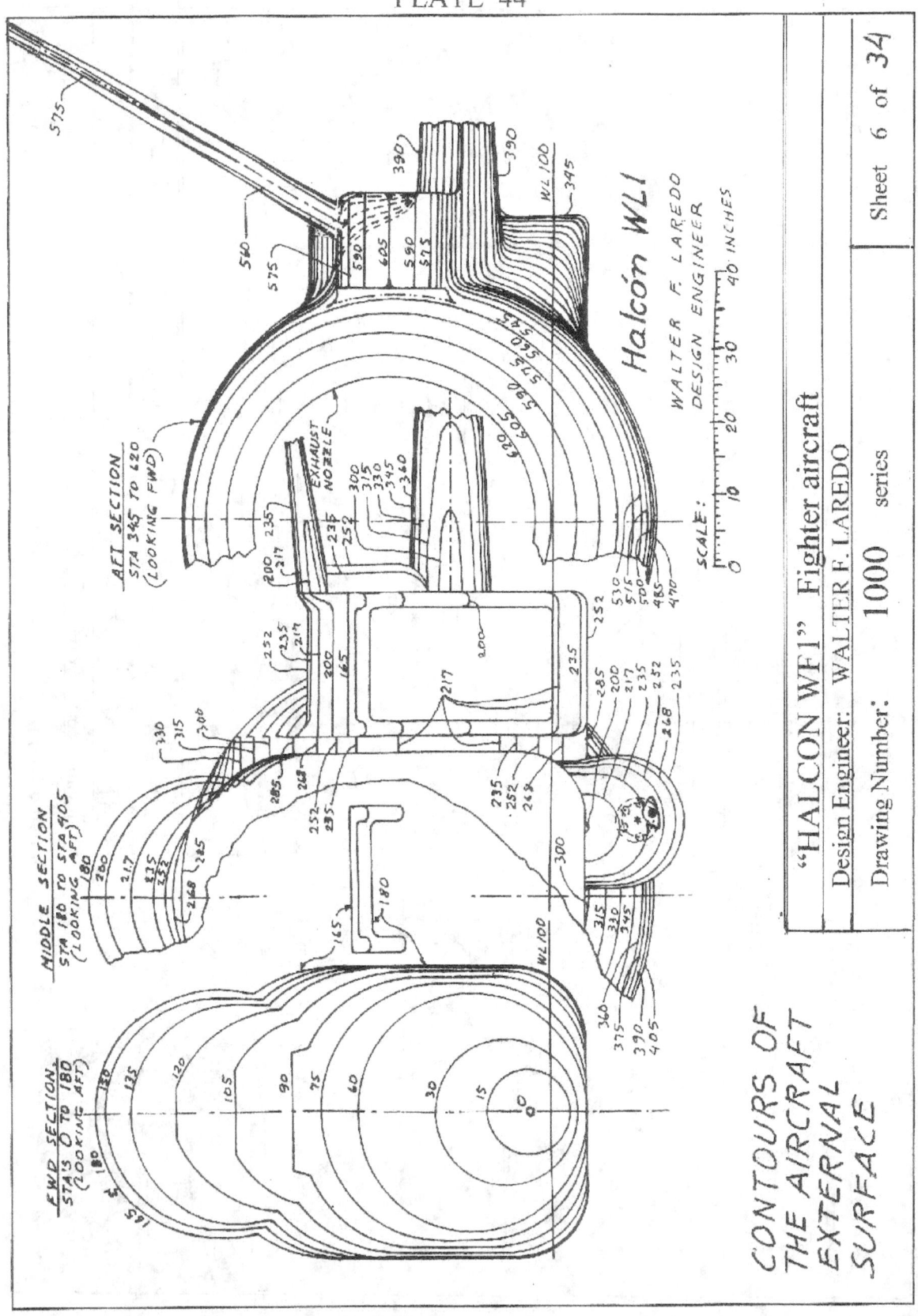

FWD SECTION
STA'S O TO 180
(LOOKING AFT)

MIDDLE SECTION
STA 180 TO STA 405
(LOOKING AFT)

AFT SECTION
STA 345 TO 420
(LOOKING FWD)

EXHAUST NOZZLE

SCALE:
0 10 20 30 40 INCHES

Halcón WL1

WALTER F. LAREDO
DESIGN ENGINEER

CONTOURS OF
THE AIRCRAFT
EXTERNAL
SURFACE

"HALCON WF1" Fighter aircraft

Design Engineer: WALTER F. LAREDO

Drawing Number: 1000 series

Sheet 6 of 34

PLATE 45

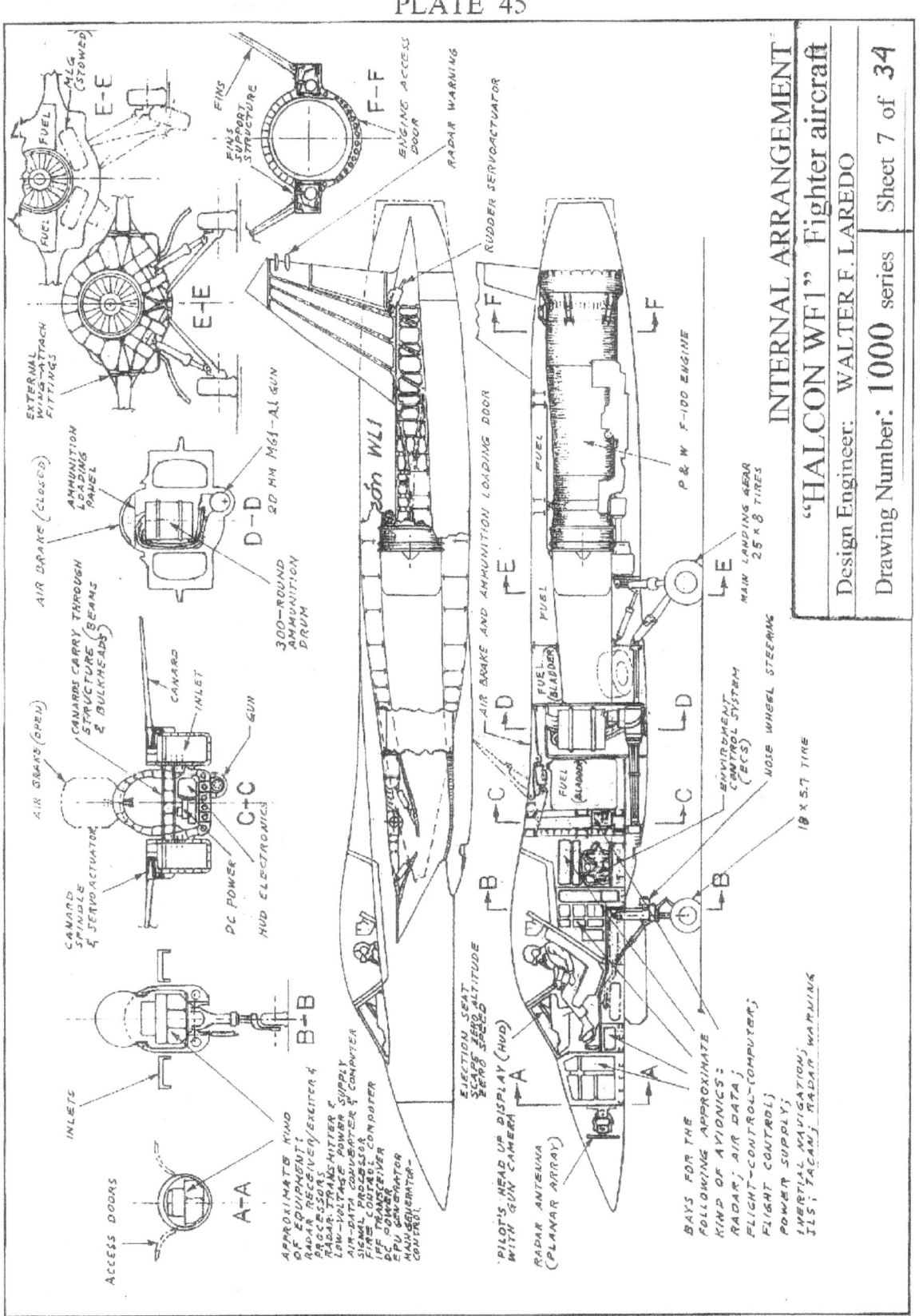

INTERNAL ARRANGEMENT

"HALCON WF1" Fighter aircraft

Design Engineer: WALTER F. LAREDO

Drawing Number: 1000 series | Sheet 7 of 34

PLATE 46

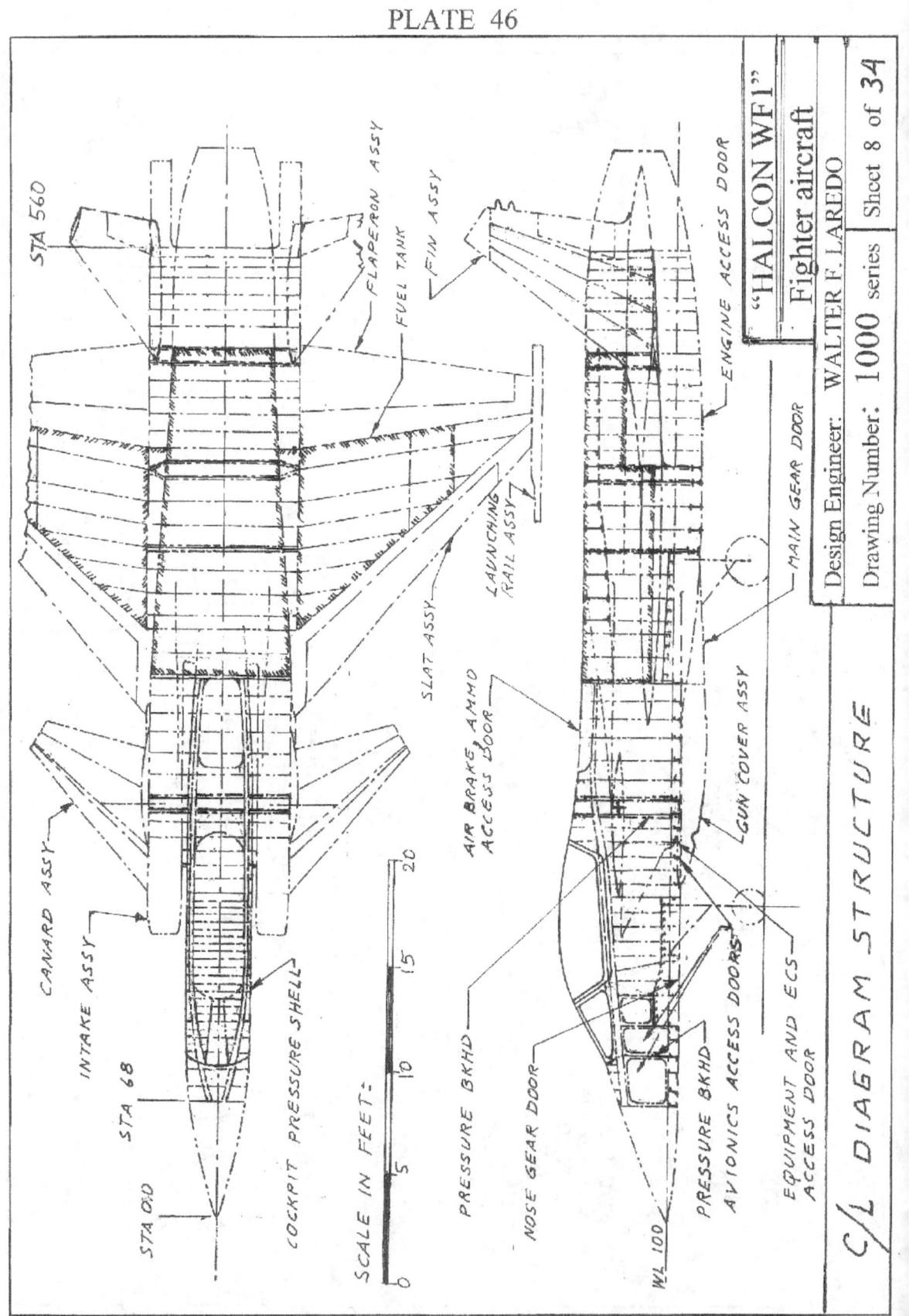

STA 560

FLAPERON ASSY

FUEL TANK

FIN ASSY

STA 0.0

STA 68

INTAKE ASSY

CANARD ASSY

COCKPIT PRESSURE SHELL

SCALE IN FEET:

0   5   10   15   20

LAUNCHING RAIL ASSY

SLAT ASSY

AIR BRAKE, AMMO ACCESS DOOR

PRESSURE BKHD

NOSE GEAR DOOR

ENGINE ACCESS DOOR

MAIN GEAR DOOR

GUN COVER ASSY

WL 100

PRESSURE BKHD

AVIONICS ACCESS DOORS

EQUIPMENT AND ECS ACCESS DOOR

C/L DIAGRAM STRUCTURE

"HALCON WF 1"
Fighter aircraft

Design Engineer: WALTER F. LAREDO

Drawing Number: 1000 series | Sheet 8 of 34

PLATE 47

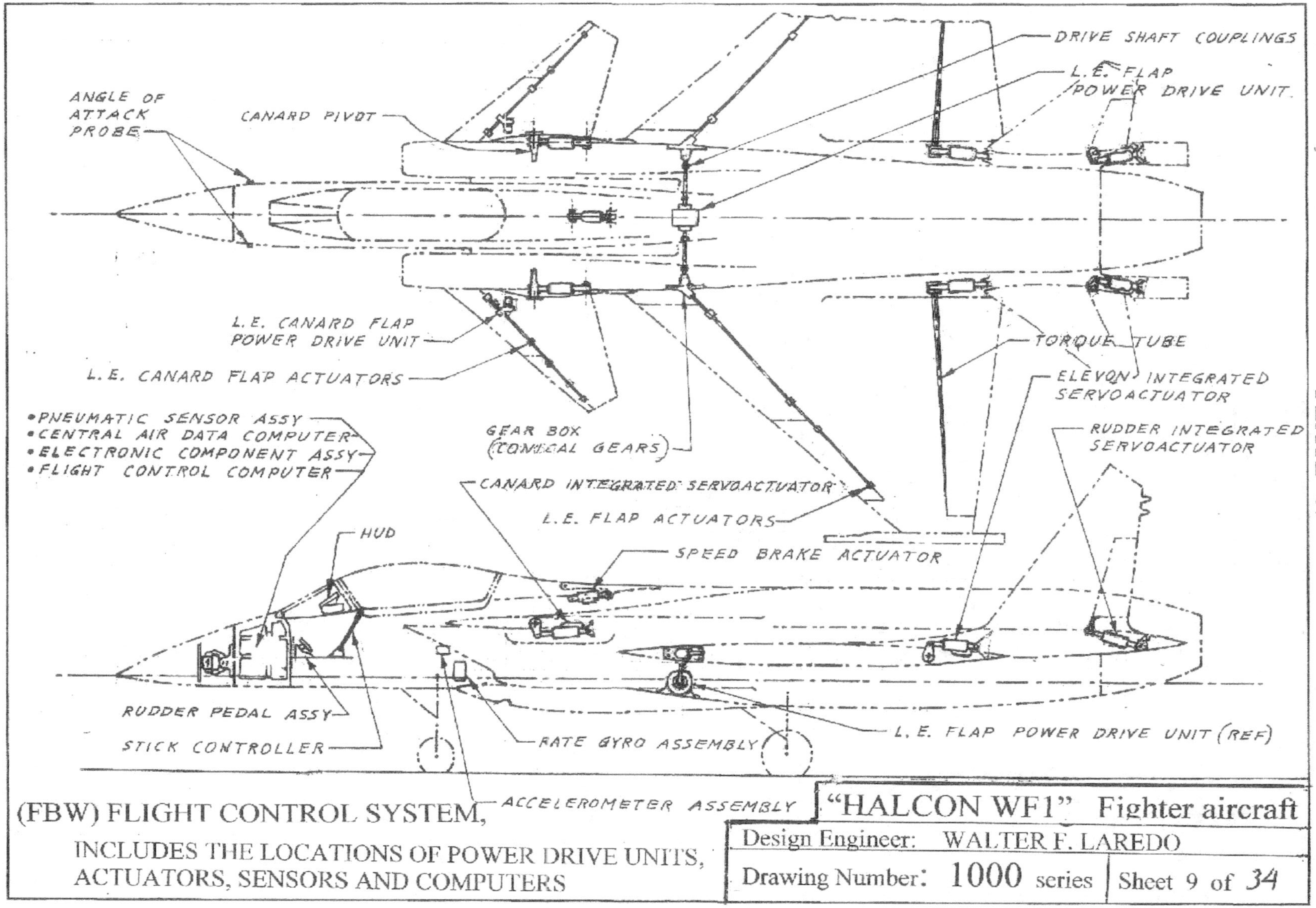

DRIVE SHAFT COUPLINGS

L.E. FLAP POWER DRIVE UNIT.

TORQUE TUBE

ELEVON INTEGRATED SERVOACTUATOR

RUDDER INTEGRATED SERVOACTUATOR

L.E. FLAP POWER DRIVE UNIT (REF)

CANARD PIVOT

L.E. CANARD FLAP POWER DRIVE UNIT

L.E. CANARD FLAP ACTUATORS

ANGLE OF ATTACK PROBE

• PNEUMATIC SENSOR ASSY
• CENTRAL AIR DATA COMPUTER
• ELECTRONIC COMPONENT ASSY
• FLIGHT CONTROL COMPUTER

GEAR BOX (CONICAL GEARS)

CANARD INTEGRATED SERVOACTUATOR

L.E. FLAP ACTUATORS

SPEED BRAKE ACTUATOR

RATE GYRO ASSEMBLY

ACCELEROMETER ASSEMBLY

HUD

RUDDER PEDAL ASSY

STICK CONTROLLER

"HALCON WF1" Fighter aircraft

| Design Engineer: | WALTER F. LAREDO | |
| Drawing Number: | 1000 series | Sheet 9 of 34 |

(FBW) FLIGHT CONTROL SYSTEM,
INCLUDES THE LOCATIONS OF POWER DRIVE UNITS,
ACTUATORS, SENSORS AND COMPUTERS

PLATE 48

AREA-RULE METHOD, MACH = 2.3

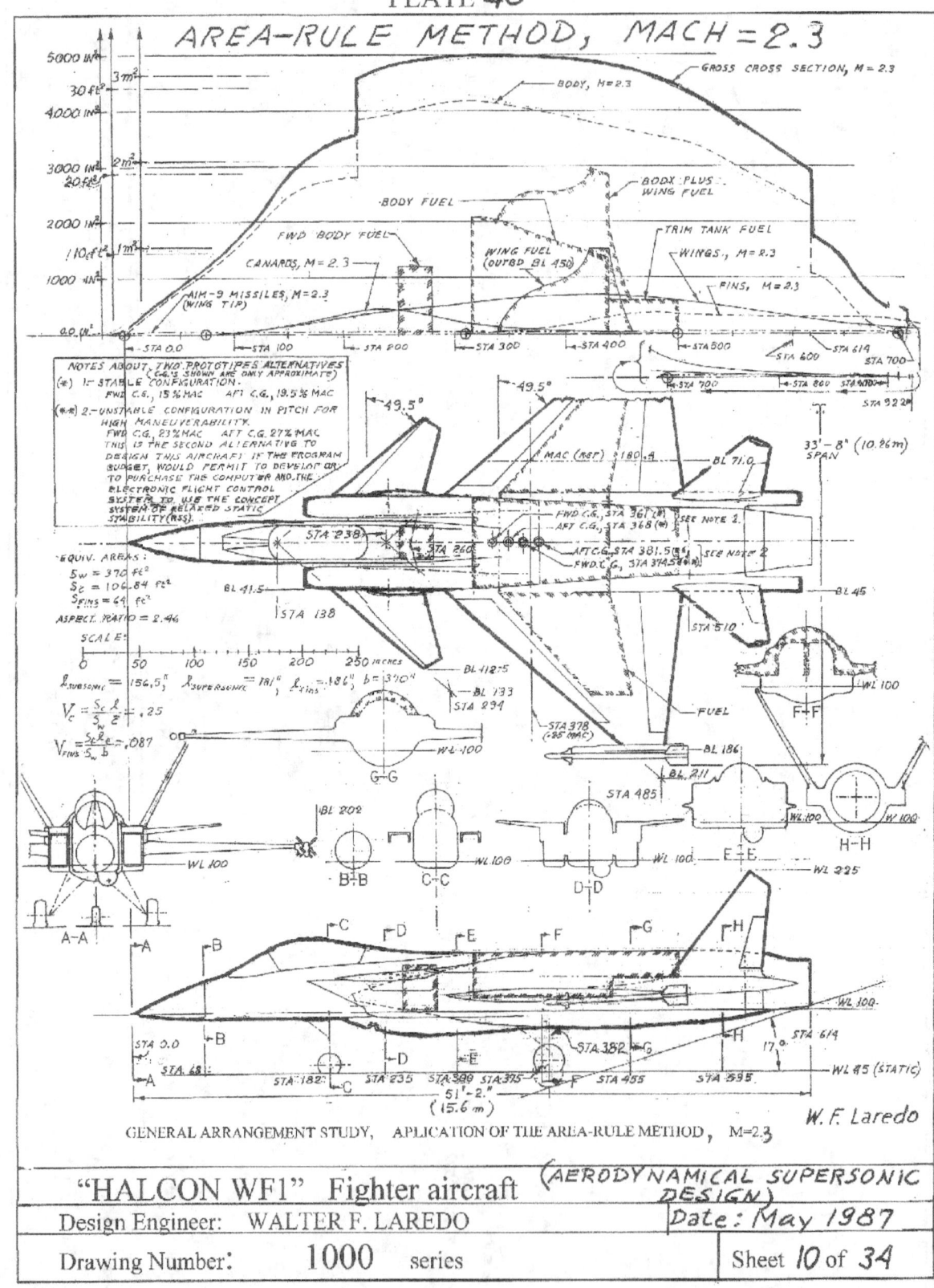

GENERAL ARRANGEMENT STUDY, APLICATION OF THE AREA-RULE METHOD, M=2.3

W. F. Laredo

"HALCON WF1" Fighter aircraft	(AERODYNAMICAL SUPERSONIC DESIGN)	
Design Engineer: WALTER F. LAREDO	Date: May 1987	
Drawing Number: 1000 series	Sheet 10 of 34	

PLATE 49

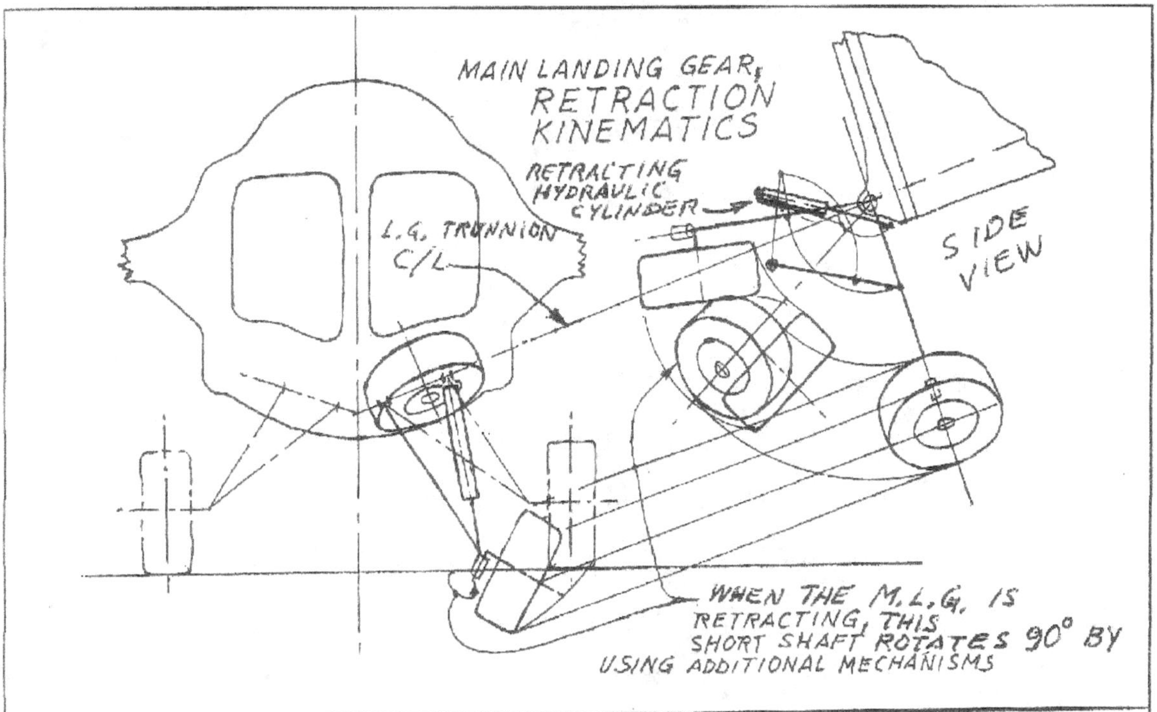

MAIN LANDING GEAR,
RETRACTION
KINEMATICS

RETRACTING
HYDRAULIC
CYLINDER

L.G. TRUNNION
C/L

SIDE
VIEW

WHEN THE M.L.G. IS
RETRACTING, THIS
SHORT SHAFT ROTATES 90° BY
USING ADDITIONAL MECHANISMS

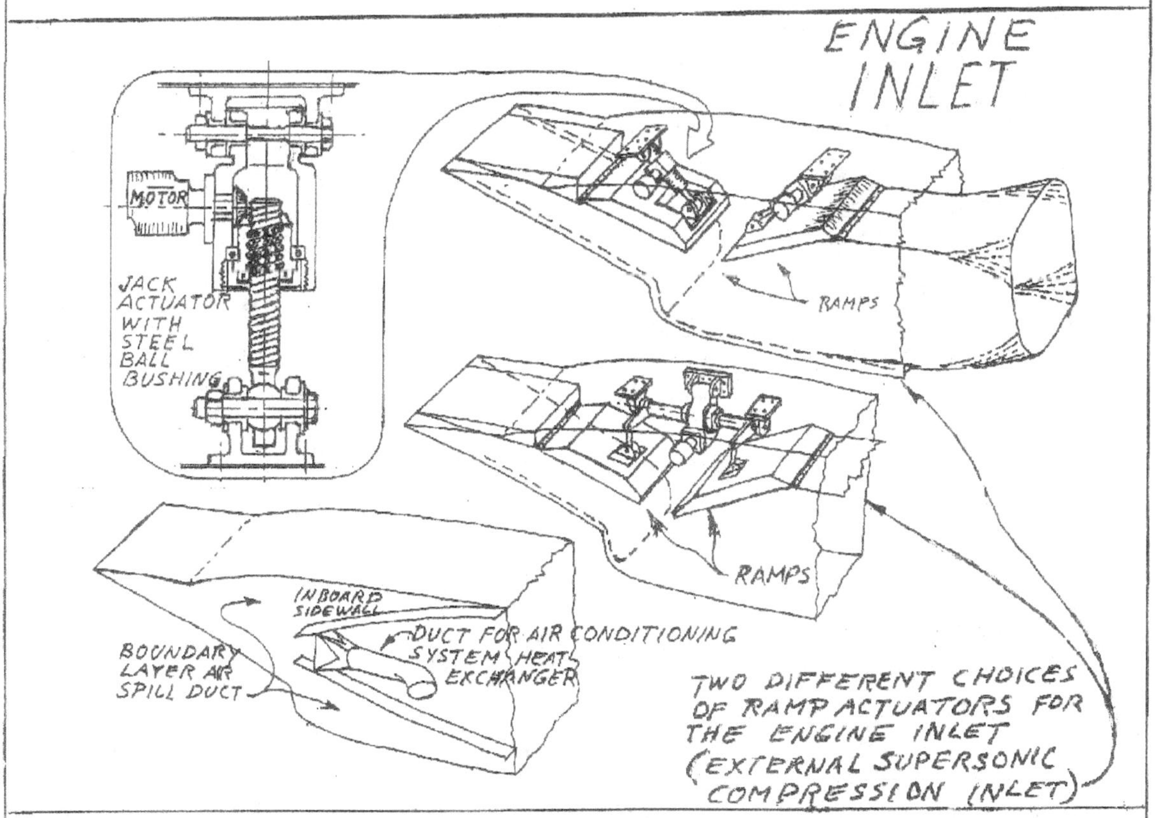

ENGINE
INLET

MOTOR

JACK
ACTUATOR
WITH
STEEL
BALL
BUSHING

RAMPS

RAMPS

INBOARD
SIDEWALL

DUCT FOR AIR CONDITIONING
SYSTEM HEAT
EXCHANGER

BOUNDARY
LAYER AIR
SPILL DUCT

TWO DIFFERENT CHOICES
OF RAMP ACTUATORS FOR
THE ENGINE INLET
(EXTERNAL SUPERSONIC
COMPRESSION INLET)

"HALCON WF1" Fighter aircraft

Design Engineer: WALTER F. LAREDO

Drawing Number: 1000   series

Sheet 11 of 34

PLATE 50

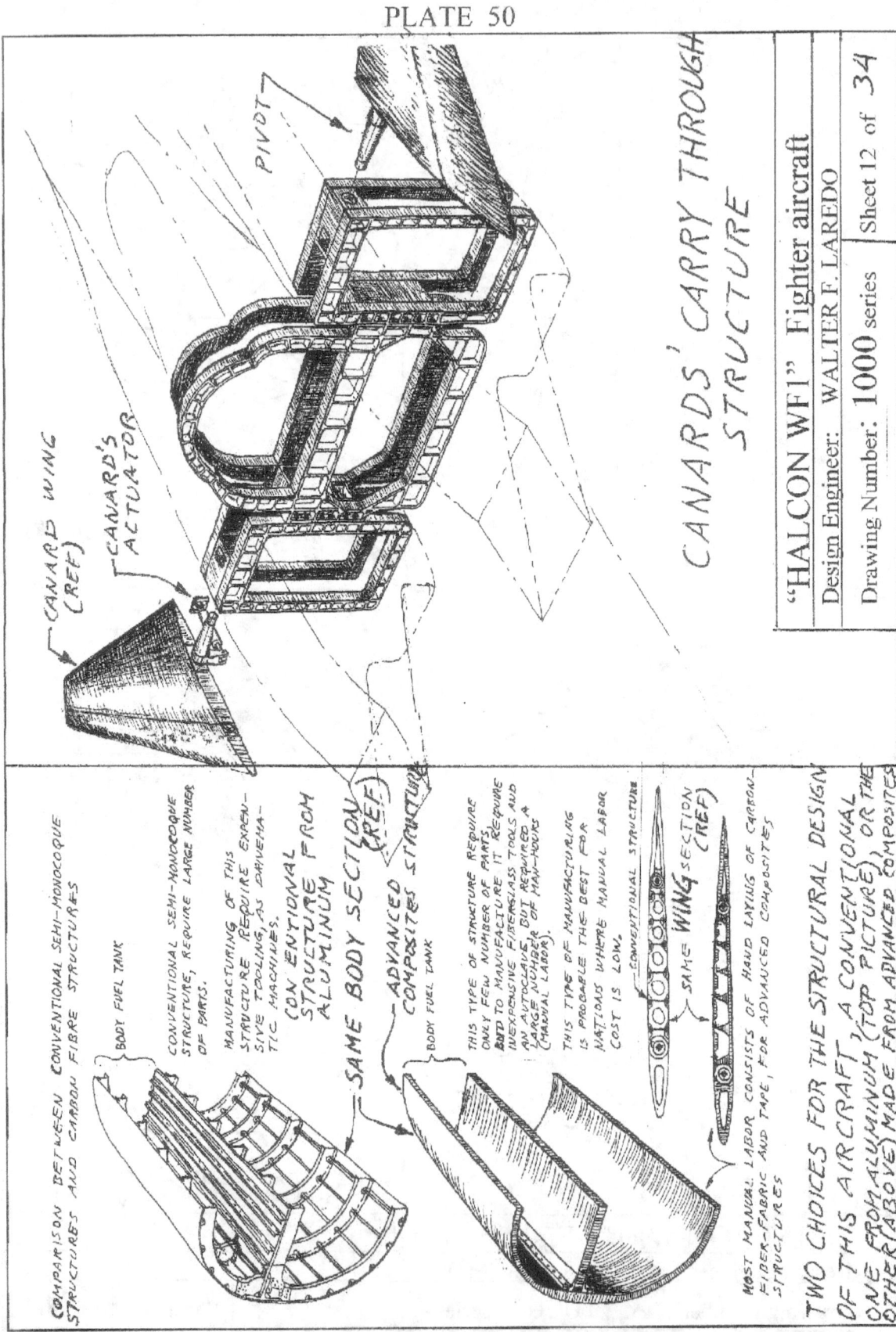

CANARD WING (REF)

CANARD'S ACTUATOR

PIVOT

CANARDS' CARRY THROUGH STRUCTURE

COMPARISON BETWEEN CONVENTIONAL SEMI-MONOCOQUE STRUCTURES AND CARBON FIBRE STRUCTURES

BODY FUEL TANK

CONVENTIONAL SEMI-MONOCOQUE STRUCTURE, REQUIRE LARGE NUMBER OF PARTS.

MANUFACTURING OF THIS STRUCTURE REQUIRE EXPENSIVE TOOLING, AS DRIVEMATIC MACHINES.

CONVENTIONAL STRUCTURE FROM ALUMINUM

SAME BODY SECTION (REF)

ADVANCED COMPOSITES STRUCTURE

BODY FUEL TANK

THIS TYPE OF STRUCTURE REQUIRE ONLY FEW NUMBER OF PARTS, BUT TO MANUFACTURE IT REQUIRE INEXPENSIVE FIBERGLASS TOOLS AND AN AUTOCLAVE, BUT REQUIRED A LARGE NUMBER OF MAN-HOURS (MANUAL LABOR).

THIS TYPE OF MANUFACTURING IS PROBABLE THE BEST FOR NATIONS WHERE MANUAL LABOR COST IS LOW.

CONVENTIONAL STRUCTURE

SAME WING SECTION (REF)

MOST MANUAL LABOR CONSISTS OF HAND LAYING OF CARBON-FIBER-FABRIC AND TAPE, FOR ADVANCED COMPOSITES STRUCTURES

TWO CHOICES FOR THE STRUCTURAL DESIGN OF THIS AIRCRAFT. A CONVENTIONAL ONE FROM ALUMINUM (TOP PICTURE), OR THE OTHER (ABOVE) MADE FROM ADVANCED COMPOSITES

"HALCON WF1" Fighter aircraft

Design Engineer: WALTER F. LAREDO

Drawing Number: 1000 series | Sheet 12 of 34

PLATE 51

"HALCON WF1", AN ADVANCED LIGHTWEIGHT FIGHTER
(FOR MORE DETAIL SEE "PROPOSAL FOR THE DEVELOPMENT OF THE HALCON WL1", PREPARED BY WALTER F. LAREDO)

The Halcon WF1 is a high performance lightweight advanced fighter Mach 2.3, designed with the latest-state- of-the-art. An aircraft developed as a new-generation lightweight supersonic combat aircraft, using new technologies, considering its high degree of readiness, mobility and striking power.

In this decade of the 1980's the Halcon WF1 will be develop as a competitor to other modern fighters in process of development as the Soviet's MIG 2000 and MIG 29, the multinational European fighter EFA, French's Rafale, Sweden's Gripen and the American Advanced Tactical Fighter (ATF). Also competitor to existing aircraft as the Mirage III, Mirage 2000, MIG 21 and F-16.

Maintenance of the Halcon WF1 will be easy, because its good accessibility through many inspections doors, at suitable location of equipment.

This aircraft configuration is basically a Delta wing with sweptback Canards, which spigots are supported by bushings inserted in the main titanium bulkhead, a structure across the fuselage and the two-dimensional air inlets. This aircraft have two vertical stabilizers. Large portions of the aircraft structure is made from advanced composites materials, in the form of graphite fibers imbedded in epoxy resin, as well as Kevlar fibers in zones requiring impact resistance.

This aircraft will be design basically for the Pratt & Whitney F100-PW-100 engine (Max. Sea Level Static Thrust = 23000 lb.), other alternate engines are the F100-PW-200 and The F100-PW-220 (both, Max. Sea Level Static Thrust 25000 lb.), engines currently in use by the F-15 and the F-16 aircraft, another alternate engine with slightly different characteristics and flight envelopes would be the General Electric F404 currently used in the F-18 with an afterburner thrust of 16000 lb. at Sea Level, Standard day. In this aircraft also could be installed equivalent European engines with a minimum of engine mountings modification. In most Halcons WL1 aircraft will use the low bypass 20 years old F100-PW-100 engine with the following characteristics:
Length (including nozzle), 190 in (4.83 m)
Weight (bare), 2310 lb. (1050 kg)
Bypass Ratio, 0.71
Afterburner Thrust, Sea Level Static, 23000 lb. (uninstalled).
Sea Level Thrust Specific Fuel Consumption (TSFC) of 2.48 lb. fuel/lb. thrust per hour with afterburner.

The Halcon WF1 aircraft, with the F100-PW-100 engine, have a thrust : weight ratio of 1.04 at combat weight. This aircraft have a combination of the four most wanted

"HALCON WF1" Fighter aircraft		
Design Engineer: WALTER F. LAREDO		
Drawing number: 1000 series	Sheet 13 of 34	

PLATE 52

characteristics, as high top speed, low sensitivity to turbulent air, and great maneuverability at typical combat speeds, plus exceptional low speed performance for short takeoff and landing.

For pilots, the maneuverability of changing the flight paths mainly in the pitch plane is considered of primary importance. The Halcon WF1 will be designed as a highly maneuvering fighter with short Turn Radius and a high Sustained Turn Rate, so that it could hold a turn without loosing speed and slip inside the adversary's arc to get in his tail.
During the studies and the design phase of the Halcon WF1 will be charted its Energy Maneuverability as plots of Turn Rate vs. Velocity for given altitudes as well as for small changes in configuration.

The pilot will have all around excellent visibility. The propulsion system will have lateral inlets of special design to obtain "LOW DISTORTION FLOW" and good "PRESSURE RECOVERY" at high angles of attack during air combat and maneuvers.

The cockpit Display System will have one Head Up Display (HUD), where vital information is presented in the line of sight of the pilot, as he looks into the forward direction. The cockpit instrument panel located next to the radar display will have either conventional instruments or more modern head down electronic display as preferred by most fighter pilots.

The Halcon WF1 as a modern supersonic aircraft will use hydraulic servoactuators, controlled by ALL ELECTRONIC FLY-BY-WIRE FLIGHT-CONTROL SYSTEM COMMAND AND STABILITY-AUGMENTATION SYSTEM. SINCE THE HALCON WF1 CENTER OF GRAVITY IS CLOSE TO THE CENTER OF PRESSURE, RELAXED STABILITY DESIGN FOR MINIMUM TRIM DRAG, WAS POSSIBLE TO REDUCE THE SIZE OF THE ELEVATORS AND FLAPERONS.

Unstable full fly-by-wire aircraft, with new generation of computers and more advanced radar, allowed using the concept of relaxed stability for minimum trim drag. Its maneuverability and controllability together will comprise good agility, hence an agile aircraft. Vital for survival in the air defense role are its maneuverability and turning capability, advantages to which should be add its full all-weather, and all-altitude defense capability.

Infrared-seeking missiles have made the tail chase of enemy aircraft unnecessary. The Halcon WF1 when followed by a seeking enemy missile, as in some modern fighters, will be alerted by their tail warning system. The basic Halcon WF1 will be armed with the following weapons:
- Infrared and radar homing air-to-air missiles as the American AIM-9 Sidewinder also could be adapted to use equivalent European missiles.
- A build-in, high performance M-61 six barrel, 20 mm cannon with 300 rounds of ammunition.

The aircraft have provisions to be used for other roles, as for ground attack, carrying laser guided missiles, bombs or anti-ship missiles.

"HALCON WF1" Fighter aircraft		
Design Engineer:	WALTER F. LAREDO	
Drawing number:	1000 series	Sheet 14 of 34

PLATE 53

The main landing gear will be designed to resist an impact at a descent rate of 13 ft per second (3.9 m/sec), good for direct short landing without flare. It will use either, conventional or carbon brakes, probably with an antiskid system.

Its modular design facilitates major repairs, re-design, and quick replacement of equipment. Major aircraft modifications and adaptations may happens sometime during the next twenty years, such as future variable cycle engines with two dimensional vectoring nozzles, better avionics, better radar and other advanced systems.

DESIGN AND DEVELOPMENTAL PHASES FOR THE HALCON WL1 AIRCRAFT

Testing with wind tunnel models will require hundreds or thousands of hours. By using CAD-CAM equipment could be produced all the drawings required to build the airplane. Applying the Area Rule for Mach 2.3 created the external geometrical design. For the development of this aircraft was used several engineering specialties as Aerodynamics, Aerodynamic Stability and Control, Structures (Stress, Aeroelasticity, Fatigue and Flutter), Electronics, Flight Control System. Materials and Processes, Electric and Hydraulic Systems, Armament and Operational equipment, Cockpit Display System, Fuel System, Propulsion, Fire Extinguishing System, Environmental Control System and Miscellaneous.

In addition to the early data obtained from wind tunnel models, the full size actual flying demonstrator will provide excellent aerodynamic representation of the flow field around the aircraft. Flying tests will provide more realistic data concerning different flight conditions, than can be obtain from wind tunnel models. An additional and important wind tunnel model test would be the launching and dropping of various stores shapes from the aircraft model and later from the actual flight tests of the demonstrator.

The ejection seat will be operable from zero to an equivalent air speed of 690 mph at zero altitude. The pilot will control the aircraft with a central conventional stick. A disadvantage of the side-stick control as in the F-16 is that it increases the frontal cross section area of the cockpit design.

The Halcon WF1 will be pressurizing with bleed air from the engine, air that will be cool with a combination of cycle machine with ram air radiators.

Fuel tanks should be inert, either, Halon or nitrogen gas.

The Maneuverability and Controllability of this aircraft will comprise Agility, with high Sustained Turn Rate, able to hold a "turn without loosing speed" and slip inside the adversary's arc to get on his tail. Good Energy Maneuverability will be charter as plots of "Turn Rate vs. Velocity" for a given Altitude and Configuration. Also will be charted the Energy Maneuver Performance in terms of specific excess power (Ps), and persistence (fuel flow).

"HALCON WF1" Fighter aircraft		
Design Engineer: WALTER F. LAREDO		
Drawing number: 1000 series	Sheet 15 of 34	

PLATE 54

HALCON WL1, PERFORMANCE AND CHARACTERISTICS

Primary role: Lightweight fighter. Secondary role; attack

DIMENSIONS:

Wing span	33 ft 8 in	10.26 m
Max. length	51 ft 2 in	15.60 m
Wheel track	8 ft 4 in	2.54 m
Wheel base	16 ft 8 in	5.08 m
Height (static)	15 ft 2 in	4.62 m

WEIGHTS:

Empty	14058 lb	6376.7 Kg
Loaded (clean), fighting mission	22200 lb	10070 Kg
Max. takeoff gross weight	32000 lb	14515 Kg
Payload, fighting mission	1800 lb	816.5 Kg
Max. armament load	6000 lb	2722 Kg
Internal fuel	9060 lb	4110 Kg
70 % internal fuel, fighting mission	6342 lb	2877 Kg
Total fuel capacity, includes external tanks	11242 lb	5099 Kg

WING:

Wing area	370 sq ft	34.37 sq. m.
Aspect Ratio	2.46	
Sweep Leading Edge	49 degrees	
MAC	181 in	4.60 m
T/C SOB / T/C TIP	5 % / 3.5 %	
Wing Angle of incidence and the Dihedral	To be determinate	

CANARD WINGS:

Area	46 sq ft	4.27 sq. m.
Area, plus fuselage lift		
Contributing sect.	106.8 sq ft	9.93 sq. m.
$Vc = (Lc)(Sc) / (c)(Sw)$		
Vc (for Sc=46 sq ft)	0.107	
Vc (for Sc=106.8 sq ft)	0.25	
Aspect Ratio	3.32	
T/C SOB / T/C tip	4.5 % / 3 %	
Sweep L.E.	49 degrees	

VERTICAL TAILS:

Sweep L.E.	36 degrees	
Area (both fins)	64 sq ft	5.95 sq. m.
$Vv = (Lv)(Sv) / (b)(Sw)$	= 0.087	
T/C root / T/C tip	4.5 % / 3.5 %	

"HALCON WF1" Fighter aircraft

Design Engineer: WALTER F. LAREDO

Drawing Number: **1000 series**		Sheet 16 of 34

PLATE 55

WING LOADING (for 22200 lb G.W.)   60 lb/sq ft   293 Kg/sq.m.
UNINSTALLED THRUST/WEIGHT (SLS)   1.04
LOAD FACTOR (for 22200 lb G.W.)   +7.33 and –3   G's limits

PERFORMANCE:
    SPEEDS
    Max. Mach No. at Sea Level (interceptor)   1.2
    Max. Mach No. at 40000 ft (interceptor)   2.2
    Cruise Mach No. at 36000 ft   0.85
    Takeoff safety speed (22200 lb G.W.)   140 mph   225 kph
    Takeoff safety speed (Max. TOW)   168 mph   270.4 kph
    Approach speed   142 mph   228.5 kph
    Landing (touchdown)   135 mph   217 kph

    CEILING
    Service ceiling   62000 ft   18898 m.
    Absolute ceiling (no weapons)   90000 ft   27432 m.

    CLIMB
    Rate of Climb (with afterburner)   To be determinate
    Initial Climb   48000 ft/min   14630 m/min
    RANGE
    Combat radius, internal fuel (20 min. reserve)   400 miles   644 Km
    Interceptor radius (70 % internal fuel)   280 miles   450 Km
    Interceptor radius, high speed interception
        with full Afterburner, Mach 2.2   84 miles   135 Km
    Ferry range, without refueling in the air   1800 miles   2897 Km

TAKEOFF AND LANDING DISTANCES
    Takeoff distance (clean aircraft)   1400 ft   427 m
    Takeoff distance at Max. TOW   3800 ft   1158 m
    Landing distance at SL (no flare)   1600 ft   488 m
    Landing distance at ST (conventional)   2600 ft   792.5 m
    Rate of sink (at landing)   14 ft/sec   4.27 m/sec

TURING RATE AT 15000 FT
    Sustaining turning rate at M.8   13.4 deg/sec
    Sustaining turning rate at M.9   13.5 deg/sec
    Instantaneous turning rate at M.5   21.6 deg/sec
    Instantaneous turning rate at M.9   17.4 deg/sec

SPECIFIC EXCESS POWER, at M.8, 1G and GW=22200 lb   (GW=10070 Kg)
    At Sea Level, P_s = 863 ft/sec   (263 m/sec)
    At 15000 ft (4572 m), P_s = 654 ft/sec   (199 m/sec)

"HALCON WF1" Fighter aircraft

Design Engineer: WALTER F. LAREDO

Drawing Number: 1000 series   Sheet 17 of 34

PLATE 56

PERFORMANCE CALCULATIONS

AIR TO AIR FIGHTING MISSION WITH A TAKEOFF GROSS WEIGHT OF 22200 LB (70 % INTERNAL FUEL),
BASIC FLIGHT MISSION TO DESIGN THE HALCON WF1 AIRCRAFT
Aircraft designed with a limit load factor $n = 7.33$ (subsonic) and $n = 6.5$ (supersonic)

To find the following, for Mach numbers 0.9, 0.8 & 0.5 and an altitude of 15000 Ft:

- Sustained Load Factor
- Sustained Turn Rate
- Instantaneous Load Factor
- Instantaneous Turn Rate
- Acceleration Level Flight
- Excess of Specific Power for :
  1. Mach 0.9 & 0.8 at an altitude of 15000 Ft
  2. Mach 0.8 at sea level

WEIGHT CHANGE CALCULATION AFTER BURNING FUEL DURING EACH MISSION PHASE, PHASES 1 THROUGH 7

Phase 1:   Takeoff
Phase 2:   Climbing
Phase 3:   Cruise and Range, between 36000 Ft to 47000 Ft
Phase 4:   Dash acceleration to combat arena
Phase 5:   Combat (4 minutes, 25000 Ft, Mach 0.9)
Phase 6:   Cruise return
Phase 7:   Loiter at sea level

Calculation of the Instantaneous Turn Rate, Continuous Turn Rate and its Radius for a safe load factor.

Calculate Total Takeoff Distance at sea level
Calculate Total Landing Distance with 50 % internal fuel.

"HALCON WF1" Fighter aircraft

Design Engineer: WALTER F. LAREDO

Drawing Number: **1000 series**   Sheet 18 of 34

PLATE 57

FOR THE Halcón WL1 FLYING AT AN ALTITUDE OF 15000 FT, TO FIND FOR EACH ONE OF THE FOLLOWING MACH NUMBERS 0.9, 0.8 AND 0.5, THE FOLLOWING:

- SUSTAINED LOAD FACTOR = n_{sust}
- SUSTAINED TURN RATE = $\dot{\psi}_{sust}$
- INSTANTANEOUS LOAD FACTOR = n_{inst}
- INSTANTANEOUS TURN RATE = $\dot{\psi}_{inst}$
- ACCELERATION LEVEL FLIGHTS $\frac{dV}{dt}$ FOR $n=5$

DATA

ALTITUDE = 15000 ft
$P = 1199.17$ Lb/ft^2
$\rho = .0015$ Lb·sec^2/ft^4
$a = 1057.23$ ft/sec
$\gamma = C_p/C_V = 1.40$
$S_w = 340$ ft^2
$C_{P_0} = .018$ (MACH NUMBERS 0.9, 0.8 AND 0.5) DATA FROM SIMILAR AIRPLANES
$K = .16$ (MACH NOS. 0.9, 0.8 AND 0.5) DATA FROM SIMILAR AIRCRAFT
$W = 22200$ Lb
$C_{L_{max}} = 0.8$ FOR SERIES 64, 65 AND 66 AIRFOILS, 5% T/C, NO FLAPS DEPLOYED

MACH NUMBER 0.9, ALTITUD 15000 ft

F-100-PW ENGINE, INSTALLED THRUST WITH MAXIMUM AFTERBURNER = $T_{max} = 20000$ Lbs
$V = 951.54$ ft/sec
$\frac{1}{2}\rho V^2 = q = 677.04$ Lb/ft^2

SUSTAINED LOAD FACTOR
$$n_{sust} = \frac{q}{W/S_w}\sqrt{\frac{1}{K}\left(\frac{T_{max}}{q S_w} - C_{P_0}\right)} = 7.0150$$

SUSTAINED TURN RATE
$$\dot{\psi} = \frac{g\sqrt{n_{sust}^2 - 1}}{V} \times (57.3) = 13.46 \ \tfrac{DEG}{SEC}$$

MACH NUMBER 0.5, ALTITUDE 15000 ft

F-100-PW ENGINE INSTALLED THRUST WITH MAXIMUM AFTERBURNING = $T_{max} = 15000$ Lb
$V = 528.62$ ft/sec
$q = \frac{1}{2}\rho V^2 = 208.97$ Lb/ft^2

n SUSTAINED:
$$n_{sust} = \frac{q}{W/S_w}\sqrt{\frac{1}{K}\left(\frac{T_{max}}{q S_w} - C_{P_0}\right)} = 3.65$$

SUSTAINED TURN RATE:
$$\dot{\psi} = \frac{g\sqrt{n_{sust}^2 - 1}}{V} \times (57.3) = 12.26 \ \tfrac{DEG}{SEC}$$

n INSTANTANEOUS:
$C_{L_{max}} = 1.8$ WITH LEADING EDGE FLAPS DEFLECTED DOWN
$$n_{inst} = \frac{q}{W/S} C_{L_{max}} = 6.27$$

INSTANTANEOUS TURN RATE:
$$\dot{\psi}_{inst} = \frac{g\sqrt{n_{ind}^2 - 1}}{V} \times (57.3) = 21.6 \ \tfrac{DEG}{SEC}$$

n INSTANTANEOUS:
$C_{L_{max}} = .8$
$$n_{inst} = \frac{q}{W/S} C_{L_{max}} = 9.03$$

INSTANTANEOUS TURN RATE:
$$\dot{\psi}_{inst} = \frac{g\sqrt{n_{inst}^2 - 1}}{V} \times (57.3) = 17.39 \ \tfrac{DEGREES}{SEC}$$

MACH NUMBER 0.8, ALTITUDE 15000 FT

F-100-PW ENGINE, INSTALLED THRUST WITH MAXIMUM AFTERBURNER = $T_{max} = 18800$ Lb
$V = 845.78$ ft/sec
$q = \frac{1}{2}\rho V^2 = 539.93$ Lb/ft^2

n SUSTAINED:
$$n_{sust} = \frac{q}{W/S_w}\sqrt{\frac{1}{K}\left(\frac{T_{max}}{q S_w} - C_{P_0}\right)} = 6.21$$

SUSTAINED TURN RATE:
$$\dot{\psi} = \frac{g\sqrt{n_{sust}^2 - 1}}{V} \times (57.3) = 13.36 \ \tfrac{DEG}{SEC}$$

n INSTANTANEOUS:
$C_{L_{max}} = .8$
$$n_{inst} = \frac{q}{W/S} C_{L_{max}} = 7.8$$

INSTANTANEOUS TURN RATE:
$$\dot{\psi}_{inst} = \frac{g\sqrt{n_{inst}^2 - 1}}{V} \times (57.3) = 15.55 \ \tfrac{DEG}{SEC}$$

TO FIND THE EXCESS SPECIFIC POWER $P_s = \frac{dh_e}{dt}$

FOR MACH 0.9, $n=5$ AT AN ALTITUDE OF 15000 ft
$C_L = \frac{nW}{qS} = .9431$
$C_D = C_{P_0} + KC_L^2 = .049$
$D = \frac{1}{2}\rho M^2 S_w C_D = 12379.44$ Lbs
$P_s = \frac{dh_e}{dt} = \frac{V(T-D)}{W} = 326.62$ ft/sec

FOR MACH 0.8, $n=5$ AT AN ALTITUDE OF 15000 ft
$C_L = \frac{nW}{qS} = .5556$
$C_D = C_{P_0} + KC_L^2 = .0673$
$D = \frac{1}{2}\rho M^2 S_w C_D = 5478.26$ Lbs
$P_s = \frac{dh_e}{dt} = \frac{V(T-D)}{W} = 208.71$ ft/sec

FOR MACH .8 $n=1$ AT AN ALTITUDE OF 15000 ft
$C_L = \frac{1W}{qS} = .11112$
$C_D = C_{P_0} + KC_L^2 = .02$
$D = \frac{1}{2}\rho M^2 S_w C_D = 1628$ Lbs
$P_s = \frac{dh_e}{dt} = \frac{V(T-D)}{W} = 654.2$ ft/sec

AND THE ACCELERATION IN LEVEL FLIGHT $\frac{dV}{dt}$ IS
$\frac{dV}{dt} = \frac{g P_s}{V} = 24.9$ ft/sec^2

FOR MACH .8 $n=1$ AT SEA LEVEL
INSTALLED THRUST WITH AFTERBURNER = 28000 Lbs
$C_L = \frac{nW}{qS} = .0633$ then $C_D = C_{P_0} + KC_L^2 = .01864$
$D = \frac{1}{2}\rho M^2 S_w C_D = 6539$ Lbs
$P_s = \frac{dh_e}{dt} = \frac{V(T-D)}{W} = \boxed{863 \ \text{ft/sec} = \text{INSTANTANEOUS RATE OF CLIMB}}$

"HALCON WF1"   Fighter aircraft

Design Engineer: WALTER F. LAREDO

Drawing Number: **1000** series

PLATE 58

COMPARISON OF SUSTAINED TURNING RATE PERFORMANCE AND THE INSTANTANEOUS TURNING RATE PERFORMANCE, BETWEEN THE FOLLOWING AIRCRAFTS, INCLUDING THE HALCON WF1.

AIRCRAFT →	MC DONNELL DOUGLAS F-15 EAGLE	Halcón WF1	MIG-21	GENERAL DYNAMICS F-16 FALCON	NORTHROP F-5A	LAVI	MC DONNELL DOUGLAS F-4 PHANTOM
SUSTAINED TURNING RATE AT MACH NUMBER .8	11.8 DEG/SEC	13.36 DEG/SEC				13.2 DEG/SEC	
SUSTAINED TURNING RATE AT M, .9		13.46 DEG/SEC	7.5 DEG/SEC	12.8 DEG/SEC	7.8 DEG/SEC		9 DEG/SEC
INSTANTANEOUS TURNING RATE AT M, .9	14.1 DEG/SEC	17.39 DEG/SEC	13.4 DEG/SEC	17.3 DEG/SEC	14 DEG/SEC		13.5 DEG/SEC
INSTANTANEOUS TURNING RATE AT M, 5	16.5 DEG/SEC	21.6 DEG/SEC	11.1 DEG/SEC	15.6 DEG/SEC	11.4 DEG/SEC		7.8 DEG/SEC
INST. TURNING RATE AT M, .8		15.55 DEG/SEC				24.3 DEG/SEC	

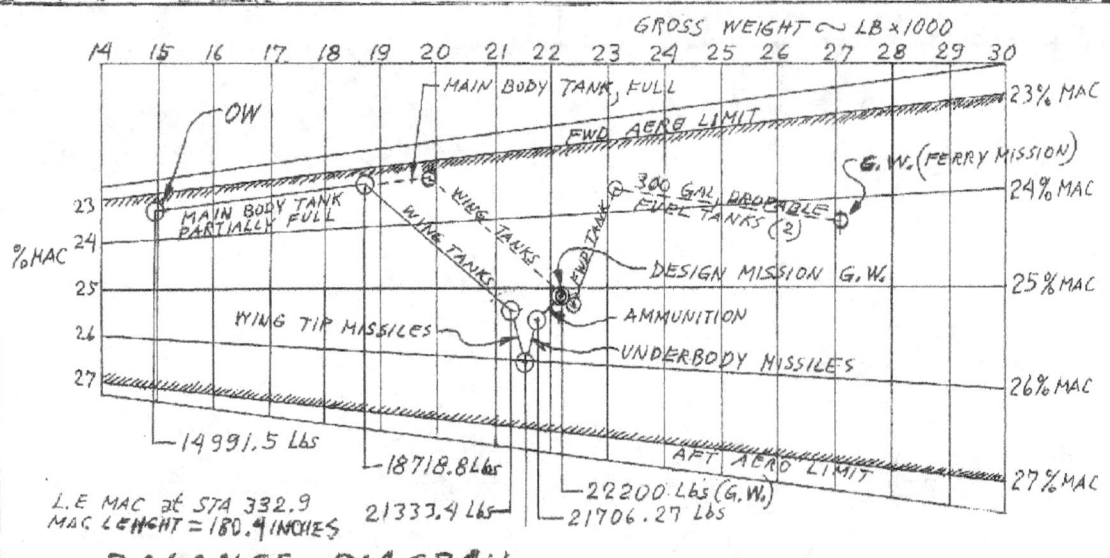

GROSS WEIGHT ~ LB × 1000

BALANCE DIAGRAM
AIRCRAFT CONFIGURATION USING ARTIFICIAL GRAVITY

"HALCON WF1" Fighter aircraft

Design Engineer: WALTER F. LAREDO

Drawing Number: 1000 series

Sheet 20 of 34

PLATE 59

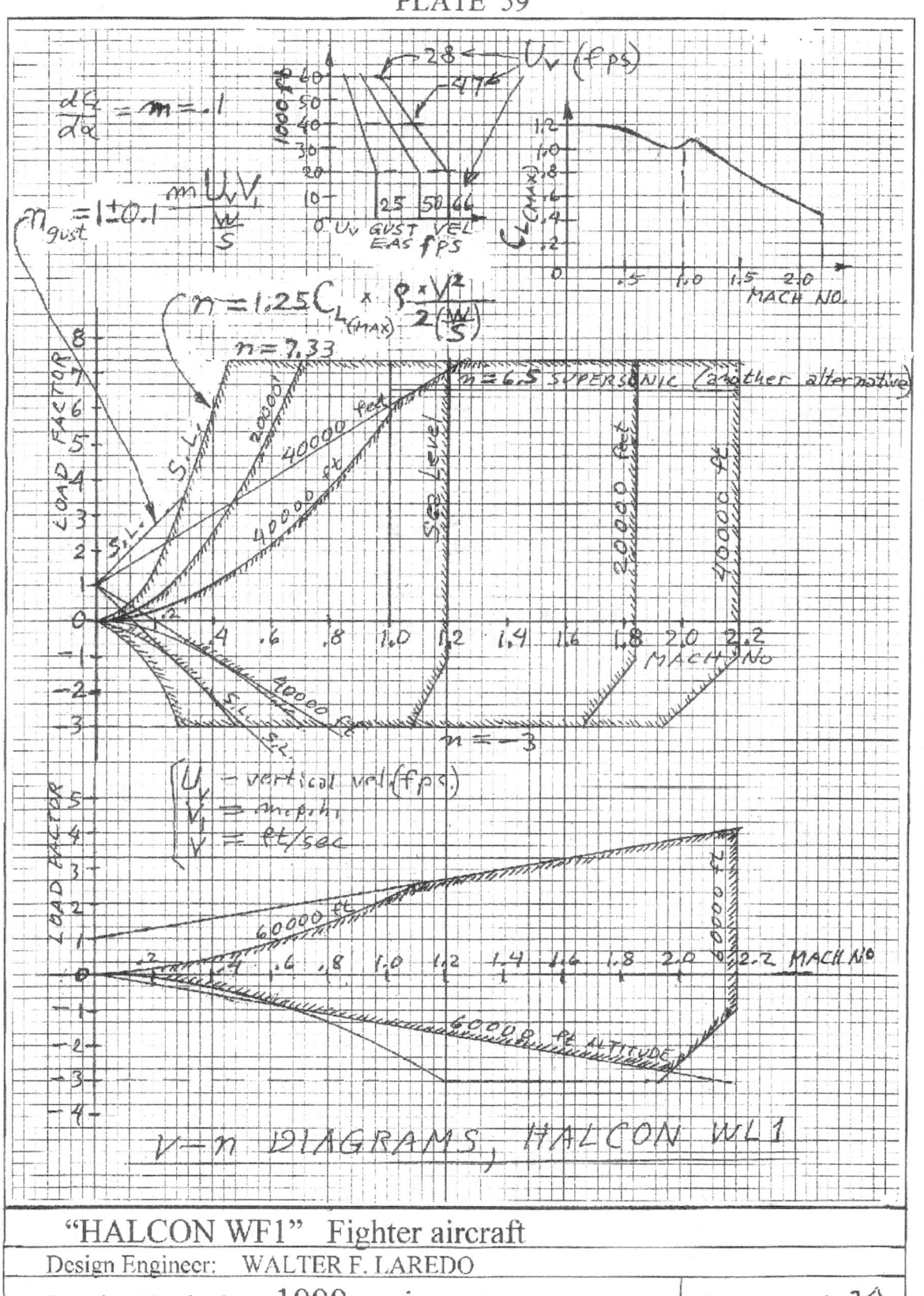

V–n DIAGRAMS, HALCON WF1

"HALCON WF1" Fighter aircraft

Design Engineer: WALTER F. LAREDO

Drawing Number: 1000 series

Sheet 21 of 34

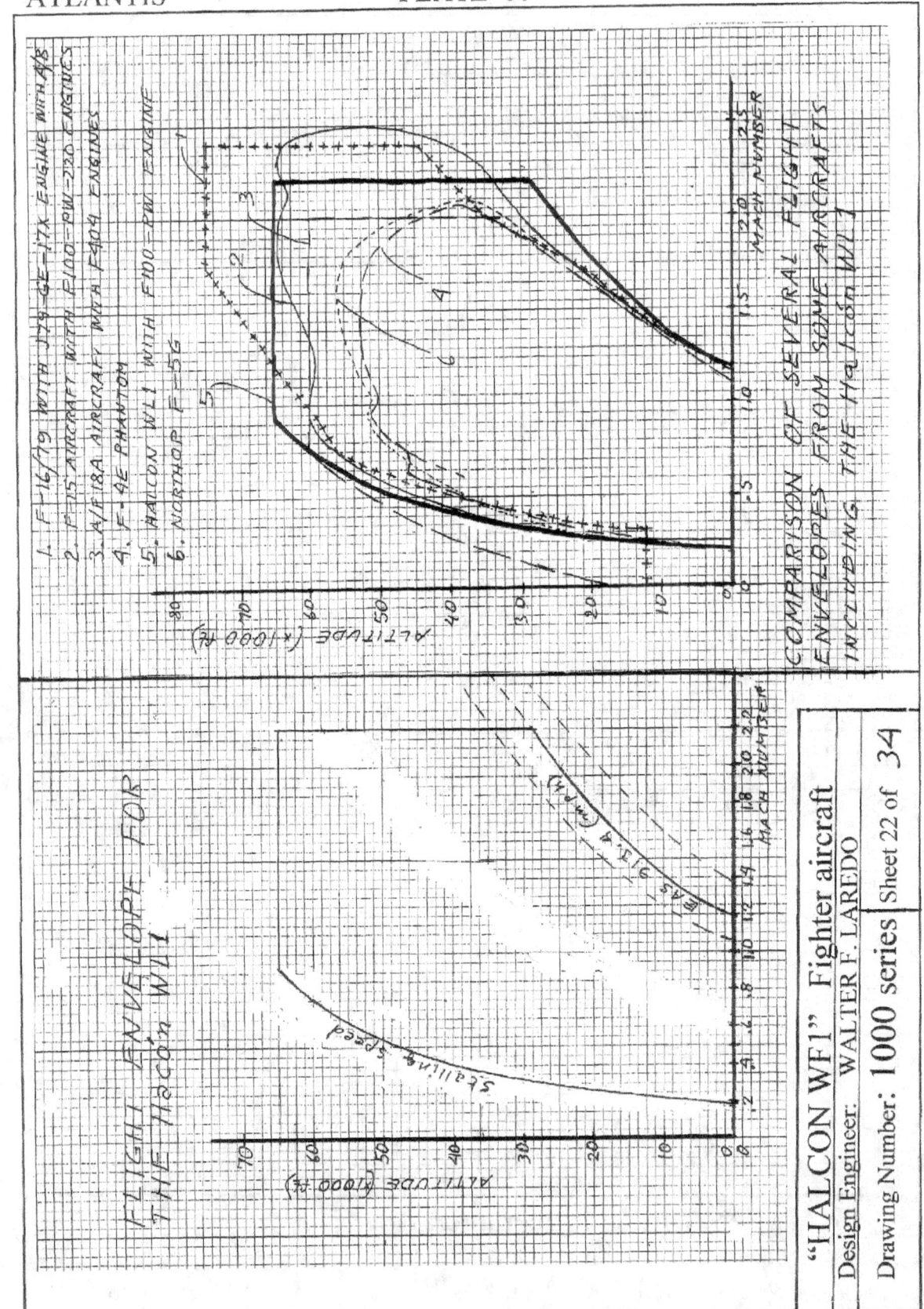

1. F-16/79 WITH J79 GE-17X ENGINE WITH A/B
2. F-05 AIRCRAFT WITH F100-PW-220 ENGINES
3. A/F18A AIRCRAFT WITH F404 ENGINES
4. F-4E PHANTOM
5. HALCON WL1 WITH F100-PW ENGINE
6. NORTHOP F-5G

COMPARISON OF SEVERAL FLIGHT ENVELOPES FROM SOME AIRCRAFTS INCLUDING THE HALCON WL1

ALTITUDE (×1000 ft)

MACH NUMBER

FLIGHT ENVELOPE FOR THE Halcon WL1

ALTITUDE (×1000 ft)

MACH NUMBER

Stalling speed

"HALCON WF1" Fighter aircraft

Design Engineer: WALTER F. LAREDO | Sheet 22 of 34

Drawing Number: 1000 series

PLATE 61

HALCON WF1
AIR TO AIR FIGHTING MISSION

WITH THE GROSS WEIGHT (70% INTERNAL FUEL) OF 28,200 LBS, THE AIRCRAFT IS GOING TO BE DESIGNED WITH A LIMIT LOAD FACTOR, $n = 7.3$

$W_1 = W_{T_O} = 22,200$ Lbs.

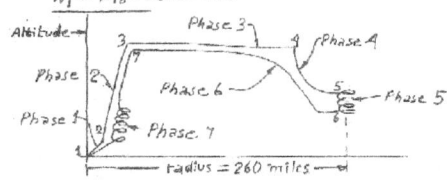

Fixed Weights:

Pilot plus gear   = 200 lbs
4 AIM missiles plus racks = 944 lbs
M-61 cannon plus accessories = 485 Lbs
300 rounds of 20mm shells = 171.43 Lbs
                           1800.43 Lbs

$W_{fixed} = 1800$ lbs

Radius using 70% fill fuel tanks = 280 miles
Max. speed for this mission = M 1.6
cruising between 36000 ft and 47000 ft

PHASE 1

$W_{T_O} = W_1 =$ Gross weight before start and before start and before takeoff. The aircraft burns 3% of W_{T_O} fuel

$\dfrac{W_2}{W_1} = .97$

PHASE 2 Climbing, from fig. below $\dfrac{W_3}{W_2} = .975$

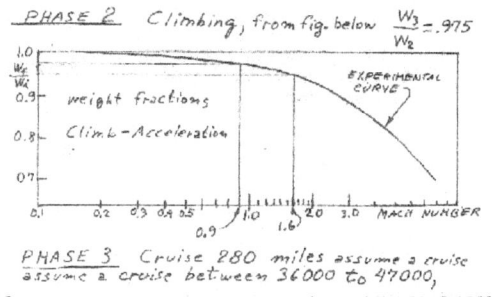

weight fractions
Climb-Acceleration
EXPERIMENTAL CURVE

PHASE 3 Cruise 280 miles assume a cruise assume a cruise between 36000 to 47000,

FROM PICTURES AT LEFT FOR MACH NUMBER .9

ALTITUDE: 36089
F-100 ENGINE
TSFC FOR PARTIAL POWER SETTINGS (NON AFTER-BURNING)

$TSFC = .91$ lb/lb-hr

and

$C_{D_0} = .018$

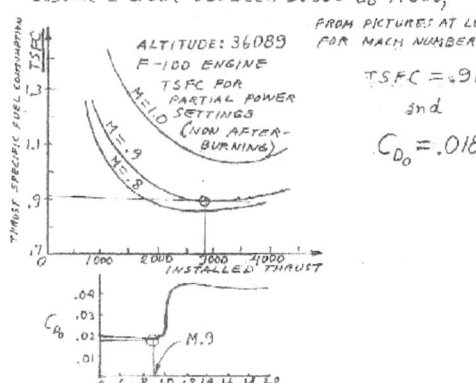

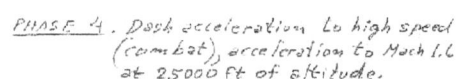

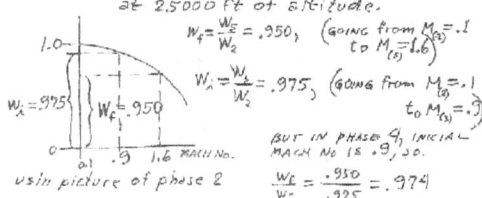

M.9

Aspect ratio $AR = 2.46$

$\bar{e} = .8$ is used for swept thin wing aircraft, with thicknesses from 3% to 8% chord.

$K = \dfrac{1}{\pi A R \bar{e}} = .1617$ subsonic

Range mission for jet aircraft	$C_L = \sqrt{C_{D_0}/3K}$	← constant altitude conditions
	$C_L = \sqrt{C_{D_0}/2K}$	← constant throttle setting condition (cruise climb)
Endurance mission for jet aircraft	$C_L = \sqrt{C_{D_0}/K}$	← minimum thrust required

$C_L = \sqrt{C_{D_0}/2K}$

$\dfrac{L}{D} = \dfrac{C_L}{C_D} = \dfrac{C_L}{C_{D_0} + K C_L^2} = \dfrac{\sqrt{C_{D_0}/2K}}{C_{D_0} + K(C_{D_0}/2K)} = \sqrt{\dfrac{8}{9}}\cdot\left(\dfrac{L}{D}\right)_{max}$

where $\left(\dfrac{L}{D}\right)_{max} = \dfrac{1}{2\sqrt{C_{D_0}K}} = 9.27$

So $\dfrac{L}{D} = \sqrt{\dfrac{8}{9}}\,(9.27) = 8.74$

To find $\dfrac{W_3}{W_4}$ using the Breguet Range equation

Range $= \left(\dfrac{V}{TSFC}\right)\left(\dfrac{L}{D}\right)\left(\ln\dfrac{W_3}{W_4}\right)$

$\ln\dfrac{W_3}{W_4} = \dfrac{(R)(TSFC)}{(V)\left(\dfrac{L}{D}\right)} = \dfrac{(280)(.91)}{(685)(8.74)} = .04256$

$\dfrac{W_3}{W_4} = e^{(.04256)} = 1.05$

or $\boxed{\dfrac{W_4}{W_3} = .953}$

PHASE 4. Dash acceleration to high speed (combat) acceleration to Mach 1.6 at 25000 ft of altitude.

$W_4 = \dfrac{W_E}{W_2} = .950$, (GOING FROM $M_{(s)} = .1$ to $M_{(s)} = 1.6$)

$W_1 = \dfrac{W_3}{W_2} = .975$, (GOING FROM $M = .1$ to $M_{(s)} = .9$)

BUT IN PHASE 4, INICIAL MACH NO IS .9, SO.

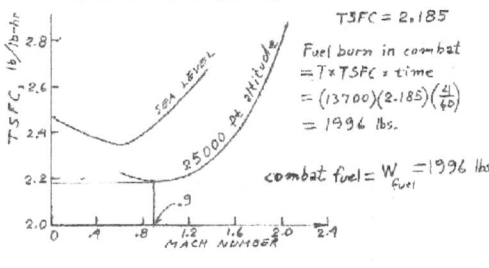

using picture of phase 2

$\dfrac{W_E}{W_2} = \dfrac{.950}{.975} = .974$

PHASE 5. FOUR MINUTES OF COMBAT TURNS AT 25000 ft of altitude and Mach Number = .9, use diagram about Installed Thrust vs. Mach Number for the F-100 Engine.

Maximum thrust with afterburner at 25000 ft = 13700 Lbs

$TSFC = 2.185$

Fuel burn in combat
$= T \times TSFC \times$ time
$= (13700)(2.185)\left(\dfrac{4}{60}\right)$
$= 1996$ lbs.

combat fuel $= W_{fuel} = 1996$ lbs

SEA LEVEL
25000 ft altitude
.9

"HALCON WF1" Fighter aircraft

Design Engineer: WALTER F. LAREDO

Drawing Number: **1000** series

PLATE 62

PHASE 6 Cruise return, same as phase 3

$$\frac{W_6}{W_7} = \frac{W_1}{W_4} = 1.05 \qquad \frac{W_7}{W_6} = .9524$$

PHASE 7 Loiter at sea level, 20 minutes

To find C_L from range and endurance chart in phase 3.

$$C_L = \sqrt{C_{D_0}/K}$$
$$C_D = C_{D_0} + K C_L^2$$

so $\dfrac{L}{D} = \dfrac{C_L}{C_D} = \dfrac{\sqrt{C_{D_0}/K}}{C_{D_0} + K C_L^2} = \dfrac{\sqrt{C_{D_0}/K}}{C_{D_0} + K(C_{D_0}/K)} = \dfrac{1}{2\sqrt{C_{D_0}\,K}}$

$$= 9.2678$$

$$Endurance = E = \frac{L}{D} \frac{1}{TSFC} \ln \frac{W_7}{W_8}$$

From F-100 engine data, for low subsonic flight $TSFC = .84$

$$E = \frac{20}{60} = (9.2678)\left(\frac{1}{.84}\right) \ln \frac{W_7}{W_8}$$

$$\frac{W_7}{W_8} = 1.03 \quad or \quad \frac{W_8}{W_7} = .970$$

In cruise back use W_6 as the new initial weight.

• The weight at landing is W_8.

$$W_8 = \left(\frac{W_7}{W_6}\right)\left(\frac{W_8}{W_7}\right) W_6 = (.9524)(.970)(16612.57) = 15350 \; Lb$$

• Fuel weight required for the mission, W_{fuel}

$$W_{fuel\atop mission} = W_{T0} - W_8 - missile - ammo = 5982.6 \; Lbs$$

The total fuel required, includes 5% reserve fuel plus 1% trapped fuel.

$$W_{fuel} = 1.06 \; W_{fuel\atop mission} = 6341.5 \; Lbs$$

• Empty weight

$$W_{(empty)} = W_{T0} - W_{fuel} - W_{fixed} = 22200 - 6341.5 - 1880$$
$$= 14058.5 \; Lbs \;\; (I)$$

Comparing this result with the one obtained by using a formula which was obtained from historical data and applies when $\frac{T}{W_0} > 0.9$ and $\frac{W}{S} < 70$

$$W_{(empty)\atop with\;conventional\;structure} = 1.605 \left(W_{TAKEOFF}\right)^{0.916} = 15372 \; Lbs$$

some parts of the aircraft structure will be made using advanced composites materials, so the aircraft will be 10% lighter.

$$W_{empty\atop (advanced\;technology\;materials)} = (15372)(.99) = 13835 \; Lbs \;\; (II)$$

$(I) > (II)$, it is acceptable

• So this mission relation between W_{empty} and W_{T0} is feasible.

TO FIND:

• THE AIRCRAFT WEIGHT AT THE BEGINNING OF COMBAT (W_5).
• THE WEIGHT AT THE END OF COMBAT (W_6).
• LANDING WEIGHT (W_8).
• MISSION FUEL WEIGHT.
• AIRCRAFT EMPTY WEIGHT
• FEASIBILITY OF TAKEOFF WEIGHT VS. EMPTY WEIGHT

MISSION PHASES

Phase 1, $\frac{W_1}{W_1} = .97 \; Lbs$
Phase 2, $\frac{W_2}{W_1} = .975 \; Lbs$
Phase 3, $\frac{W_3}{W_2} = .9524 \; Lbs$ AIRCRAFT TAKEOFF WEIGHT $= W_0 = W_1$
Phase 4, $\frac{W_4}{W_3} = .974 \; Lbs$ $= 22200 \; Lb$
Phase 5, $W_{fuel\atop combat} = 1996 \; Lbs$
Phase 6, $\frac{W_7}{W_6} = .9524 \; Lbs$
Phase 7, $\frac{W_8}{W_7} = .970 \; Lbs$

• AIRCRAFT WEIGHT AT THE BEGINNING OF COMBAT.

$$W_5 = \left(\frac{W_1}{W_1}\right)\left(\frac{W_2}{W_1}\right)\left(\frac{W_3}{W_2}\right)\left(\frac{W_4}{W_3}\right) W_1 = (.877)(22200) = 19476 \; Lbs$$

• WEIGHT AT THE END OF COMBAT $= W_6 = W_5 - $ EXPENDABLE FIXED WEIGHTS

EXPENDABLE FIXED WEIGHTS

FUEL (4 MINUTES OF COMBAT) ..... 1996 Lbs
4 AIM MISSILES .... 696 Lbs
300 ROUNDS OF 20 MM AMMUNITION ..... 171.43 Lbs
                 2863.43 Lbs

$$W_6 = 19476 \; Lbs - 2863.9326 = 16612.57 \; Lbs$$

HALCON WF1

AT the end of phase 4 (Dash operation) the aircraft will be flying horizontally, at the following conditions:
altitude = 25000 ft.
Mach number = 1.6, $(V = 1625 \; ft/sec)$
Max. Thrust $\approx 21000 \; Lb$
mass weight $= (.877)(W) = 19476 \; Lb$
$C_D = .04092$ (from similar aircraft) supersonic co.
$\gamma = \frac{C_P}{C_V} = 1.4$ for air
$p = $ pressure at 25000 ft (air)
$S = 370 \; ft^2$

$$D = \frac{1}{2}\gamma P M^2 S \, C_D = 21046 \; Lb$$

$$P_s = \frac{dh_e}{dt} = \frac{V(T-D)}{W} = \frac{(1625)(21000 - 21046)}{19476}$$

$$P_s = -3.85 \; ft/sec$$

desacceleration in horizontal flight $= \frac{gh}{V} = ft/sec$

$$\frac{gh}{V} = \frac{(32.2)(-3.85)}{(1625)} = -.076 \approx 0 \; (negligible)$$

so that the airplane will be on a level flight.
its maximum sustained turn rate $\dot\psi$ and n are:

$$n_{sustained} = \frac{q}{W/S} \sqrt{\frac{1}{K}\left(\frac{T_{max}}{q \, S} - C_{D_0}\right)} \quad where:$$
     $M = 1.8$
     $V = 1625 \; ft/sec$
     $\rho = .001065 \; slugs/ft^3$
$$= 1.0189$$
     $K = .25$
     $C_{D_0} = .04$
$$\dot\psi = \frac{g\sqrt{n^2-1}}{V} = .039 \; rad/sec = .22 \; degrees/sec$$
     $S = 370$
     $T_{max} = 21000 \; Lb$
$R \approx \infty$, straight line flight

Calculate $\dot\psi$ and n for .9M, at an altitude of 25000 ft.

$$n_{sustained} = \frac{q}{W/S}\sqrt{\frac{1}{K}\left(\frac{T_{max}}{q \, S} - C_{D_0}\right)}$$
     $V = 1015.74 \; ft/sec$
     $\rho = .001065 \; slugs/ft^3$
$$= 7.81 \; which \; is \; a \; high \; load \; factor$$
     $K = .16$
for the pilot and the aircraft above the limit strength of the structure.
     $C_{D_0} = .018$
     $T_{max} = 21000 \; Lb$

$$\dot\psi = \frac{g\sqrt{n^2-1}}{V} = .238 \; rad/sec = 13.7 \; degrees/sec$$

"HALCON WF1" Fighter aircraft

Design Engineer: WALTER F. LAREDO

Drawing Number: **1000 series**

PLATE 63

$Radius = \dfrac{V^2}{g\sqrt{n^2-1}} = 4250\ ft$

$wing\ tilt\ angle,\quad \phi = arc\ cos\left(\dfrac{1}{n}\right) \approx 82.2°$

To calculate the instantaneous turn rate for $C_{L_{max}} = .8$

$n_{inst} = \dfrac{q\,C_{L_{max}}}{W/S} = 8.35$

$\dot\psi = \dfrac{g\sqrt{n^2-1}}{V}(57.3) = 15\ degrees\ per\ second$

The m g's of the sustained turn rate and the instantaneous turn rate are too high and could damage or destroy the aircraft structure. So using a lower value of $n = 5$, is calculated the new Thrust setting (T).

$P_s = \dfrac{dh_e}{dt} = 0 = V\left[\dfrac{T}{W} - \dfrac{q\,C_D}{W/S} - \dfrac{k}{q}\,n^2\,\dfrac{W}{S}\right]$

$T = 11589\ lbs$

max. sustained turn rate $\dot\psi = \dfrac{g\sqrt{n^2-1}}{V}(57.3) = 8.3\ degrees/sec$

$Radius = \dfrac{V^2}{g\sqrt{n^2-1}} = 6540\ feet$

HALCON WL1

TOTAL TAKEOFF DISTANCE AT SEA LEVEL

mission: Air to air combat, 4 AIM missiles, M-61 300 RNDS 20 mm, partially fill internal fuel tanks.
Takeoff weight = 22200 Lbs

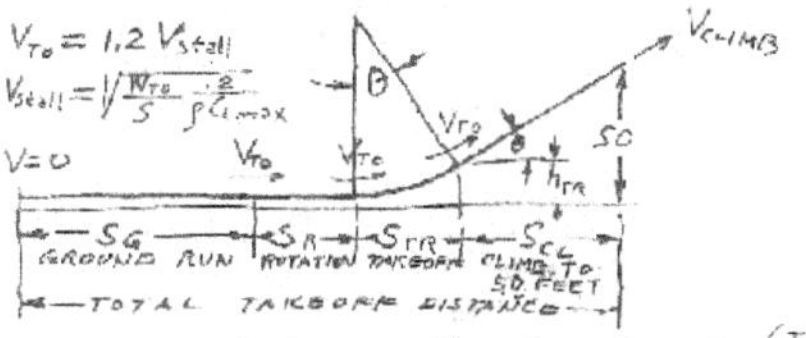

$V_{TO} = 1.2\ V_{stall}$

$V_{stall} = \sqrt{\dfrac{W_{TO}}{S}\cdot\dfrac{2}{\rho\,C_{L_{max}}}}$

$V = 0$

$\boxed{Total\ Takeoff\ distance = S_G + S_R + S_{TR} + S_{CL}}\quad (I)$

$S_G = \left(\dfrac{1}{2}V_{TO}^2\right)\dfrac{1}{a}$, acceleration a is at $.7V_{TO}$

$S_R = t\,V_{TO}$, $t = 3$, (3 seconds of rotation)

$S_{TR} = R\sin\theta$; $R = \dfrac{V_{TO}^2}{n\,g}$

$n = \dfrac{L}{W} = 1.15$ $\Big\}(*)$

$\sin\theta = \dfrac{T-D}{W_{TO}}$

$S_{CL} = \dfrac{50 - h_{TR}}{\tan\theta}$, where $h_{TR} = R(1-\cos\theta)$

$(*)\quad L = W_{TO} + (centrifugal\ force) = W_{TO} + \dfrac{W_{TO}}{g}\cdot\dfrac{V_{TO}^2}{R}$ divide by W_{TO}

$n = \dfrac{L}{W_{TO}} = 1 + \dfrac{V_{TO}^2}{Rg} = \dfrac{\frac12\rho(1.2 V_{TO})^2 S_w(.8\,C_{L_{max}})}{\frac12\rho(V_{stall}^2) S_w\,C_{L_{max}}}$

$n = 1.15$ the airplane is rotated to an angle of attack such that $C_L = .8\ C_{L_{max}}$

EQUATION (I) BECOME:

TOTAL TAKEOFF DISTANCE =

(II)

$\dfrac{1.44\,\dfrac{W_{TO}}{S_w}}{g\,\rho\,C_{L_{max}}\left[\dfrac{T}{W_{TO}} - \dfrac{D}{W_{TO}} - \mu\left(1 - \dfrac{L_a}{W_{TO}}\right)\right]} + 3V_{TO} + \left(\dfrac{V_{TO}^2}{1.15\,g}\right)\left(\dfrac{T-D}{W_{TO}}\right) + \dfrac{50 - h_{TR}}{\tan\theta}$

where:

$W_{TO} = 22200\ lbs$

$S_w = 370\ ft^2$

$\rho = .002377$

$g = 32.2$

$\mu = .025$

$T = 20000\ lbs$, F-100 engine (installed)

$C_{L_{max}} = 1.7$ considered deflected flaps and ground effect

$AR = 2.46$

$e = .8$ for swept thin wings aircraft

$k = .162$, from $k = \dfrac{1}{\pi\cdot AR\cdot e}$

$V_{stall} = 172.3\ ft/sec$ where $V_{stall} = \sqrt{\dfrac{W_{TO}}{S_w}\cdot\dfrac{2}{\rho\,C_{L_{max}}}}$

$V_{TO} = 206.8\ ft/sec$ where $V_{TO} = 1.2\ V_{stall}$

$L_a = 4422.9\ lbs$ = Average lift in ground running where $V = .7V_{TO}$

$L_a = \frac12\rho V^2 S_w\,C_{L_a}$

$D = 3163.76\ lbs$

from $D = \frac12\rho V_n^2 S_w\,C_D$

(see next page for C_D)

$C_D = .2746$ from $C_D = \left[C_{D_0} + k\,C_L^2\right] + \Delta C_{D_{flaps}} + \Delta C_{D_{gear}} + etc.$

where:

$\begin{cases} C_{D_0} = .016 \text{ (for supersonic fighters in subsonic cruise } C_{D_0}\text{ from }.014\text{ to }.022)\\[4pt] C_{L_a} = .42 \text{ (flaps deployed and } \alpha_{wing} = 0)\\[4pt] \Delta C_{D_{flaps}} = .04 \text{ from charts}\\[4pt] \Delta C_{D_{gear}} = .19 \quad \Delta C_{D_{gear}} \text{ was obtained from:} \end{cases}$

$D = C_{D_{gear}}\cdot S_w\cdot q = \Delta f\cdot A_f\cdot q$

A_{gear} is frontal area = $4.7\ ft^2$

and $\Delta f_{gear} = 15$ from diagram shown below

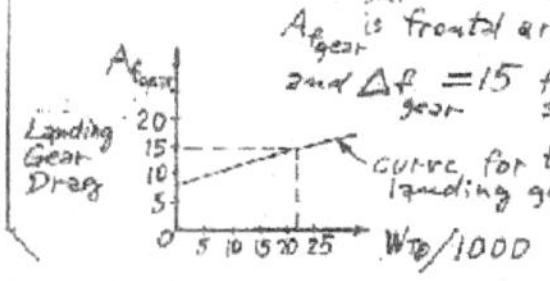

Replacing all the indicated values from Equation (II)

$S_G = 1024.25\ ft$

$S_f = 620.4\ ft$

$S_{re} = 771.82\ ft$

The value of S_{TR} should be decrease to 336.11 ft because $h_{TR} = 296\ ft$, which is higher than the required 50 ft. obstacle.

$\boxed{TOTAL\ TAKEOFF\ DISTANCE = 1981\ ft}$

The actual Takeoff Distance will be by far less, because in this calculation was not considered, the additional lifting by the canards.

"HALCON WF1" Fighter aircraft

Design Engineer: WALTER F. LAREDO	
Drawing Number: 1000 series	Sheet 25 of 34

PLATE 64

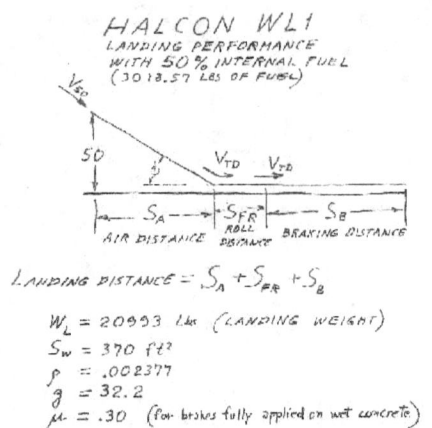

HALCON WL1
LANDING PERFORMANCE
WITH 50% INTERNAL FUEL
(3013.57 LBS OF FUEL)

AIR DISTANCE   ROLL DISTANCE   BRAKING DISTANCE

$LANDING\ DISTANCE = S_A + S_{FR} + S_B$

$W_L = 20993$ Lbs (LANDING WEIGHT)
$S_w = 370$ ft²
$\rho = .002377$
$g = 32.2$
$\mu = .30$ (for brakes fully applied on wet concrete)
$C_{L_{max}} = 1.73$ (with T.F and L.E. flaps in landing condition)
$V_{stall} = \sqrt{\frac{W_L}{S_w} \cdot \frac{2}{\rho C_{L_{max}}}} = 166$ ft/sec.
$V_{50} = (1.3) V_{stall} = 215.8$ ft/sec.
$V_{TD} = (1.15) V_{stall} = 190.9$ ft/sec. (Touchdown vel.)
$AR = 2.46$
$\bar{e} = .8$
$K = \frac{1}{\pi \cdot AR \cdot \bar{e}} = .162$
$L = W_L = 20993$ Lbs

$C_{D_0} = .016$
$C_{L(SA)} = (.8) C_{L_{max}} = 1.73$
$C_{L_G} = .42$
$\Delta C_{D_{flaps}} = .041$
$\Delta C_{D_{gear}} = .19$

$C_{D(SA)} = \left[C_{D_0} + K C_{L(SA)}^2 \right] + \Delta C_{D_{flaps}} + \Delta C_{D_{gear}} = .7318$

$V_{(SA)} = 208$ where $V_{50} > V_{(SA)} > V_{TD}$

$D_{(SA)} = \frac{1}{2} \rho V_{(SA)}^2 S_w C_{D(SA)} = 13923.$ Lbs.

$S_A = \frac{L}{D_{(SA)}} \cdot \left[\frac{V_{50}^2 - V_{TD}^2}{2g} + 50 \right] = 312.5$ ft

$S_{FR} = t V_{TD} = (3)(190.9) = 572.7$ ft   ($t = 3$ seconds)

$S_B = \frac{W_L}{g \mu \rho S_w \left(\frac{C_D}{\mu} - C_{L_G} \right)} \ln \left[1 + \frac{\rho S_w}{W_L} \left(\frac{C_D}{\mu} - C_{L_G} \right) V_{TD}^2 \right]$

where: $\bar{C}_D = \left[C_{D_0} + K C_{L_G}^2 \right] + \Delta C_{D_{flap}} + \Delta C_{D_{gear}} = .2758$

so $S_B = 2805$

$TOTAL\ LANDING\ DISTANCE = S_A + S_{FR} + S_B = \underline{3690\ ft}$

The actual landing distance will be considerable
less than indicated; because in this calculation was not
considered neither the air brake flap (above the fuselage), nor
the canards which will increase lift reducing the
approach speed.

"HALCON WF1" Fighter aircraft

Design Engineer: WALTER F. LAREDO

Drawing Number: 1000 series

Sheet 26 of 34

PLATE 65

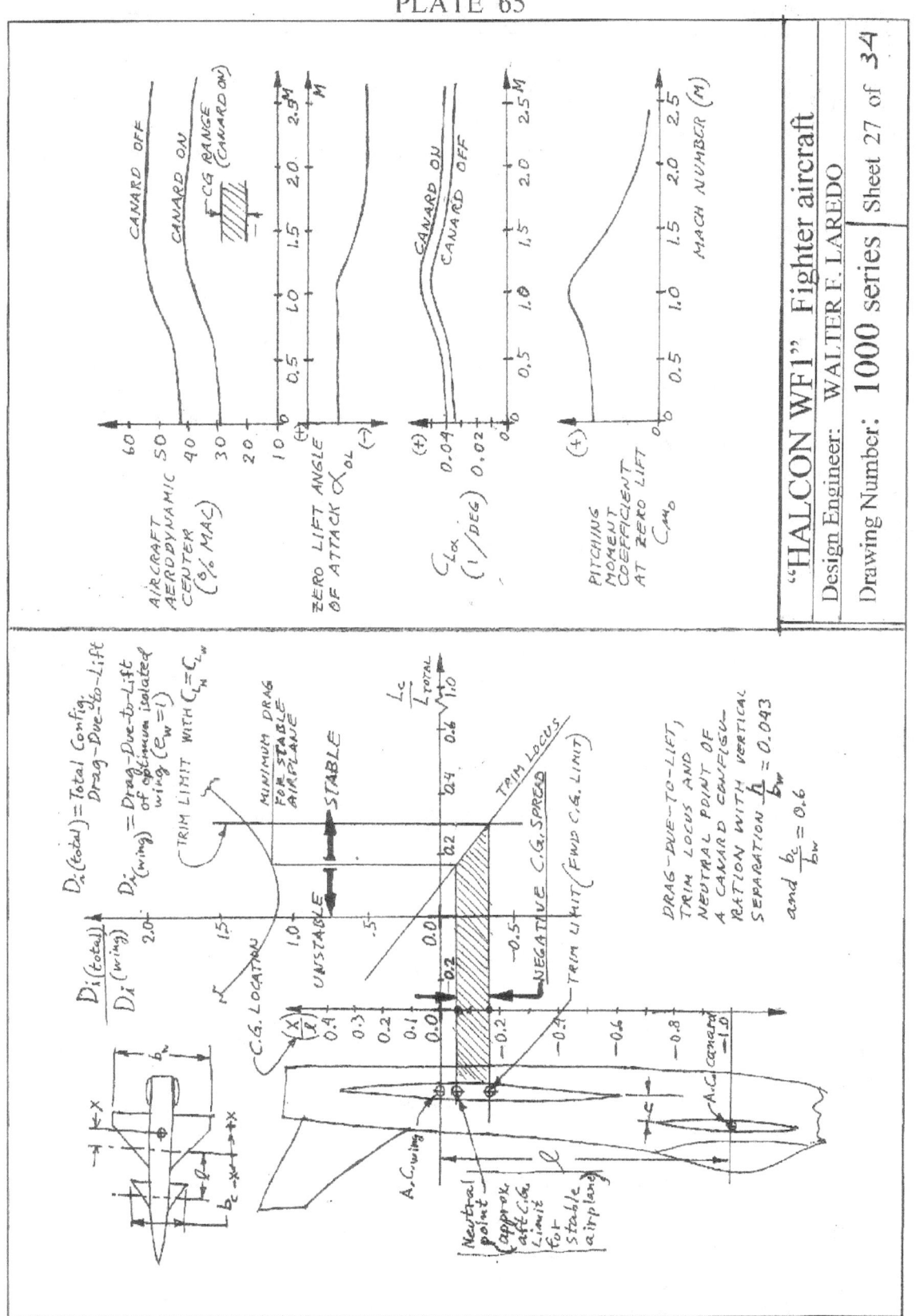

"HALCON WF1" Fighter aircraft

Design Engineer: WALTER F. LAREDO

Drawing Number: 1000 series | Sheet 27 of 34

PLATE 66

STABILITY AND CONTROL

The delta wing is relatively large, reducing wing loading providing better low-speed performance, and permitting higher turn-rates at high altitude. The wing has almost no camber, but leading-edge flaps (very rare in deltas) and large trailing-edge elevons, which drops as flaps during maneuvers, can produce more lift per unit area than the inefficient conventional delta wing.

The Halcon WL1 aircraft will use a sophisticated electronic control system, using a concept called Relaxed Static Stability (RSS), such instability will require active controls, the aircraft will employ canard surfaces for gust alleviation, maneuvering, and trimmed lift enhancement. A canard configuration offers advantages of less rearward aerodynamic center travel than a conventional tail arrangement; also the wing-canard combination is of load sharing. The wing and the canard variable camber controlled by the computer will be use to optimize the body-wing lift to drag ratio, at lift coefficients up to maximum sustained maneuver levels. Beyond that point the wing flaps will be deflected for stability and control demands.

The Halcon WL1 will have an autopilot and use triple-redundant, digital fly-by-wire flight-control system command and stability-augmentation. This aircraft also will carry a mechanical backup for the flight control systems.

The addition of canards, slightly ahead of and above the wing, increases the lift available at a given angle of attack. Canards allows the aircraft to operate over a greater range of angles of attack, and reduces stability because the center of lift and center of gravity are moved closer together. A very important purpose of the canards is to improve maneuverability in combat, also to operate from shorter runways.

The aircraft has an electronic fly-by-wire flight control system, which allows the aircraft center of gravity range to be extended further aft than on a conventional aircraft. This will produce less basic stability. Without the FBW system, it could not be possible to position the aircraft center of gravity behind the center of pressure to give a reduced or negative static margin.

For the halcon all-electronic fly-by-wire (FBW) Flight-control system is used, where electrical circuits replace the conventional hydromechanical system with linkages and cables.

The control configured vehicle (CCV) concept is related to the ratio of the aircraft balance to the aircraft static longitudinal stability, allowing the CG to be moved further aft than in conventional configurations. With the CG near the center of pressure, drag is reduced especially at high load factors and at supersonic speeds, this aft CG location also reduces trim drag. Because the center of gravity is so close to the center of pressure, therefore is used relaxed stability design, which allowed reducing de size of canards and flaperons.
As in all jet fighters, the control surfaces of the Halcon WF1 will be actuated by servos, using. CCV (Control Configured Vehicle) technology, which uses FBW (Fly-by-Wire) Flight Control

"HALCON WF1" Fighter aircraft

| Design Engineer: | WALTER F. LAREDO | |
| Drawing number | 1000 series | Sheet 28 of 34 |

PLATE 67

System. A flight control system with instant response to pilot input, which computer, will accept, refuse or modify the pilot's command to prevent the aircraft from exceeding its maneuvering structural limits, then the FBW will send the pilot input directly to servomotors that operates the control surfaces. The movable surfaces are continuously adjusted to let the aircraft follows it's required trajectory. The unstable aircraft is prevented from divergent accelerations by continuous positive control. CCV technology is ideal for this type of aircraft configuration, because it reduces the severe trim drag. The system will use a Flight Control Computer and a Central Air Data Computer. The Fly-by-Wire Control System will be integrated with the engine control and link with the weapons system, in order to relieve the pilot from many monitoring tasks, letting him to concentrate in dogfighting.

Electronic wire pulses command small, light, high-pressure (preferable 4000 psi) actuators for the control surfaces. Computer control of elevons and all-moving canard independent from the pilot maintains the aircraft stability.

The short coupled canard configuration of the Halcon WF1 will have good longitudinal aerodynamic and lateral-directional characteristics. The wing efficiency is highly increased by the canard, particularly at high angles of attack, by deflecting the airflow on the wing; the canard also will allow large movements of the center of gravity, which could be handle by a simple autopilot.

On most fighters today roll rate is automatically limited when the aircraft flies below a certain speed, pulling g's, this prevents an inertial coupling from raising angle of attack to where the aircraft becomes unstable in yaw. On the other hand if there were no roll limiter, pilots would put on a margin of safety, ending up with problems. The designers would like to increase roll rates while maintaining high angle of attack stability.

For the HALCON WF1 will be develop charts with a variety family of curves to represent the approximate performance of this aircraft, for example one important family of curves are the dog-house shaped curves, where the left wall is determined by MAXIMUM LIFT, the roof by MAXIMUM G's, and the right wall by MAXIMUM SPEED. These curves will be used to compare one fighter aircraft to another by superposing their doghouse curves. Other family of curves will show TURNS AT CONSTANT G's, when speed is lost or gained, and SUSTAINED, WHICH IS THE MAXIMUM TURN RATE AT ANY CONSTANT SPEED.

Approximate values for the wing-body-canard configuration of the Halcon WF1, using the concept of Relaxed Static Stability will be as indicated below, but need to be recalculated:

Transonic aerodynamic center shift from the subsonic...... 12 % MAC
Subsonic aerodynamic center at.................................32 % MAC
Supersonic aerodynamic center, varies from.........40 % to 44 % MAC
If a 5 % MAC static margin is designed too,
      the aft C.G. limit will be...............................27 % MAC
At supersonic speeds with the C.G. at the aft limit,
      the static margin will be usually less than.............17 %

Low stability couples with positive values of (Cm,o), pitching moment at zero lift will result in "GOOD MANEUVER CAPABILITY AND LOW TRIM DRAG". At high dynamic pressure conditions, the airplane will trim at a negative angle of attack, probably caused by the large pitching moment at zero lift (Cm,o). This aircraft will require a new, more accurate analysis of Stability Control, including trimming and other flight conditions. Will be created charts for canard and elevon trim requirements, which must cover most of the flight envelope, curves that will reflect the "Natural Longitudinal Static Stability" of the airplane, assuming no thrust effect.

"HALCON WF1". Fighter aircraft	.
Design Engineer: WALTER F. LAREDO	.
Drawing number 1000 series	Sheet 29 of 34 .

PLATE 68

Just for a preliminary estimation could be assume that the wing's elevons are fix and that the airplane could employ the canards for trimming and maneuvering, although wing leading edge flaps and elevons will be also use to decamber the wing at supersonic speed.

The lift curve slope per degree of α, and the pitching moment coefficient at zero lift is maximum at Mach 1. The zero lift angle of attack increases negatively as Mach number increases, the aerodynamic center shift aft by 12 % MAC at transonic and supersonic speeds. At the aft C.G. the airplane will be unstable over most of the flight envelope, (the rate of change of the canard angle is positive and the rate of change of the wing's elevon angle is negative). Subsonically the aft center of gravity is 5 % ahead of the aerodynamic center, and the airplane is still unstable. Supersonically the aft center of gravity will be in between 13 and 17 % ahead of the aerodynamic center and the configuration will become more stable, but not too stable because this unstability is caused by the rate of change of (Cm,o) with respect to Mach number, δ Cm/ δ Mach is some small negative number at supersonic speeds. This is a destabilizing effect because the elevon must deflect in a negative direction, as speed is increased to compensate for it, and the canards must deflect increasing its angle of attack. At low supersonic stability levels, the speed effect is dominating static stability characteristics. The canard's area was designed for the forward C.G. at transonic and supersonic speeds.

Optimum (Cm,o) for an aircraft with relaxed stability is different than for a conventional aircraft. Canard and elevon deflection required to maneuver at the aft C.G. condition, for an additional "g" required that additional canard angles are small, and the required wing elevon angles are smaller yet, positive or negative. This is one of the primary benefits of RELAXED STABILITY.

The increase in vertical tail area by using two vertical tails will give to the Halcon WF1 aircraft a good lateral directional stability for a variety of angles of attack as long as they are not too large. Fins better exposed to the air stream increasing its rudder effectiveness. The redundancy of using two vertical tails increases survivability. If there will be some decrease in wing-body directional stability at Mach numbers greater than 2, then this problem can be alleviated by small design changes in the wing Strake.

The Roll Power is provided from differentially operated elevons, with one second or less time to bank from 0 to 90 degrees, could exist throughout most of the flight envelope. Data that will be gathered from the wind tunnel test plus the prototype flying tests, Data used to reprogram the software of the flight control system until the aircraft could perform in very efficient way under all kinds of flight conditions.

STRUCTURAL ARRANGEMENT

Early in this proposal, the primary structure of this aircraft was designed from aluminum lithium alloy, and the rest including the secondary structure from advanced composites. However this concept changed, because it seems more promising that in order to reduce weight and to reduce the number of parts, the whole aircraft should be build from advanced composites, using the latest state-of-the-art in structural design, considering this material is good for long fatigue life. Materials ways of arrangement would be selected to limit crack growth/time.

Materials using carbon fibers, kevlar (an aramid fiber) and fiberglass, all them imbedded in an epoxy resin matrix would be use in honeycomb sandwich shell form or in laminar form.

Both, conventional and advanced composite structures will be designed with sufficient load path redundancy to prevent catastrophic failure due to a 23-mm shell hit on the primary airframe structure.

This structure will be analyze for several load conditions using Matrix computer analysis of structures to find Static and Dynamic loads, Stresses, deflection and flutter.

WING:

The wing structure consists of the wing box, leading edge and trailing edge flaps. The leading edge and the trailing edge flaps are made from full depth honeycomb because of its thinness. A honeycomb sandwich structure made from advanced composites would be lighter than a similar conventional from aluminum alloy. The composite wing box structure is multispar, where at the wing-body join, each spar end is bolted to its corresponding extension from a body frame, so wing spars became extensions of body frames. The wing box section outboard BL 45 serves as support structure for the variable camber leading and trailing edge components, the wing box is also an integral fuel tank. Integral stringers stiffen the skin to prevent buckling prior to limit load. For a wing box from advanced

"HALCON WF1" Fighter aircraft	
Design Engineer: WALTER F. LAREDO	
Drawing number 1000 series	Sheet 30 of 34

PLATE 69

composites, the skin ply orientation are + & - 45 degrees Gp/Ep to carry shear with some 0.0 degrees added to improve their crack stopping capability.

VERTICAL FINS:

The fins are assembled structures of several honeycomb core skins sandwich components, which includes spars, ribs to support the rudder hinges and a root rib to contribute the transfer of loads from the fin to the body support structure. The fins are full depth honeycomb due to its thinness.
The skin fiber orientation are + & - 45 degrees Gr/Ep to carry shear loads.

FOREBODY

The forebody contains the pressurized cockpit, avionics packages and radar system, part of the avionics are located below the pressure deck and other part behind the pilot's seat, they are accessible through a non-structural access door. The radar system is contained within the radome, which is mounted on Sta. 68 bulkhead. A load-carrying frame at Sta. 188 supports the nose landing gear.

AFT-BODY

A load carrying frame, Sta.390 supports the main landing gear. The aft body structure supports the wing and fin spars, which load carrying frames are located in between Sta's 364 & 440 and Sta's 510 & 575 respectively, it also serves as the support structure for the M61 gun.

The ammunition storage and the environmental control equipment are located in the non-pressurized area of the aft-body and are accessible through non-structural doors. The aft-body includes the engine intake duct, body fuel tanks, two external semi-submerged missile systems and the mechanical equipment.

The duct penetrates the body frames and becomes part of the body structure. The ducts, the outer skin and the body side rib provides the boundaries for the body fuel tank between Sta's 315 and 500. Two AIM-9 missiles or similar are stored in external conformal pockets located at the lower outboard corner of the inlets external surfaces. The fuselage have four pairs of fail-safe longerons, to carry fuselage bending loads prior to limit load, designed by pairs for redundancy to avoid catastrophic failure due to a shell hit. For the fighter aircraft version with conventional aluminum structure, the skins are stiffened by close spaced fuselage frames, but in the advanced composites sandwich structural version, the skin faces (internal and external) are stiffened by the honeycomb core, the fiber orientation in such skins are primarily + & - 45 degrees to carry the shear loads.

The upper temperature limit for the 305 degrees F cure for graphite-epoxy systems was established to be 250 to 275 degrees F. The 275 degrees F temperature is the maximum temperature encountered during any flight condition.

For the basic aircraft, its engine will be supported at four points, the front mount is a vertical link and a rail, which is hung at the top of intercostals; this support reacts only vertical loads allowing the engine to expand. The other two mounts are trunnions that reacts engine axial, vertical and torsion loads, while allowing engine radial expansion, this radial expansion is allowed by the trunnions sliding inside spherical bearings. Side loads at the rear mount probable will be reached by a horizontal link between the body of the engine and the aft body frames. Access to the engine equipment for normal maintenance is provided by a lower structural access door, which carry shear loads when it is lock closed, a major engine maintenance require its removal from the aircraft.

An interface between the engine and the exhaust or afterburner system is located aft of the rear mount but forward of the flameholder location and sealed by a gas tight rolling-bellows. A fan discharge air-cools all surfaces of the afterburner exhaust system, upstream of the nozzle exit plane.

PROPULSION

The Halcon WF1 will use the popular low bypass Pratt & Whitney F100-PW-100, A/B engine:

Area of each two-dimensional inlet	To be determinate
Nozzle Area	Same as in the F-15 or F-16 fighters
Length (including nozzle)	190 in (4.83 m)
Weight (bare)	2737 lb. (1242 Kg)
Sea Level Static Airflow	217 lb./sec (98 Kg/sec)
Bypass Ratio	0.71

"HALCON WF1" Fighter aircraft

Design Engineer:	WALTER F. LAREDO	
Drawing number	1000 series	Sheet 31 of 34

PLATE 70

Thrust with afterburner
Sea Level Static (uninstalled) 23 000 lb. (102 000 KN)

Specific Fuel Consumption (TSFC), Sea Level
2.48 lb. fuel/lb. thrust per second with afterburner
(70.46 mg fuel /Newton per second with reheat)

This cycle was selected as a compromise between high altitude supersonic dash and low altitude subsonic cruise mission. The Halcon WF1 will have in interceptor configuration a thrust-to-weight ratio at takeoff greater than one.

Also to this aircraft could be adapt equivalent American or European engines, where the designs of the aircraft inlet and exhaust may or may not be changed. As alternate engines are: Snecmas M53, -2, -5 & P2, F100-PW-220, TF30-P-100 and the latest F404-GE.

This aircraft will have two variable geometry air intakes, one at each side of the forward fuselage, each with two dimensional inlets for external compression with movable ramps, throat bleed system and auxiliary doors for takeoff.

The propulsion system. it will consist of the engine, the variable geometry air intakes and the exhaust system. A variety of tests will be required to cover all flight conditions. Test that can be performed by renting testing facilities in America or in Europe.

WEAPONS

The combat arena could be divided into regions having different requirements for both, the aircraft and the weapons. Flying at high altitude and speed is for fighting beyond visual range. At high cruise speed but fuel-efficient, the aircraft radar can search a broad area of the sky, and once the target is detected and identified to turn into it and launch preferable an Active Radar Homing Missile then leave. A Semiactive Radar Missile requires, staying locked into the target and for a while illuminating it with its radar. Active Radar Homing missiles are AMRAAM, Phoenix and others, and among the Semiactive Radar guided missiles is the SPARROW.

Witting visual range (3 to 5 miles), pointing all-aspect Infrared Missiles and leave, or get engaged in gun battle (perhaps maneuvering to the aircraft limit for short time). Since a decade ago, short-range, all-aspect-attack infrared seeking missiles have made the tail chase of an enemy aircraft unnecessary, and also was innovation of warning systems installed in the aircraft's tails to alert them of coming missiles.

A very maneuverable canard configuration aircraft. In addition to the internal gun and the electronic countermeasures and despite its small size the Halcon WF1 will have seven hard points for mounting weapons, it will have two at the wing tips, two under the wing and two semisubmerged on both sides of the lower fuselage. The Halcon WF1 can carry an arrangement combination of infrared seeking missiles and radar homing missiles. in interception configuration can carry six short range infrared guided missiles as the AIM-9 Sidewinders, and for the international version aircraft, the Matra R.550 Magic or any other of the same equivalent characteristics. For a Mach 2.3 interception mission it could carry two AIM-9, conformally in a semi-submerged arrangement at both sides of the fuselage to reduce drag, and to reduce also radar, infrared and optical signatures.

Pylons will allow the Halcon WF1 to carry many different weapons and sensors for its diverse missions, including dropable fuel tanks.

The aircraft will carry one internal 20 mm M61 six-barrel gun, with 300 rounds of ammunition.
Maximum armament load (attack mission)   6000 lb. (2722 Kg)
Stores, Hard Points:

Fuselage 3   (two conformal and one underneath)
Under wing   2
Wing tips 2

MATERIALS

The Halcon WF1 could be designed with either two kinds of structures, the first one from conventional aluminum alloy and the other, perhaps the best one from advanced composites materials with some titanium fittings. Some steel hardware as the landing gears to be use on both types of structures.

"HALCON WF1" Fighter aircraft		
Design Engineer: WALTER F. LAREDO		
Drawing number 1000 series		Sheet 32 of 34

PLATE 71

For both choices of structures, the wing leading edge, the wing trailing edge flaps, the canard surfaces, the fins, the landing gear doors, the inlets and the radome structural components will be made from advanced composites materials.

In the conventional structural construction of aluminum, cold be saved more weight by using aluminum-lithium alloy, however still more weight would be saved as much as 20 % by using advanced composites materials, enabling fighters to takeoff and land on runways less than 1500 ft. long. The size of autoclave for curing composite parts is important, the larger the autoclave, the larger the components to be cure and less the number of production brakes (splicing joins) to be made, which also translates into weight savings. The largest part of the aircraft structure probable is the wing skins with its bonded spars and the aft fuselage skins.

The materials for the composite parts are: carbon fibers, Kevlar, fiberglass, boron, could be use a single material or a combination of them as required, materials that could be obtained in the form of tape, imbedded in epoxy resin. For the prototype aircraft, the honeycomb sandwich panels and the skin lay-ups would be made by hand on long wood tables. The wing bottom skin and the wing spars will be bonded together and put to get cured in the autoclave, later this integral skin-spar unit will be mechanically fastened to the upper skin and sealed, since this wing is also an integral fuel tank.

The major structural splicing in the Halcon WF1 aircraft will employ small titanium attachment fittings, made by superplastic forming and diffusion bonding. Bonding of titanium fittings to carbon composites structures are more effective than the bonding of aluminum fittings to the same structure, mostly because titanium have smaller differential thermoexpansion and is good for corrosion resistance.

Canopy and windshield materials are primary polycarbonate, and for economical reasons could be build as an assembly of several parts. A single unit canopy windshield structure would be the best, but very expensive.

AVIONICS

The avionics and its characteristics, will be established by a group of experts. Here is only described the rough preliminary idea. For the Halcon WF1 aircraft probably will be choose an efficient radar system, small enough to be accommodated in the radome small room available there.

The cockpit instrumentation includes a head-up display (HUD) where vital information is presented to the pilot's line of vision. The Halcon WF1 may have either a conventional arrangement of instruments or a head-down system of electronic displays, controlled by programmable computers, the company producing these aircraft, the pilots and the engineers would decide the type of cockpit instrumentation arrangement. The pilot could allow the Digital Automatic Flight Control System, take over a broad or a narrow part of the controls, reducing the pilot's workload. This enables the pilot to concentrate on other work such as in radar observation or in dog fighting. Avionics, cockpit displays, radar, computers and the rest of all electronics could be purchase from countries that produce them. Combat avionics for the Halcon WF1 most probably would be: RADAR, HUD, UHF, ECM, UHF/IFF, RWR, and TACAN.

FUEL SYSTEM

The internal fuel capacity for this aircraft is 9060 lb. (4109 Kg).

For a fighting mission of a Gross Weight of 22200 lb. and with 70 % of internal fuel, the weight of this amount of fuel is 6342 lb. (2877 Kg).

The fuel from the wing tanks will flow first to the feed tanks located in the body, and from there to the engine. The maximum flow of the fuel system is of 40 000 lb./hr. The fuel pressure at the inlet of the engine fuel pump should be more than 5 gpsi and less than 50 gpsi. Positive pressure should be maintain in all flight conditions at the inlet of the engine driven pumps, included maneuvers and inverted flight, by electrical boost pumps mounted inside a negative "g" housing located at the bottom of the main tank.

Because fuels are volatile, all fuel tanks including the external, will be pressurized to operate the aircraft at high altitude, pressure obtained by bleeding air from the engine compressor, this air will flow through a system that includes also a dehumidifier then to the tanks.

The aircraft will have a single point ground refuel receptacle connected internally with the air refuel probe manifold.

The Halcon WF1 fuel tanks will be inert depending on threat, the aircraft will be design with either system, one that uses halon or the other that uses nitrogen to inert the fuel tanks, also a bottle will be located inside the nose gear well and will provide 30 minutes of inerting at near empty fuel tank level. Additional survivability and vulnerability

"HALCON WF1" Fighter aircraft		
Design Engineer: WALTER F. LAREDO		
Drawing number 1000 series		Sheet 33 of 34

PLATE 72

protection is giving by a self-sealing blanket between fuel and inlet duct and around the engine in the area of the body fuel tank. Some redundant fuel ducting system will be required for combat survivability.

ELECTRIC POWER GENERATION AND DIDTRIBUTION SYSTEM

Normal power generation system is achieved from the engine's takeoff shaft driving probable two separate accessories gear boxes, where in each is mounted a variable frequency (VSCF) 60 KVA main generator and a hydraulic pump. The cycle-converters convert the variable frequency output of the generator to the fix required number of 117 power of the aircraft. A central Starting/APU unit will provide engine-starting power.

There will be Generator's control breakers. Redundant AC buses will be required for combat survivability.

HYDRAULIC SYSTEM. Two or three independent hydraulic systems will supply hydraulic
power for the Flight Controls and other systems. Variable displacement, constant pressure preferable 4000-psi pumps, installed in the accessory drive units. Two pumps will supply hydraulic power to elevons, rudders, canards, landing gear retraction, steering, brakes, inlet ramps, gun drive and other systems. The wing leading edge flaps could be designed to be actuated either, hydraulically or electrically.

ENVIRONMENTAL CONTROL SYSTEM. The Environmental Control System
provides air-conditioned for the pilot, for cooling the Avionics Equipment and for the fuel tank pressurization system.

The system extracts engine bleed air from an intermediate compressor stage for all climbs and cruise conditions. Last stage compressor bleed air is required for the cockpit pressurization and equipment operation during Idle descent and other Low Power flight conditions.

Engine bleed air, powers the Air Cycle Machine through the dual nozzle drive turbine when ram air pressure is insufficient. Will be study if in addition to the ram air cooled heat exchanger will be necessary a second heat exchanger cooled by tank fuel.

SONIC ENVIRONMENT. Sonic fatigue is nor considered a problem for the wing, fuselage,
canards, fins and other critical components. In the critical structural areas, the sonic environment will not be severe, the highest sound pressure levels are confined to aft nozzles sections. followed by the fins and its neighborhood areas, but will not be a problem for the aircraft critical components.

Some time during the development of the program with the obtained data will be made a preliminary jet noise analysis for static, and takeoff thrust conditions with afterburner, then to made a drawing of the external view of the aircraft showing lines of constant sound pressure levels.

ANTI-ICING. On the Halcon WF1, anti-icing will prevent the formation of ice on the engine nose
cone, engine inlet and its guide vanes, pitot static, total temperature sensor, angle-of-attack and other probes. The engine inlet anti-icing could use regulated compressor bleed air from each engine. and 115-volt AC electrical heaters will heat the probes. Also should be study how convenient would be if most of the systems are heated electrically.

OXYGEN SYSTEM. The oxygen system for the Halcon WF1 will be designed either, as a
gaseous system or as a liquid oxygen system. Located either, in the forward fuselage or behind the pilot seat.

WEIGHT AND BALANCE. At the time I made this proposal for the Halcon WF1 aircraft, I
did a quick weight and balance estimation (see diagram). However before starting the aircraft final detail design stage, will be required to do a new precise weight and balance analysis.

ENGINE. See propulsion information in forward pages, in the last twenty years, engineers have
developed engines providing the same thrust while weighting about half as much, having 30 % fewer parts and significantly lower fuel consumption.

"HALCON WF1". Fighter aircraft		
Design Engineer: WALTER F. LAREDO		
Drawing number 1000 series		Sheet 34 of 34

PLATE 73

POWER PLANT

Two 200 lb.-thrust turbojet engines power this VSTOL, jet aircraft.
For liftoff the efflux is diverted to drive two fans housed in the wings. Each
fan augments the jet thrust by 300 per cent, hence the liftoff thrust of each
fan is 600 lb., giving a total lift of 1200 lb.

SPECIFICATIONS

Accommodation	2 tandem (as in a motorcycle)
Wing's span, (b)	23.42 ft
Wing's gross area, (Sw)	131.7 sq. ft.
Airfoil	Supercritical
Wing chord, root/ tip	6.2 ft / 4.0 ft
MAC	5.6 ft
Wing aspect Ratio, AR= (b square)/ Sw	4.16

(Continues in next page)

ADVANCED ENGINEERING PROJECTS

Name of project:

MECHANICAL FLYING HORSE, VSTOL MACHINE

Design Engineer: Walter F. Laredo   *W.Laredo*   Date: March 15, 2002

Areas of Development:
Aerodynamics ✓, Stability Control ✓, Thermodynamics ✓, Materials ✓,
Structures (Design and Analysis) ✓, Mechanisms & Systems ✓,

NOTE 1:   This is the PRELIMINARY DESIGN for a real engineering
project.

BIG TOYS FOR THE VERY RICH

Drawing Number: 1100

PLATE 74

(Continues from page 1)

Horizontal tail area (Sc)	24 sq. ft.
Vertical tail area (Sv)	8.2 sq. ft.
Tail arm for hor. tail (Lc)	8 ft.
Tail arm for ver. tail (Lv)	7.18 ft.

Tail volume:

Horizontal tail

$$Vc = (Lc \times Sc) / (M.A.C. \times Sw) \qquad 0.26$$

Vertical tail

$$Vv = (Lv \times Sv) / (b \times Sw) \qquad 0.02$$

Weight empty	420 lb.
Weight max. takeoff	840 lb.
Max. wing loading	6.38 lb. / sq. ft.
Max. power loading	2.1 lb. / thrust lb.

Ratio of Fans liftoff thrust to gross weight

$$L/W = 1200 \text{ lb. T.} / 840 \text{ lb. G. W.} \qquad 1.43$$

PERFORMANCE

Max. Level speed at SL	130 mph
Service ceiling	15 000 ft
Range	200 mile

NOTE:

SHEET NUMBERING SYSTEM FOR DRAWINGS

Airframe drawings	sheets 3 through 16
Propulsion system drawings	sheets 17 through 26

ADVANCED ENGINEERING PROJECTS

Name of project: MECHANICAL FLYING HORSE

Design Engineer: Walter F. Laredo   Date: March 15, 2002

Drawing Number: 1100   Sheet 2 of 26

PLATE 75

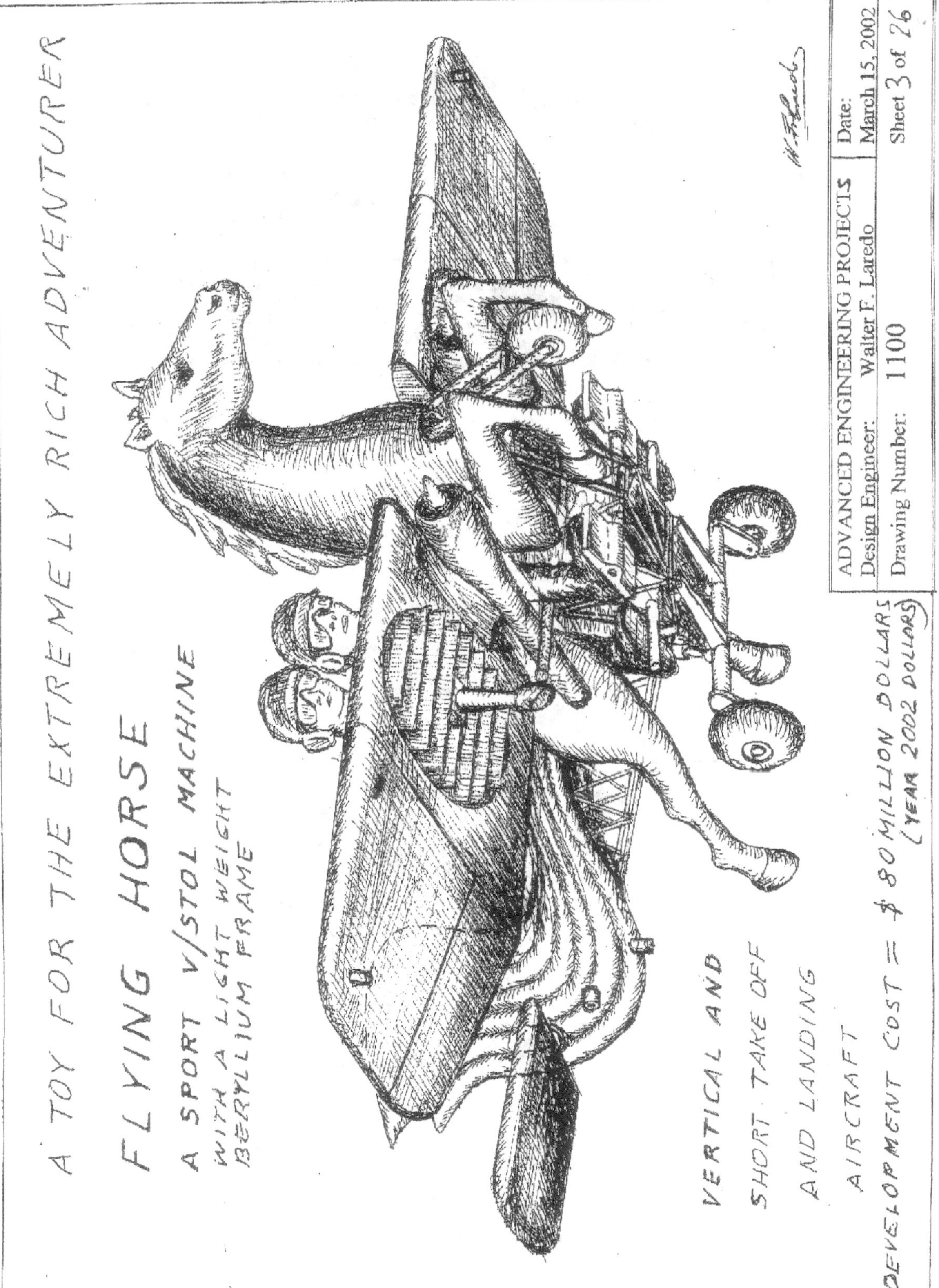

A TOY FOR THE EXTREMELY RICH ADVENTURER

FLYING HORSE

A SPORT V/STOL MACHINE
WITH A LIGHT WEIGHT
BERYLLIUM FRAME

VERTICAL AND

SHORT TAKE OFF

AND LANDING

AIRCRAFT

DEVELOPMENT COST = $ 80 MILLION DOLLARS
(YEAR 2002 DOLLARS)

ADVANCED ENGINEERING PROJECTS	Date:
Design Engineer: Walter F. Laredo	March 15, 2002
Drawing Number: 1100	Sheet 3 of 26

PLATE 76

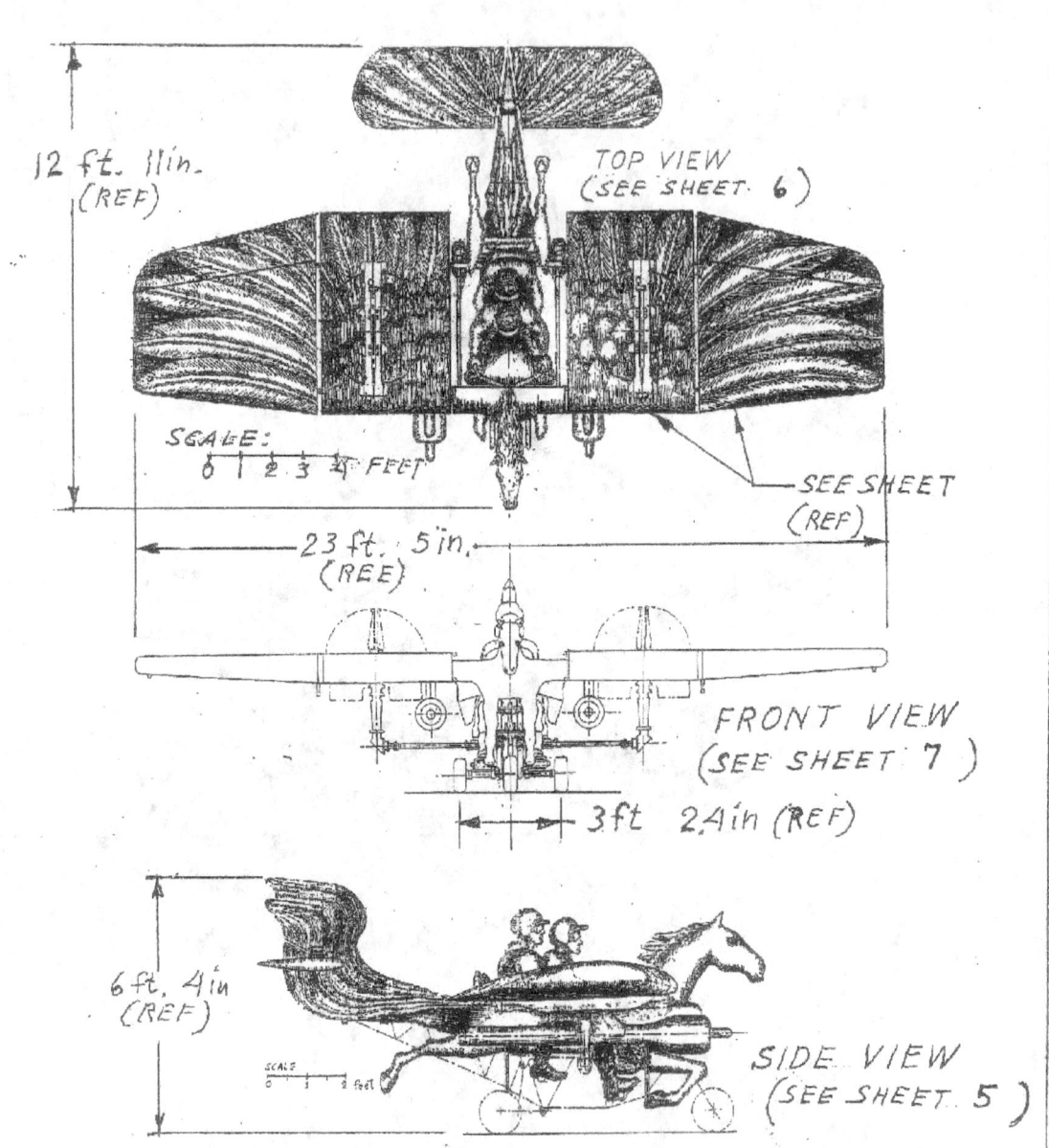

12 ft. 11in. (REF)

TOP VIEW (SEE SHEET 6)

SCALE: 0 1 2 3 4 FEET

SEE SHEET (REF)

23 ft. 5 in. (REF)

FRONT VIEW (SEE SHEET 7)

3ft 2.4in (REF)

6 ft. 4in (REF)

SCALE 0 1 2 feet

SIDE VIEW (SEE SHEET 5)

Name of project:	MECHANICAL FLYING HORSE	
ADVANCED ENGINEERING PROJECT		
Design Engineer: Walter F. Laredo		Date: March 15, 2002
Drawing Number: 1100		Sheet 4 of 26

PLATE 77

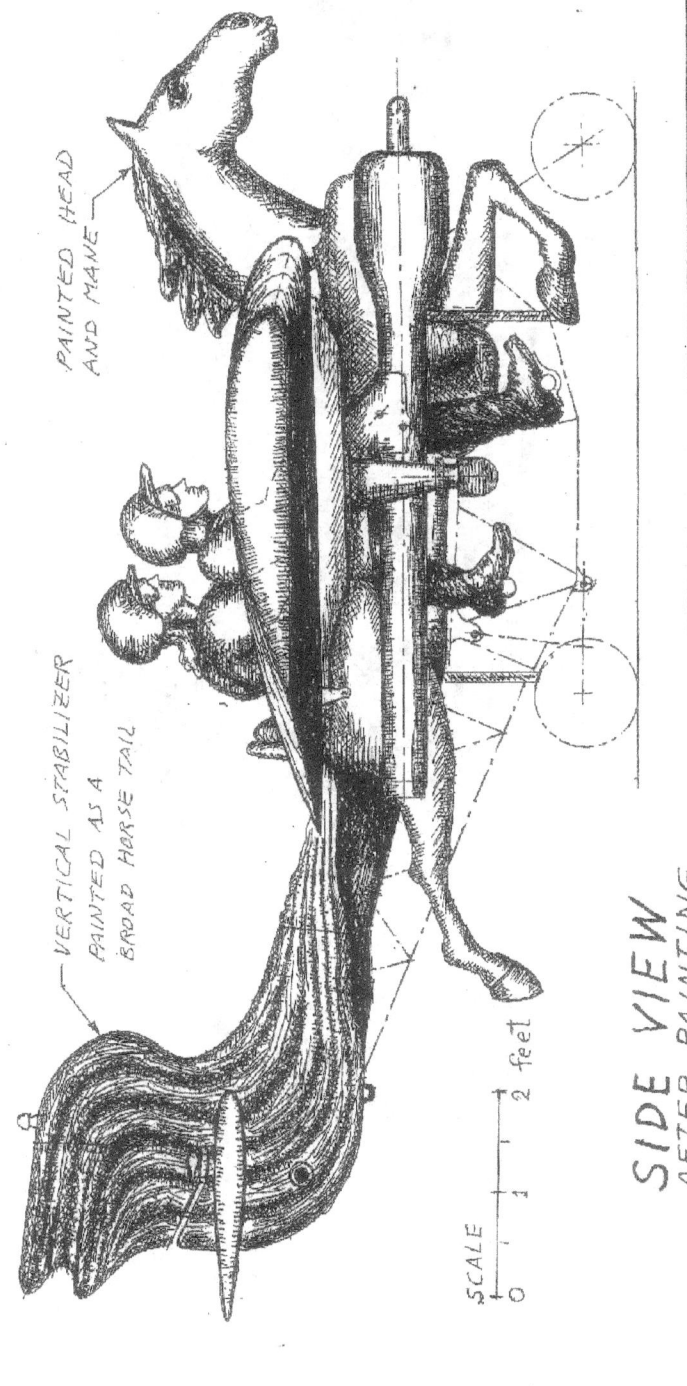

PAINTED HEAD
AND MANE

VERTICAL STABILIZER
PAINTED AS A
BROAD HORSE TAIL

SIDE VIEW
AFTER PAINTING

PAINTING OF VERTICAL TAIL IMITATES
LONG BUNDLES OF TAIL HAIR

SCALE

0   1   2 feet

Name of project:	MECHANICAL FLYING HORSE	
ADVANCED ENGINEERING PROJECT		
Design Engineer:	Walter F. Laredo	Date: March 15, 2002
Drawing Number:	1100	Sheet 5 of 26

PLATE 78

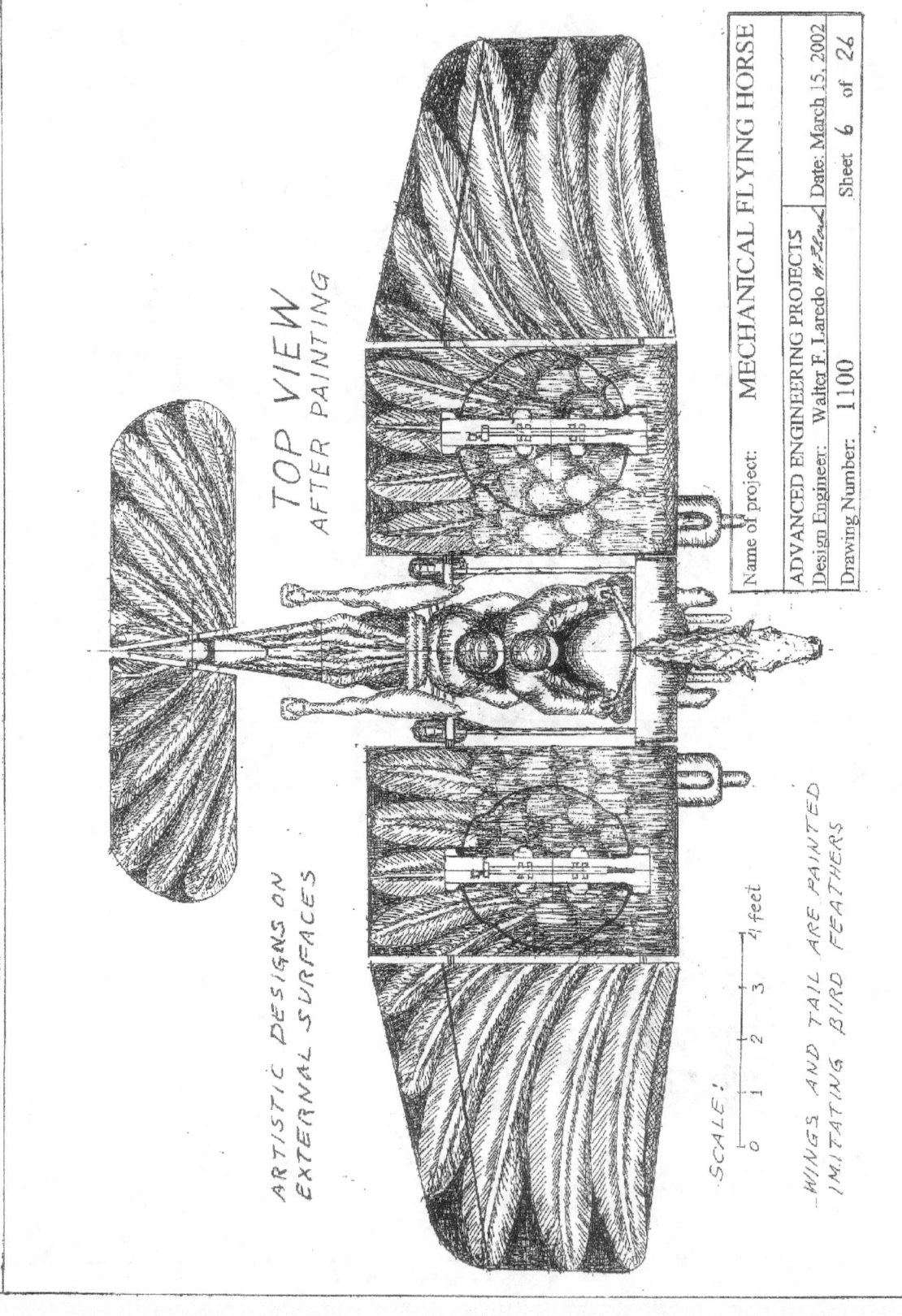

TOP VIEW
AFTER PAINTING

ARTISTIC DESIGNS ON
EXTERNAL SURFACES

SCALE:

0   1   2   3   4 feet

WINGS AND TAIL ARE PAINTED
IMITATING BIRD FEATHERS

Name of project: MECHANICAL FLYING HORSE

ADVANCED ENGINEERING PROJECTS
Design Engineer: Walter F. Laredo   Date: March 15, 2002

Drawing Number: 1100   Sheet   6   of   26

PLATE 79

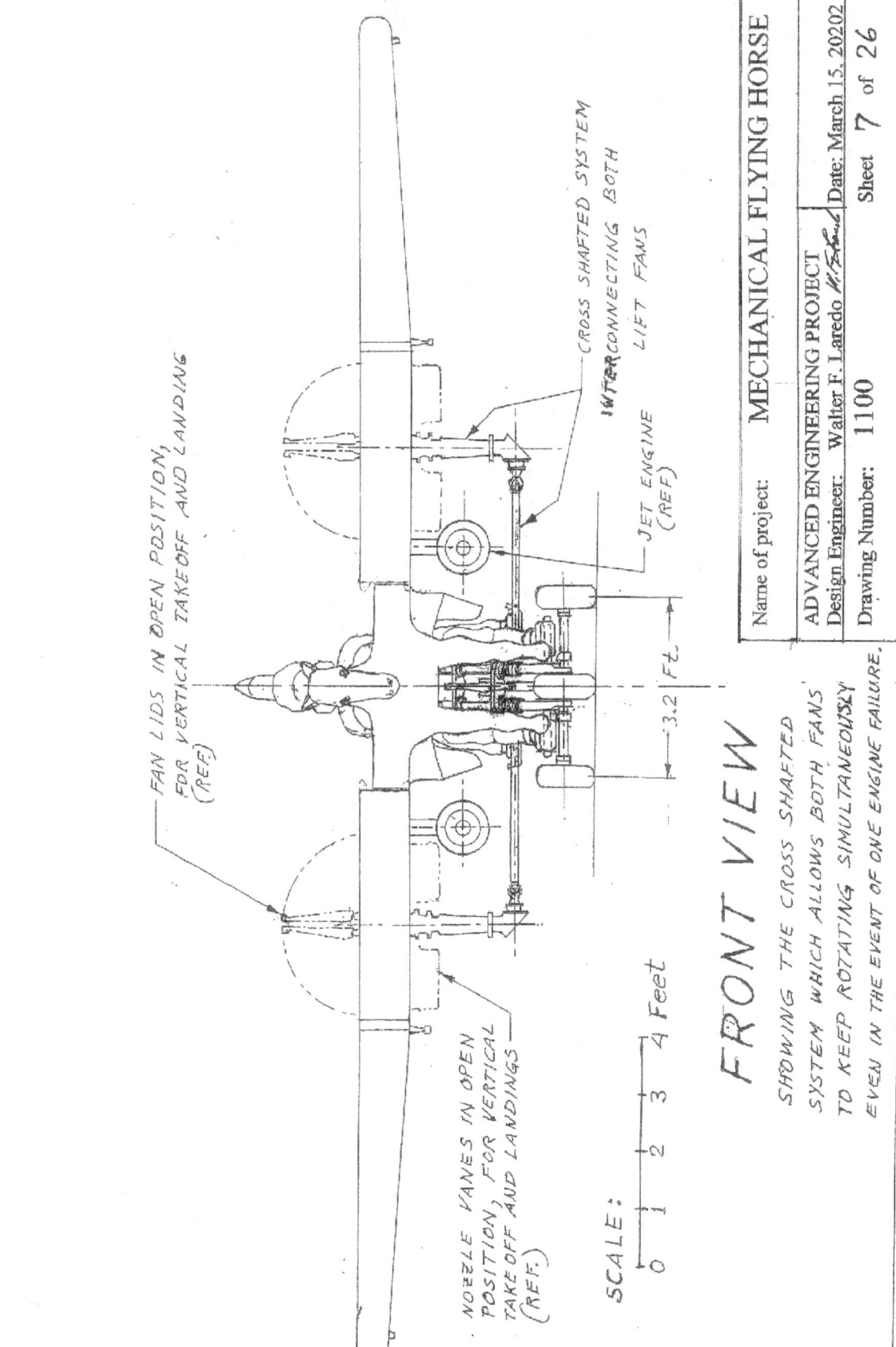

FRONT VIEW

SHOWING THE CROSS SHAFTED
SYSTEM WHICH ALLOWS BOTH FANS
TO KEEP ROTATING SIMULTANEOUSLY
EVEN IN THE EVENT OF ONE ENGINE FAILURE.

FAN LIDS IN OPEN POSITION,
FOR VERTICAL TAKEOFF AND LANDING
(REF.)

NOZZLE VANES IN OPEN
POSITION, FOR VERTICAL
TAKE OFF AND LANDINGS
(REF.)

CROSS SHAFTED SYSTEM

INTERCONNECTING BOTH
LIFT FANS

JET ENGINE
(REF.)

3.2 FT.

SCALE:
0 1 2 3 4 Feet

Name of project:	MECHANICAL FLYING HORSE
ADVANCED ENGINEERING PROJECT	
Design Engineer: Walter F. Laredo	Date: March 15, 20202
Drawing Number: 1100	Sheet 7 of 26

PLATE 80

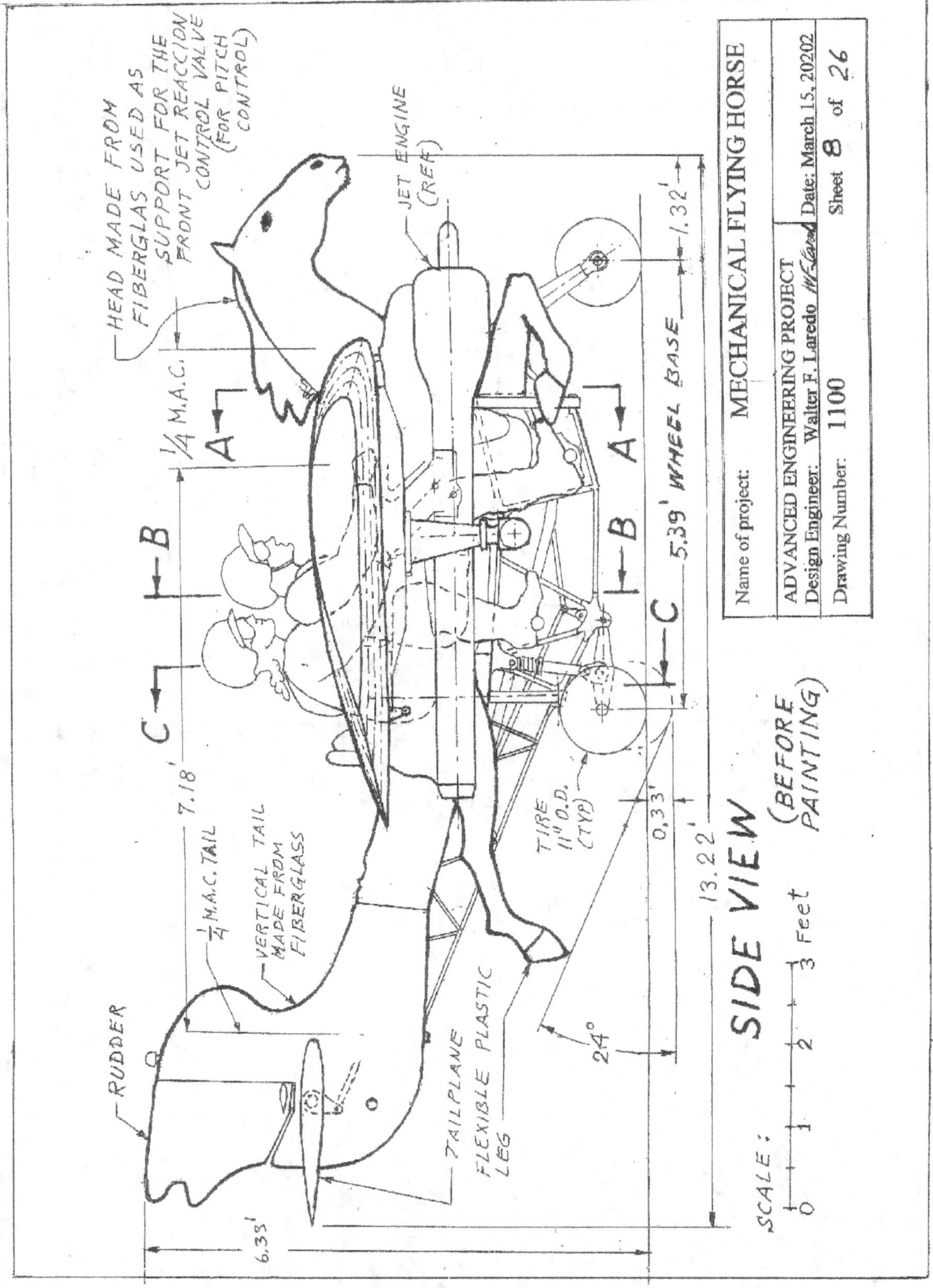

RUDDER

HEAD MADE FROM FIBERGLAS USED AS SUPPORT FOR THE FRONT JET REACCION CONTROL VALVE (FOR PITCH CONTROL)

JET ENGINE (REF.)

¼ M.A.C.

A

B

C

7.18'

¼ M.A.C. TAIL

VERTICAL TAIL MADE FROM FIBERGLASS

TAILPLANE

FLEXIBLE PLASTIC LEG

6.33'

TIRE 11" O.D. (TYP)

24°

13.22'

0.33'

1.32'

5.39' WHEEL BASE

A

B

C

SIDE VIEW (BEFORE PAINTING)

SCALE:

0 1 2 3 Feet

Name of project:	MECHANICAL FLYING HORSE	
ADVANCED ENGINEERING PROJECT		Date: March 15, 2020
Design Engineer: Walter F. Laredo W. Laredo		
Drawing Number: 1100		Sheet 8 of 26

PLATE 81

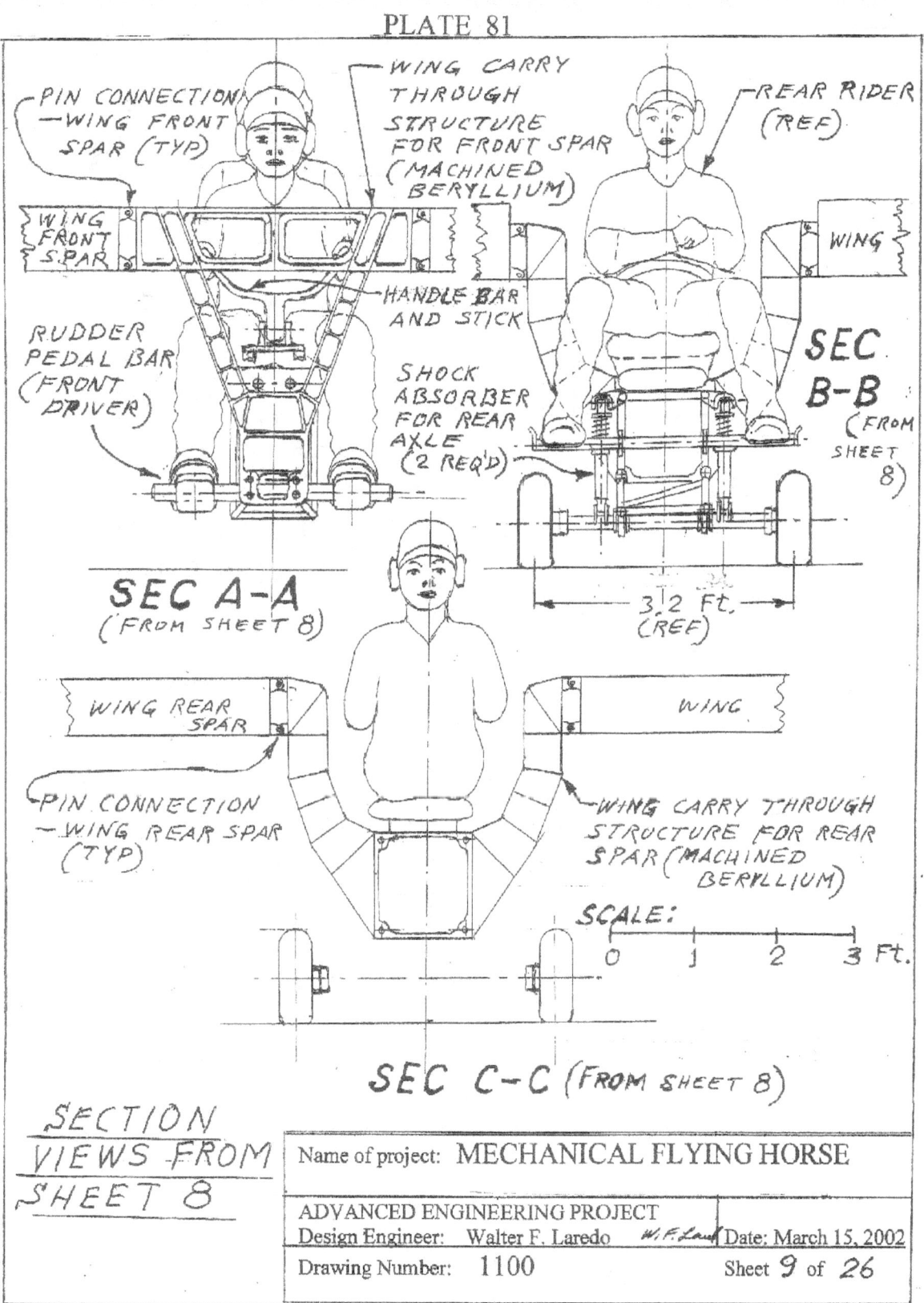

PIN CONNECTION
—WING FRONT
SPAR (TYP)

WING CARRY
THROUGH
STRUCTURE
FOR FRONT SPAR
(MACHINED
BERYLLIUM)

REAR RIDER
(REF)

WING
FRONT
SPAR

WING

RUDDER
PEDAL BAR
(FRONT
DRIVER)

HANDLE BAR
AND STICK

SHOCK
ABSORBER
FOR REAR
AXLE
(2 REQ'D)

SEC
B-B
(FROM
SHEET
8)

SEC A-A
(FROM SHEET 8)

3.2 Ft.
(REF)

WING REAR
SPAR

WING

PIN CONNECTION
—WING REAR SPAR
(TYP)

WING CARRY THROUGH
STRUCTURE FOR REAR
SPAR (MACHINED
BERYLLIUM)

SCALE:

0   1   2   3 Ft.

SEC C-C (FROM SHEET 8)

SECTION
VIEWS FROM
SHEET 8

Name of project: MECHANICAL FLYING HORSE

ADVANCED ENGINEERING PROJECT
Design Engineer: Walter F. Laredo   W.F.Laud   Date: March 15, 2002

Drawing Number: 1100   Sheet 9 of 26

PLATE 82

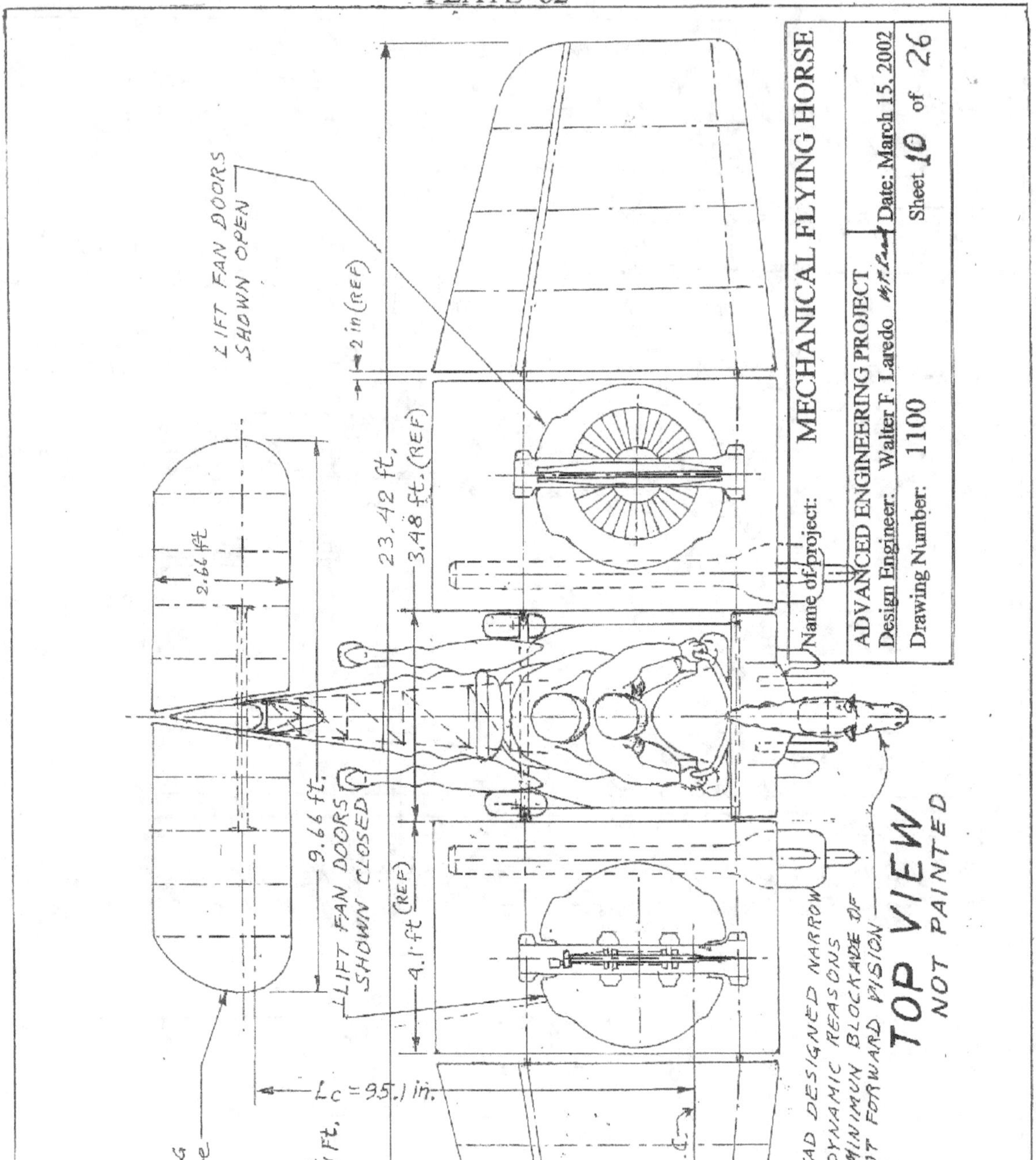

LIFT FAN DOORS SHOWN OPEN

2 in (REF)

23.42 ft.

3.48 ft. (REF)

2.66 ft

9.66 ft.

LIFT FAN DOORS SHOWN CLOSED

4.1 ft (REF)

L_C = 95.1 in.

.25 M.A.C.

ALL MOVING tail plane

SCALE:

0 1 2 3 4 ft.

HORSE HEAD DESIGNED NARROW FOR AERODYNAMIC REASONS AND FOR MINIMUN BLOCKADE OF PILOT FORWARD VISION

TOP VIEW
NOT PAINTED

MECHANICAL FLYING HORSE

Name of Project:

ADVANCED ENGINEERING PROJECT
Design Engineer: Walter F. Laredo   W.F.Laredo   Date: March 15, 2002

Drawing Number: 1100   Sheet 10 of 26

PLATE 83

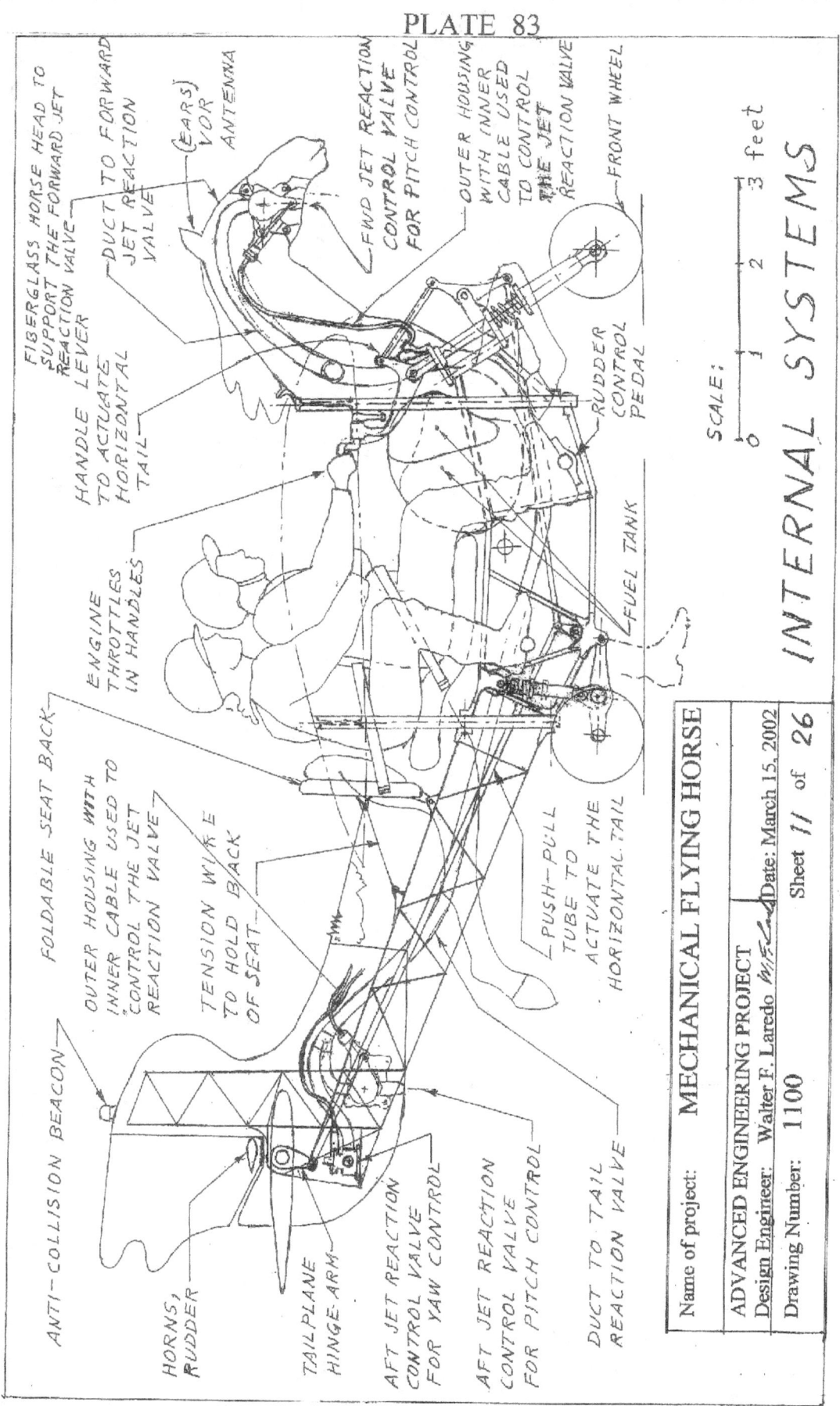

FIBERGLASS HORSE HEAD TO SUPPORT THE FORWARD JET REACTION VALVE

DUCT TO FORWARD JET REACTION VALVE

(EARS) VOR ANTENNA

FWD JET REACTION CONTROL VALVE FOR PITCH CONTROL

OUTER HOUSING WITH INNER CABLE USED TO CONTROL THE JET REACTION VALVE

FRONT WHEEL

HANDLE LEVER TO ACTUATE HORIZONTAL TAIL

RUDDER CONTROL PEDAL

SCALE:

0   1   2   3 feet

ENGINE THROTTLES IN HANDLES

FUEL TANK

FOLDABLE SEAT BACK

OUTER HOUSING WITH INNER CABLE USED TO CONTROL THE JET REACTION VALVE

TENSION WIRE TO HOLD BACK OF SEAT

PUSH-PULL TUBE TO ACTUATE THE HORIZONTAL TAIL

ANTI-COLLISION BEACON

HORNS, RUDDER

TAILPLANE HINGE ARM

AFT JET REACTION CONTROL VALVE FOR YAW CONTROL

AFT JET REACTION CONTROL VALVE FOR PITCH CONTROL

DUCT TO TAIL REACTION VALVE

INTERNAL SYSTEMS

Name of project:	MECHANICAL FLYING HORSE	
ADVANCED ENGINEERING PROJECT		
Design Engineer:	Walter F. Laredo W.F.L	Date: March 15, 2002
Drawing Number:	1100	Sheet 11 of 26

PLATE 84

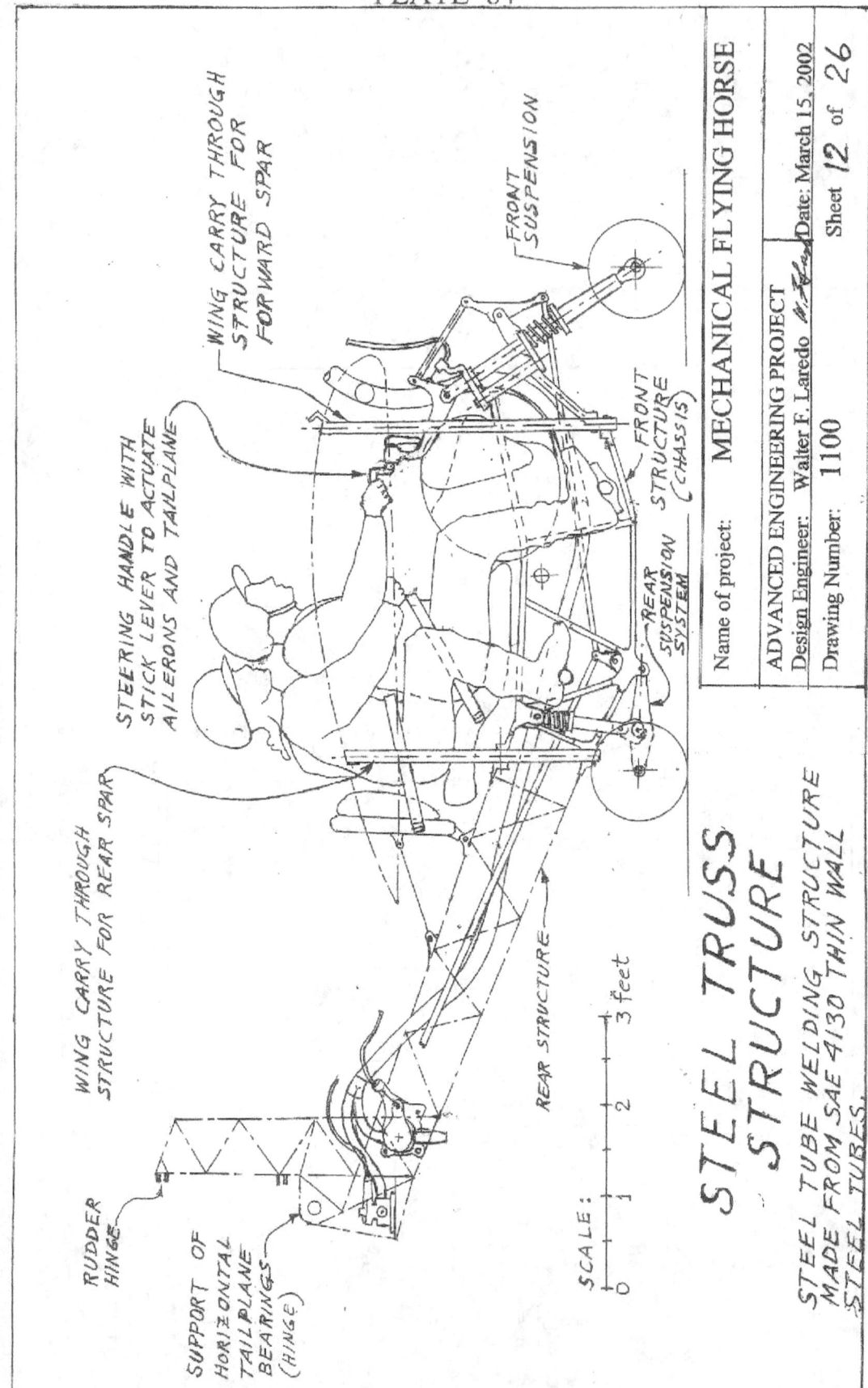

WING CARRY THROUGH STRUCTURE FOR FORWARD SPAR

STEERING HANDLE WITH STICK LEVER TO ACTUATE AILERONS AND TAILPLANE

FRONT SUSPENSION

FRONT STRUCTURE (CHASSIS)

REAR SUSPENSION SYSTEM

WING CARRY THROUGH STRUCTURE FOR REAR SPAR

RUDDER HINGE

SUPPORT OF HORIZONTAL TAILPLANE BEARINGS (HINGE)

REAR STRUCTURE

SCALE:
0   1   2   3 feet

STEEL TRUSS STRUCTURE

STEEL TUBE WELDING STRUCTURE MADE FROM SAE 4130 THIN WALL STEEL TUBES.

Name of project:	MECHANICAL FLYING HORSE
ADVANCED ENGINEERING PROJECT	
Design Engineer: Walter F. Laredo	Date: March 15, 2002
Drawing Number: 1100	Sheet 12 of 26

PLATE 85

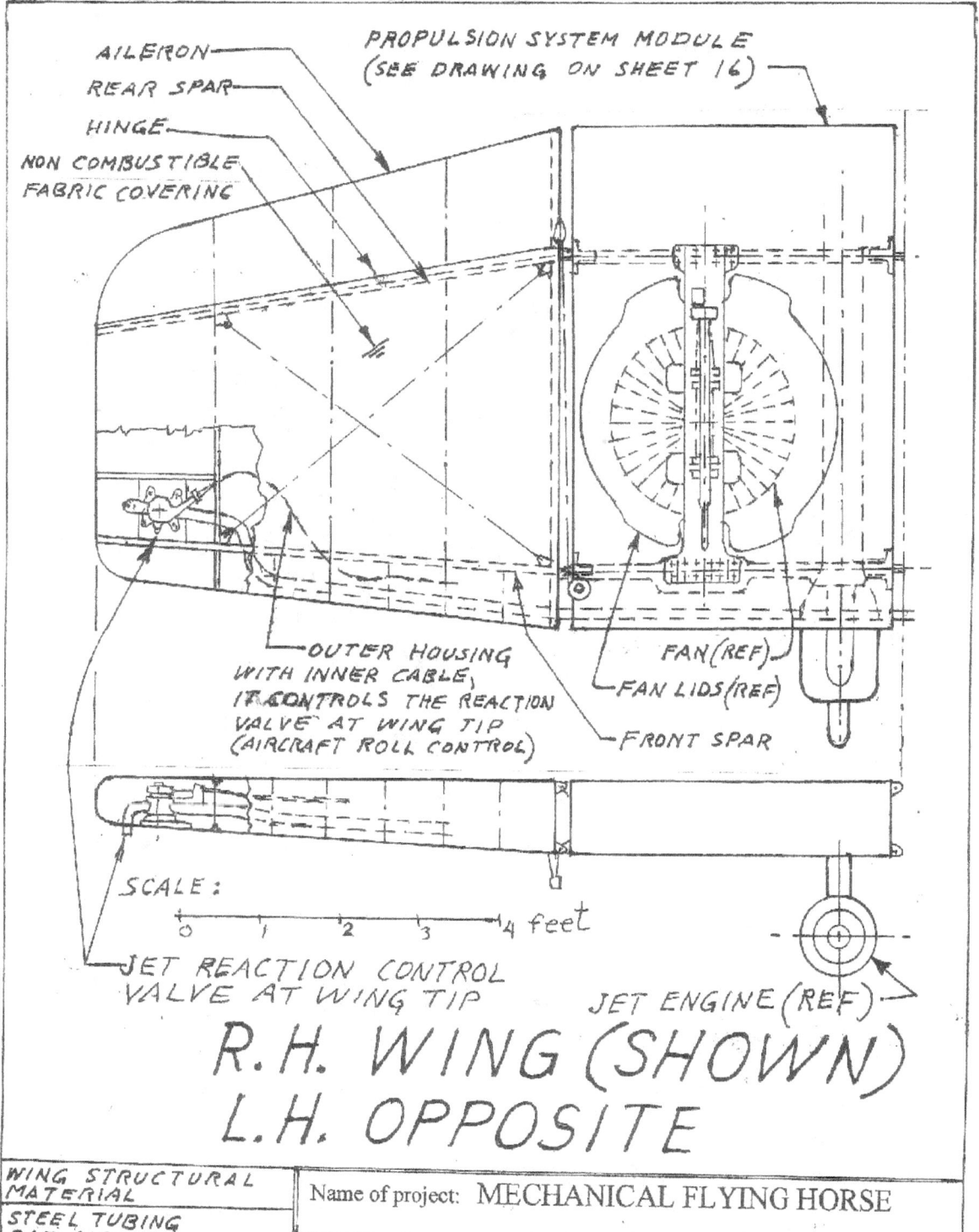

AILERON
REAR SPAR
HINGE
NON COMBUSTIBLE
FABRIC COVERING

PROPULSION SYSTEM MODULE
(SEE DRAWING ON SHEET 16)

OUTER HOUSING
WITH INNER CABLE,
IT CONTROLS THE REACTION
VALVE AT WING TIP
(AIRCRAFT ROLL CONTROL)

FAN (REF)
FAN LIDS (REF)
FRONT SPAR

SCALE:

0   1   2   3   4 feet

JET REACTION CONTROL
VALVE AT WING TIP

JET ENGINE (REF)

R.H. WING (SHOWN)
L.H. OPPOSITE

WING STRUCTURAL MATERIAL	Name of project: MECHANICAL FLYING HORSE	
STEEL TUBING SAE 4130	ADVANCED ENGINEERING PROJECT	
SPARS: 3/4 O.D. × .040" WALL	Design Engineer: Walter F. Laredo *W.F. Laredo*	Date: March 15, 2002
RIBS: 1/4 O.D. × .010" WALL	Drawing Number: 1100	Sheet 13 of 26

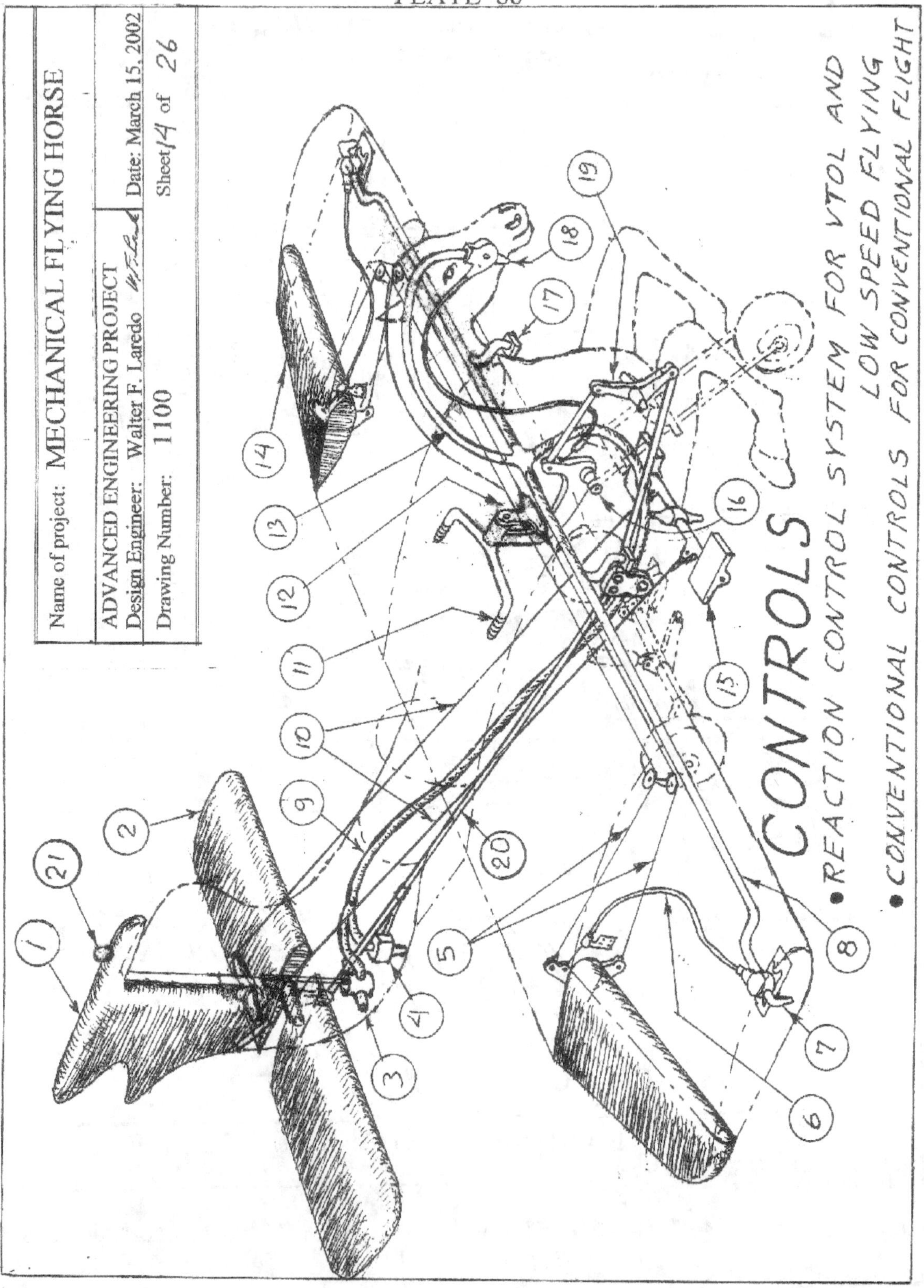

PLATE 86

CONTROLS

• REACTION CONTROL SYSTEM FOR VTOL AND LOW SPEED FLYING
• CONVENTIONAL CONTROLS FOR CONVENTIONAL FLIGHT

Name of project:	MECHANICAL FLYING HORSE	
ADVANCED ENGINEERING PROJECT		
Design Engineer:	Walter F. Laredo	Date: March 15, 2002
Drawing Number:	1100	Sheet 14 of 26

PLATE 87

(Continuation from CONTROLS, sheet 14)

1   Rudder
2   Horizontal tail
3   Yaw-control reaction jet
4   Pitch-control reaction jet
5   Cables to actuate ailerons
6   Outer housing with inner cable system controls wing tip reaction
       control valve. Inner cables are actuated by aileron movements
7   Roll-control reaction jet (wing tip)
8   Lateral duct, supplies compressed air for roll-control reaction jets
9   Duct to tail reaction control jets
10   Steel cables to actuate rudder
11   Pilot's handlebar, it steers front wheel, also actuates ailerons and tail
       plane, the throttles controls are located at handles
12   Crank, pulls ailerons control cables
13   Duct to forward control reaction jet
14   Aileron
15   Pedals, controls rudder
16   Pin, hold handlebar
17   Connector with one way valve, supplies compressed air from engine
       air bleed hole to reaction jets control system
18   Fwd. Pitch-control reaction jet
19   Rocker connected to pilot's handlebar lever, actuates horizontal tail
20   Push-pull tube, actuates horizontal tail
21   Tail strobe light.

Name of project: MECHANICAL FLYING HORSE

ADVANCED ENGINEERING PROJECTS
Design Engineer: Walter F. Laredo

Date: March 15, 2002

Drawing Number: 1100

Sheet 15 of 26

PLATE 88

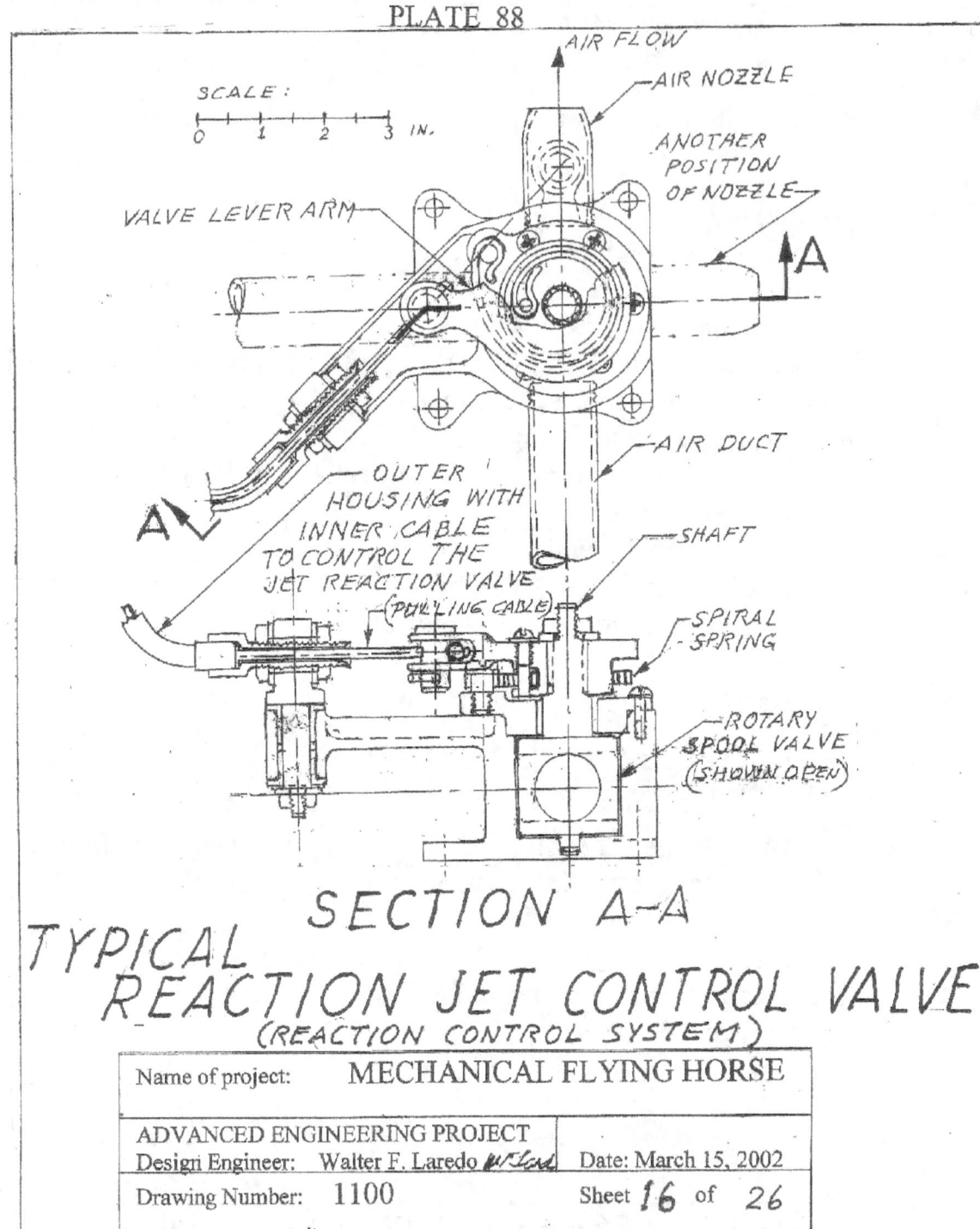

AIR FLOW

AIR NOZZLE

ANOTHER POSITION OF NOZZLE

SCALE:
0 1 2 3 IN.

VALVE LEVER ARM

A

AIR DUCT

SHAFT

OUTER HOUSING WITH INNER CABLE TO CONTROL THE JET REACTION VALVE (PULLING CABLE)

A

SPIRAL SPRING

ROTARY SPOOL VALVE (SHOWN OPEN)

SECTION A-A

TYPICAL REACTION JET CONTROL VALVE
(REACTION CONTROL SYSTEM)

Name of project:	MECHANICAL FLYING HORSE	
ADVANCED ENGINEERING PROJECT		
Design Engineer: Walter F. Laredo	Date: March 15, 2002	
Drawing Number: 1100	Sheet 16 of 26	

PLATE 89

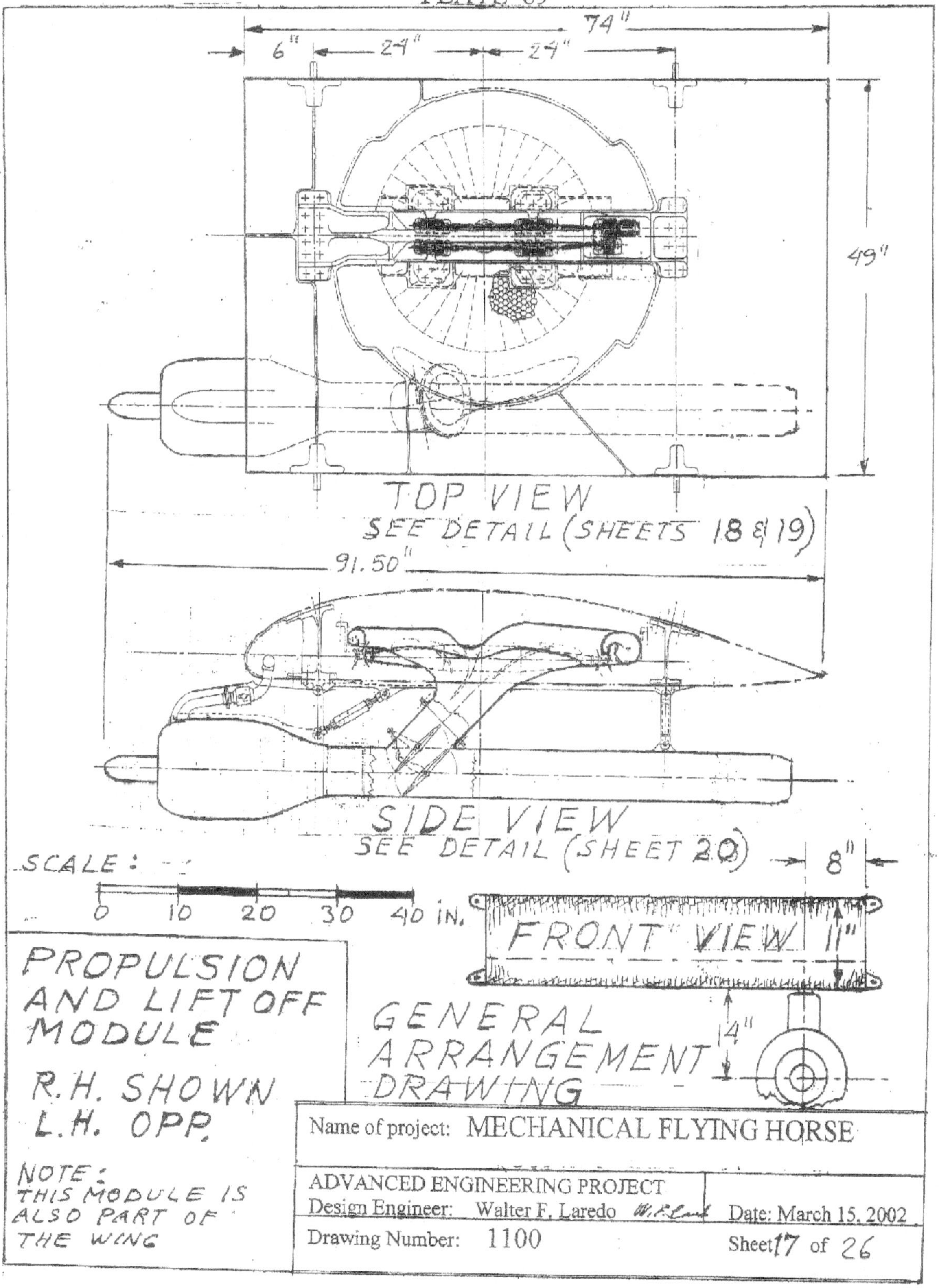

74"

6" 24" 24"

49"

TOP VIEW
SEE DETAIL (SHEETS 18 & 19)

91.50"

SIDE VIEW
SEE DETAIL (SHEET 20)

8"

SCALE:

0   10   20   30   40 IN.

FRONT VIEW   1"

PROPULSION
AND LIFT OFF
MODULE

R.H. SHOWN
L.H. OPP,

GENERAL
ARRANGEMENT
DRAWING

4"

NOTE:
THIS MODULE IS
ALSO PART OF
THE WING

Name of project:	MECHANICAL FLYING HORSE	
ADVANCED ENGINEERING PROJECT		
Design Engineer: Walter F. Laredo W. F. Laredo		Date: March 15, 2002
Drawing Number: 1100		Sheet 17 of 26

PLATE 90

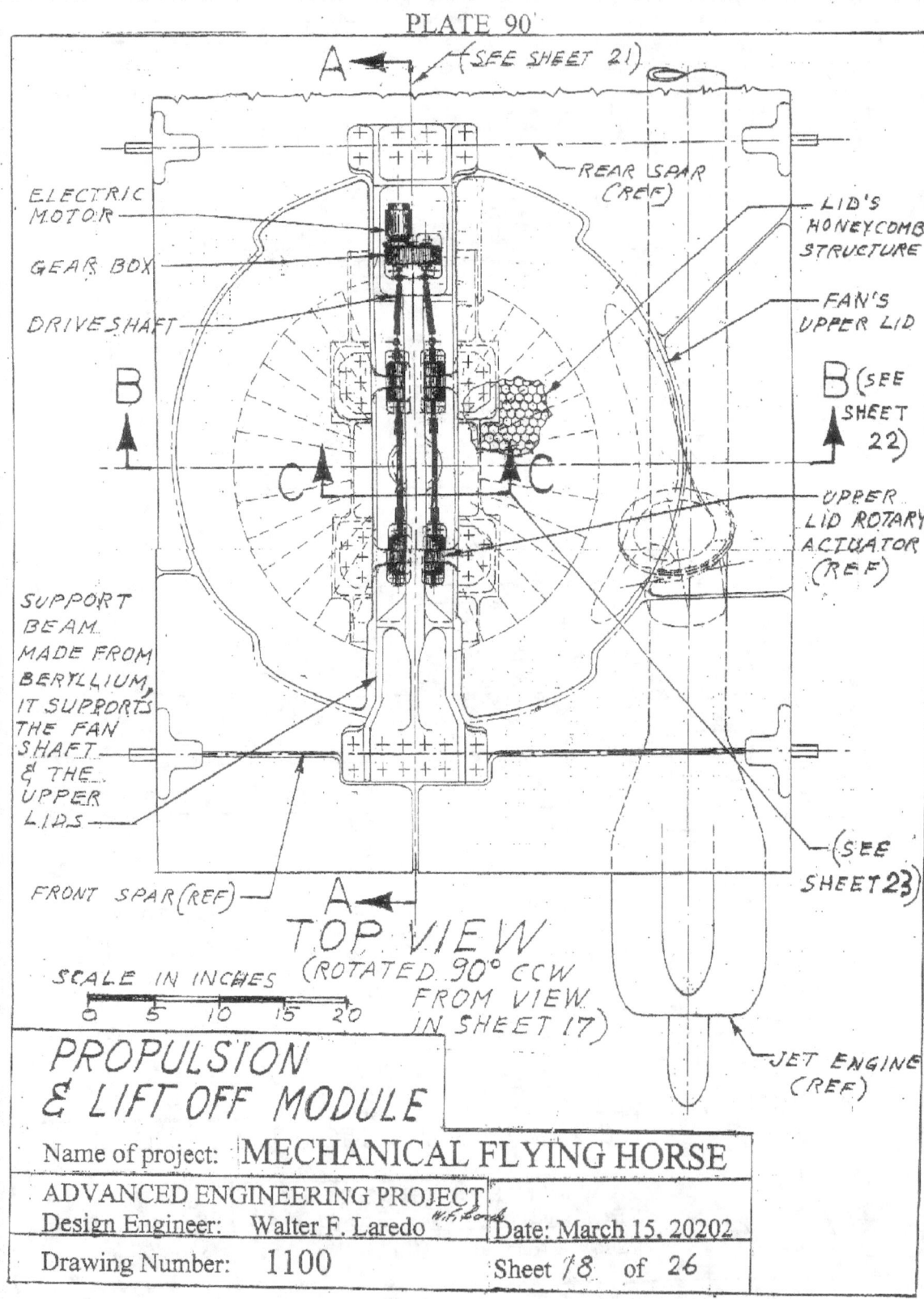

A ← (SEE SHEET 21)

ELECTRIC MOTOR

GEAR BOX

DRIVE SHAFT

B

C

REAR SPAR (REF)

LID'S HONEYCOMB STRUCTURE

FAN'S UPPER LID

B (SEE SHEET 22)

C

UPPER LID ROTARY ACTUATOR (REF)

SUPPORT BEAM MADE FROM BERYLLIUM, IT SUPPORTS THE FAN SHAFT & THE UPPER LIDS

FRONT SPAR (REF)

A ←

TOP VIEW
(ROTATED 90° CCW FROM VIEW IN SHEET 17)

(SEE SHEET 23)

JET ENGINE (REF)

SCALE IN INCHES
0   5   10   15   20

PROPULSION & LIFT OFF MODULE

Name of project: MECHANICAL FLYING HORSE

ADVANCED ENGINEERING PROJECT
Design Engineer: Walter F. Laredo   Date: March 15, 20202

Drawing Number: 1100   Sheet 18 of 26

PLATE 91

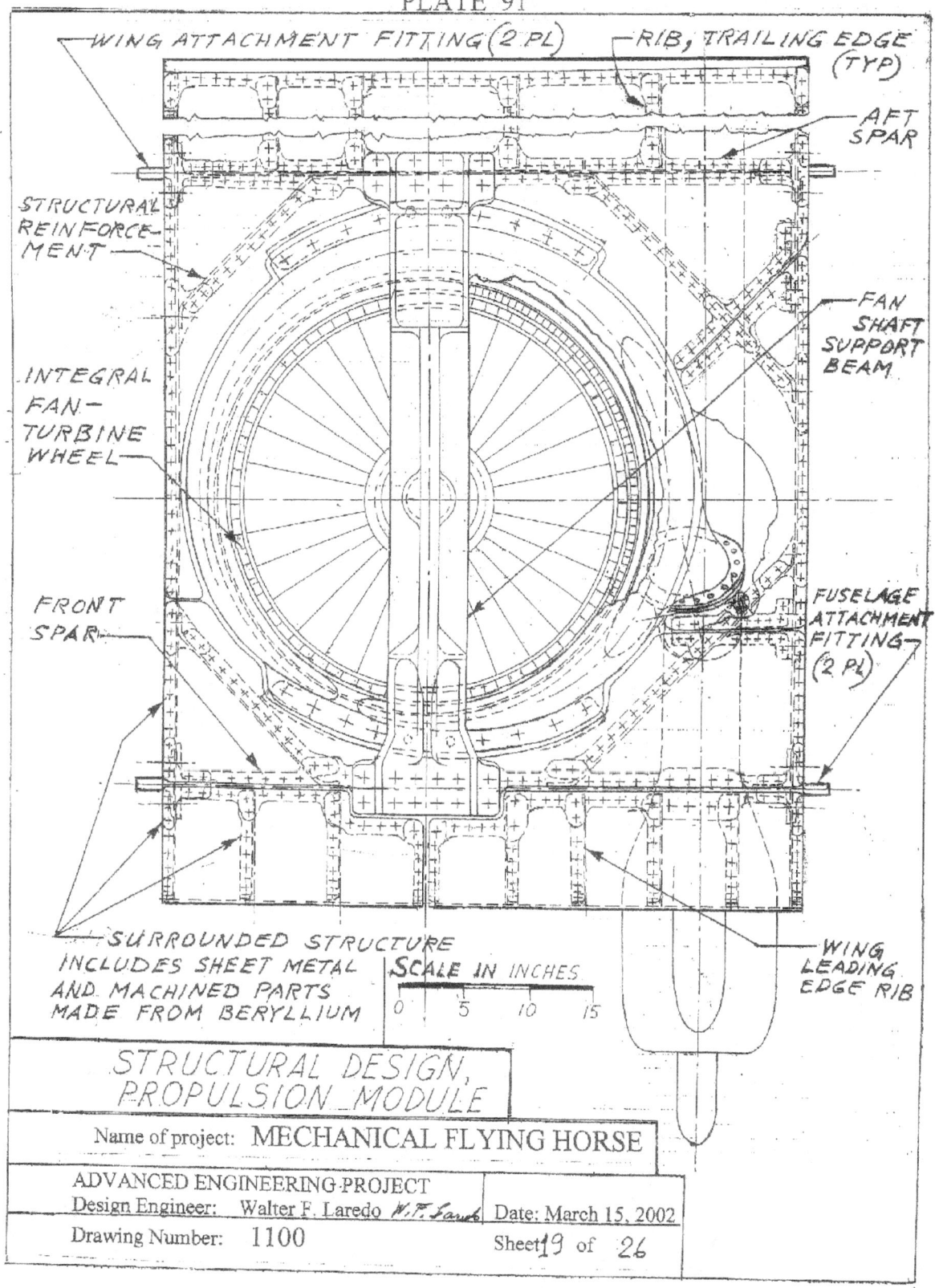

WING ATTACHMENT FITTING (2 PL)

RIB, TRAILING EDGE (TYP)

AFT SPAR

STRUCTURAL REINFORCE-MENT

FAN SHAFT SUPPORT BEAM

INTEGRAL FAN-TURBINE WHEEL

FRONT SPAR

FUSELAGE ATTACHMENT FITTING (2 PL)

SURROUNDED STRUCTURE INCLUDES SHEET METAL AND MACHINED PARTS MADE FROM BERYLLIUM

SCALE IN INCHES

0   5   10   15

WING LEADING EDGE RIB

STRUCTURAL DESIGN, PROPULSION MODULE

Name of project: MECHANICAL FLYING HORSE

ADVANCED ENGINEERING PROJECT

Design Engineer: Walter F. Laredo   *W.F. Laredo*   Date: March 15, 2002

Drawing Number: 1100   Sheet 19 of 26

PLATE 92

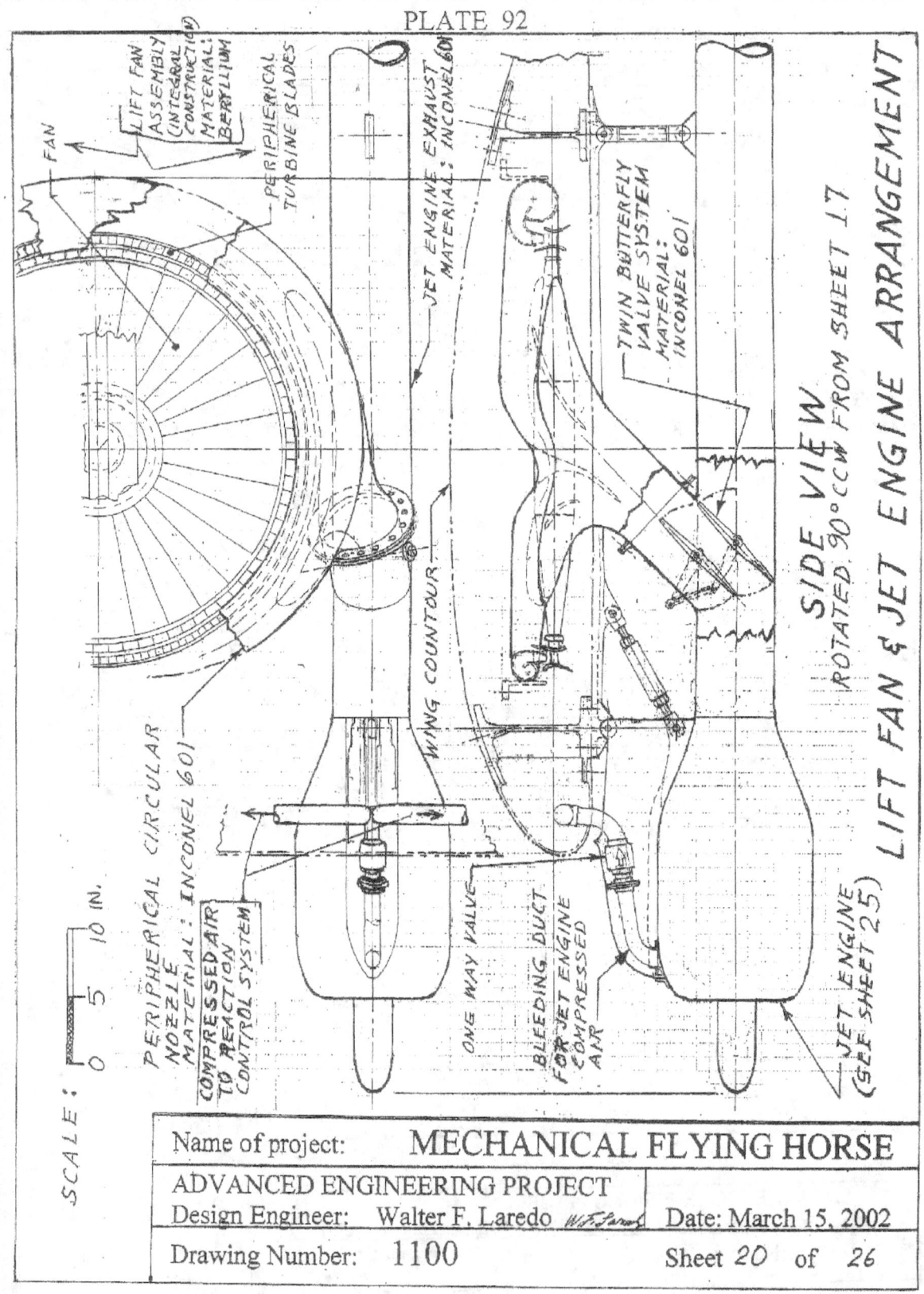

SIDE VIEW
ROTATED 90° CCW FROM SHEET 17

LIFT FAN & JET ENGINE ARRANGEMENT

FAN

LIFT FAN ASSEMBLY (INTEGRAL CONSTRUCTION) MATERIAL: BERYLLIUM

PERIPHERICAL TURBINE BLADES

JET ENGINE EXHAUST MATERIAL: INCONEL 601

TWIN BUTTERFLY VALVE SYSTEM MATERIAL: INCONEL 601

WING COUNTOUR

PERIPHERICAL CIRCULAR NOZZLE MATERIAL: INCONEL 601

COMPRESSED AIR TO REACTION CONTROL SYSTEM

ONE WAY VALVE

BLEEDING DUCT FOR JET ENGINE COMPRESSED AIR

JET ENGINE (SEE SHEET 25)

SCALE: 0   5   10 IN.

Name of project:	MECHANICAL FLYING HORSE	
ADVANCED ENGINEERING PROJECT		
Design Engineer: Walter F. Laredo		Date: March 15, 2002
Drawing Number: 1100		Sheet 20 of 26

PLATE 93

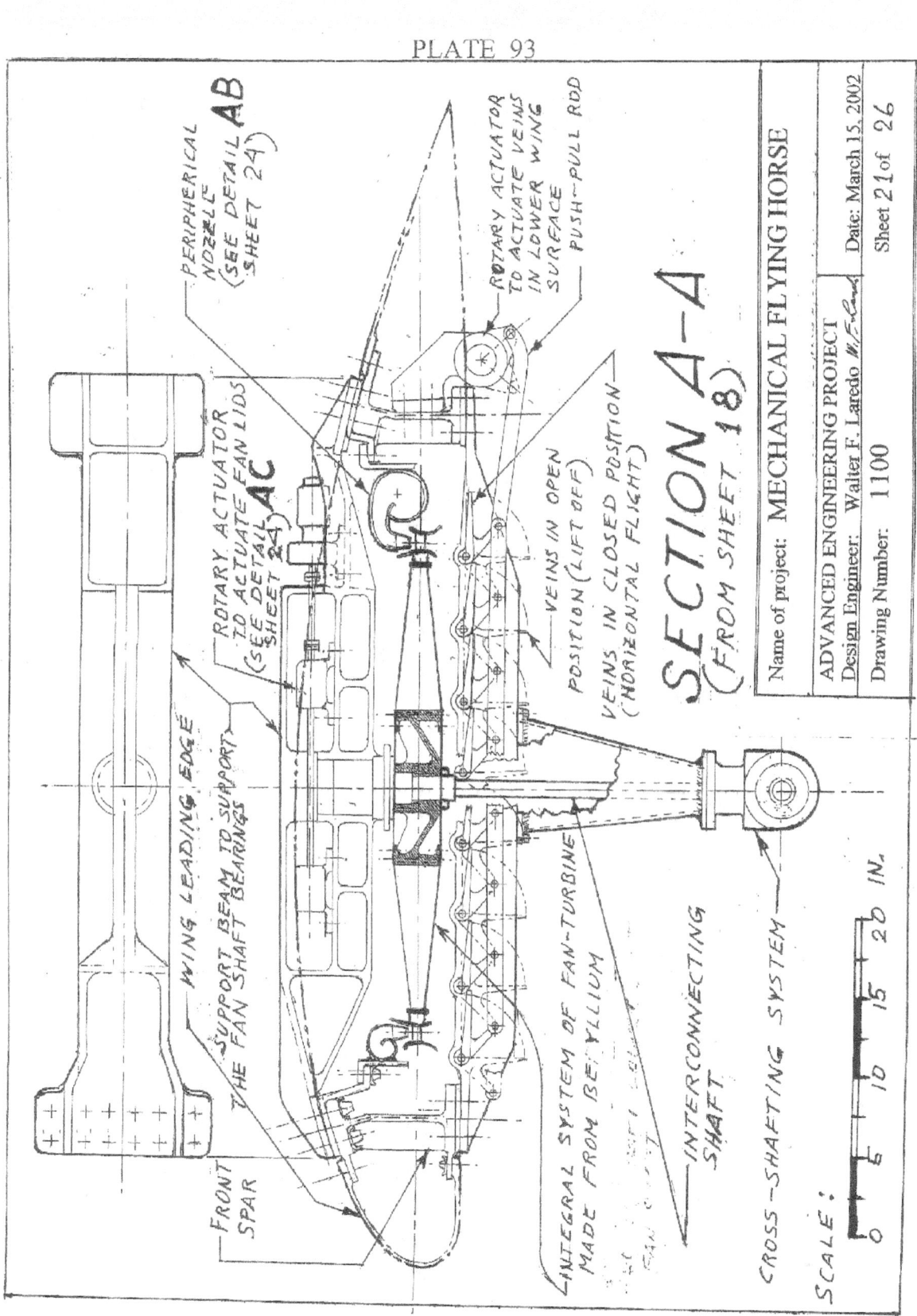

PERIPHERICAL NOZZLE (SEE DETAIL **AB** SHEET 24)

ROTARY ACTUATOR TO ACTUATE VEINS IN LOWER WING SURFACE

PUSH-PULL ROD

ROTARY ACTUATOR TO ACTUATE FAN LIDS (SEE DETAIL **AC** SHEET 24)

WING LEADING EDGE

SUPPORT BEAM TO SUPPORT THE FAN SHAFT BEARINGS

VEINS IN OPEN POSITION (LIFT OFF)

VEINS IN CLOSED POSITION (HORIZONTAL FLIGHT)

FRONT SPAR

INTEGRAL SYSTEM OF FAN-TURBINE MADE FROM BERYLLIUM

INTERCONNECTING SHAFT

CROSS-SHAFTING SYSTEM

SECTION A-A
(FROM SHEET 18)

SCALE:

0   5   10   15   20 IN.

Name of project: MECHANICAL FLYING HORSE

ADVANCED ENGINEERING PROJECT
Design Engineer: Walter F. Laredo W.F.Laredo   Date: March 15, 2002

Drawing Number: 1100   Sheet 21 of 26

PLATE 94

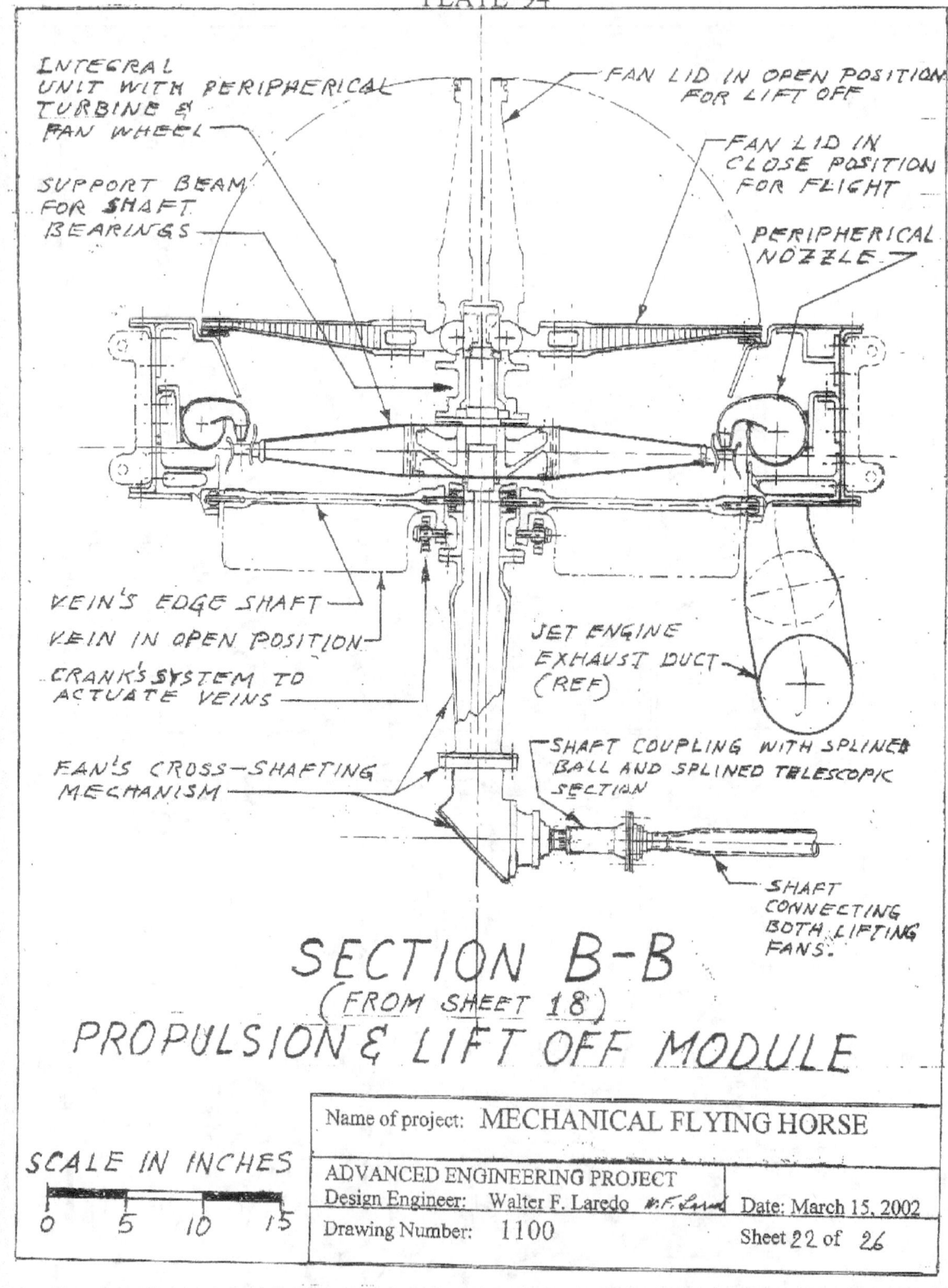

INTEGRAL UNIT WITH PERIPHERICAL TURBINE & FAN WHEEL

SUPPORT BEAM FOR SHAFT BEARINGS

FAN LID IN OPEN POSITION FOR LIFT OFF

FAN LID IN CLOSE POSITION FOR FLIGHT

PERIPHERICAL NOZZLE

VEIN'S EDGE SHAFT

VEIN IN OPEN POSITION

CRANK'S SYSTEM TO ACTUATE VEINS

JET ENGINE EXHAUST DUCT (REF)

FAN'S CROSS-SHAFTING MECHANISM

SHAFT COUPLING WITH SPLINED BALL AND SPLINED TELESCOPIC SECTION

SHAFT CONNECTING BOTH LIFTING FANS.

SECTION B-B
(FROM SHEET 18)
PROPULSION & LIFT OFF MODULE

SCALE IN INCHES

0   5   10   15

Name of project:	MECHANICAL FLYING HORSE	
ADVANCED ENGINEERING PROJECT		
Design Engineer: Walter F. Laredo *W.F. Laredo*		Date: March 15, 2002
Drawing Number: 1100		Sheet 22 of 26

PLATE 95

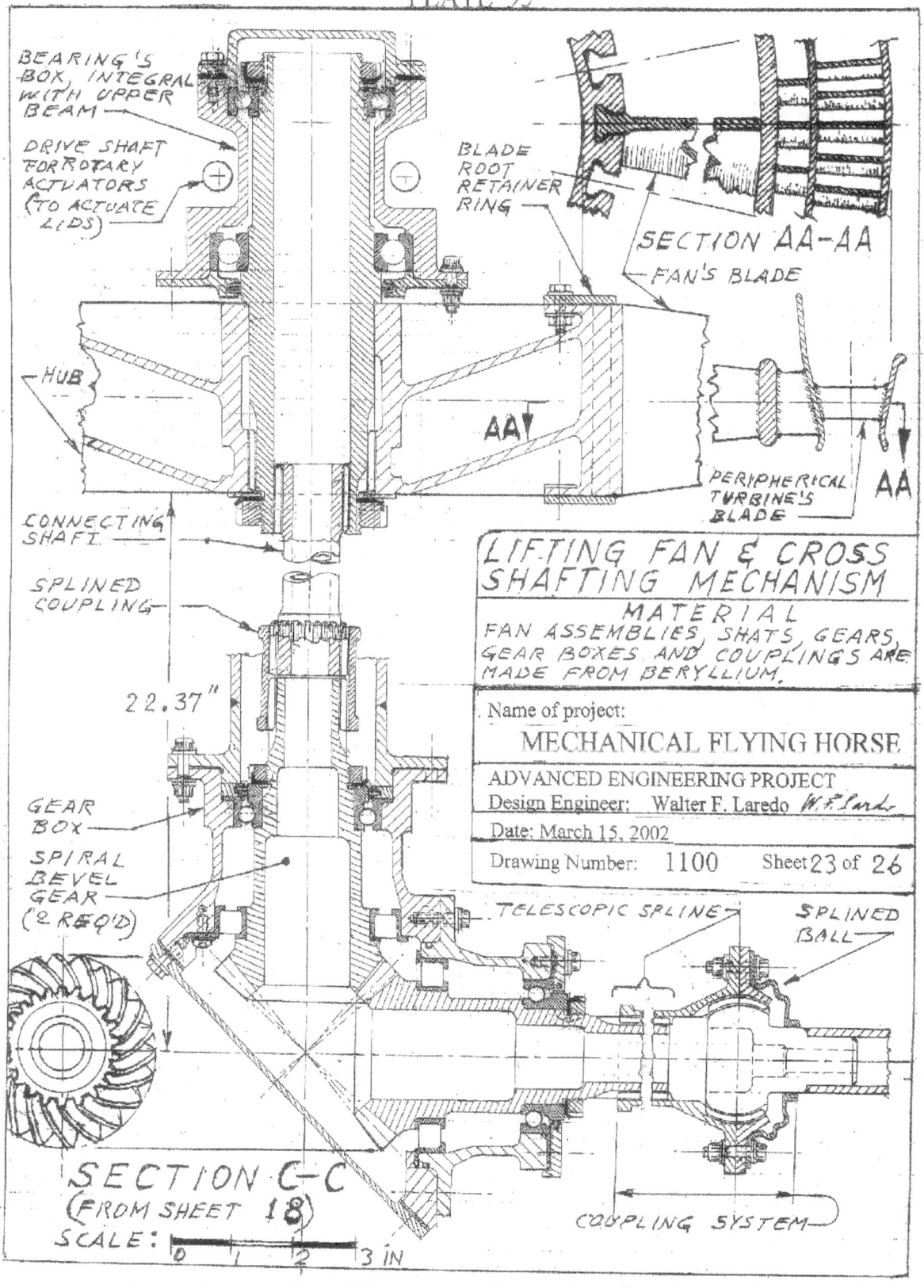

BEARING'S BOX, INTEGRAL WITH UPPER BEAM

DRIVE SHAFT FOR ROTARY ACTUATORS (TO ACTUATE LIDS)

HUB

CONNECTING SHAFT

SPLINED COUPLING

22.37"

GEAR BOX

SPIRAL BEVEL GEAR ('2 REQ'D)

BLADE ROOT RETAINER RING

SECTION AA-AA
—FAN'S BLADE

PERIPHERICAL TURBINE'S BLADE

AA

LIFTING FAN & CROSS SHAFTING MECHANISM

MATERIAL
FAN ASSEMBLIES, SHATS, GEARS, GEAR BOXES AND COUPLINGS ARE MADE FROM BERYLLIUM.

Name of project:
MECHANICAL FLYING HORSE

ADVANCED ENGINEERING PROJECT
Design Engineer: Walter F. Laredo

Date: March 15, 2002

Drawing Number: 1100   Sheet 23 of 26

TELESCOPIC SPLINE

SPLINED BALL

SECTION C-C
(FROM SHEET 18)
SCALE: 0 1 2 3 IN

COUPLING SYSTEM

PLATE 96

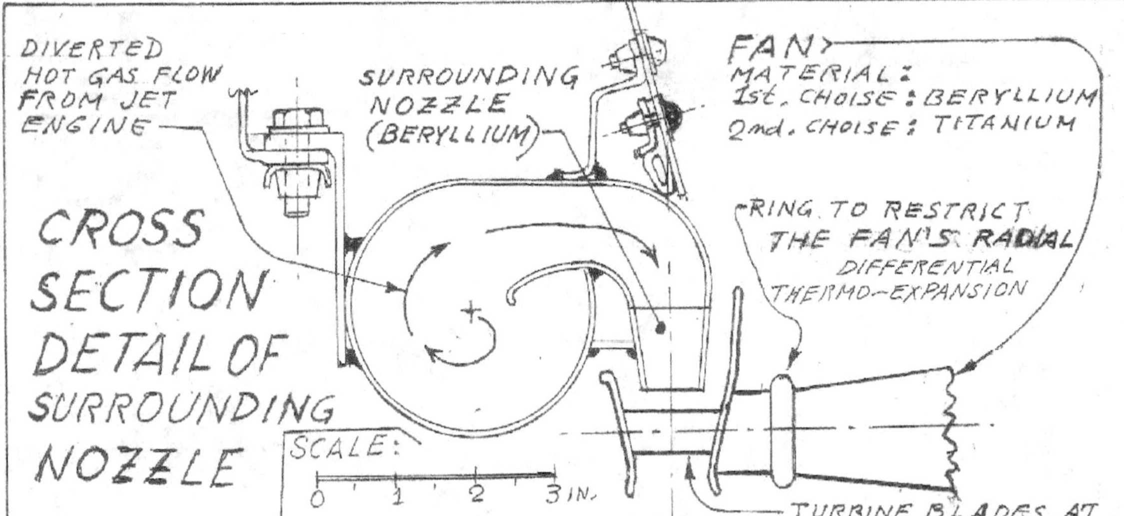

DIVERTED
HOT GAS FLOW
FROM JET
ENGINE

SURROUNDING
NOZZLE
(BERYLLIUM)

FAN>
MATERIAL:
1st. CHOISE: BERYLLIUM
2nd. CHOISE: TITANIUM

RING TO RESTRICT
THE FAN'S RADIAL
DIFFERENTIAL
THERMO-EXPANSION

CROSS
SECTION
DETAIL OF
SURROUNDING
NOZZLE

SCALE:
0 1 2 3 IN.

DETAIL AB
(FROM SHEET 21)

TURBINE BLADES AT
THE FAN'SYSTEM
PERIPHERY

FAN UPPER LID
ROTARY ACTUATOR
WITH EPICYCLIC
PLANETARY
GEARS (4 REQ'D)

FAN
UPPER LID
(BERYLLIUM)

UNIVERSAL JOIN

DRIVE
SHAFT

SCALE
IN INCHES
0 ½ 1 1½

SUPRORT
BEAM
(REF)

SHAFT
COUPLING

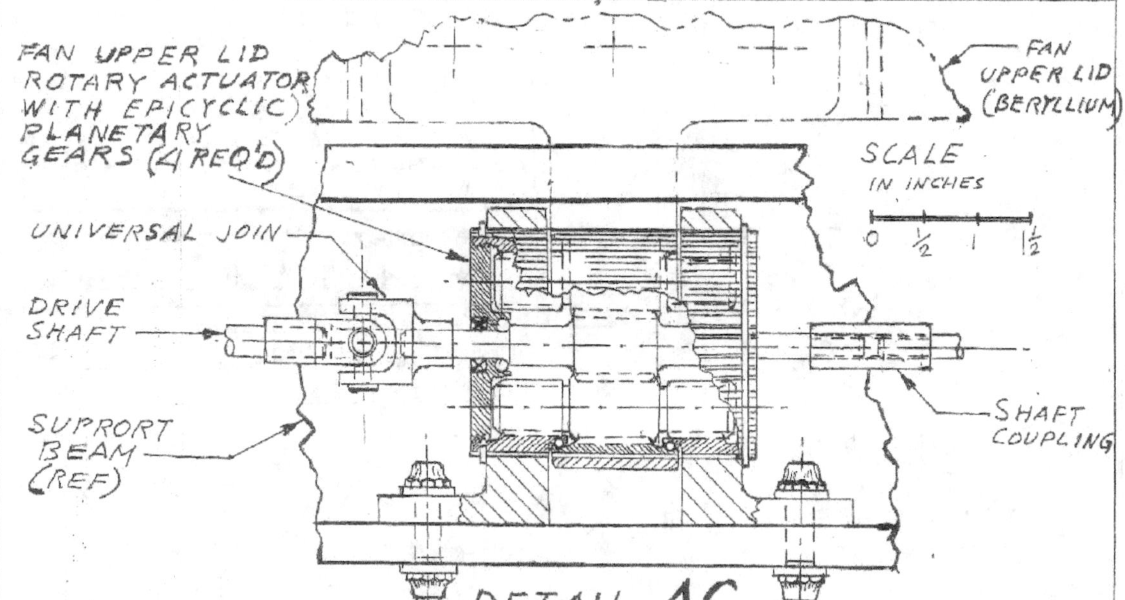

DETAIL AC
(FROM SHEET 21)

PROPULSION AND LIFT OFF MODULE

Name of project:	MECHANICAL FLYING HORSE	
ADVANCED ENGINEERING PROJECT		
Design Engineer: Walter F. Laredo		Date: March 15, 20202
Drawing Number: 1100		Sheet 24 of 26

PLATE 97

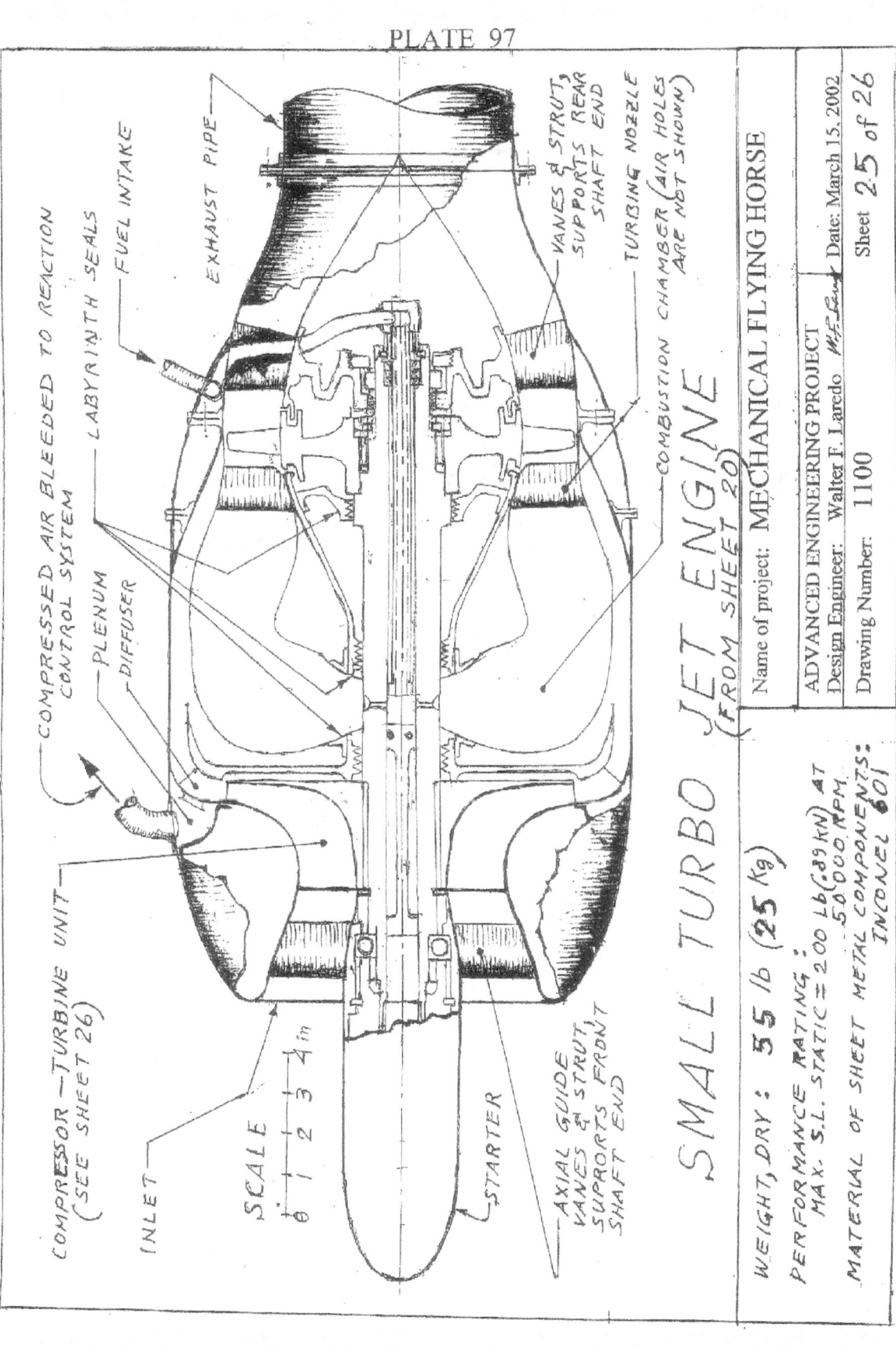

COMPRESSED AIR BLEEDED TO REACTION CONTROL SYSTEM

LABYRINTH SEALS

FUEL INTAKE

EXHAUST PIPE

VANES & STRUTS, SUPPORTS REAR SHAFT END

TURBINE NOZZLE

COMBUSTION CHAMBER (AIR HOLES ARE NOT SHOWN)

PLENUM

DIFFUSER

COMPRESSOR — TURBINE UNIT (SEE SHEET 26)

INLET

SCALE
0 1 2 3 4 in

STARTER

AXIAL GUIDE VANES & STRUT SUPPORTS FRONT SHAFT END

SMALL TURBO JET ENGINE
(FROM SHEET 20)

WEIGHT, DRY: **55** lb **(25** Kg**)**

PERFORMANCE RATING:
MAX. S.L. STATIC = 200 Lb (.89 kN) AT 50,000 R.P.M.

MATERIAL OF SHEET METAL COMPONENTS: INCONEL 601

Name of project:	MECHANICAL FLYING HORSE	
ADVANCED ENGINEERING PROJECT		
Design Engineer:	Walter F. Laredo	Date: March 15, 2002
Drawing Number:	1100	Sheet 25 of 26

PLATE 98

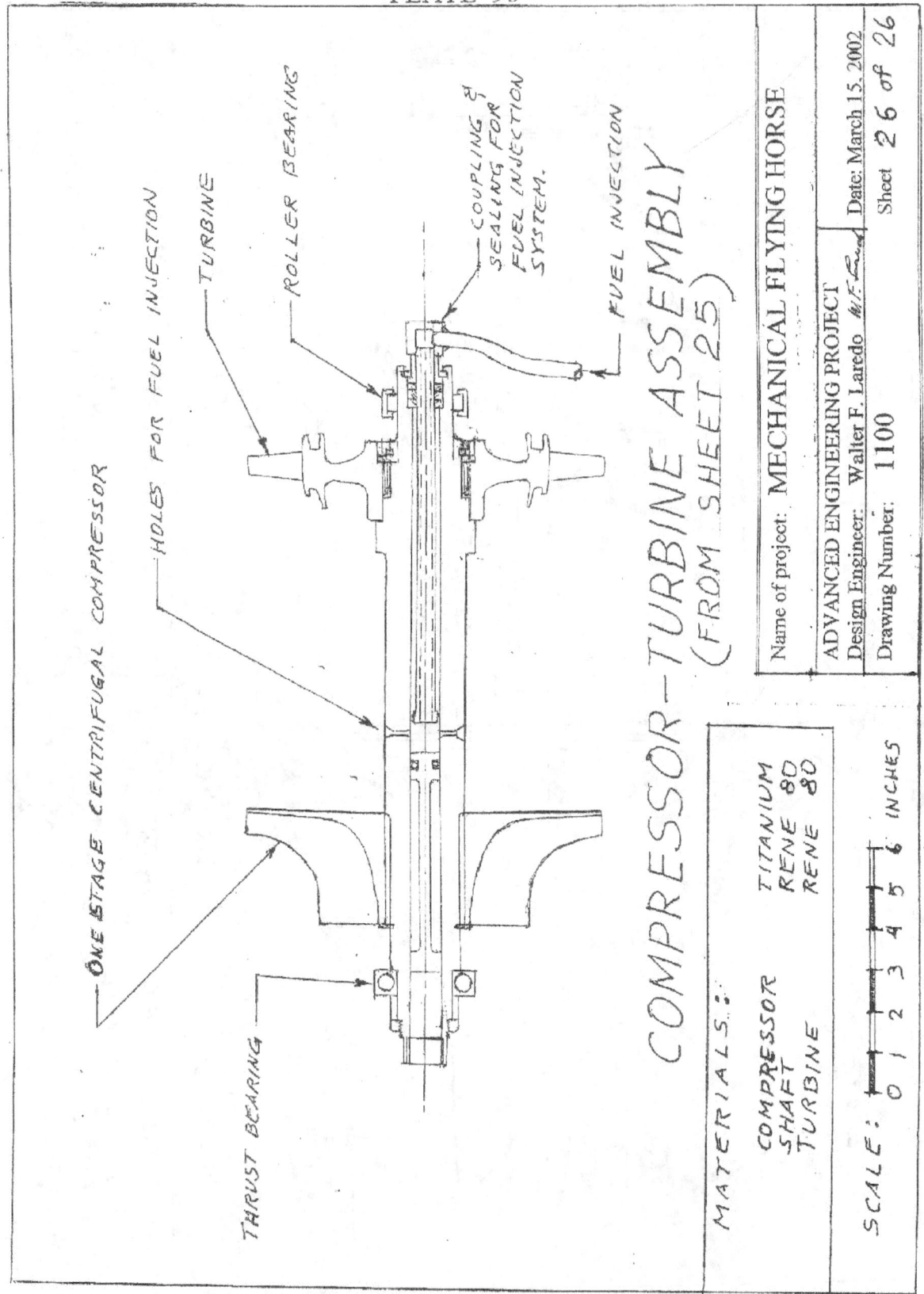

COMPRESSOR-TURBINE ASSEMBLY
(FROM SHEET 25)

ONE STAGE CENTRIFUGAL COMPRESSOR

HOLES FOR FUEL INJECTION

TURBINE

ROLLER BEARING

COUPLING & SEALING FOR FUEL INJECTION SYSTEM.

FUEL INJECTION

THRUST BEARING

MATERIALS:

COMPRESSOR   TITANIUM
SHAFT   RENE 80
TURBINE   RENE 80

SCALE: 0 1 2 3 4 5 6 INCHES

Name of project:	MECHANICAL FLYING HORSE	
ADVANCED ENGINEERING PROJECT		
Design Engineer: Walter F. Laredo W.F.Laredo		Date: March 15, 2002
Drawing Number: 1100		Sheet 26 of 26

PLATE 99

POWER PLANT (BOTH CONFIGURATIONS)
65 HP, air cooled Hirth engine,
Limited to 15 HP in road-ridden mode.

SPECIFICATIONS (WING TYPE)

Accommodation	2 tandem (as in a motorcycle)
Wing's span, (b)	24 ft 8in
Wing's gross area, (Sw)	23.3 sq. ft.
Airfoil	High lift (see Supl. Inf.)
Wing chord,	5.0 ft
MAC	5.0 ft
Wing aspect Ratio, AR= b/a	4.93
Canard Area (Sc)	25.32 sq. ft.
Canard MAC	2 ft.

(Continues in next page)

ADVANCED ENGINEERING PROJECTS

Name of project:

FLYING CAR (AIRCRAFT-CAR, VEHICLE)

PROJECT INCLUDES THE FOLLOWING TWO CONFIGURATIONS:
    I. RETRACTABLE WING TYPE
    II. POWERED PARACHUTE TYPE

Design Engineer: Walter F. Laredo *W.F.Laredo* | Date: March 15, 2002

Areas of Development:
Aerodynamics ✓, Stability Control ✓, Thermodynamics ✓, Materials ✓,
Structures (Design and Analysis) ✓, Mechanisms & Systems ✓,

NOTE: This is the PRELIMINARY DESIGN for a real engineering project.

Drawing Number: 1200 | Sheet 1 of 18

PLATE 100

(Continues from page 1)

Vertical tail area (Sv)	12 sq. ft.
Canard arm (Lc)	7 ft. 4in
Tail arm for ver. tail (Lv)	11 ft. 3in
Vertical tail volume:	
$Vv = (Lv \times Sv) / (b \times Sw)$	0.044
Weight empty	515 lb.
Weight max. takeoff	910 lb.
Max. wing loading	7.38 lb. / sq. ft.
Max. power loading	14.0 lb. / HP

PERFORMANCE (WING TYPE CONFIGURATION)

Max. flying level speed at SL	130 mph
Stall speed	46 mph
Max. Road ridden speed in car mode	30 mph
Service ceiling	15 000 ft
Range	250 miles

PERFORMANCE (POWERED PARACHUTE TYPE)

Cruise speed	33 mph
Stall speed	26 mph

NOTE:

SHEET NUMBERING SYSTEM FOR DRAWINGS

Airframe drawings	sheets 3 through 13
Propulsion system drawings	sheets 14 through 18

ADVANCED ENGINEERING PROJECTS

Name of project:

FLYING CAR

(INCLUDES TWO DIFFERENT CONFIGURATIONS)

Design Engineer: Walter F. Laredo *W.F.Laredo*	Date: March 15, 2002
Drawing Number: 1200	Sheet 2 of 18

PLATE 101

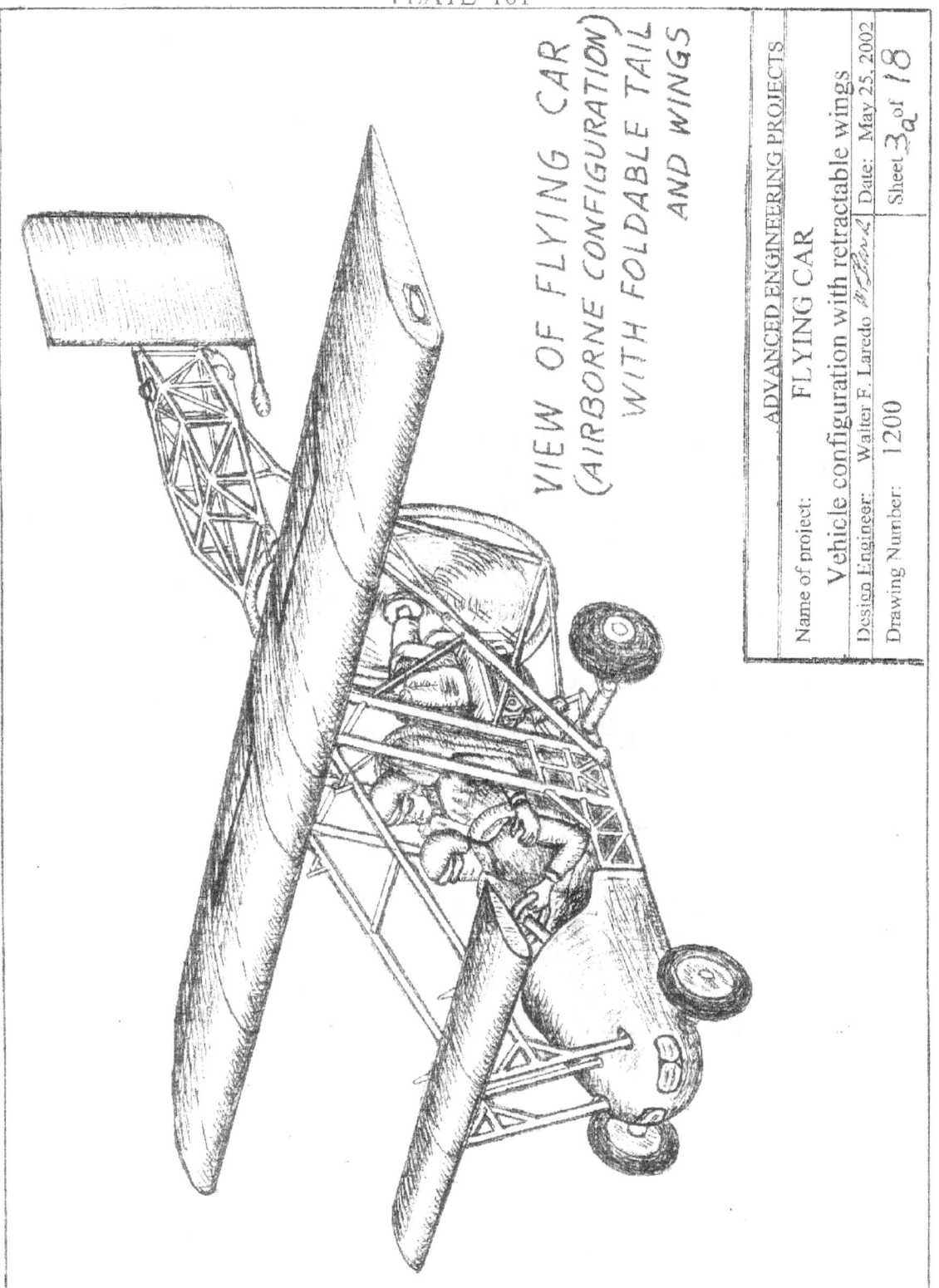

VIEW OF FLYING CAR
(AIRBORNE CONFIGURATION)
WITH FOLDABLE TAIL
AND WINGS

ADVANCED ENGINEERING PROJECTS

FLYING CAR

Name of project:

Vehicle configuration with retractable wings

| Design Engineer: | Walter F. Laredo | Date: | May 25, 2002 |
| Drawing Number: | 1200 | | Sheet 3a of 18 |

PLATE 102

STORED RUDDER

FOLDED WING

FLYING CAR
IN ROAD
RIDING MODE

STORED
CANARD
WINGS

ADVANCED ENGINEERING PROJECTS		
FLYING CAR		
Name of project:		
Vehicle configuration with retractable wings		
Design Engineer:	Walter F. Laredo *W. Laredo*	Date: May 25. 2002
Drawing Number: 1200		Sheet 3 b of 18

PLATE 103

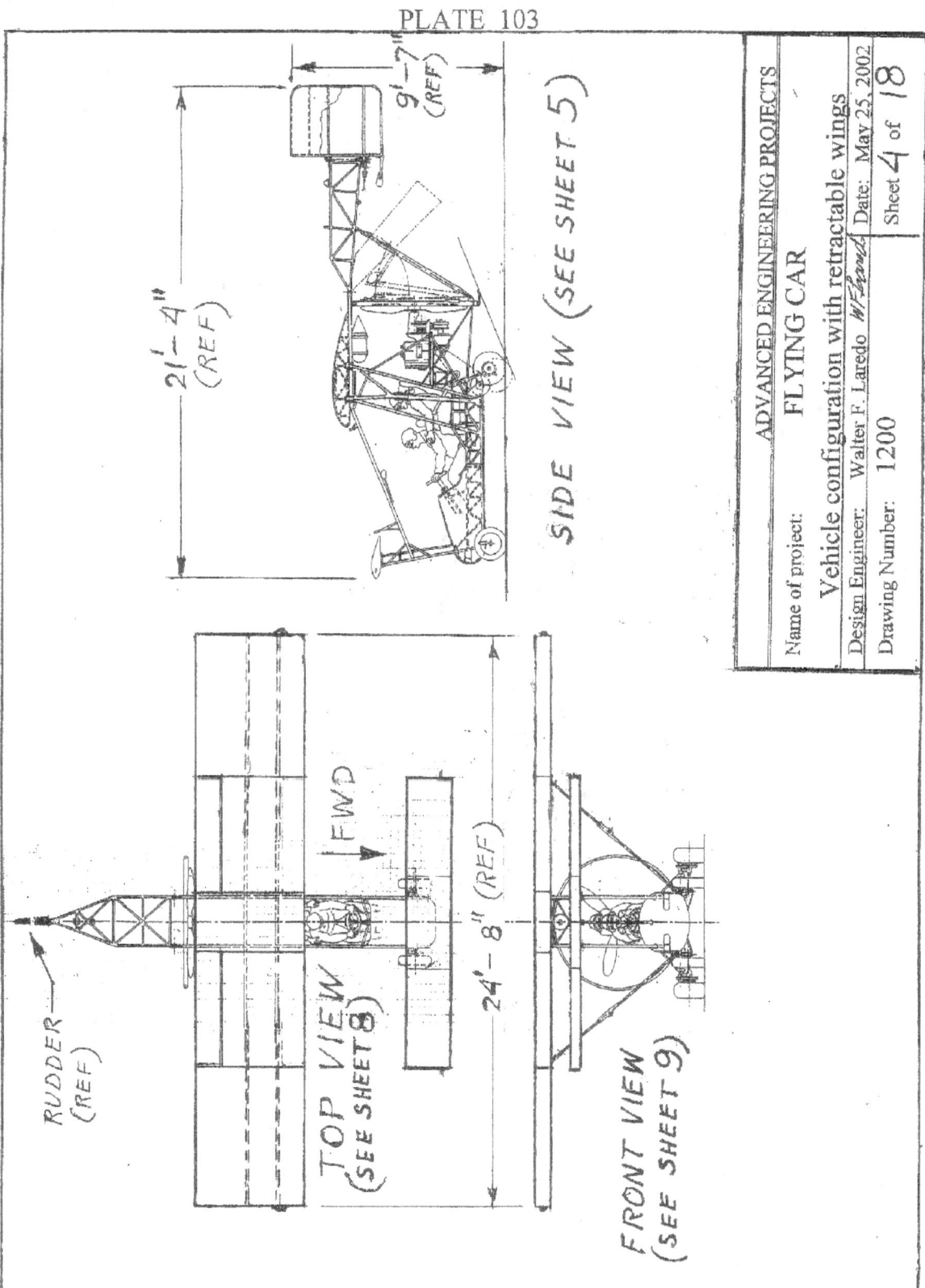

SIDE VIEW (SEE SHEET 5)

$21'-4''$
(REF)

$9'-7''$
(REF)

RUDDER
(REF)

FWD

TOP VIEW
(SEE SHEET 8)

$24'-8''$ (REF)

FRONT VIEW
(SEE SHEET 9)

ADVANCED ENGINEERING PROJECTS		
Name of project:	FLYING CAR	
	Vehicle configuration with retractable wings	
Design Engineer:	Walter F. Laredo *W.F.Laredo*	Date: May 25, 2002
Drawing Number:	1200	Sheet 4 of 18

PLATE 104

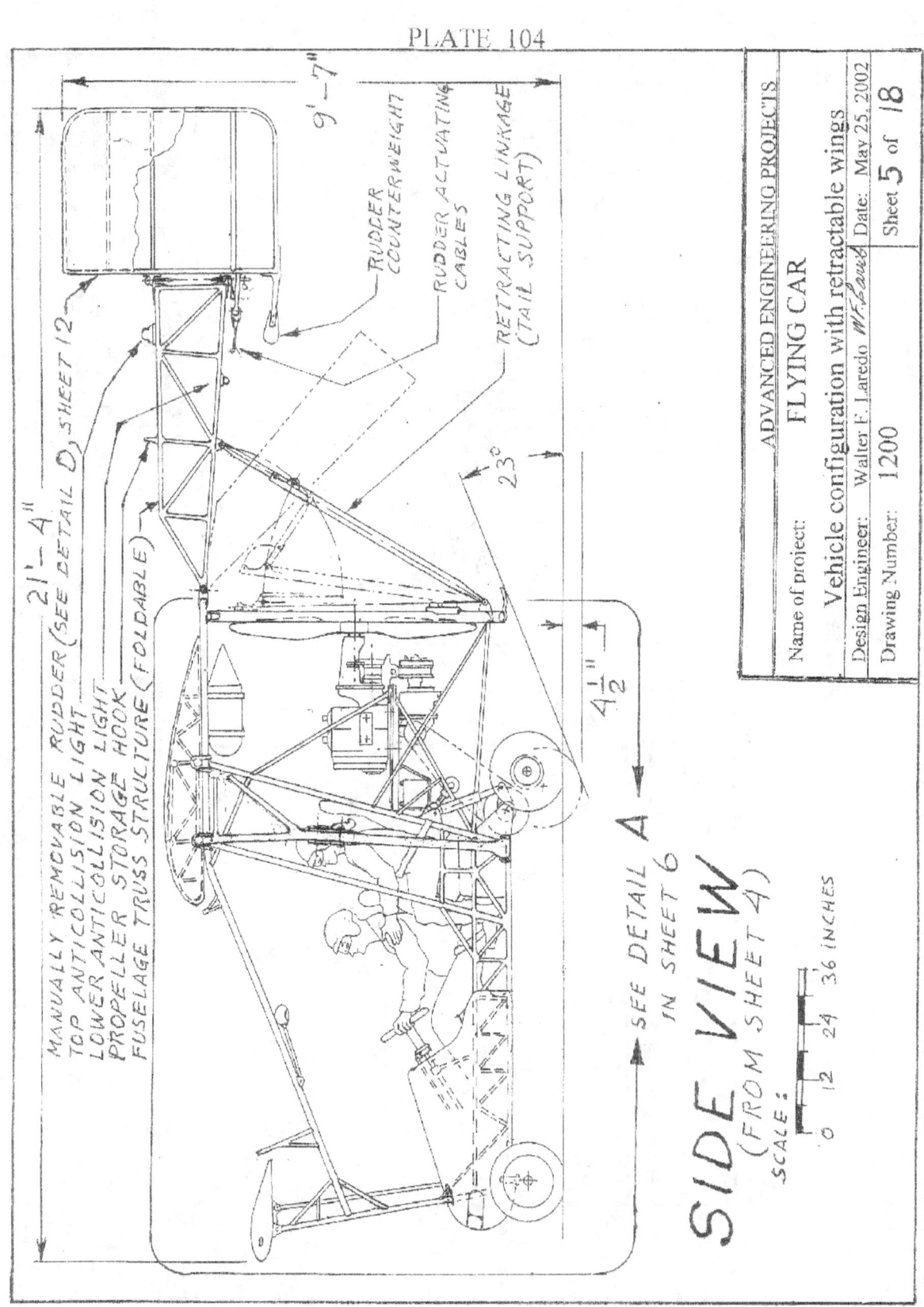

SIDE VIEW
(FROM SHEET 4)
SCALE:

0   12   24   36 INCHES

SEE DETAIL A
IN SHEET 6

MANUALLY REMOVABLE RUDDER (SEE DETAIL D, SHEET 12)
TOP ANTICOLLISION LIGHT
LOWER ANTICOLLISION LIGHT
PROPELLER STORAGE HOOK
FUSELAGE TRUSS STRUCTURE (FOLDABLE)

RUDDER COUNTERWEIGHT
RUDDER ACTIVATING CABLES
RETRACTING LINKAGE (TAIL SUPPORT)

21'–4"
9'–7"
23°
$4\frac{1}{2}$"

Name of project:	ADVANCED ENGINEERING PROJECTS
	FLYING CAR
	Vehicle configuration with retractable wings
Design Engineer: Walter F. Laredo *W.F. Laredo*	Date: May 25, 2002
Drawing Number: 1200	Sheet 5 of 18

PLATE 105

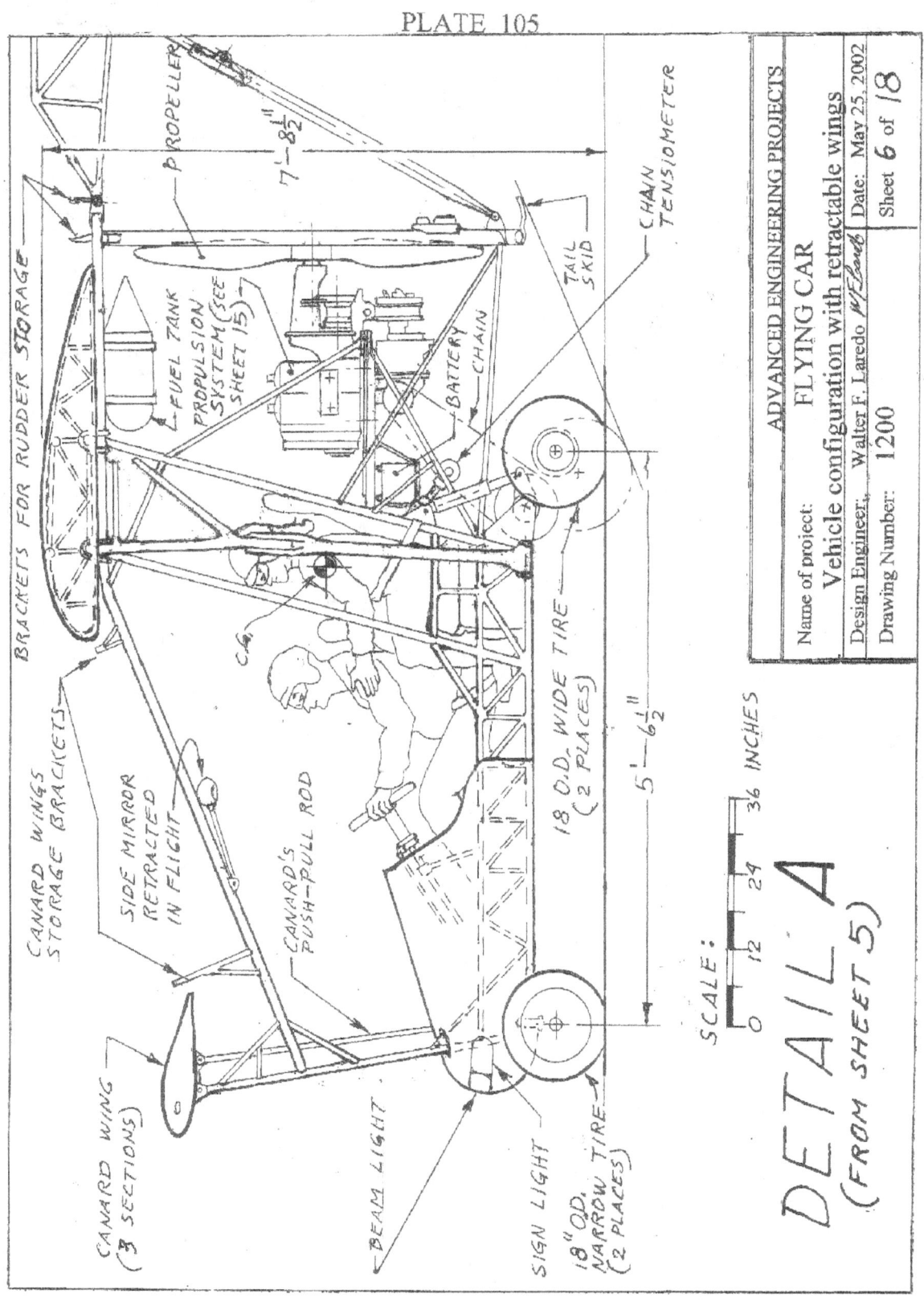

DETAIL A
(FROM SHEET 5)

SCALE:

0   12   24   36 INCHES

PROPELLER

7'—8½"

FUEL TANK

PROPULSION SYSTEM (SEE SHEET 15)

BATTERY

CHAIN

CHAIN TENSIOMETER

TAIL SKID

18 O.D. WIDE TIRE (2 PLACES)

5'—6½"

BRACKETS FOR RUDDER STORAGE

CANARD WINGS STORAGE BRACKETS

SIDE MIRROR RETRACTED IN FLIGHT

CANARD'S PUSH-PULL ROD

CANARD WING (3 SECTIONS)

BEAM LIGHT

SIGN LIGHT

18" O.D. NARROW TIRE (2 PLACES)

ADVANCED ENGINEERING PROJECTS		
Name of project:	FLYING CAR	
	Vehicle configuration with retractable wings	
Design Engineer:	Walter F. Laredo W.F.Laredo	Date: May 25, 2002
Drawing Number:	1200	Sheet 6 of 18

PLATE 106

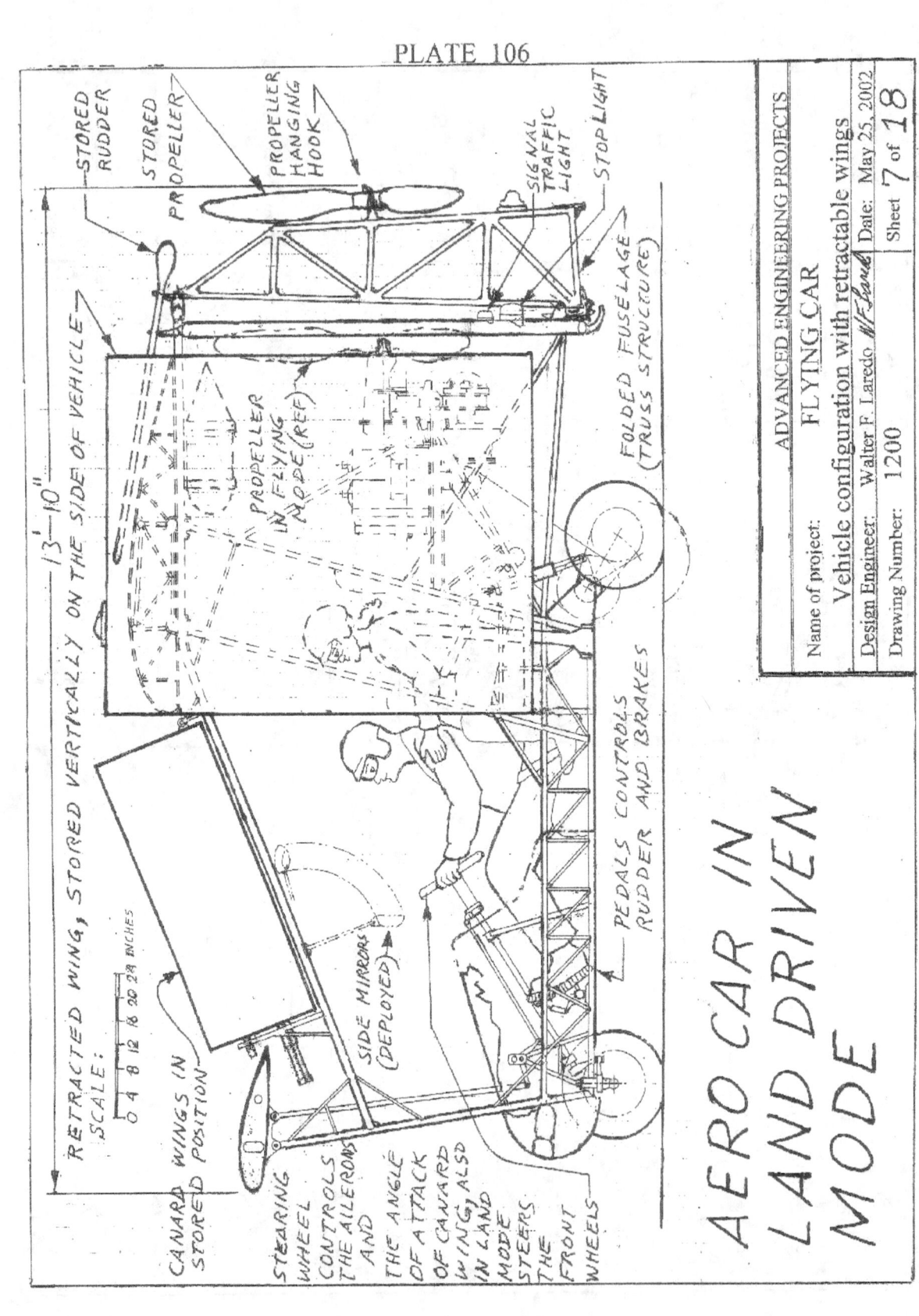

AERO CAR IN LAND DRIVEN MODE

STORED RUDDER

STORED PROPELLER

PROPELLER HANGING HOOK

SIGNAL TRAFFIC LIGHT

STOP LIGHT

RETRACTED WING, STORED VERTICALLY ON THE SIDE OF VEHICLE

13'-10"

SCALE:
0 4 8 12 16 20 24 INCHES

PROPELLER IN FLYING MODE (REF)

FOLDED FUSELAGE (TRUSS STRUCTURE)

CANARD WINGS IN STORED POSITION

SIDE MIRRORS (DEPLOYED)

PEDALS CONTROLS RUDDER AND BRAKES

STEERING WHEEL CONTROLS THE AILERONS AND THE ANGLE OF ATTACK OF CANARD WING, ALSO IN LAND MODE STEERS THE FRONT WHEELS

Name of project:
ADVANCED ENGINEERING PROJECTS
FLYING CAR
Vehicle configuration with retractable wings

Design Engineer: Walter F. Laredo   Date: May 25, 2002

Drawing Number: 1200   Sheet 7 of 18

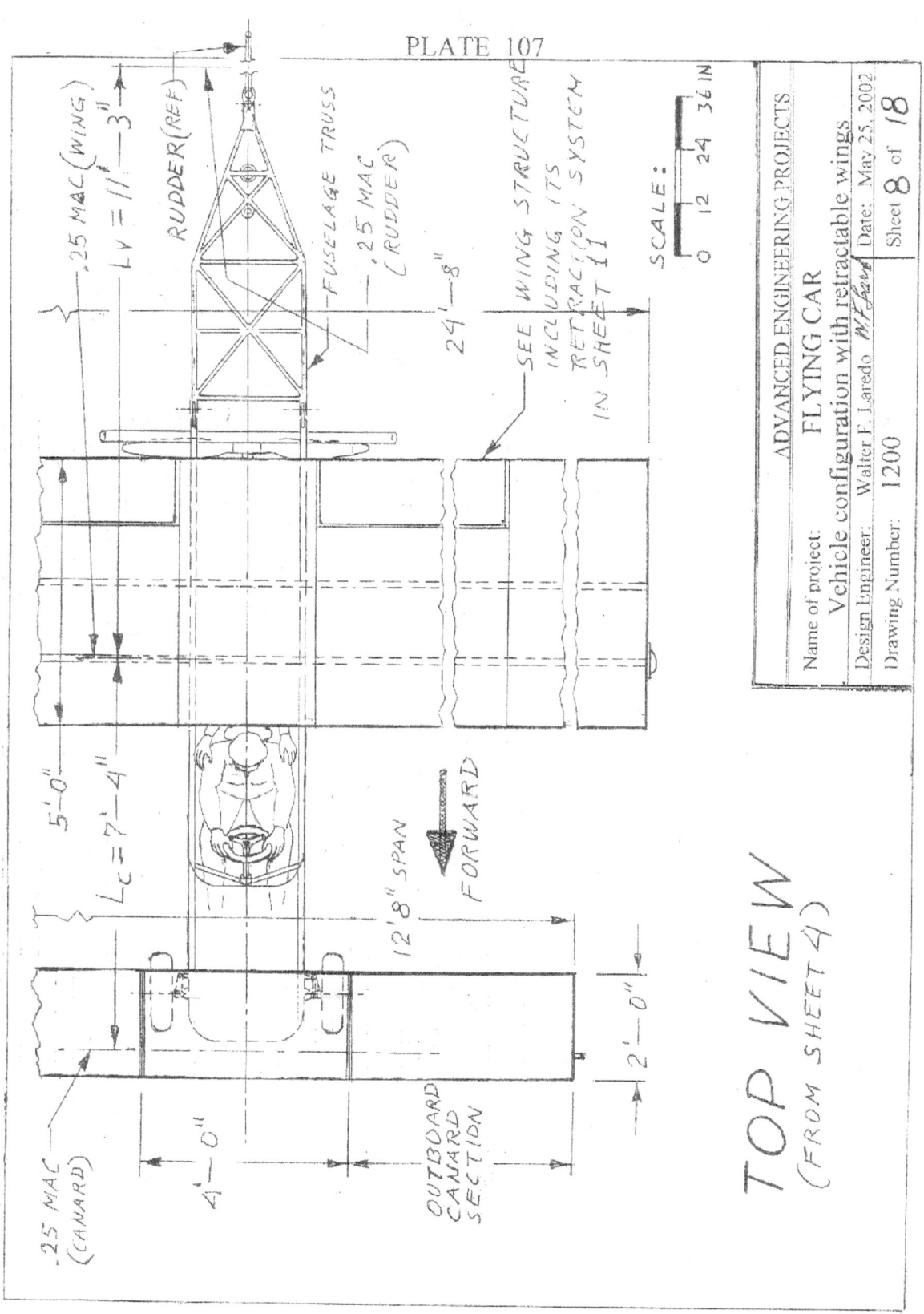

PLATE 107

TOP VIEW
(FROM SHEET 4)

.25 MAC (CANARD)

RUDDER (REF.)

.25 MAC (WING)

LV = 11'—3"

FUSELAGE TRUSS

.25 MAC (RUDDER)

24'—8"

SEE WING STRUCTURE INCLUDING ITS RETRACTION SYSTEM IN SHEET 11

SCALE:

0   12   24   36 IN

5'—0"

$L_c = 7'—4"$

12'8" SPAN

FORWARD

4'—0"

2'—0"

OUTBOARD CANARD SECTION

ADVANCED ENGINEERING PROJECTS
FLYING CAR
Vehicle configuration with retractable wings

Name of project:

Design Engineer: Walter F. Laredo   W. F. Laredo   Date: May 25, 2002

Drawing Number: 1200   Sheet 8 of 18

PLATE 108

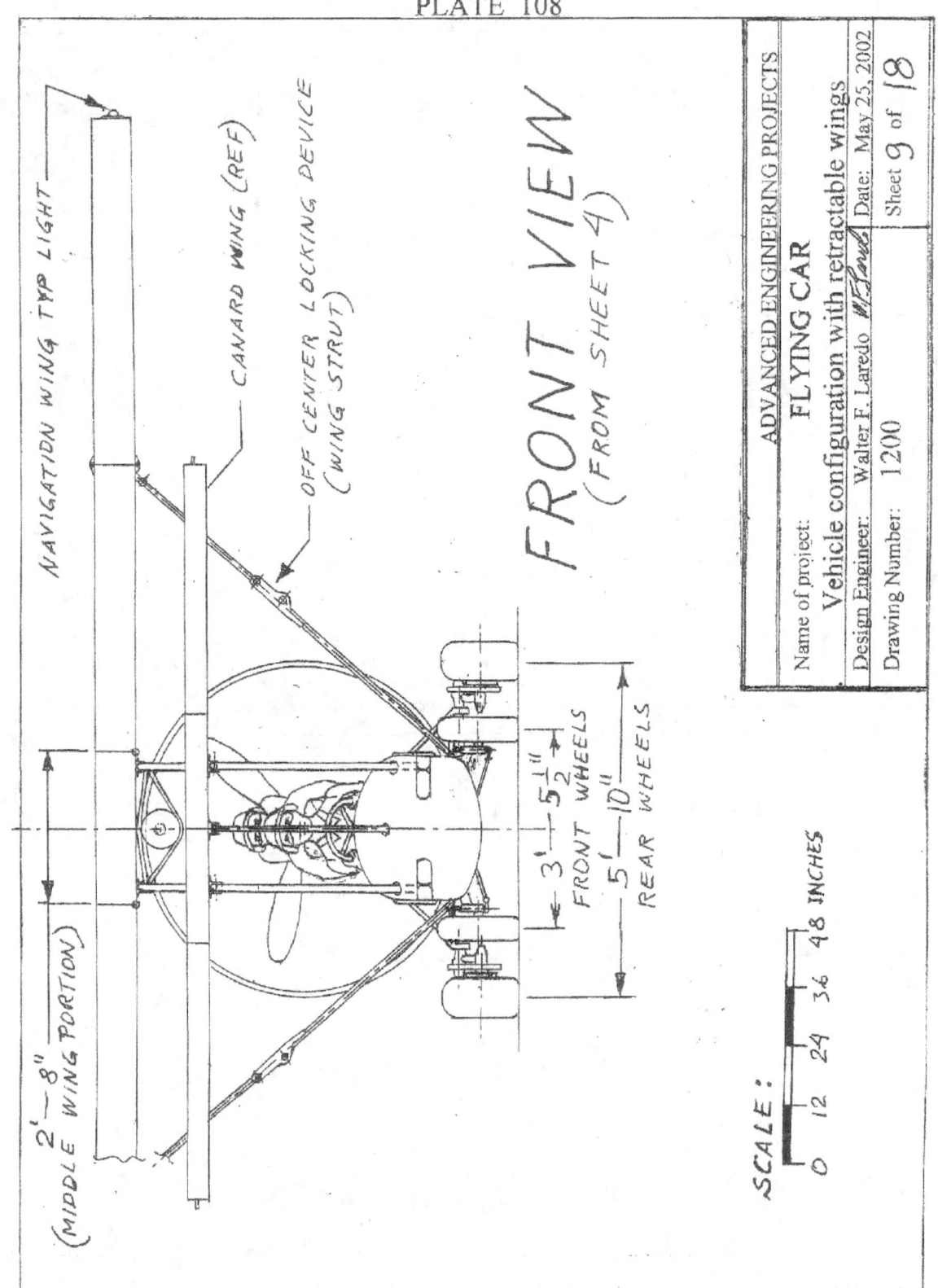

NAVIGATION WING TYP LIGHT

CANARD WING (REF)

OFF CENTER LOCKING DEVICE
(WING STRUT)

FRONT VIEW
(FROM SHEET 4)

2' — 8"
(MIDDLE WING PORTION)

3' — 5½"
FRONT WHEELS

5' — 10"
REAR WHEELS

SCALE :

0   12   24   36   48 INCHES

ADVANCED ENGINEERING PROJECTS

Name of project: FLYING CAR
Vehicle configuration with retractable wings

Design Engineer: Walter F. Laredo   W.F. Laredo   Date: May 25, 2002

Drawing Number: 1200   Sheet 9 of 18

PLATE 109

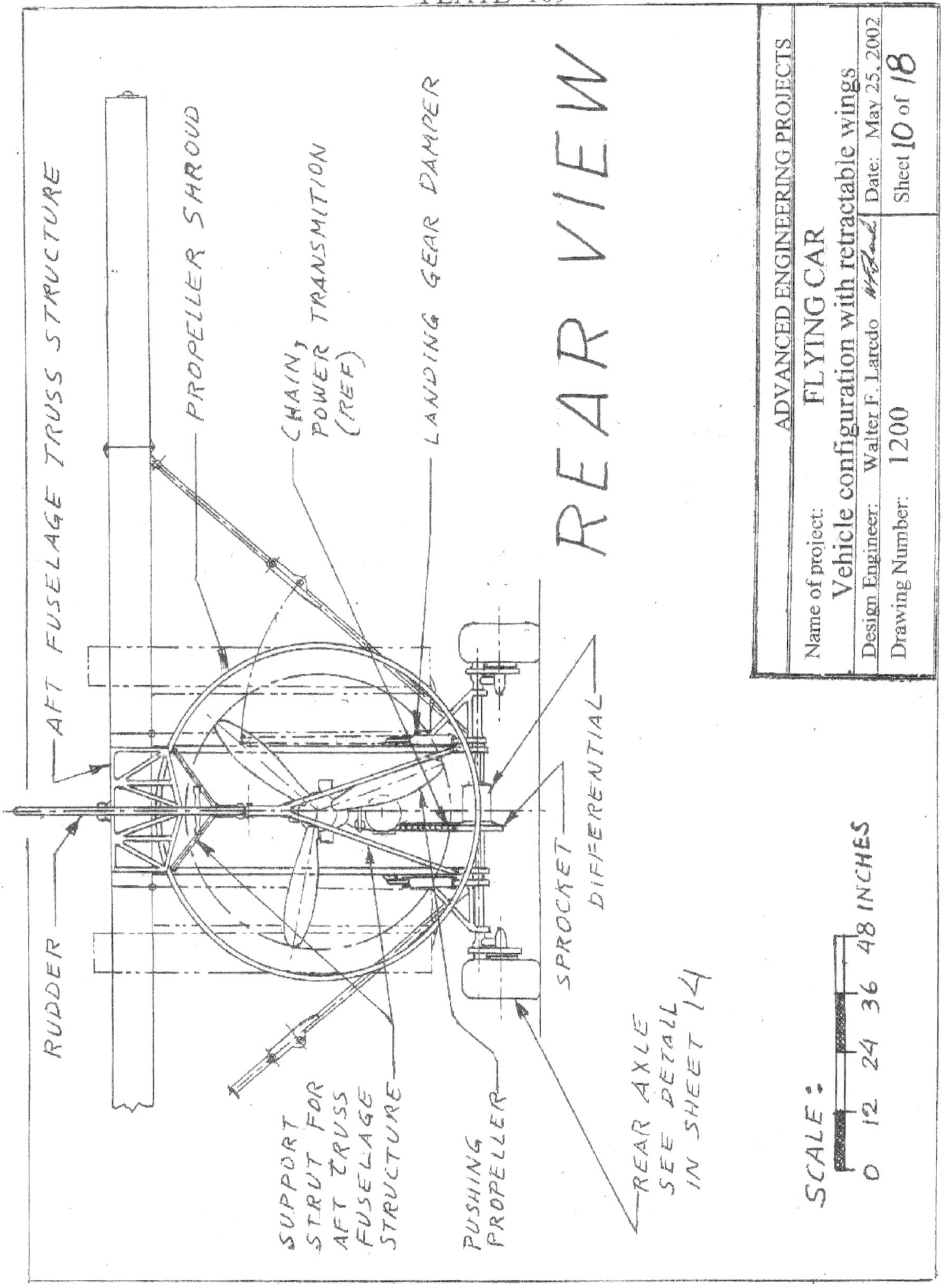

REAR VIEW

RUDDER

AFT FUSELAGE TRUSS STRUCTURE

PROPELLER SHROUD

CHAIN, POWER TRANSMITION (REF)

LANDING GEAR DAMPER

SUPPORT STRUT FOR AFT TRUSS FUSELAGE STRUCTURE

PUSHING PROPELLER

SPROCKET

DIFFERENTIAL

REAR AXLE SEE DETAIL IN SHEET 14

SCALE :

0   12   24   36   48 INCHES

ADVANCED ENGINEERING PROJECTS		
Name of project:	FLYING CAR	
Vehicle configuration with retractable wings		
Design Engineer:	Walter F. Laredo	Date: May 25, 2002
Drawing Number:	1200	Sheet 10 of 18

PLATE 110

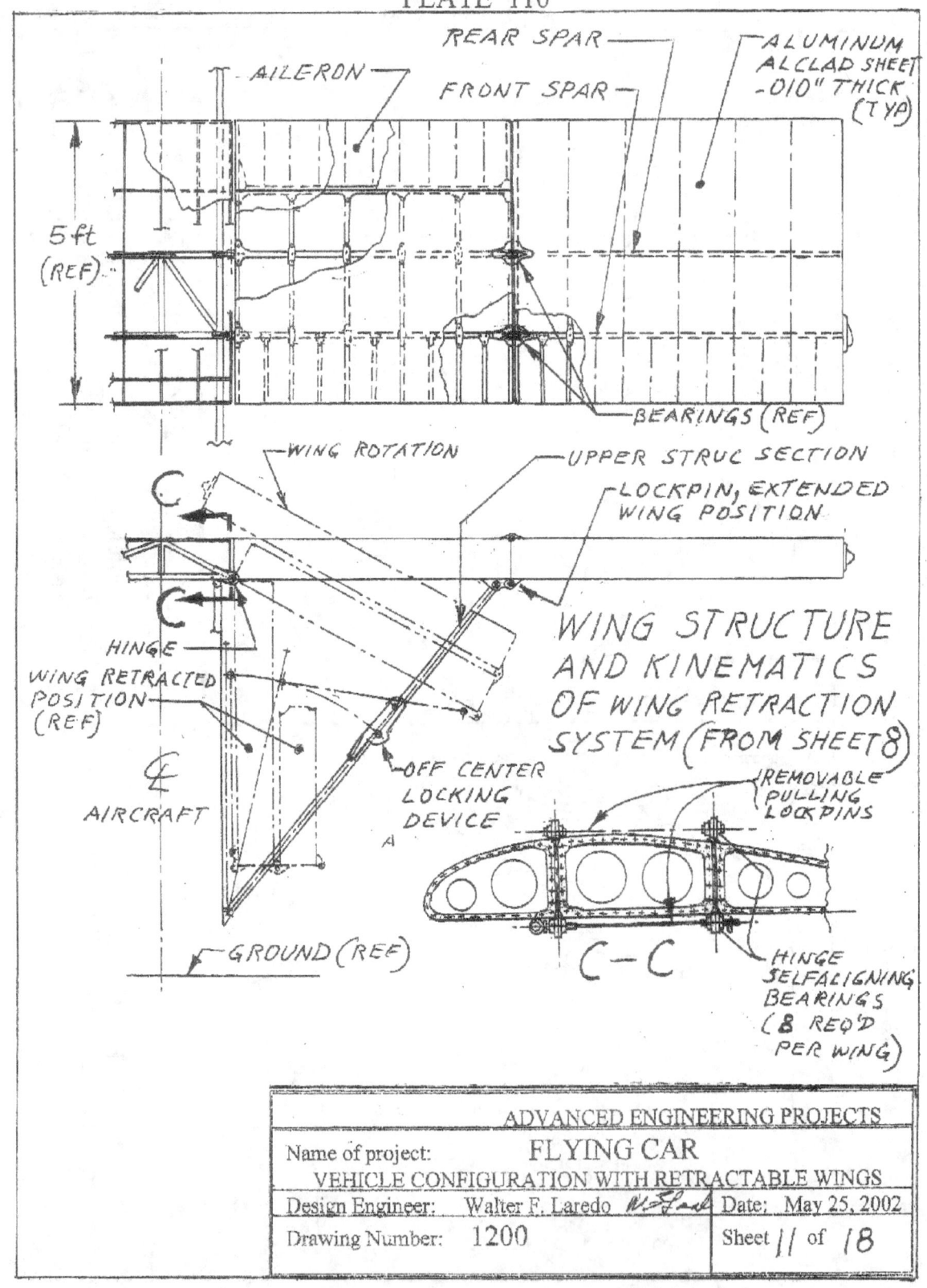

AILERON
REAR SPAR
FRONT SPAR
ALUMINUM ALCLAD SHEET -.010" THICK (TYP)
5 ft (REF)
BEARINGS (REF)
WING ROTATION
UPPER STRUC SECTION
LOCKPIN, EXTENDED WING POSITION
HINGE
WING RETRACTED POSITION (REF)
OFF CENTER LOCKING DEVICE
AIRCRAFT
GROUND (REF)

WING STRUCTURE AND KINEMATICS OF WING RETRACTION SYSTEM (FROM SHEET 8)

REMOVABLE PULLING LOCKPINS
C—C
HINGE SELFALIGNING BEARINGS (8 REQ'D PER WING)

ADVANCED ENGINEERING PROJECTS		
Name of project:	FLYING CAR	
VEHICLE CONFIGURATION WITH RETRACTABLE WINGS		
Design Engineer: Walter F. Laredo	Date: May 25, 2002	
Drawing Number: 1200	Sheet 11 of 18	

PLATE 111

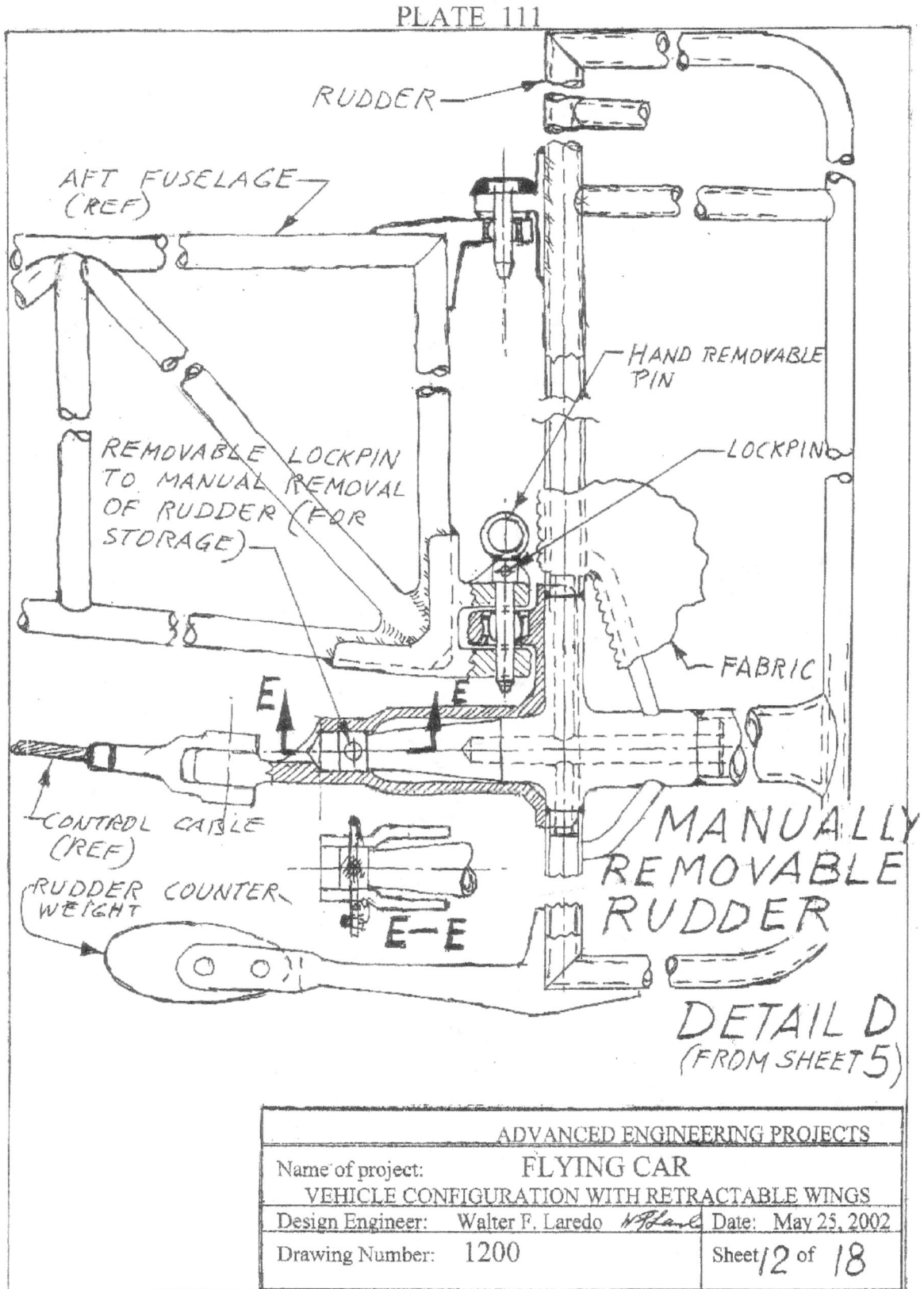

RUDDER

AFT FUSELAGE
(REF)

HAND REMOVABLE
PIN

REMOVABLE LOCKPIN
TO MANUAL REMOVAL
OF RUDDER (FOR
STORAGE)

LOCKPIN

FABRIC

CONTROL CABLE
(REF)

RUDDER COUNTER
WEIGHT

E—E

MANUALLY
REMOVABLE
RUDDER

DETAIL D
(FROM SHEET 5)

ADVANCED ENGINEERING PROJECTS		
Name of project:	FLYING CAR	
VEHICLE CONFIGURATION WITH RETRACTABLE WINGS		
Design Engineer: Walter F. Laredo	Date: May 25, 2002	
Drawing Number: 1200	Sheet 12 of 18	

PLATE 112

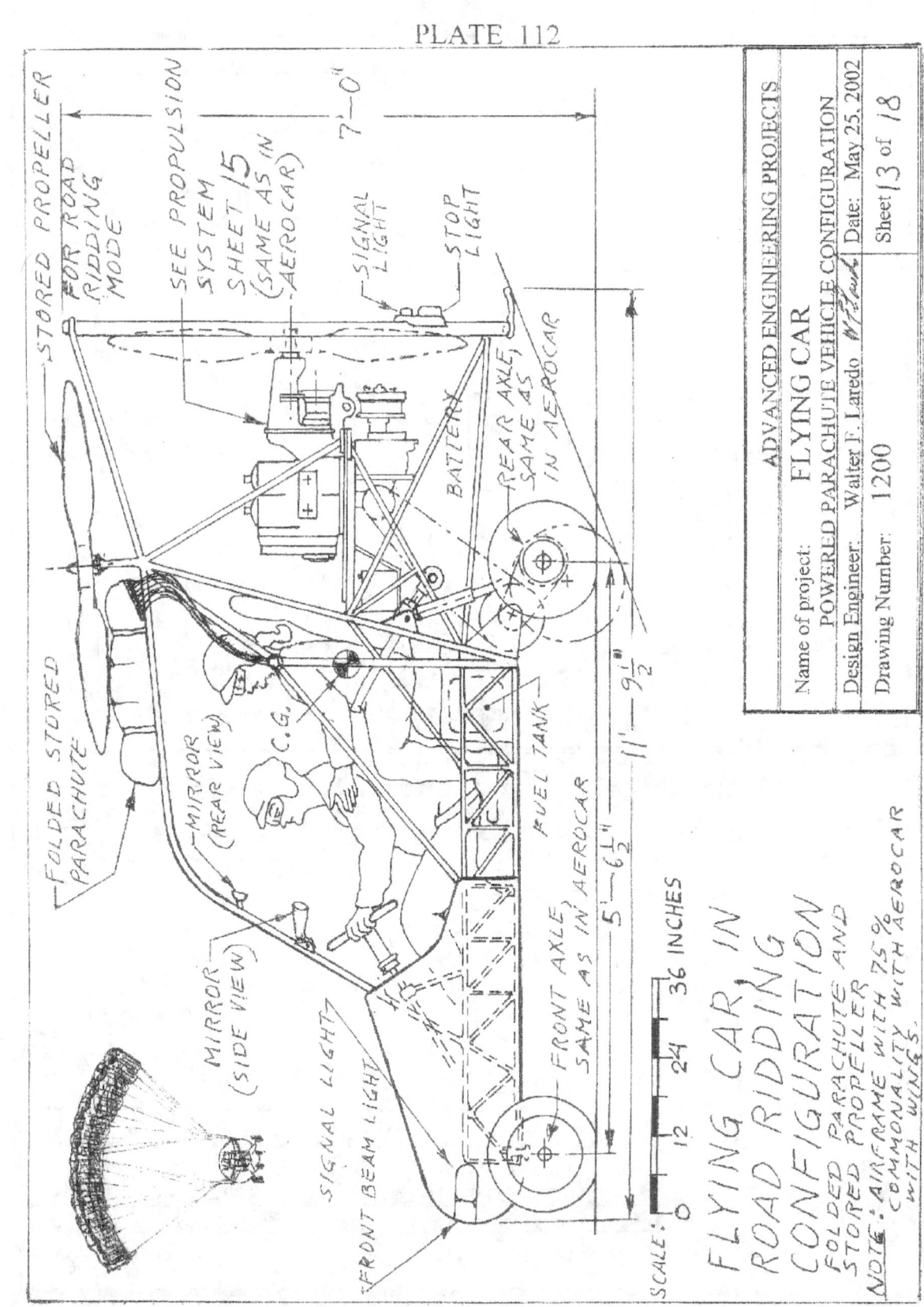

FLYING CAR IN ROAD RIDDING CONFIGURATION

FOLDED PARACHUTE AND STORED PROPELLER

NOTE: AIRFRAME WITH 75% COMMONALITY WITH AEROCAR WITH WINGS.

STORED PROPELLER

FOR ROAD RIDDING MODE

SEE PROPULSION SYSTEM SHEET 15 (SAME AS IN AEROCAR)

7'—0"

SIGNAL LIGHT

STOP LIGHT

REAR AXLE, SAME AS IN AEROCAR

BATTERY

FOLDED STORED PARACHUTE

MIRROR (REAR VIEW)

C.G.

FUEL TANK

FRONT AXLE, SAME AS IN AEROCAR

5'—6½"

11'—9½"

MIRROR (SIDE VIEW)

SIGNAL LIGHT

FRONT BEAM LIGHT

SCALE:

0   12   24   36 INCHES

	ADVANCED ENGINEERING PROJECTS		
Name of project:	POWERED PARACHUTE VEHICLE CONFIGURATION		
	FLYING CAR		
Design Engineer:	Walter F. Laredo	W Laredo	Date: May 25, 2002
Drawing Number:	1200		Sheet 13 of 18

PLATE 113

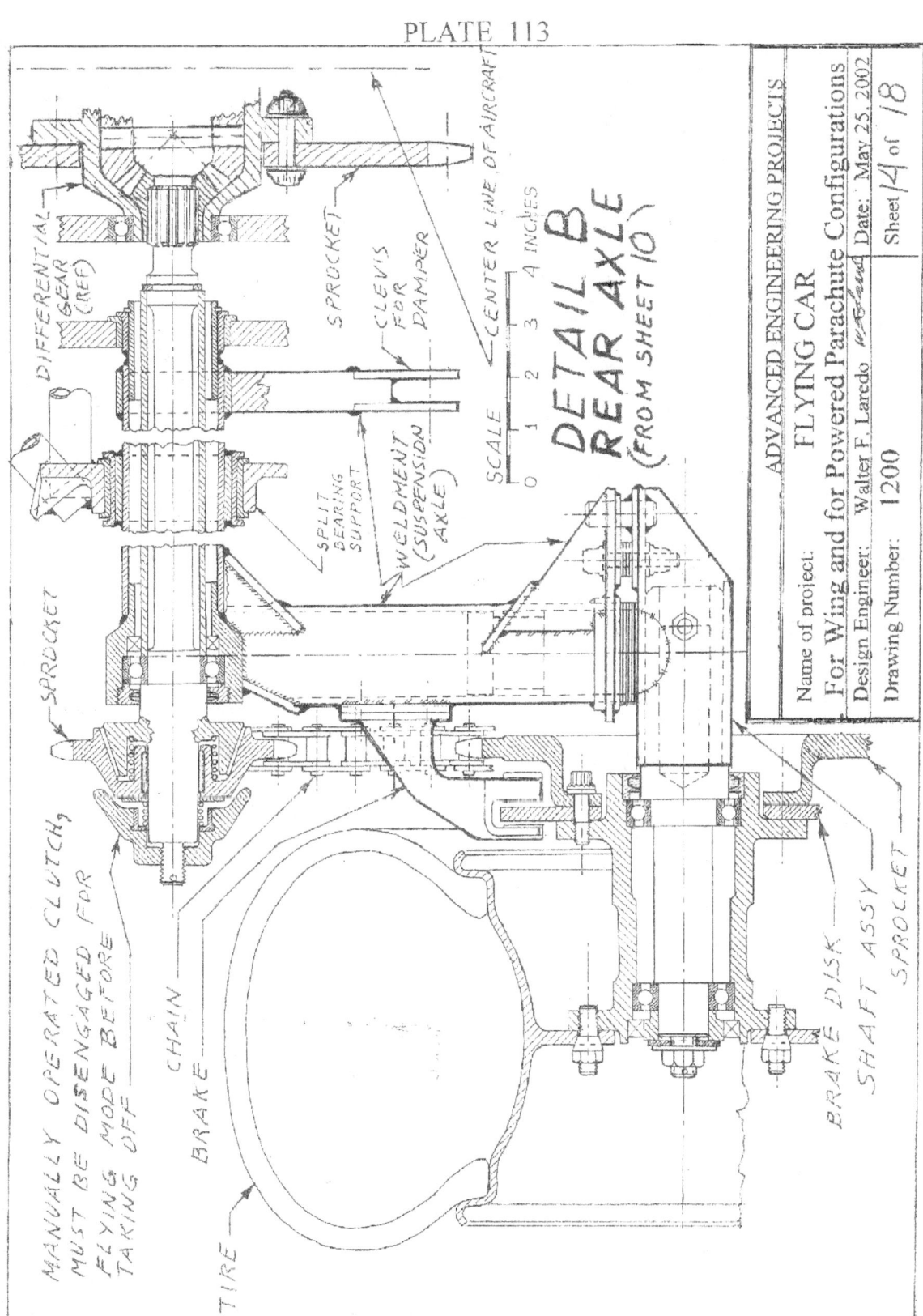

DETAIL B
REAR AXLE
(FROM SHEET 10)

SCALE
0 1 2 3 4 INCHES

DIFFERENTIAL GEAR (REF)

SPROCKET

CLEVIS FOR DAMPER

CENTER LINE OF AIRCRAFT

SPLIT BEARING SUPPORT

WELDMENT (SUSPENSION AXLE)

SPROCKET

MANUALLY OPERATED CLUTCH, MUST BE DISENGAGED FOR FLYING MODE BEFORE TAKING OFF

CHAIN

BRAKE

TIRE

BRAKE DISK

SHAFT ASSY

SPROCKET

ADVANCED ENGINEERING PROJECTS		
Name of project:		
For Wing and for Powered Parachute Configurations		
FLYING CAR		
Design Engineer: Walter F. Laredo	W.F. Laredo	Date: May 25, 2002
Drawing Number: 1200		Sheet 14 of 18

PLATE 114

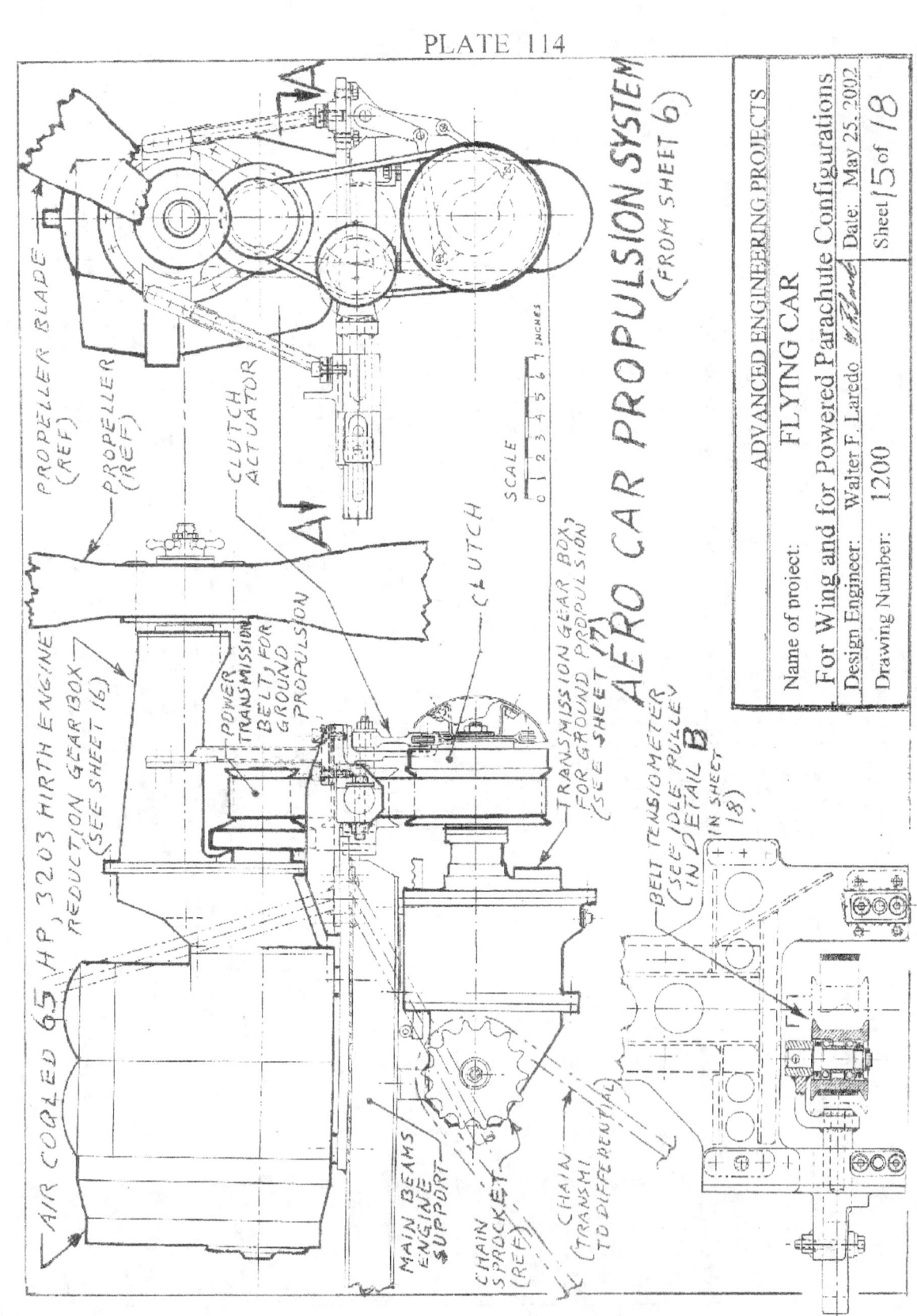

AERO CAR PROPULSION SYSTEM
(FROM SHEET 6)

PROPELLER BLADE (REF)

PROPELLER (REF)

CLUTCH ACTUATOR

CLUTCH

SCALE
0 1 2 3 4 5 6 7 INCHES

AIR COOLED 65 HP, 3203 HIRTH ENGINE

REDUCTION GEARBOX (SEE SHEET 16)

POWER TRANSMISSION BELT FOR GROUND PROPULSION

TRANSMISSION GEAR BOX FOR GROUND PROPULSION (SEE SHEET 7)

BELT TENSIOMETER (SEE IDLE PULLEY IN DETAIL B IN SHEET 18)

MAIN BEAMS ENGINE SUPPORT

CHAIN SPROCKET (REF)

CHAIN (TRANSM) TO DIFFERENTIAL

ADVANCED ENGINEERING PROJECTS
FLYING CAR
For Wing and for Powered Parachute Configurations

Name of project:		
Design Engineer:	Walter F. Laredo	Date: May 25, 2002
Drawing Number:	1200	Sheet 15 of 18

PLATE 115

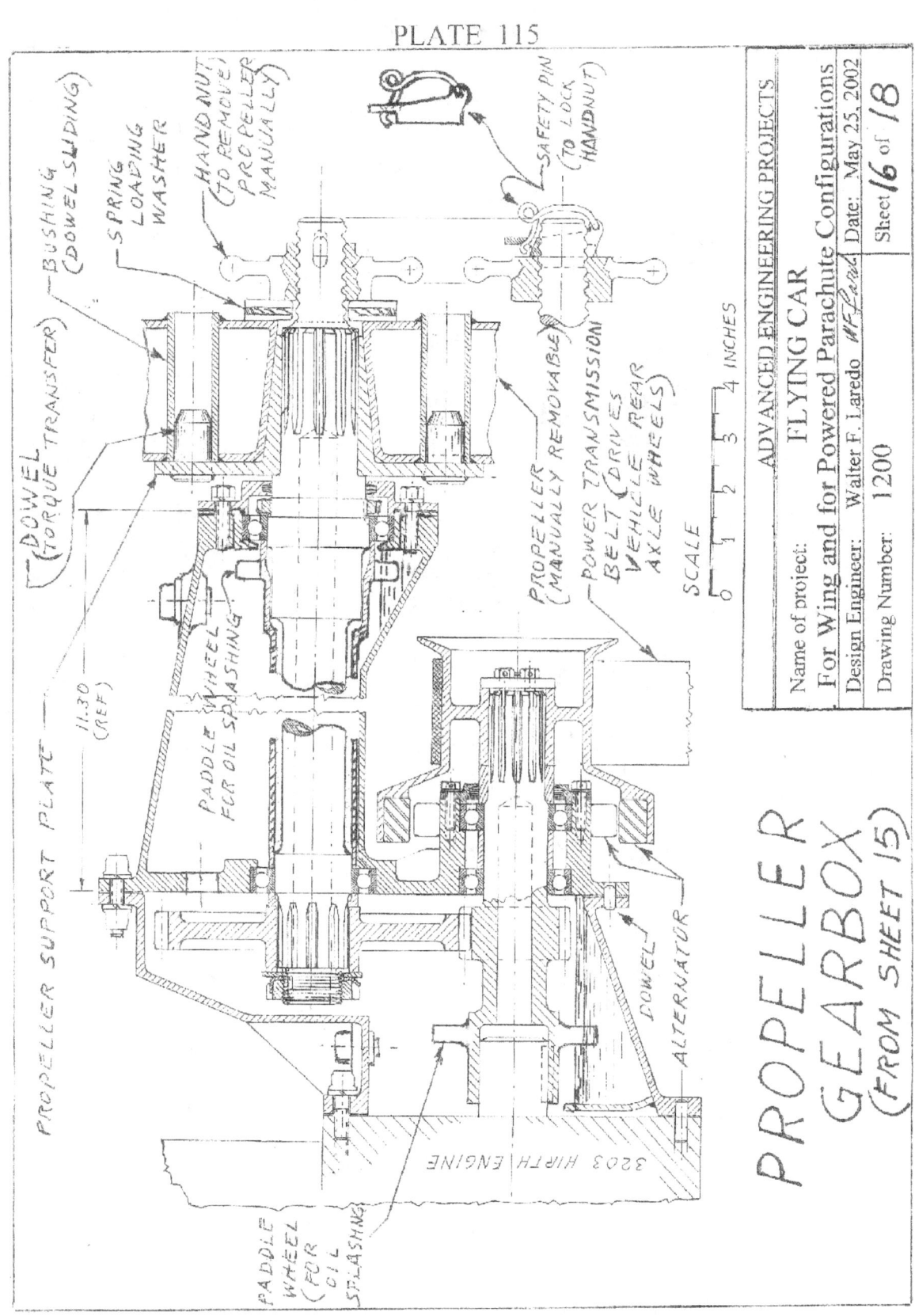

PLATE 116

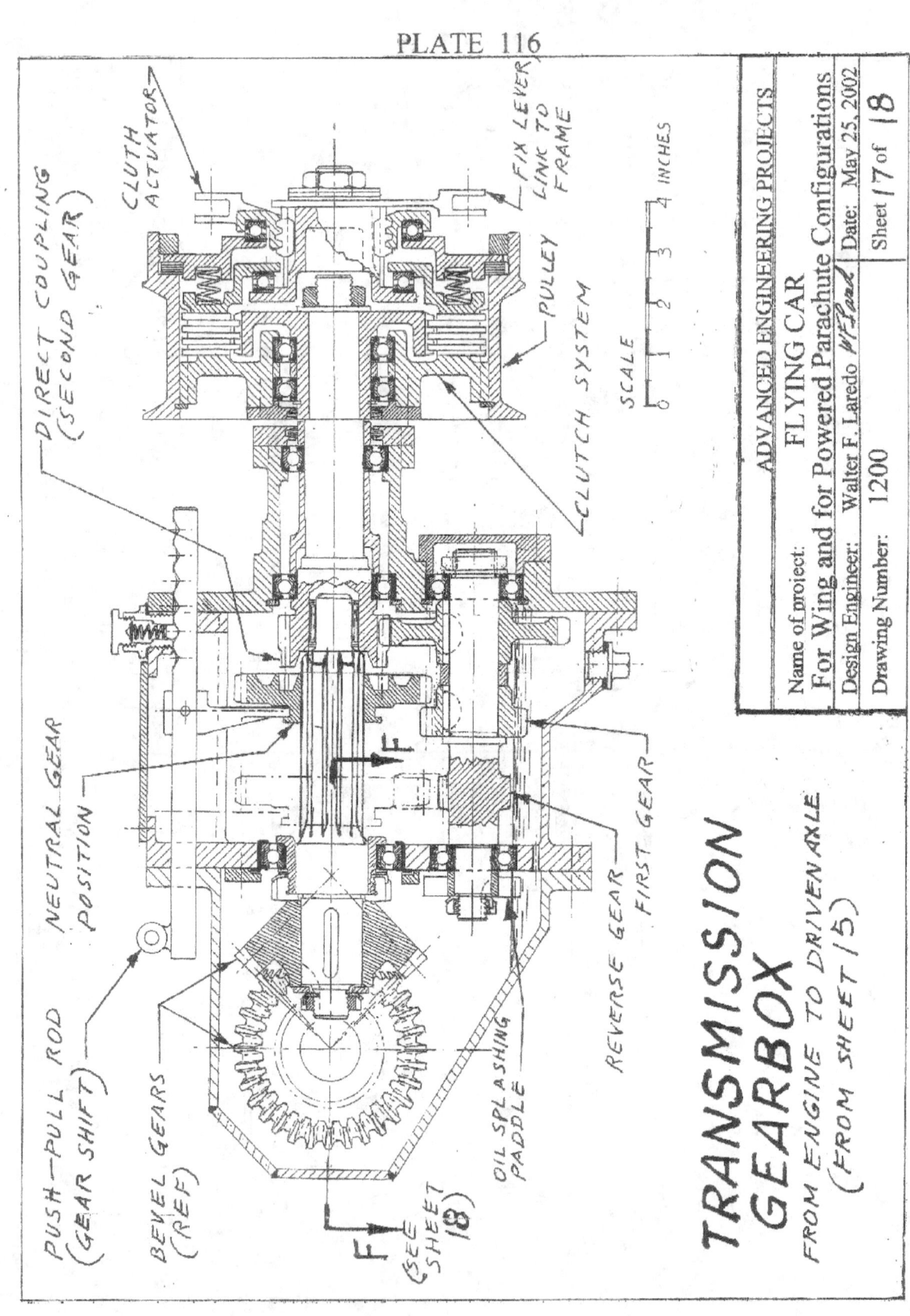

DIRECT COUPLING
(SECOND GEAR)

CLUTH ACTUATOR

FIX LEVER LINK TO FRAME

PULLEY

CLUTCH SYSTEM

PUSH-PULL ROD
(GEAR SHIFT)

NEUTRAL GEAR POSITION

BEVEL GEARS
(REF)

F
(SEE SHEET 18)

OIL SPLASHING PADDLE

REVERSE GEAR

FIRST GEAR

SCALE

0 1 2 3 4 INCHES

TRANSMISSION GEARBOX

FROM ENGINE TO DRIVEN AXLE
(FROM SHEET 15)

ADVANCED ENGINEERING PROJECTS

Name of project: FLYING CAR
For Wing and for Powered Parachute Configurations
Design Engineer: Walter F. Laredo   Date: May 25, 2002
Drawing Number: 1200   Sheet 17 of 18

PLATE 117

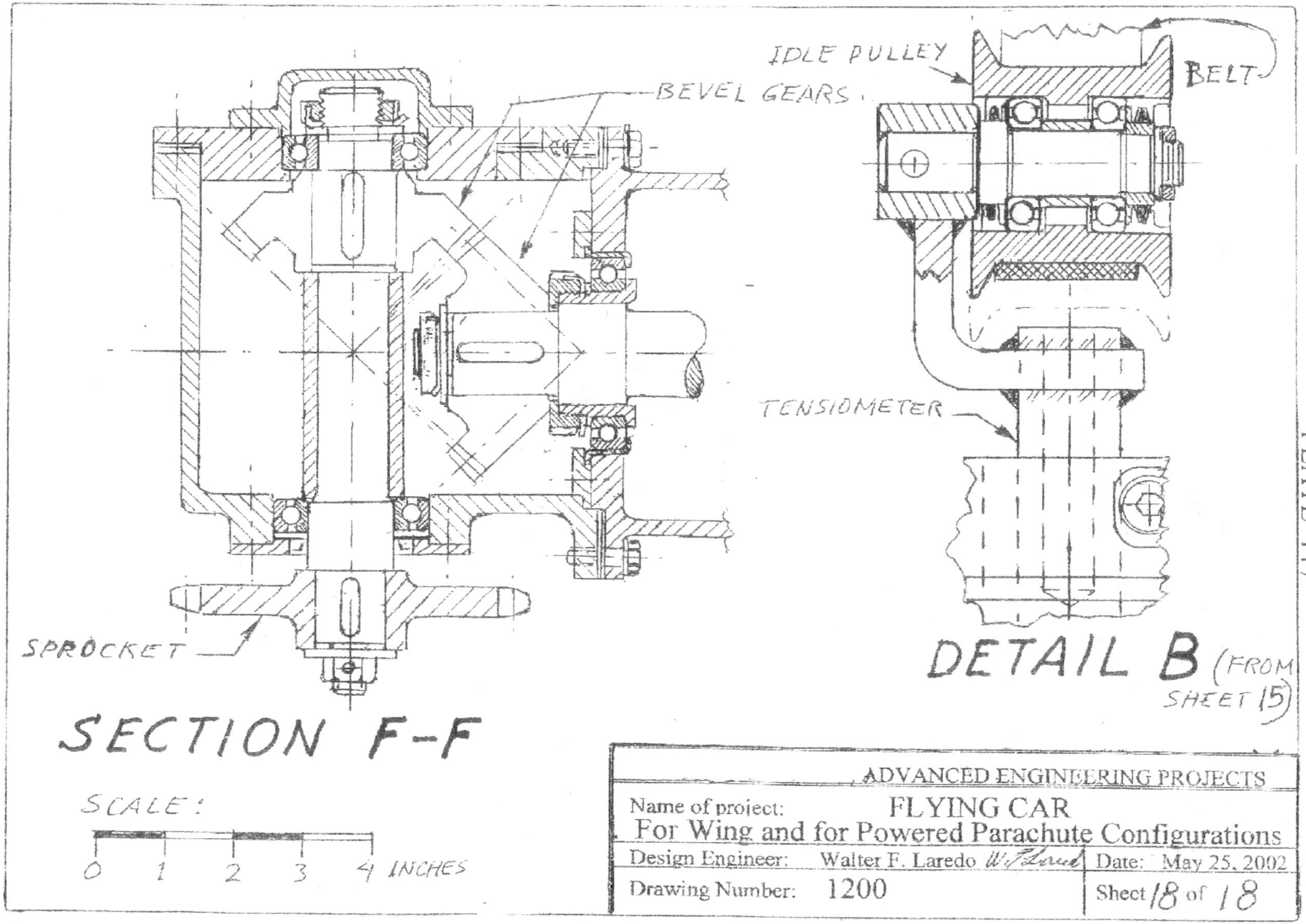

BELT

IDLE PULLEY

BEVEL GEARS

TENSIOMETER

DETAIL B (FROM SHEET 15)

SPROCKET

SECTION F-F

SCALE:

0 1 2 3 4 INCHES

ADVANCED ENGINEERING PROJECTS

FLYING CAR

Name of project: For Wing and for Powered Parachute Configurations

Design Engineer: Walter F. Laredo W. Laredo Date: May 25, 2002

Drawing Number: 1200

Sheet 18 of 18

PLATE 118

Sheet No.	Name	Quantity	Material
Sheet 17	ARM, FOREARM & HAND ASSEMBLY	R.H. (1) L.H. (1)	STEEL
Sheets 14 & 16	ROBOT'S FOOT	R.H. (1) L.H. (1)	STEEL
Sheets 13 & 15	ROBOT HIP AND PIVOTS	1	STEEL
Sheet 12	HYDRAULIC POWER SYSTEM	R.H. (1) L.H. (1)	—
Sheet 11	LEG SYSTEM ASSEMBLY	RIGHT HAND (1) LEFT HAND (1)	STEEL
Sheet 10	ROBOT'S STRUCTURAL ARRANGEMENT	1	—
Sheet 9	ROBOT'S GENERAL ARRANGEMENT	1	—
Sheet 8	WALKING CONTROL SYSTEM	1	—

ADVANCED ENGINEERING PROJECTS

Name of project:

GIANT ROBOT WORKER

MECHANICAL BEAST USED TO TRANSPORT HEAVY CARGO, BY
CARRYING IT THROUGH MOUNTAINS WITH NO ROADS.
CARGO INCLUDES BOULDERS AND MARBLE COLUMNS, TO BE USE TO
BUILD TEMPLES AND BUILDINGS .

Design Engineer: Walter F. Laredo	Date: August 8, 2002

Areas of Development: Stability Control ✔, Thermodynamics __, Materials ✔
Structures (Design and Analysis) ✔, Mechanisms & Systems ✔ .

NOTE: This is the PRELIMINARY DESIGN for a real engineering project.

Drawing Number: 1300	Sheet 1 of 17

PLATE 119

W. F. Laredo
8-8-03

ADVANCED ENGINEERING PROJECTS		
Name of project:	**GIANT ROBOT**	
It can transport heavy structures above mountains with no roads		
Design Engineer: Walter F. Laredo		Date: August 8, 2002
Drawing Number: 1300		Sheet 2 of 17

PLATE 120

GIANT WALKING ROBOT
STABILITY AND CONTROL SYSTEM

This robot operates automatically by its own electronic brain, however if necessary it could be operated by a human operator seating inside its head, but always with the help of the robot's computers.

The robot has three laser scanners, each act as a special eye, the one located on top of the robot's head is for distant scanning, 25 feet and beyond, it have 340 degrees field of view. It is also a laser range finder, which warns the robot's electronic brain of approaching dangers as trees, rocky walls or precipice edges. The other two, are short-range laser scanners located one forward of the robot's belly and the other behind the robot's hip. Both facing down to scan the ground, one scans few feet in front of the robot and the other few feet behind, scanning a piece of terrain not larger than the robot's footprint, the area where the robot will step next.

With an artificial intelligence of an insect, this robot has an artificial nervous system, which works as follows:

After scanning a miniature terrain either forward or behind the robot, the laser scanner transfers this 3D mapping information to the robot's brain, which with this DATA will order one of the legs to move, it depends, one step forward or one aft. Then the leg will move first up, then as it moves forward will rotate its foot to the proper angle in order that its underneath surface will match exactly the same average slope of the terrain it will step next.

The kinematics sequence of each robot's step begins at the moment when the robot raises its leg followed by a forward movement. When this leg is still in the air the foot rotates an angle that should match the same average slope of the terrain where it will step next. As the robot steps on this terrain, thick hairy sensors sticking out from its legs would sense the terrain characteristics such as softness, hardness, humid or dry or muddy. After that step, the robot will stand on both feet, one foot forward and the other behind, then without moving its feet the robot body will move (Continues in next sheet)

ADVANCED ENGINEERING PROJECTS

Name of project: **GIANT ROBOT**

It can transport heavy structures above mountains with no roads

Design Engineer: Walter F. Laredo	Date: August 8, 2002
Drawing Number: 1300	Sheet 3 of 17

PLATE 121

(Continues from preceding sheet)
forward, then rise vertically, with the body the position of the center of gravity will rise along a stable vertical theoretical line.

Here CG means center of gravity of a system that includes robot plus cargo or payload. The CG position is constantly corrected, in order for the robot to remain stable. This is done by the robot's stability and control system, which consist of a computer, a guidance system and load or pressure sensors located under the robot feet. The robot's feet are broad and rectangular, underneath each of the four-foot corner's there is a protrusion, all used to support the robot's weight and inside each protrusion there is a load sensor. When the robot stands on one foot only, its electronic brain receives signs of the feet load sensors. With this DATA, stability is computed several times per second and the robot position with its center of gravity is changed continuously by the Hydraulic power system, until the robot weight will be evenly distributed on the four protrusions of the standing foot. This stable condition will happen only if the vertical projection of the center of gravity falls inside the footprint or inside the area made by the four load sensors.

Each step of the robot represents a complete cycle of a set of operations, after this cycle was completed; the robot is ready to start the step, a new cycle, operation that could be repeated over and over.

The robot walking kinematics and its stability are controlled by the stability control and guidance systems, which consist of computers, load distribution, terrain pressure sensors, laser scanners range finders, a guidance system with gyros and two accelerometers. The guidance system considers the robot's platform as the theoretical horizontal plane of reference. Another plane of reference is a vertical one used for navigation purposes along the robot's trip, it extends from the point of the robot's departure to the point of its arrival.

The robot feet are designed in such way, when the robot is standing on both feet, the toes of one foot penetrates the toes spaces of the other foot, as interlocked combs, so both feet rest at the same time on the same surface or on a single footprint. Something similar as if both feet blended into a single-foot resting on a single footprint, a very stable structure.

ADVANCED ENGINEERING PROJECTS

Name of project: GIANT ROBOT

It can transport heavy structures above mountains with no roads

Design Engineer: Walter F. Laredo	Date: August 8, 2002
Drawing Number: 1300	Sheet 4 of 17

PLATE 122

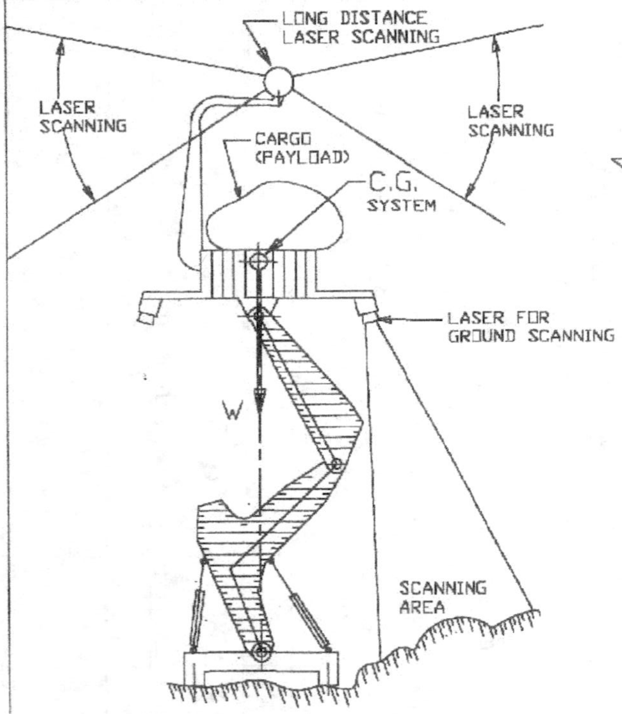

ENVELOPE FOR STABLE
C.G. LOCATIONS

C.G. UPPER POSITION

C.G. LOWER POSITION

STABLE C.G. TRAVEL RANGE (TILTED UP TERRAIN)

FORWARD

VERTICAL LINE FROM C.G. TO FOOTPRINT SURFACE

TILTED TERRAIN

HORIZONTAL TERRAIN

C.G.

W

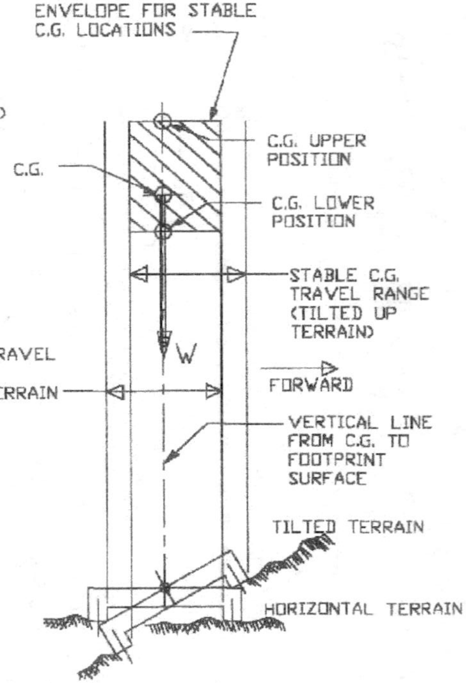

DIAGRAM OF ENVELOPE
FOR STABLE C.G. LOCATIONS
SIDE VIEW
(FRONT VIEW SIMILAR)

C.G. (CENTER OF GRAVITY OF SYSTEM, THE SYSTEM INCLUDES ROBOT AND CARGO)

STABLE C.G. TRAVEL RANGE FOR HORIZONTAL TERRAIN

SCANNING AREA

LEG

FOOT

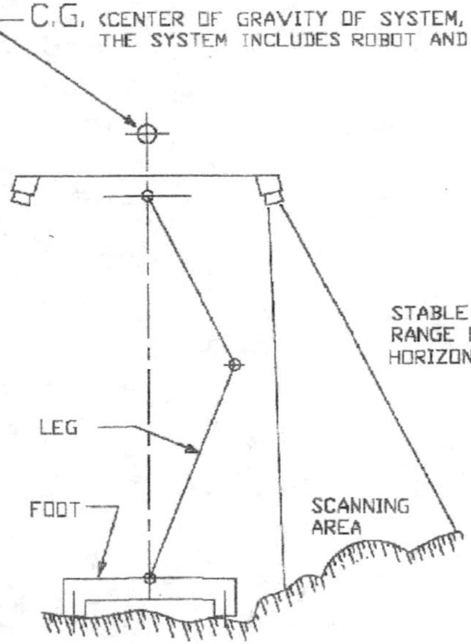

DIAGRAM OF
GIANT ROBOT
(SIDE VIEW)

LONG DISTANCE LASER SCANNING

LASER SCANNING

LASER FOR GROUND SCANNING

C.G. SYSTEM

CARGO (PAYLOAD)

SCANNING AREA

W

LASER SCANNING

ROBOT CONFIGURATION
(SIDE VIEW)

W = WEIGHT OF ROBOT + WEIGHT OF PAYLOAD

STABILITY AND
CONTROL

FOR ROBOT WALKING KINEMATICS
SEE NEXT TWO PAGES

ADVANCED ENGINEERING PROJECTS	
Name of project: GIANT ROBOT	
IT CAN TRANSPORT HEAVY STRUCTURES ABOVE MOUNTAINS WITH NO ROADS	
Design Engineer: WALTER F. LAREDO	Date: August 8, 2002
Drawing Number: 1300	Sheet 5 of 17

PLATE 123

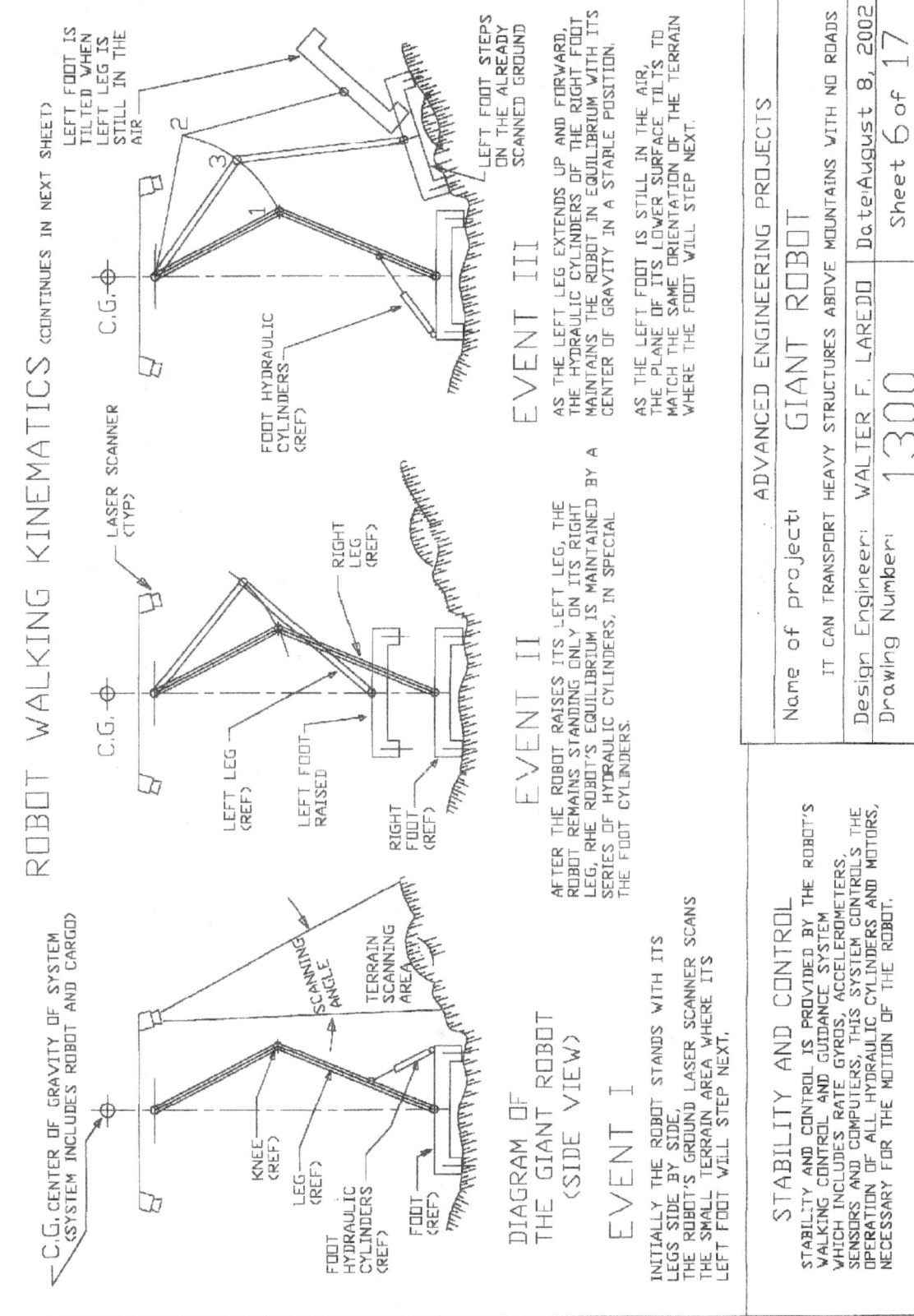

ROBOT WALKING KINEMATICS (CONTINUES IN NEXT SHEET)

C.G. CENTER OF GRAVITY OF SYSTEM
(SYSTEM INCLUDES ROBOT AND CARGO)

C.G.

C.G.

C.G.

LEFT FOOT IS TILTED WHEN LEFT LEG IS STILL IN THE AIR

LEFT FOOT STEPS ON THE ALREADY SCANNED GROUND

LASER SCANNER (TYP)

FOOT HYDRAULIC CYLINDERS (REF)

RIGHT LEG (REF)

LEFT LEG (REF)

LEFT FOOT RAISED

RIGHT FOOT (REF)

SCANNING ANGLE

TERRAIN SCANNING AREA

KNEE (REF)

LEG (REF)

FOOT HYDRAULIC CYLINDERS (REF)

FOOT (REF)

DIAGRAM OF THE GIANT ROBOT (SIDE VIEW)

EVENT I

INITIALLY THE ROBOT STANDS WITH ITS LEGS SIDE BY SIDE.
THE ROBOT'S GROUND LASER SCANNER SCANS THE SMALL TERRAIN AREA WHERE ITS LEFT FOOT WILL STEP NEXT.

EVENT II

AFTER THE ROBOT RAISES ITS LEFT LEG, THE ROBOT REMAINS STANDING ONLY ON ITS RIGHT LEG, THE ROBOT'S EQUILIBRIUM IS MAINTAINED BY A SERIES OF HYDRAULIC CYLINDERS, IN SPECIAL THE FOOT CYLINDERS.

EVENT III

AS THE LEFT LEG EXTENDS UP AND FORWARD, THE HYDRAULIC CYLINDERS OF THE RIGHT FOOT MAINTAINS THE ROBOT IN EQUILIBRIUM WITH ITS CENTER OF GRAVITY IN A STABLE POSITION.

AS THE LEFT FOOT IS STILL IN THE AIR, THE PLANE OF ITS LOWER SURFACE TILTS TO MATCH THE SAME ORIENTATION OF THE TERRAIN WHERE THE FOOT WILL STEP NEXT.

STABILITY AND CONTROL

STABILITY AND CONTROL IS PROVIDED BY THE ROBOT'S WALKING CONTROL AND GUIDANCE SYSTEM WHICH INCLUDES RATE GYROS, ACCELEROMETERS, SENSORS AND COMPUTERS, THIS SYSTEM CONTROLS THE OPERATION OF ALL HYDRAULIC CYLINDERS AND MOTORS, NECESSARY FOR THE MOTION OF THE ROBOT.

ADVANCED ENGINEERING PROJECTS		
Name of project: GIANT ROBOT		
IT CAN TRANSPORT HEAVY STRUCTURES ABOVE MOUNTAINS WITH NO ROADS		
Design Engineer: WALTER F. LAREDO	Date: August 8, 2002	
Drawing Number: 1300	Sheet 6 of 17	

PLATE 124

ROBOT WALKING KINEMATICS (CONTINUATION OF PRIOR SHEET)

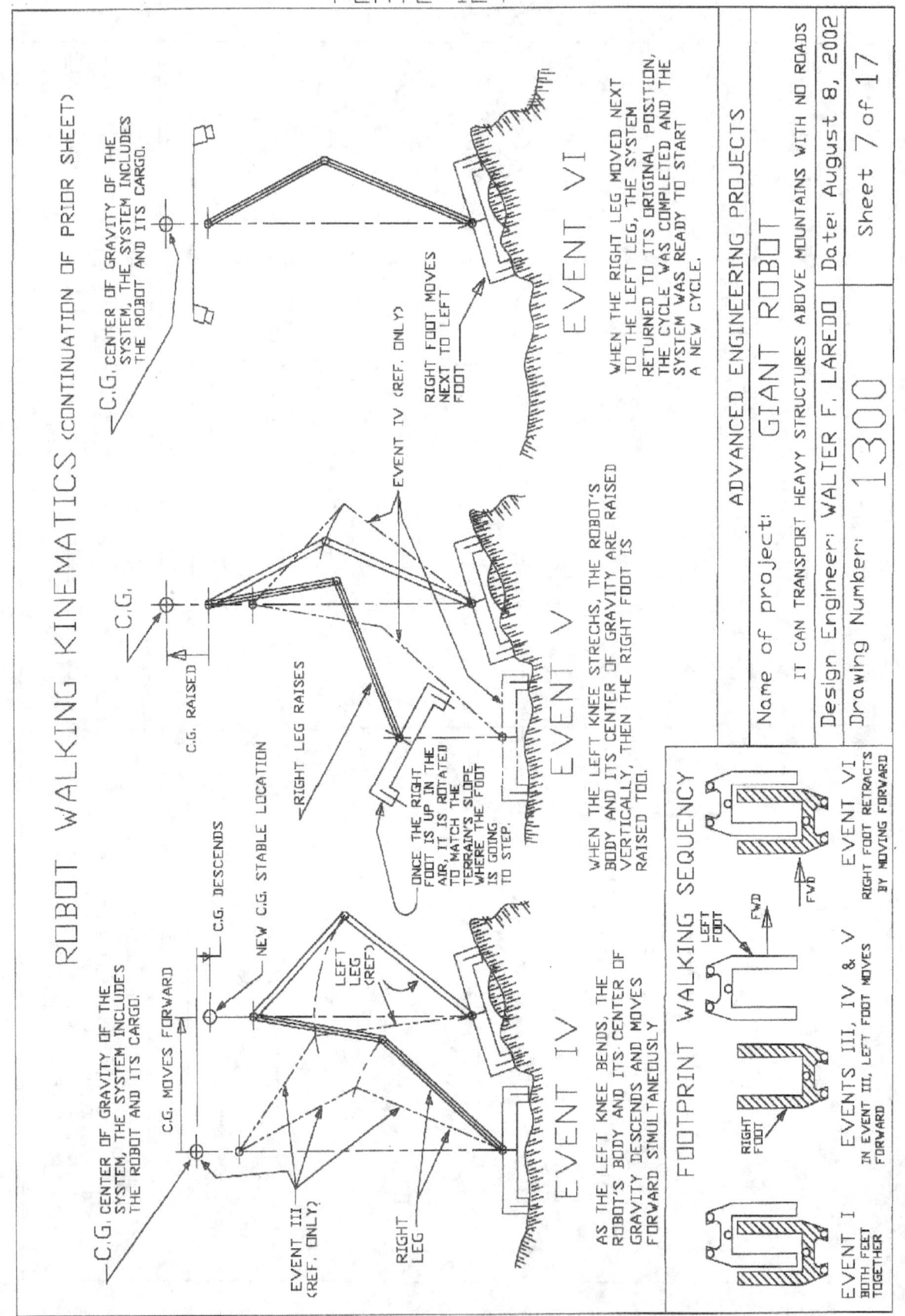

C.G. CENTER OF GRAVITY OF THE SYSTEM, THE SYSTEM INCLUDES THE ROBOT AND ITS CARGO.

C.G. CENTER OF GRAVITY OF THE SYSTEM, THE SYSTEM INCLUDES THE ROBOT AND ITS CARGO.

C.G.

C.G. RAISED

RIGHT LEG RAISES

C.G. DESCENDS

C.G. MOVES FORWARD

NEW C.G. STABLE LOCATION

EVENT IV (REF. ONLY)

RIGHT FOOT MOVES NEXT TO LEFT FOOT

EVENT VI

WHEN THE RIGHT LEG MOVED NEXT TO THE LEFT LEG, THE SYSTEM RETURNED TO ITS ORIGINAL POSITION, THE CYCLE WAS COMPLETED AND THE SYSTEM WAS READY TO START A NEW CYCLE.

ONCE THE RIGHT FOOT IS UP IN THE AIR, IT IS ROTATED TO MATCH THE TERRAIN'S SLOPE WHERE THE FOOT IS GOING TO STEP.

EVENT V

WHEN THE LEFT KNEE STRECHS, THE ROBOT'S BODY AND ITS CENTER OF GRAVITY ARE RAISED VERTICALLY, THEN THE RIGHT FOOT IS RAISED TOO.

EVENT III (REF. ONLY)

RIGHT LEG

LEFT LEG (REF.)

EVENT IV

AS THE LEFT KNEE BENDS, THE ROBOT'S BODY AND ITS CENTER OF GRAVITY DESCENDS AND MOVES FORWARD SIMULTANEOUSLY.

FOOTPRINT WALKING SEQUENCY

LEFT FOOT

RIGHT FOOT

FWD

FWD

EVENT VI
RIGHT FOOT RETRACTS BY MOVING FORWARD

EVENTS III, IV & V
IN EVENT III, LEFT FOOT MOVES FORWARD

EVENT I
BOTH FEET TOGETHER

ADVANCED ENGINEERING PROJECTS		
Name of project:	GIANT ROBOT	
	IT CAN TRANSPORT HEAVY STRUCTURES ABOVE MOUNTAINS WITH NO ROADS	
Design Engineer: WALTER F. LAREDO	Date: August 8, 2002	
Drawing Number: 1300		Sheet 7 of 17

PLATE 125

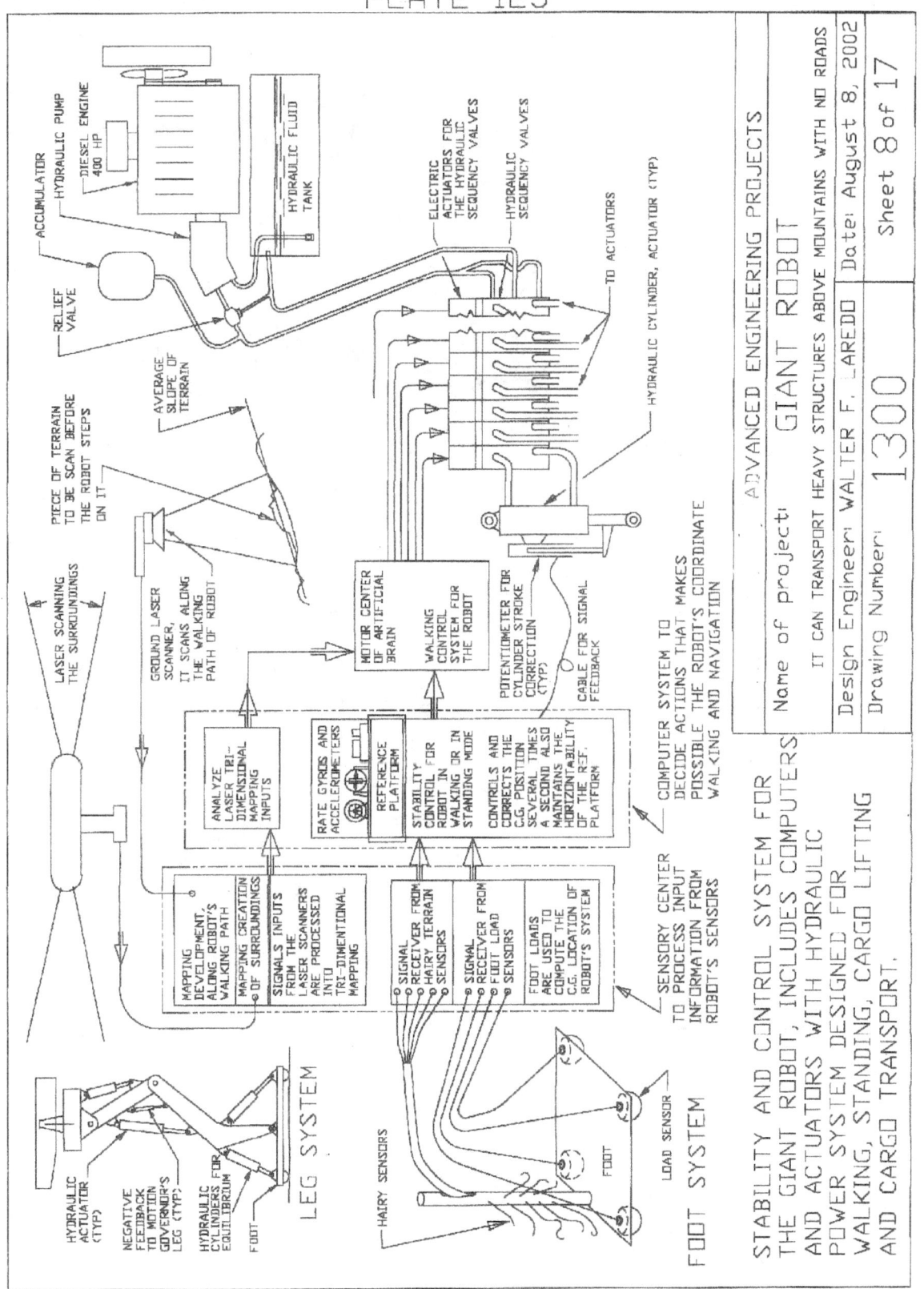

Name of project: **GIANT ROBOT**

IT CAN TRANSPORT HEAVY STRUCTURES ABOVE MOUNTAINS WITH NO ROADS

Design Engineer: WALTER F. LAREDO   Date: August 8, 2002

Drawing Number: **1300**   Sheet 8 of 17

ADVANCED ENGINEERING PROJECTS

STABILITY AND CONTROL SYSTEM FOR THE GIANT ROBOT, INCLUDES COMPUTERS AND ACTUATORS WITH HYDRAULIC POWER SYSTEM DESIGNED FOR WALKING, STANDING, CARGO LIFTING AND CARGO TRANSPORT.

LEG SYSTEM

FOOT SYSTEM

PLATE 126

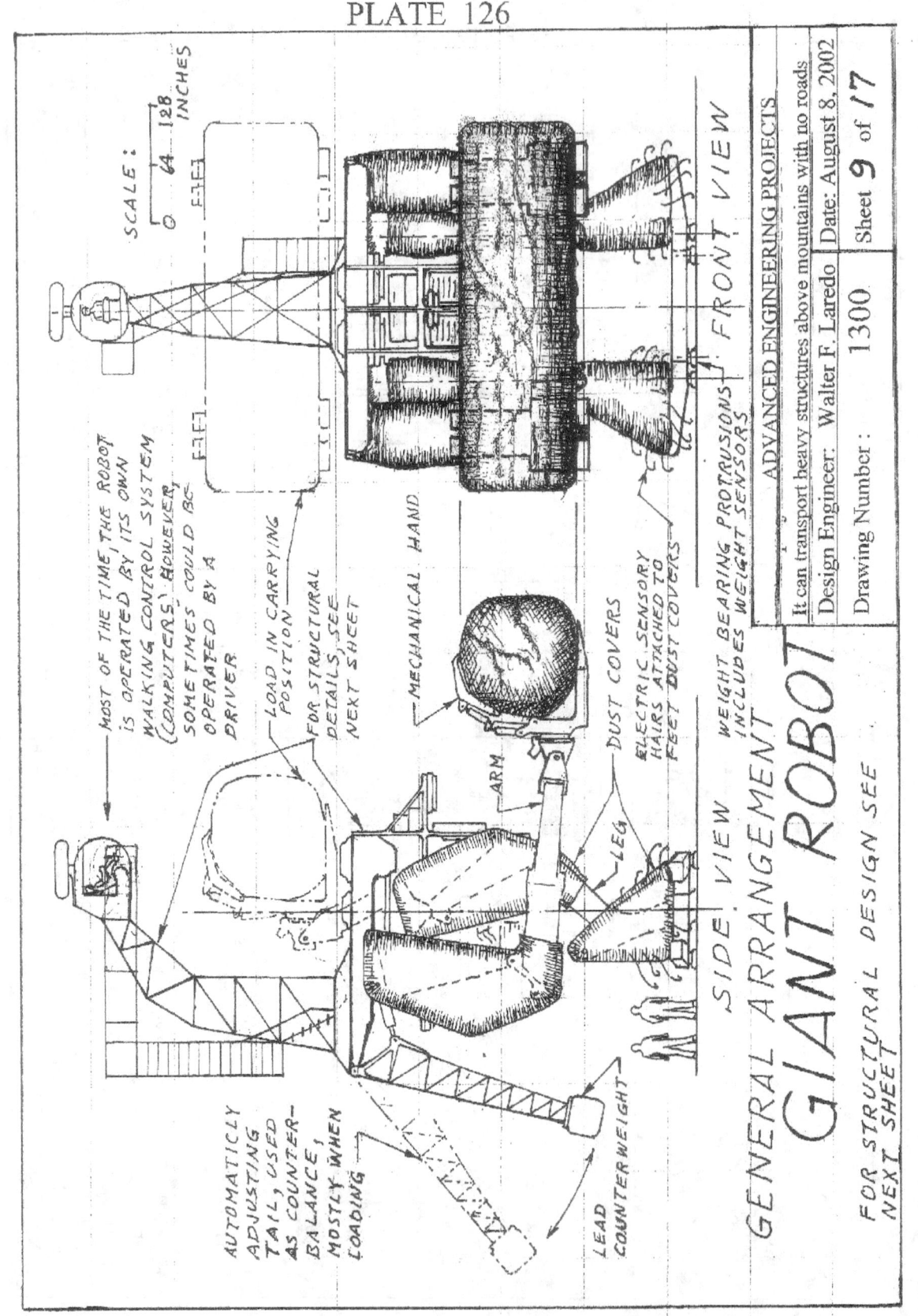

SCALE:

0 64 128 INCHES

MOST OF THE TIME THE ROBOT IS OPERATED BY ITS OWN WALKING CONTROL SYSTEM (COMPUTERS) HOWEVER, SOMETIMES COULD BE OPERATED BY A DRIVER

LOAD IN CARRYING POSITION

FOR STRUCTURAL DETAILS, SEE NEXT SHEET

MECHANICAL HAND

ARM

LEG

DUST COVERS

ELECTRIC SENSORY HAIRS ATTACHED TO FEET DUST COVERS

FRONT VIEW

AUTOMATICLY ADJUSTING TAIL, USED AS COUNTER-BALANCE, MOSTLY WHEN LOADING

LEAD COUNTERWEIGHT

SIDE VIEW   WEIGHT   WEIGHT BEARING PROTRUSIONS INCLUDES WEIGHT SENSORS

GENERAL ARRANGEMENT

GIANT ROBOT

FOR STRUCTURAL DESIGN SEE NEXT SHEET

ADVANCED ENGINEERING PROJECTS		
It can transport heavy structures above mountains with no roads		
Design Engineer: Walter F. Laredo	Date: August 8, 2002	
Drawing Number: 1300	Sheet **9** of *17*

PLATE 127

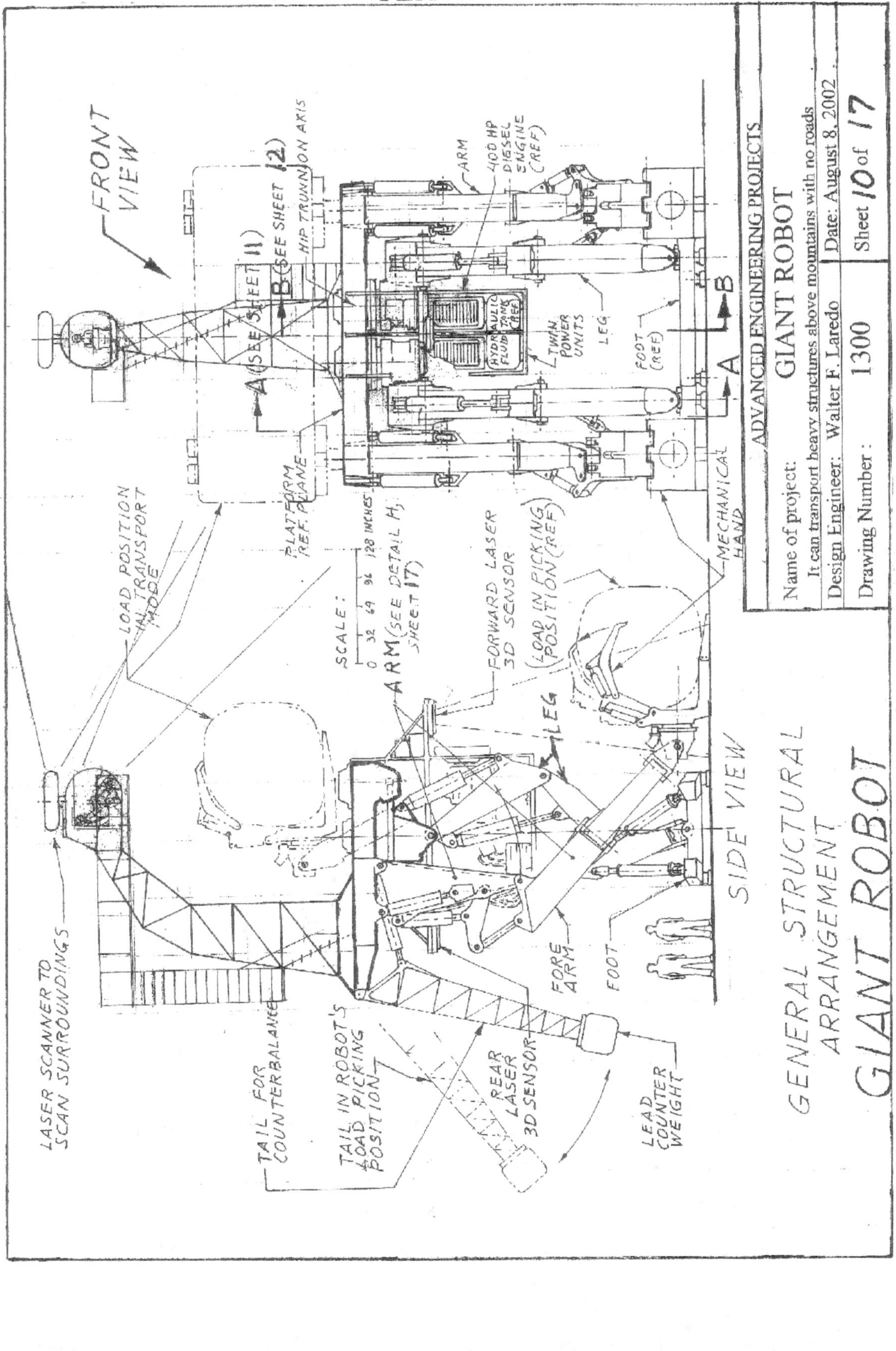

FRONT VIEW

A (SEE SHEET 11)
B (SEE SHEET 12)
HIP TRUNNION AXIS

ARM
400 HP DIESEL ENGINE (REF)

HYDRAULIC FLUID TANK (REF)

TWIN POWER UNITS

LEG

FOOT (REF)

A — A
B — B

LOAD POSITION IN TRANSPORT MODE

PLATFORM REF. PLANE

SCALE:
0 32 64 96 128 INCHES

ARM (SEE DETAIL H, SHEET 17)

FORWARD LASER 3D SENSOR

LOAD IN PICKING POSITION (REF)

LEG

MECHANICAL HAND

FORE ARM

FOOT

LASER SCANNER TO SCAN SURROUNDINGS

TAIL FOR COUNTERBALANCE

TAIL IN ROBOT'S LOAD PICKING POSITION

REAR LASER 3D SENSOR

LEAD COUNTER WEIGHT

SIDE VIEW

GENERAL STRUCTURAL ARRANGEMENT

GIANT ROBOT

ADVANCED ENGINEERING PROJECTS		
Name of project:	GIANT ROBOT	
It can transport heavy structures above mountains with no roads		
Design Engineer: Walter F. Laredo	Date: August 8, 2002	
Drawing Number: 1300	Sheet 10 of 17	

PLATE 128

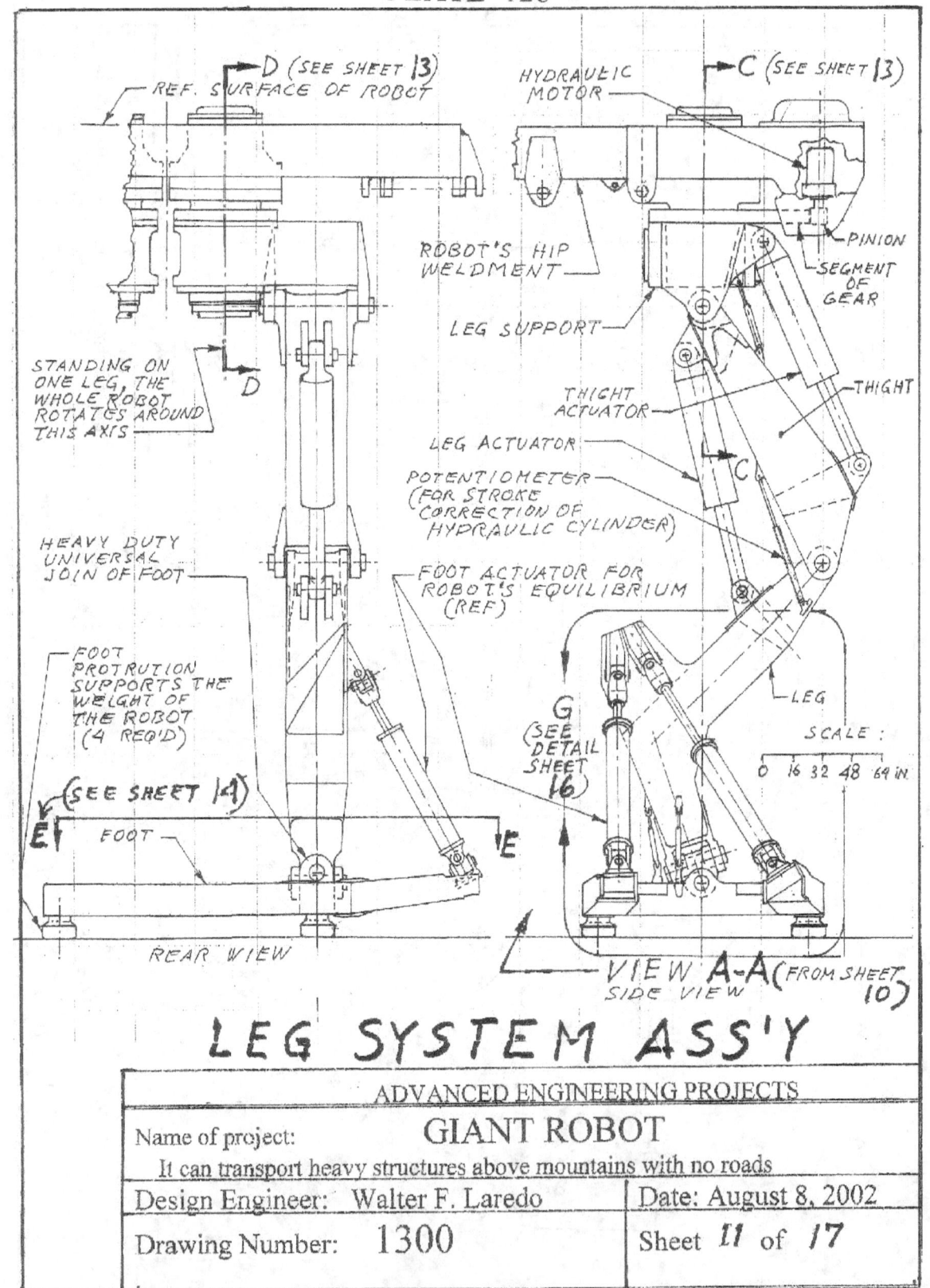

D (SEE SHEET 13)

REF. SURFACE OF ROBOT

HYDRAULIC MOTOR

C (SEE SHEET 13)

PINION

SEGMENT OF GEAR

ROBOT'S HIP WELDMENT

LEG SUPPORT

THIGHT

D

STANDING ON ONE LEG, THE WHOLE ROBOT ROTATES AROUND THIS AXIS

THIGHT ACTUATOR

LEG ACTUATOR

POTENTIOMETER (FOR STROKE CORRECTION OF HYDRAULIC CYLINDER)

HEAVY DUTY UNIVERSAL JOIN OF FOOT

FOOT ACTUATOR FOR ROBOT'S EQUILIBRIUM (REF)

LEG

C

FOOT PROTRUTION SUPPORTS THE WEIGHT OF THE ROBOT (4 REQ'D)

G (SEE DETAIL SHEET 16)

SCALE:

0 16 32 48 64 IN.

(SEE SHEET 14)

E

FOOT

E

REAR VIEW

VIEW A-A (FROM SHEET 10)
SIDE VIEW

LEG SYSTEM ASS'Y

ADVANCED ENGINEERING PROJECTS	
Name of project: **GIANT ROBOT**	
It can transport heavy structures above mountains with no roads	
Design Engineer: Walter F. Laredo	Date: August 8, 2002
Drawing Number: 1300	Sheet 11 of 17

PLATE 129

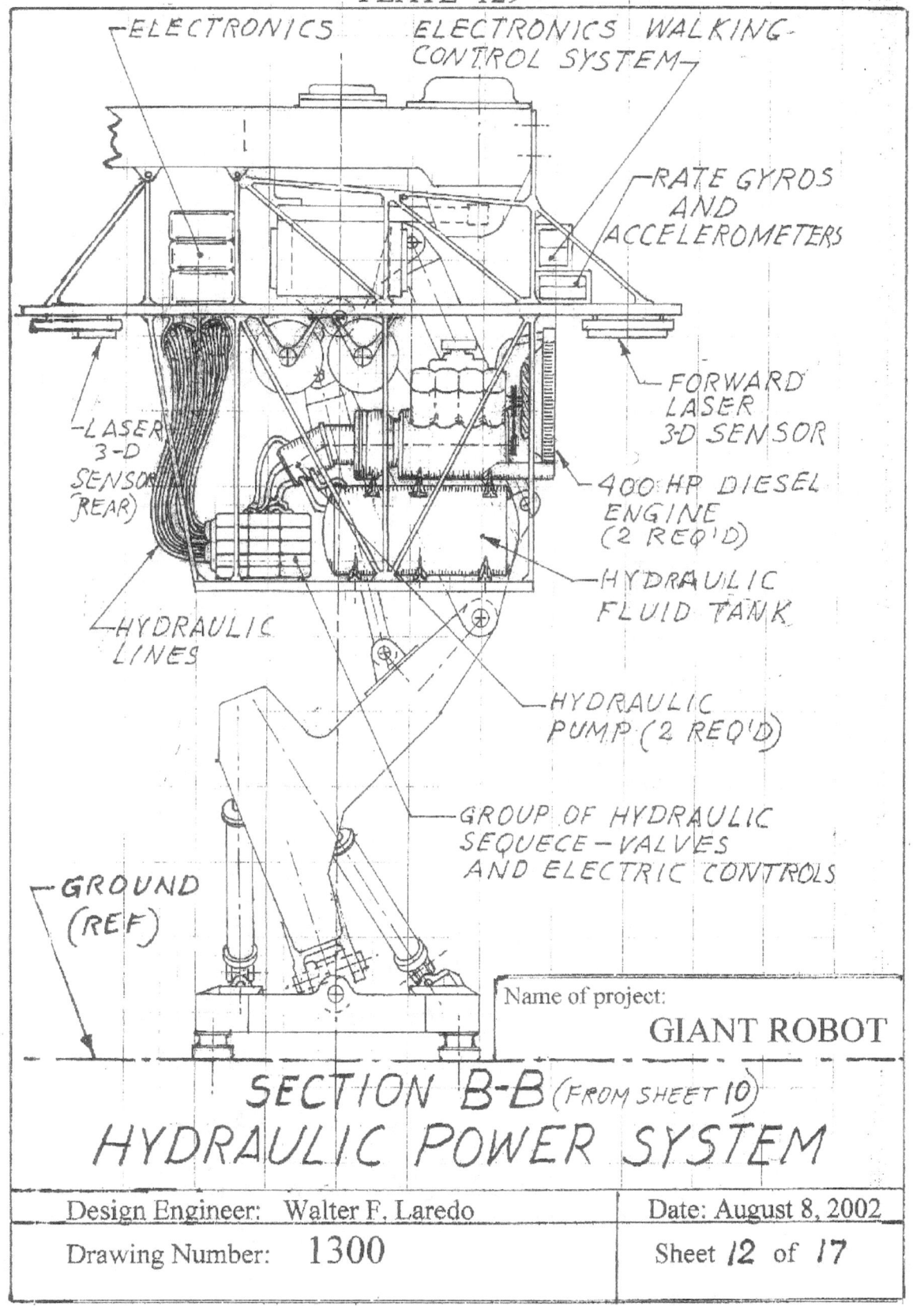

- ELECTRONICS
- ELECTRONICS WALKING CONTROL SYSTEM
- RATE GYROS AND ACCELEROMETERS
- FORWARD LASER 3-D SENSOR
- LASER 3-D SENSOR (REAR)
- 400 HP DIESEL ENGINE (2 REQ'D)
- HYDRAULIC FLUID TANK
- HYDRAULIC LINES
- HYDRAULIC PUMP (2 REQ'D)
- GROUP OF HYDRAULIC SEQUECE - VALVES AND ELECTRIC CONTROLS
- GROUND (REF)

Name of project:

GIANT ROBOT

SECTION B-B (FROM SHEET 10)
HYDRAULIC POWER SYSTEM

Design Engineer: Walter F. Laredo	Date: August 8, 2002
Drawing Number: 1300	Sheet 12 of 17

PLATE 130

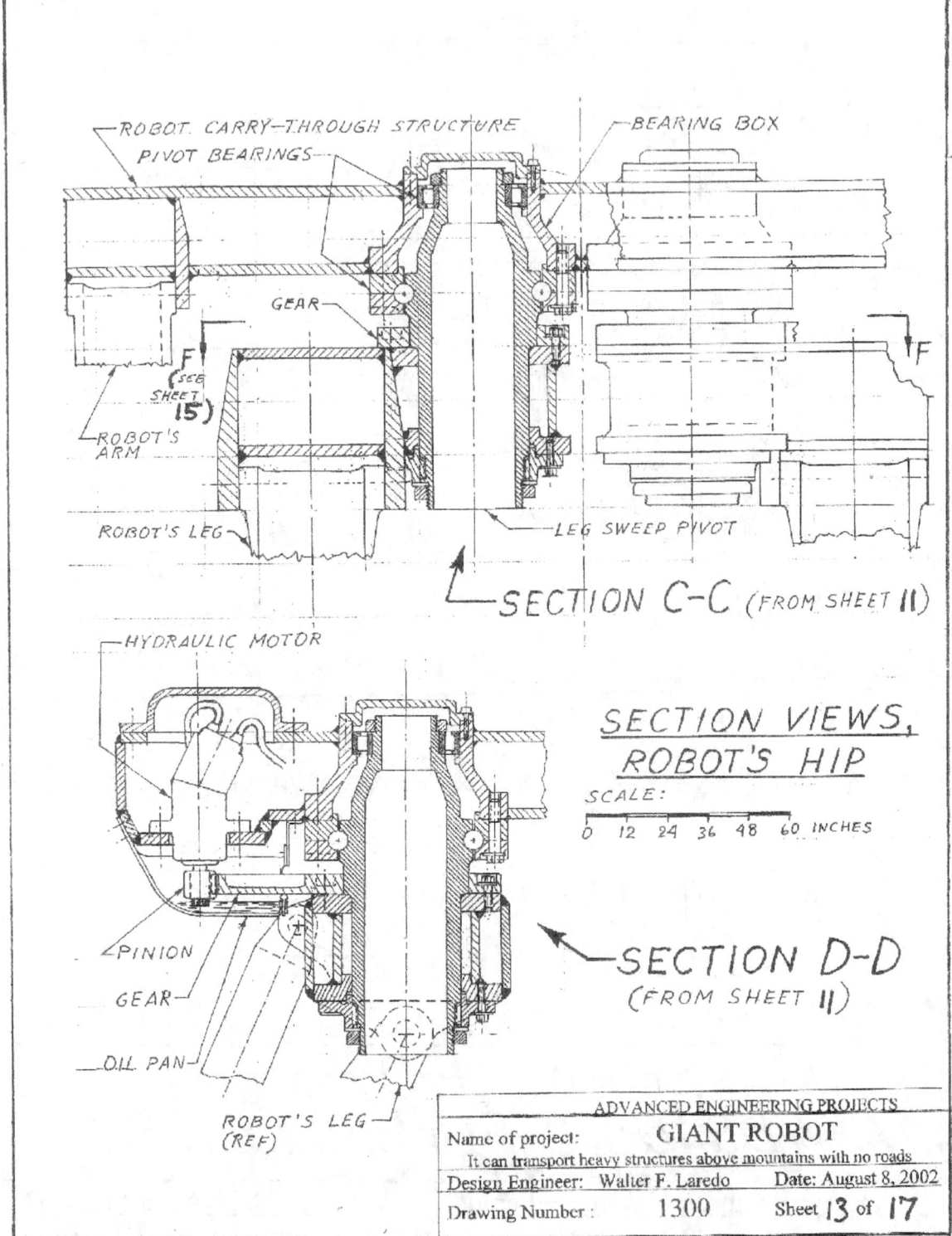

ROBOT. CARRY-THROUGH STRUCTURE

PIVOT BEARINGS

BEARING BOX

GEAR

F (SEE SHEET 15)

ROBOT'S ARM

ROBOT'S LEG

LEG SWEEP PIVOT

F

SECTION C–C (FROM SHEET 11)

HYDRAULIC MOTOR

SECTION VIEWS, ROBOT'S HIP

SCALE:

0   12   24   36   48   60 INCHES

PINION

GEAR

OIL PAN

SECTION D-D

(FROM SHEET 11)

ROBOT'S LEG (REF)

ADVANCED ENGINEERING PROJECTS

Name of project:	GIANT ROBOT	
It can transport heavy structures above mountains with no roads		
Design Engineer: Walter F. Laredo	Date: August 8, 2002	
Drawing Number :	1300	Sheet 13 of 17

PLATE 131

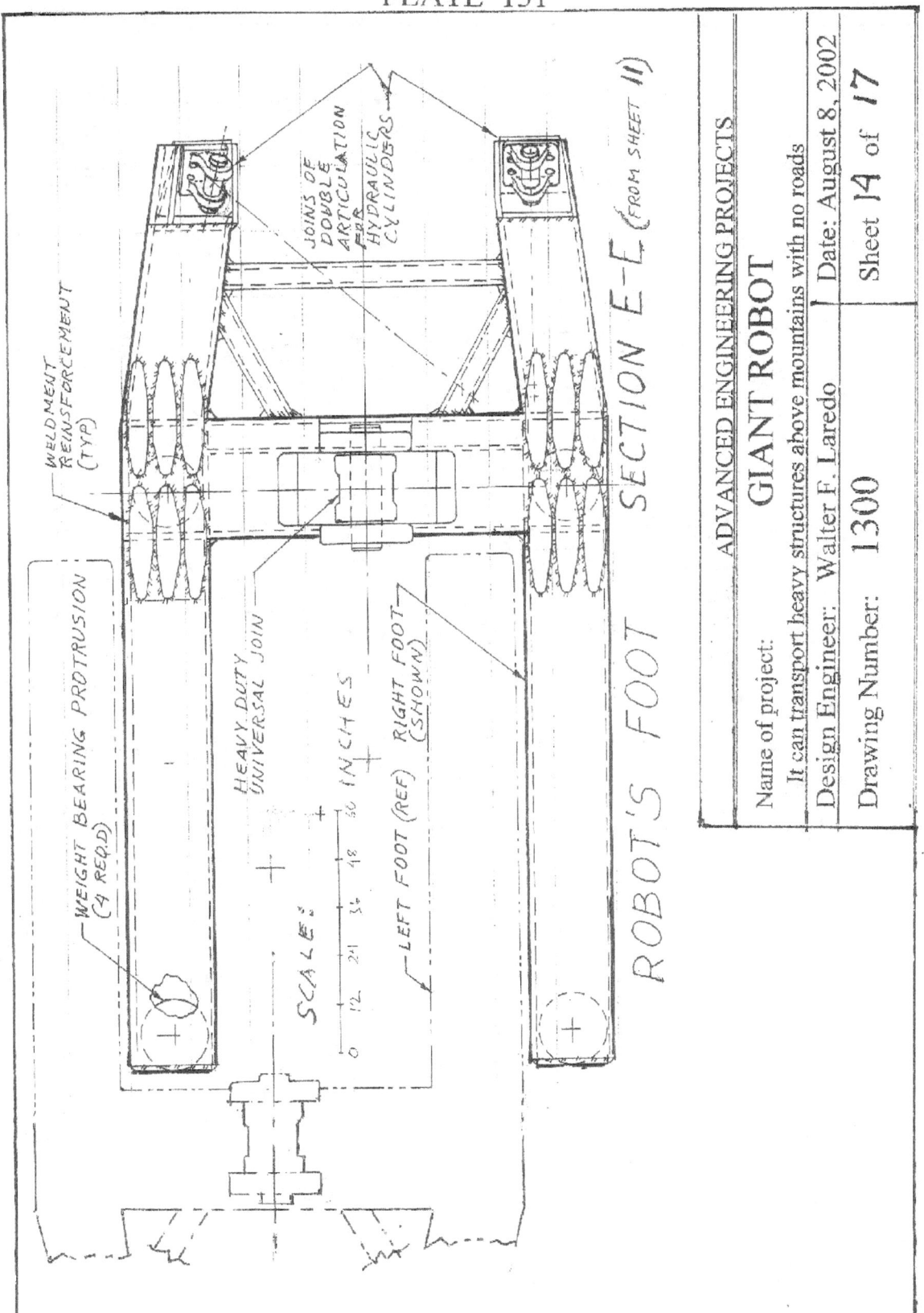

JOINS OF
DOUBLE
ARTICULATION
FOR
HYDRAULIC
CYLINDERS

WELDMENT
REINSFORCEMENT
(TYP)

WEIGHT BEARING PROTRUSION
(4 REQ.D)

HEAVY DUTY
UNIVERSAL JOIN

LEFT FOOT (REF) RIGHT FOOT
(SHOWN)

SCALE:

0   12   24   36   48   60 INCHES

SECTION E-E (FROM SHEET 11)

ROBOT'S FOOT

ADVANCED ENGINEERING PROJECTS		
GIANT ROBOT		
Name of project:		
It can transport heavy structures above mountains with no roads		
Design Engineer: Walter F. Laredo	Date: August 8, 2002	
Drawing Number: 1300	Sheet 14 of 17	

PLATE 132

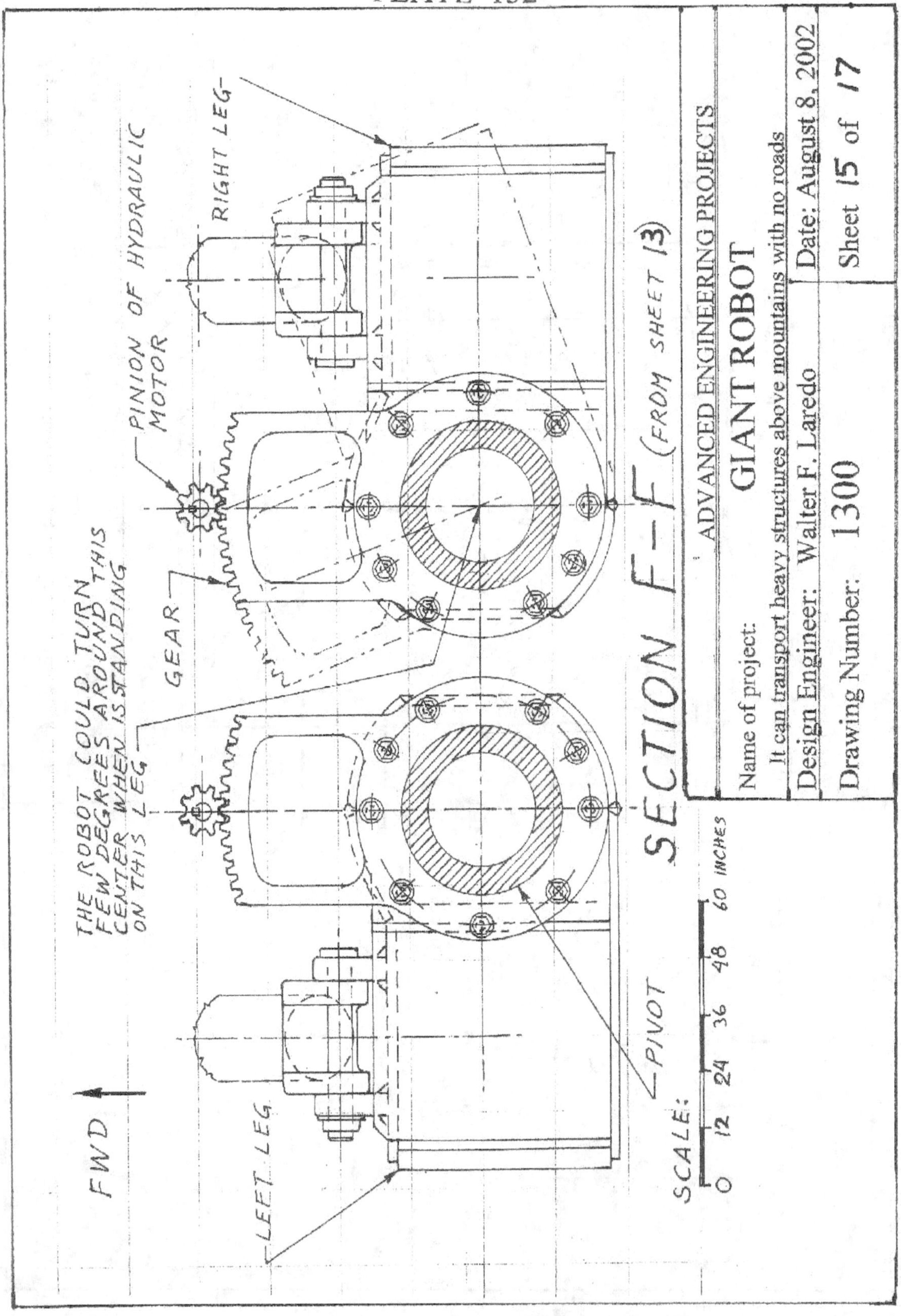

FWD

THE ROBOT COULD TURN
FEW DEGREES AROUND THIS
CENTER WHEN IS STANDING
ON THIS LEG

PINION OF HYDRAULIC
MOTOR

RIGHT LEG

GEAR

LEFT LEG

PIVOT

SCALE:

0   12   24   36   48   60 INCHES

SECTION F–F (FROM SHEET 13)

ADVANCED ENGINEERING PROJECTS		
GIANT ROBOT		
Name of project:		
It can transport heavy structures above mountains with no roads		
Design Engineer: Walter F. Laredo		Date: August 8, 2002
Drawing Number: 1300		Sheet 15 of 17

PLATE 133

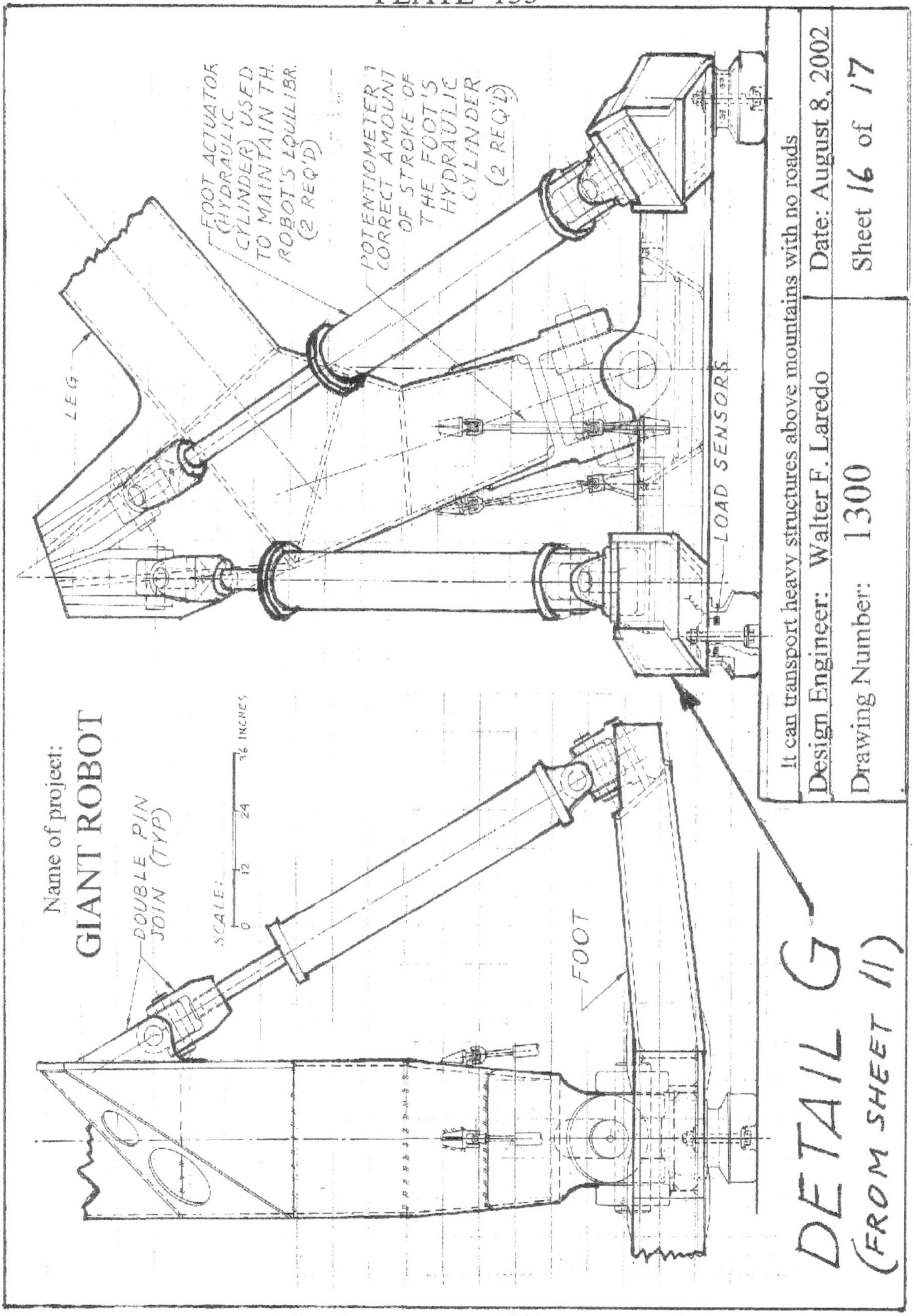

Name of project:
GIANT ROBOT

DOUBLE PIN
JOIN (TYP)

SCALE:

0   12   24   36 INCHES

FOOT ACTUATOR
(HYDRAULIC
CYLINDER) USED
TO MAINTAIN TH
ROBOT'S EQUILIBR.
(2 REQ'D)

POTENTIOMETER
CORRECT AMOUNT
OF STROKE OF
THE FOOT'S
HYDRAULIC
CYLINDER
(2 REQ'D)

LEG

FOOT

LOAD SENSORS

DETAIL G
(FROM SHEET 11)

It can transport heavy structures above mountains with no roads	
Design Engineer: Walter F. Laredo	Date: August 8, 2002
Drawing Number: 1300	Sheet 16 of 17

PLATE 134

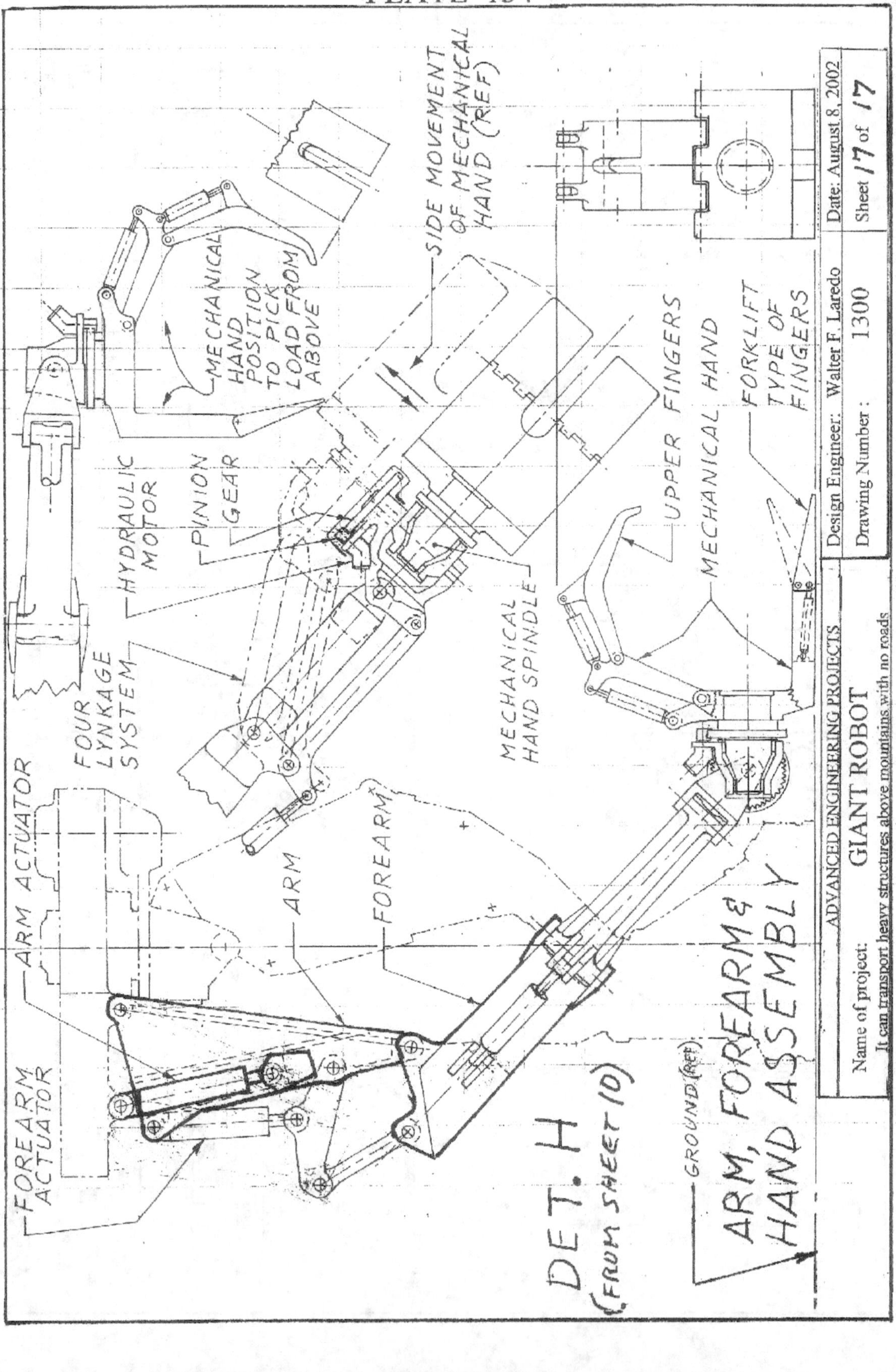

SIDE MOVEMENT OF MECHANICAL HAND (REF)

MECHANICAL HAND POSITION TO PICK LOAD FROM ABOVE

HYDRAULIC MOTOR

PINION GEAR

FOUR LYNKAGE SYSTEM

ARM ACTUATOR

MECHANICAL HAND SPINDLE

ARM

FOREARM

UPPER FINGERS

MECHANICAL HAND

FORKLIFT TYPE OF FINGERS

FOREARM ACTUATOR

DET. H.
(FROM SHEET 10)

GROUND (REF)

ARM, FOREARM & HAND ASSEMBLY

ADVANCED ENGINEERING PROJECTS

GIANT ROBOT

Design Engineer: Walter F. Laredo	Date: August 8, 2002
Drawing Number: 1300	Sheet 17 of 17

Name of project:

It can transport heavy structures above mountains with no roads

THE MARS SPACESHIP

The Mars Spaceship will be designed to be a transport for round trips to planet Mars. However in the 21st Century it could become the standard type of transport for interplanetary travel, including trips not only to Mars, but also to our Moon, to the asteroids, the Jupiter moons, and Saturn's Titan. Depending from the length of time required for a particular voyage, the ship should be fit with the proper type of propulsion system, which are interchangeable.

26 pages of preliminary designs follows this sheet, those are real designs draw to scale, which may be used as a guidance in order to develop the final design step for a real engineering project including construction drawings.

Testing of the ship components should be included in the main program, where testing will be performed on the ground as well as on space, where some modules of the International Space Station would become laboratories for this testing purpose.

Sometimes long continuous running testing as long as three years of duration, will be required for fatigue and for durability of some structural-mechanical components, including the lubricated gears and bearings running in the vacuum, as the ones in the hubs of the major centrifugal rotating bodies of the spaceship.

The centrifugal bodies for the Mars spaceship could be tested for three years in space by mounting them on a strong support shaft attached to one side of the existent space station. Later it may become a permanent part of the International Space Station used as a graviting environment for the off of duty crew. For the actual Mars-spaceship should be build an identical assembly of centrifugal bodies which should contain in its inside, the living quarters, a small hospital and the hydrophonic gardens.

PLATE 135

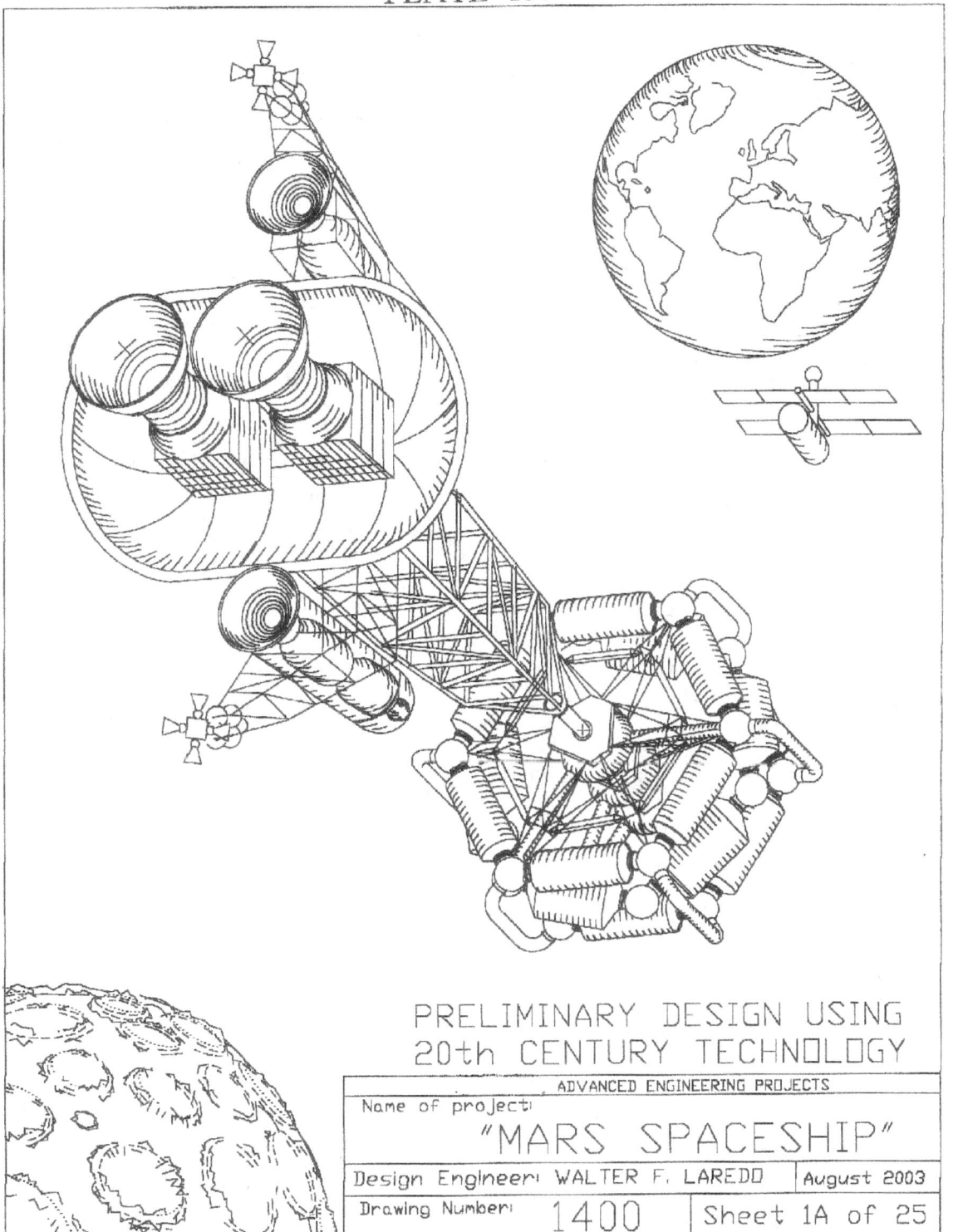

PRELIMINARY DESIGN USING
20th CENTURY TECHNOLOGY

ADVANCED ENGINEERING PROJECTS		
Name of project: "MARS SPACESHIP"		
Design Engineer: WALTER F. LAREDO		August 2003
Drawing Number: 1400		Sheet 1A of 25

PLATE 136

MARS SPACESHIP

VEHICLE GROSS WEIGHT: 1,000 METRIC TONS

PERFORMANCE

THE CRUISING SPEED OF THIS SHIP WITH RESPECT TO THE EARTH WILL VARY ACCORDING TO THE TYPE OF PROPULSION SYSTEM USED, FROM 38,000 MPH TO ABOVE. THE DESIRED ONE WAY TRANSIT TIME BETWEEN EARTH AND MARS WILL BE LESS THAN FIVE MONTHS, HOWEVER IT COULD TAKE MORE TIME DEPENDING OF THE THRUST OF THE PROPULSION SYSTEM USED BY THIS SPACECRAFT. A PROPULSION SYSTEM WITH A SPECIFIC IMPULSE OF 4,000 SECONDS OR MORE WILL BE DESIRABLE TO EXPLORE PART OF THIS SOLAR SYSTEM, HOWEVER A LESS SPECIFIC IMPULSE WOULD BE ACCEPTABLE AS MORE REALISTIC.

THE ACCELERATION AFTER DEPARTING FROM AN ORBIT AROUND A CELESTIAL BODY AND THE DECELERATION WHEN APPROACHING AND ARRIVING TO AN ORBIT OF ANOTHER ONE DEPENDS ON THE THRUST PROVIDED BY THE TYPE OF SPACE PROPULSION SYSTEM THAT WOULD BE USED ON THIS SPACESHIP. THE NUCLEAR FUSION PROPULSION SYSTEM WOULD BE THE BEST, HOWEVER THE DEVELOPMENT OF VARIOUS TYPES OF ELECTRICAL PROPULSION SYSTEM SEEMS MORE FEASABLE FOR THE NEAR FUTURE.

SOME PROPULSION INFORMATION IS SHOWN IN SHEET 20 FROM THIS PRELIMINARY DESIGN PACKAGE.

TEST FLIGHTS

PRIOR TO THE VOYAGE TO MARS, IT WOULD BE NECESSARY TO PERFORM TWO SHORT TEST FLIGHTS, DEPARTING EITHER FROM A SPACE STATION, OR DIRECTLY FROM AN EARTH PARKING ORBIT TO AN ORBIT AROUND THE MOON AND BACK. THIS IS A SPACECRAFT THAT WOULD TRAVEL AT HIGH INTERPLANETARY CRUISING SPEED HOWEVER ITS ACCELERATION AFTER DEPARTING FROM A PARKING ORBIT AND ITS DECELERATION WHEN APPROACHING TO THE OTHER PLANET ORBIT, WOULD BE RELATIVELY SLOW DUE TO ITS PROPULSION SYSTEM WITH A SMALL THRUST, BUT WITH THE GREAT ADVANTAGE OF ACCELERATING FOR LONG PERIODS OF TIME.

ARTIFICIAL GRAVITY

THE CIRCULAR ROTATING BODIES OF THIS SPACESHIP (WITH AN AVERAGE RADIUS OF 40 FEET) WILL PROVIDE TO THE CREW AND PASSENGERS INSIDE IT WITH A CENTRIFUGAL FORCE (ARTIFICIAL GRAVITY) EQUIVALENT TO SOMEWHERE BETWEEN ONE THIRD AND ONE QUARTER OF THE ACCELERATION OF GRAVITY AT THE EARTH'S SURFACE. A CENTRIPETAL FORCE IS PRODUCED BY ROTATING THESE BODIES BETWEEN 5 AND 4.38 RPM RESPECTIVELY. THE ROTATION OF THESE BODIES WILL START FROM ZERO AND SLOWLY BUILD UP TO THE FULL RPM'S IN FOUR HOURS, AN OPERATION THAT SHOULD BE PERFORMED PRIOR TO BOARDING THE SHIP.

NOTES:

FOR CONSTRUCTION MATERIALS, INSULATION INSTALLATION, STRUCTURAL ASSEMBLY, INSTALLATION OF SYSTEMS, MATERIALS PROCESSES, ETC., SEE THE ACTUAL PROPOSAL BOOK.

FLIGHT PROFILE FOR THE "MARS SPACESHIP" MISSION

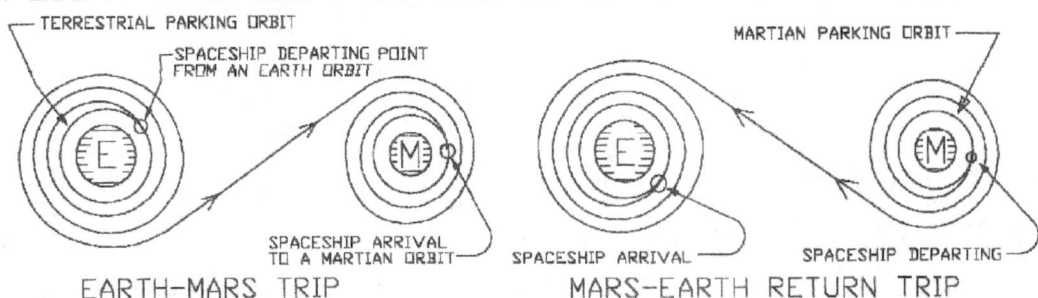

EARTH-MARS TRIP   MARS-EARTH RETURN TRIP

ADVANCED ENGINEERING PROJECTS		
Name of project: "MARS SPACESHIP"		
Design Engineer: WALTER F. LAREDO	Date: August 2003	
Drawing Number: 1400	Sheet 1B of 25	

PLATE 137

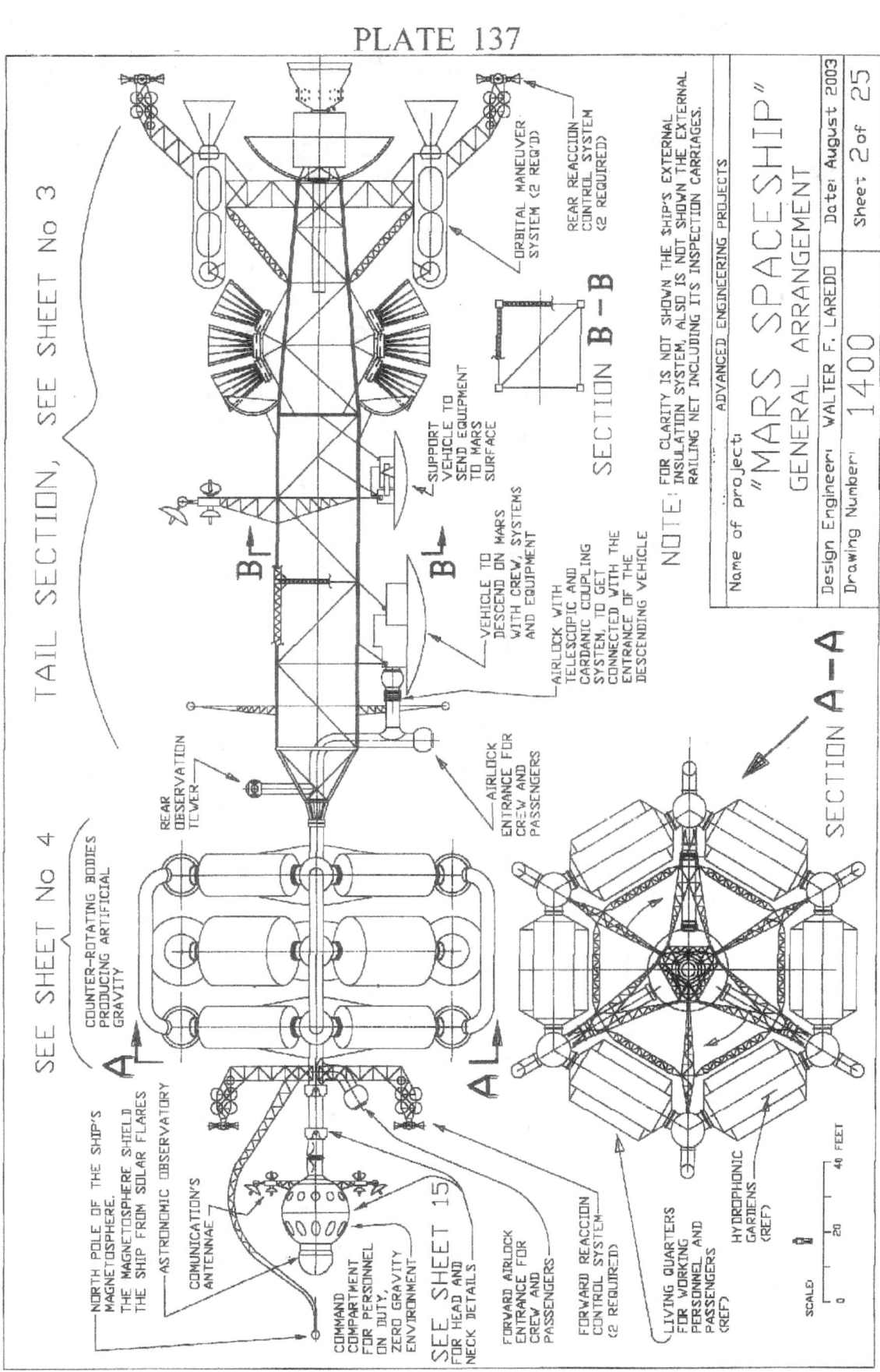

TAIL SECTION, SEE SHEET No 3

SEE SHEET No 4

COUNTER-ROTATING BODIES PRODUCING ARTIFICIAL GRAVITY

NORTH POLE OF THE SHIP'S MAGNETOSPHERE. THE MAGNETOSPHERE SHIELD THE SHIP FROM SOLAR FLARES

ASTRONOMIC OBSERVATORY

COMUNICATION'S ANTENNAE

COMMAND COMPARTMENT FOR PERSONNEL ON DUTY, ZERO GRAVITY ENVIRONMENT

SEE SHEET 15 FOR HEAD AND NECK DETAILS

FORWARD AIRLOCK ENTRANCE FOR CREW AND PASSENGERS

FORWARD REACCION CONTROL SYSTEM (2 REQUIRED)

LIVING QUARTERS FOR WORKING PERSONNEL AND PASSENGERS (REF)

HYDROPHONIC GARDENS (REF)

SCALE

0   20   40 FEET

SECTION A—A

REAR OBSERVATION TOWER

SUPPORT VEHICLE TO SEND EQUIPMENT TO MARS SURFACE

VEHICLE TO DESCEND ON MARS WITH CREW, SYSTEMS AND EQUIPMENT

AIRLOCK WITH TELESCOPIC AND CARDANIC COUPLING SYSTEM, TO GET CONNECTED WITH THE ENTRANCE OF THE DESCENDING VEHICLE

AIRLOCK ENTRANCE FOR CREW AND PASSENGERS

ORBITAL MANEUVER SYSTEM (2 REQ'D)

REAR REACCION CONTROL SYSTEM (2 REQUIRED)

SECTION B—B

B

B

A

A

NOTE: FOR CLARITY IS NOT SHOWN THE SHIP'S EXTERNAL INSULATION SYSTEM, ALSO IS NOT SHOWN THE EXTERNAL RAILING NET INCLUDING ITS INSPECTION CARRIAGES.

ADVANCED ENGINEERING PROJECTS

Name of project: "MARS SPACESHIP" GENERAL ARRANGEMENT

| Design Engineer: WALTER F. LAREDO | Date: August 2003 |
| Drawing Number: 1400 | Sheet: 2 of 25 |

PLATE 138

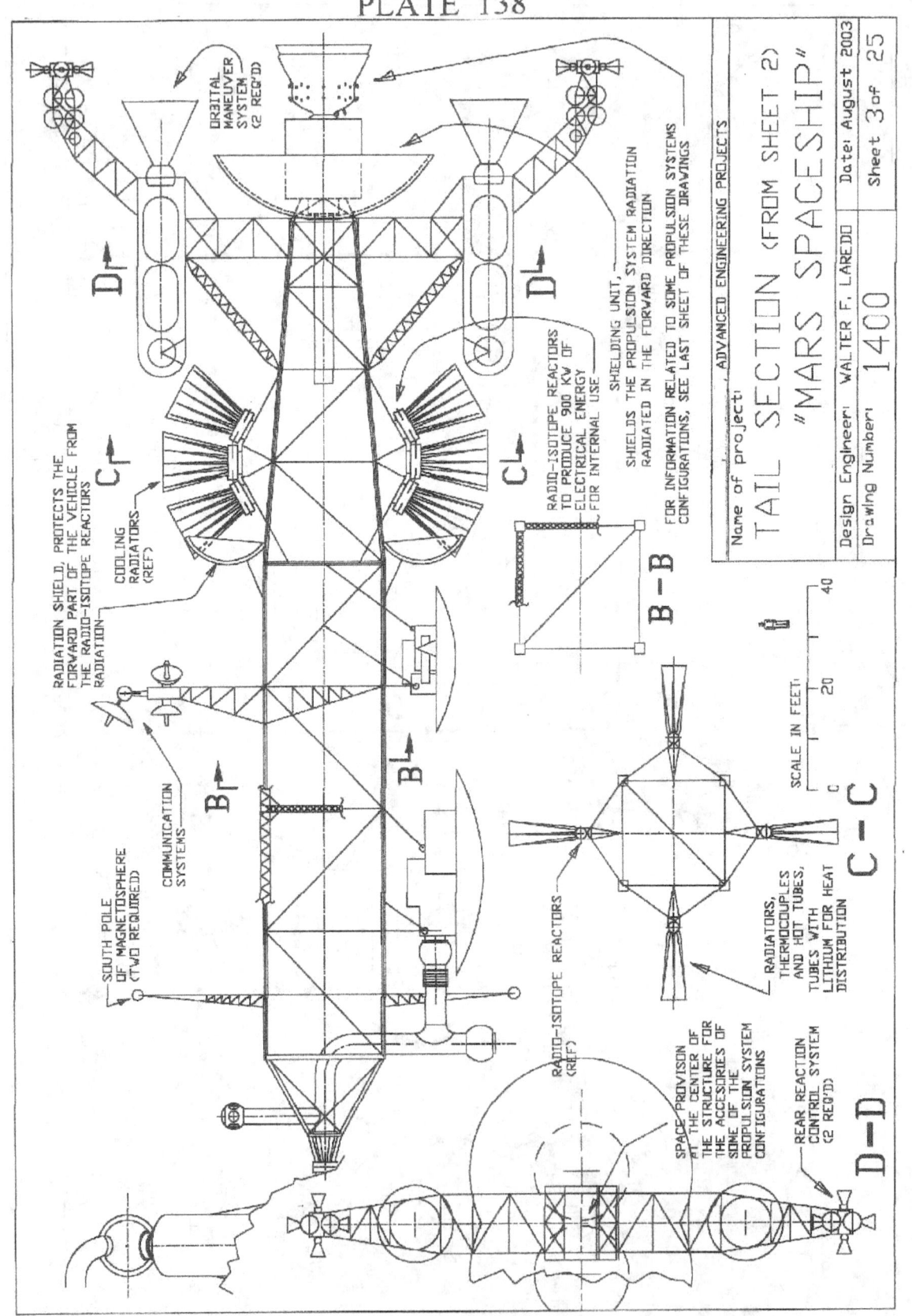

ORBITAL MANEUVER SYSTEM (2 REQ'D)

RADIATION SHIELD, PROTECTS THE FORWARD PART OF THE VEHICLE FROM THE RADIO-ISOTOPE REACTORS RADIATION

COOLING RADIATORS (REF)

SOUTH POLE OF MAGNETOSPHERE (TWO REQUIRED)

COMMUNICATION SYSTEMS

RADIO-ISOTOPE REACTORS TO PRODUCE 900 KW OF ELECTRICAL ENERGY FOR INTERNAL USE

SHIELDING UNIT, SHIELDS THE PROPULSION SYSTEM RADIATION RADIATED IN THE FORWARD DIRECTION

FOR INFORMATION RELATED TO SOME PROPULSION SYSTEMS CONFIGURATIONS, SEE LAST SHEET OF THESE DRAWINGS

B — B

SCALE IN FEET:
0   20   40

C — C

RADIATORS, THERMOCOUPLES AND HOT TUBES, TUBES WITH LITHIUM FOR HEAT DISTRIBUTION

RADIO-ISOTOPE REACTORS (REF)

SPACE PROVISION AT THE CENTER OF THE STRUCTURE FOR THE ACCESORIES OF SOME OF THE PROPULSION SYSTEM CONFIGURATIONS

REAR REACTION CONTROL SYSTEM (2 REQ'D)

D — D

Name of project: ADVANCED ENGINEERING PROJECTS

TAIL SECTION (FROM SHEET 2) "MARS SPACESHIP"

| Design Engineer: WALTER F. LAREDO | Date: August 2003 |
| Drawing Number: 1400 | Sheet 3 of 25 |

PLATE 139

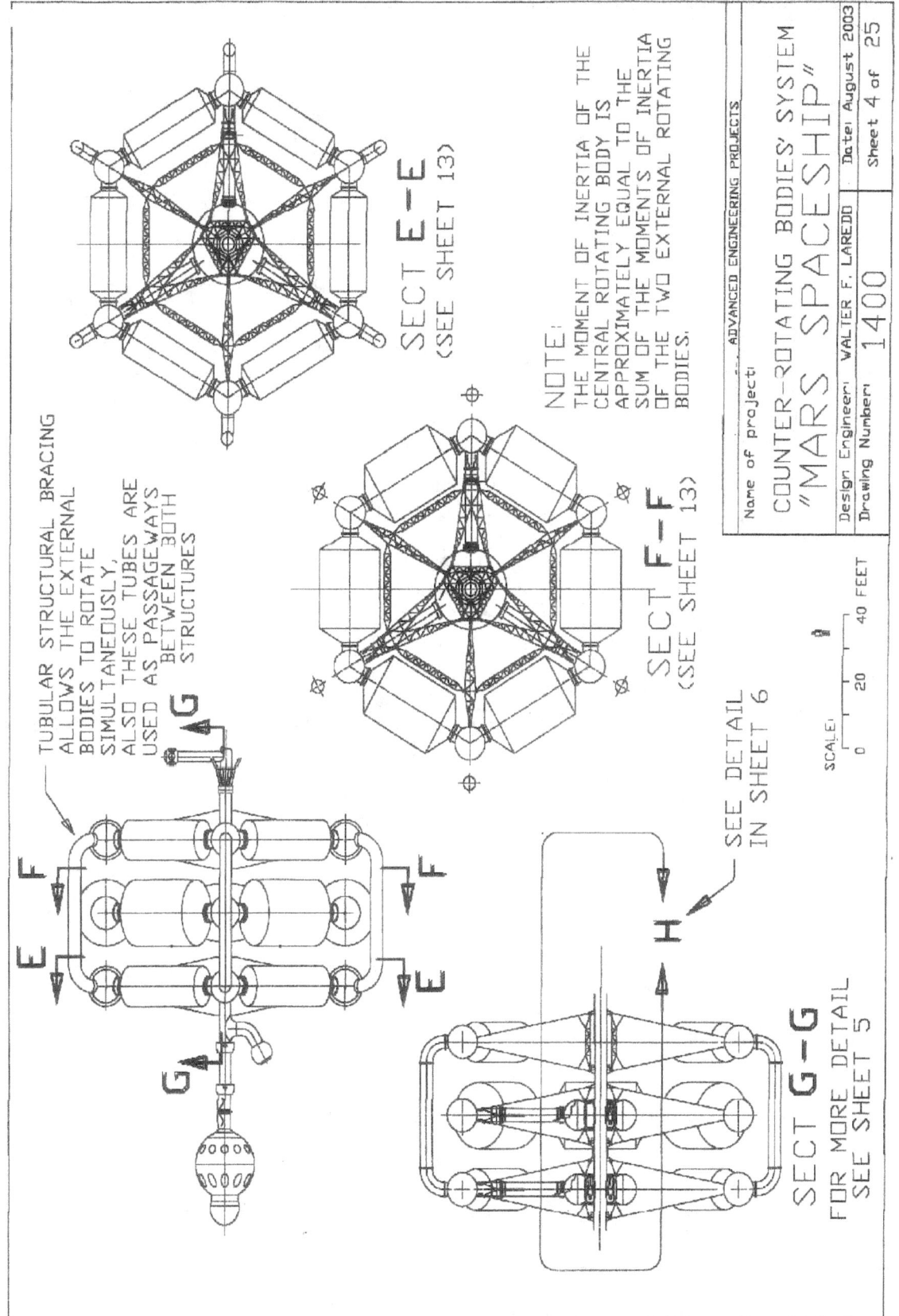

SECT E-E
(SEE SHEET 13)

NOTE:
THE MOMENT OF INERTIA OF THE
CENTRAL ROTATING BODY IS
APPROXIMATELY EQUAL TO THE
SUM OF THE MOMENTS OF INERTIA
OF THE TWO EXTERNAL ROTATING
BODIES.

SECT F-F
(SEE SHEET 13)

TUBULAR STRUCTURAL BRACING
ALLOWS THE EXTERNAL
BODIES TO ROTATE
SIMULTANEOUSLY,
ALSO THESE TUBES ARE
USED AS PASSAGEWAYS
BETWEEN BOTH
STRUCTURES

SEE DETAIL
IN SHEET 6

SCALE:
0   20   40 FEET

SECT G-G
FOR MORE DETAIL
SEE SHEET 5

Name of project: ADVANCED ENGINEERING PROJECTS

COUNTER-ROTATING BODIES' SYSTEM
"MARS SPACESHIP"

Design Engineer: WALTER F. LAREDO   Date: August 2003
Drawing Number: 1400   Sheet 4 of 25

PLATE 140

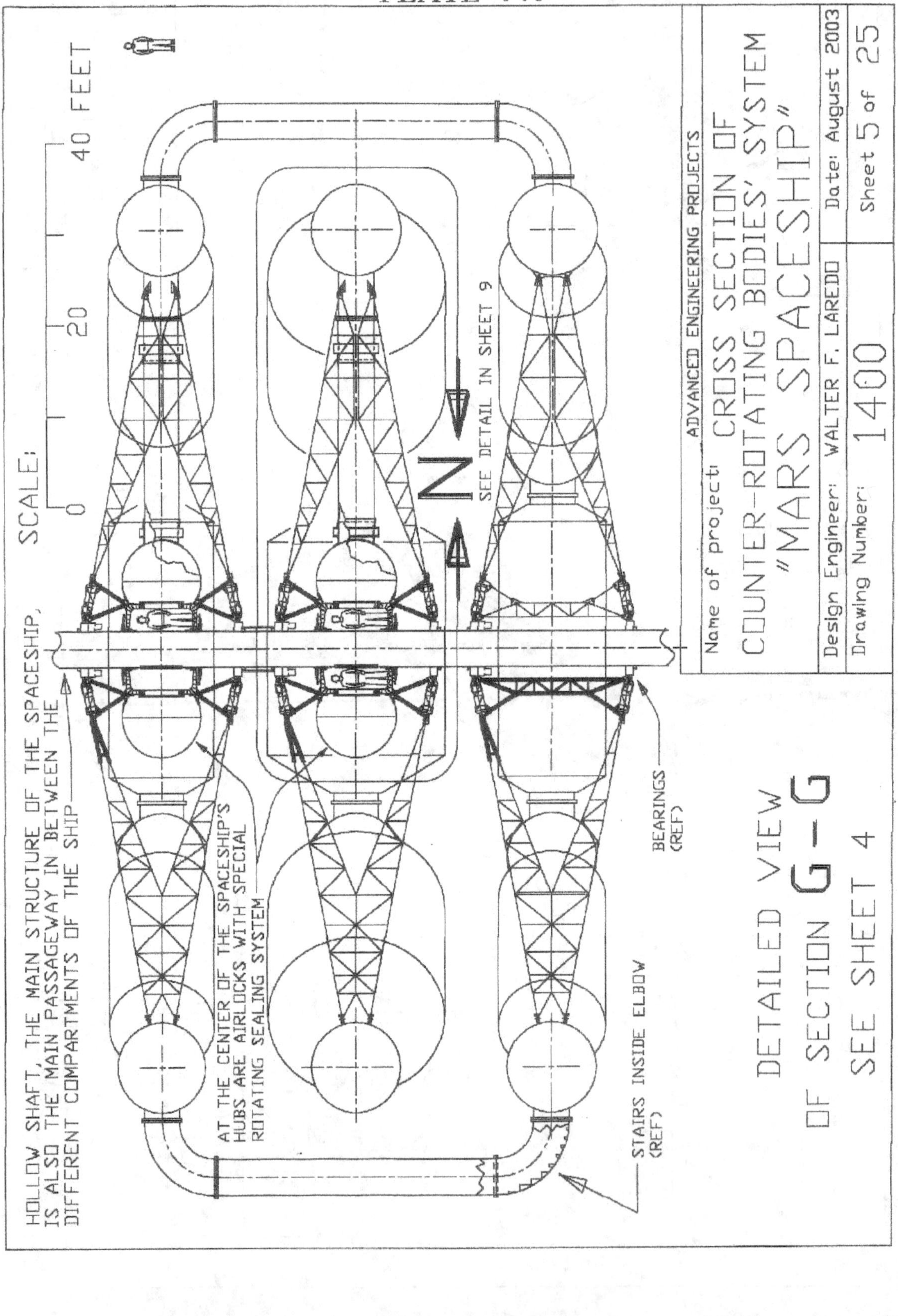

SCALE:

0   20   40 FEET

HOLLOW SHAFT, THE MAIN STRUCTURE OF THE SPACESHIP, IS ALSO THE MAIN PASSAGEWAY IN BETWEEN THE DIFFERENT COMPARTMENTS OF THE SHIP

AT THE CENTER OF THE SPACESHIP'S HUBS ARE AIRLOCKS WITH SPECIAL ROTATING SEALING SYSTEM

SEE DETAIL IN SHEET 9

N

BEARINGS (REF)

STAIRS INSIDE ELBOW (REF)

DETAILED VIEW OF SECTION G-G SEE SHEET 4

ADVANCED ENGINEERING PROJECTS

Name of project: CROSS SECTION OF COUNTER-ROTATING BODIES' SYSTEM "MARS SPACESHIP"

Design Engineer: WALTER F. LAREDO

Drawing Number: 1400

Date: August 2003

Sheet 5 of 25

PLATE 141

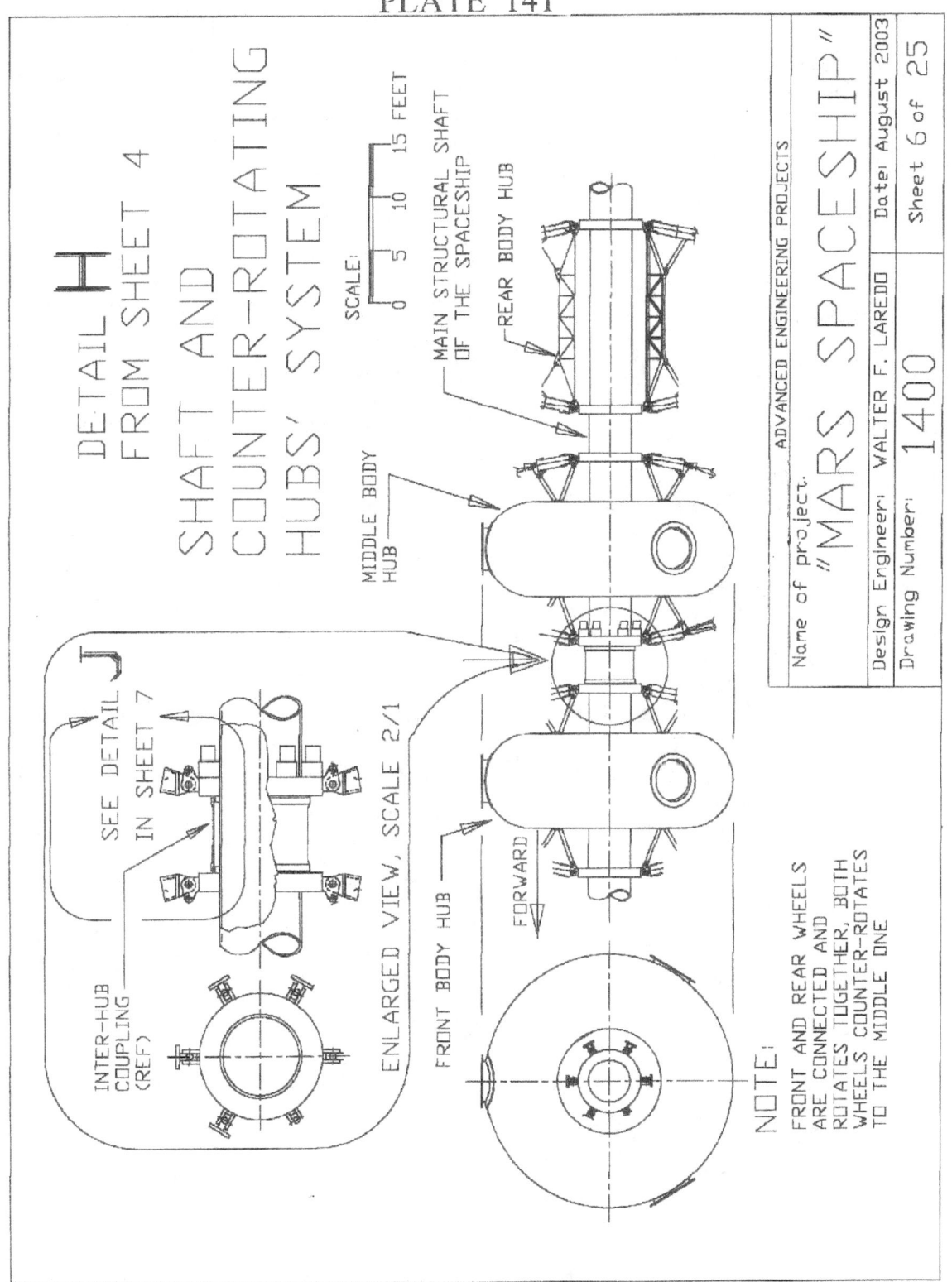

DETAIL H
FROM SHEET 4

SHAFT AND
COUNTER-ROTATING
HUBS' SYSTEM

SCALE:

0   5   10   15 FEET

MAIN STRUCTURAL SHAFT
OF THE SPACESHIP

REAR BODY HUB

MIDDLE BODY
HUB

SEE DETAIL J
IN SHEET 7

INTER-HUB
COUPLING
(REF)

ENLARGED VIEW, SCALE 2/1

FRONT BODY HUB

FORWARD

NOTE:
FRONT AND REAR WHEELS
ARE CONNECTED AND
ROTATES TOGETHER, BOTH
WHEELS COUNTER-ROTATES
TO THE MIDDLE ONE

Name of project. ADVANCED ENGINEERING PROJECTS

"MARS SPACESHIP"

Design Engineer: WALTER F. LAREDO   Date: August 2003

Drawing Number: 1400   Sheet 6 of 25

PLATE 142

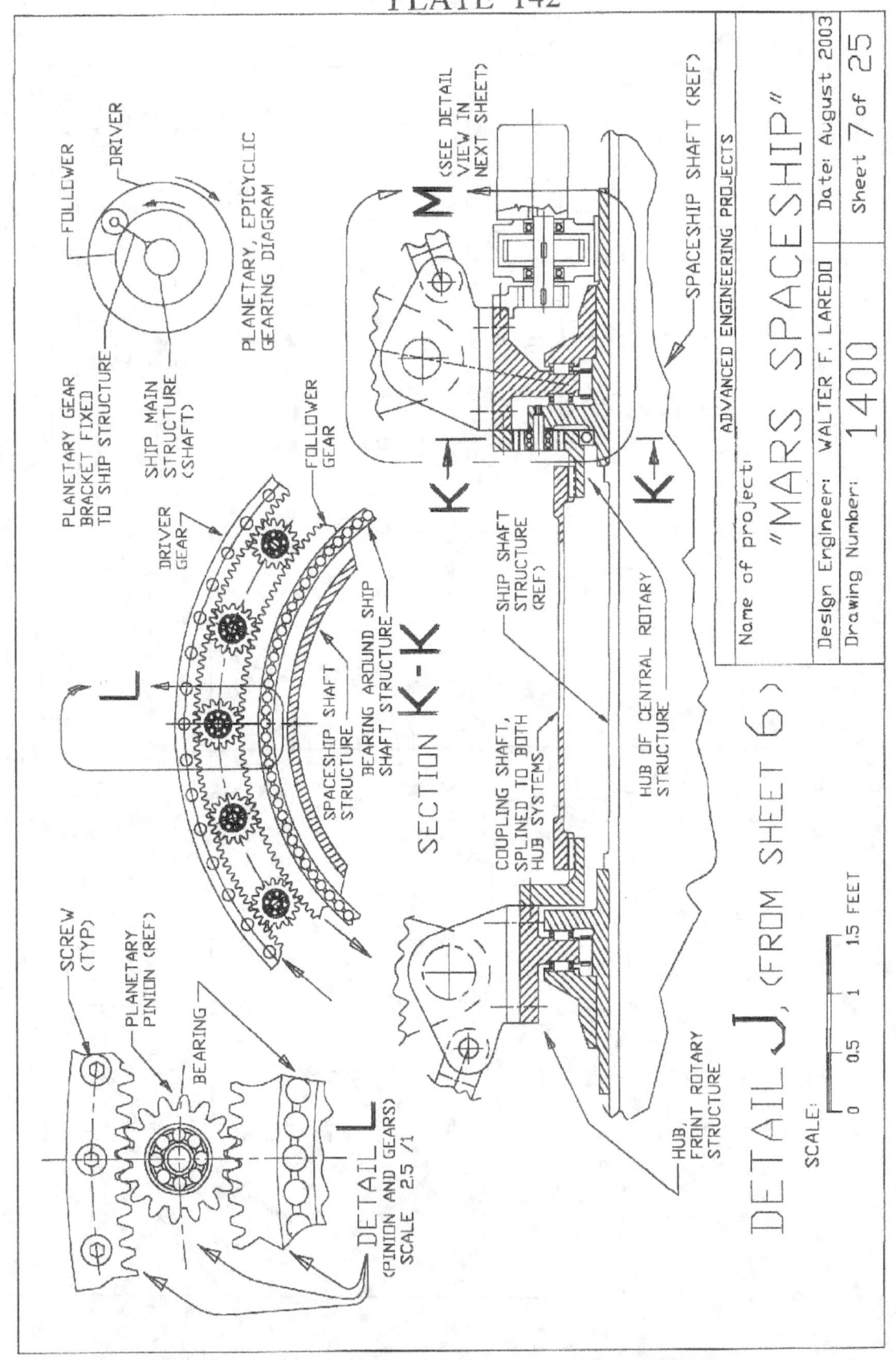

FOLLOWER

DRIVER

PLANETARY, EPICYCLIC GEARING DIAGRAM

PLANETARY GEAR BRACKET FIXED TO SHIP STRUCTURE

SHIP MAIN STRUCTURE (SHAFT)

DRIVER GEAR

FOLLOWER GEAR

SPACESHIP SHAFT STRUCTURE

BEARING AROUND SHIP SHAFT STRUCTURE

SECTION K-K

SCREW (TYP)

PLANETARY PINION (REF)

BEARING

DETAIL L
(PINION AND GEARS)
SCALE 2.5 /1

M (SEE DETAIL VIEW IN NEXT SHEET)

SPACESHIP SHAFT (REF)

SHIP SHAFT STRUCTURE (REF)

COUPLING SHAFT, SPLINED TO BOTH HUB SYSTEMS

HUB OF CENTRAL ROTARY STRUCTURE

HUB, FRONT ROTARY STRUCTURE

DETAIL J (FROM SHEET 6)

SCALE:

0   0.5   1   1.5 FEET

Name of project: "MARS SPACESHIP"

ADVANCED ENGINEERING PROJECTS

| Design Engineer: WALTER F. LAREDO | Date: August 2003 |
| Drawing Number: 1400 | Sheet 7 of 25 |

PLATE 143

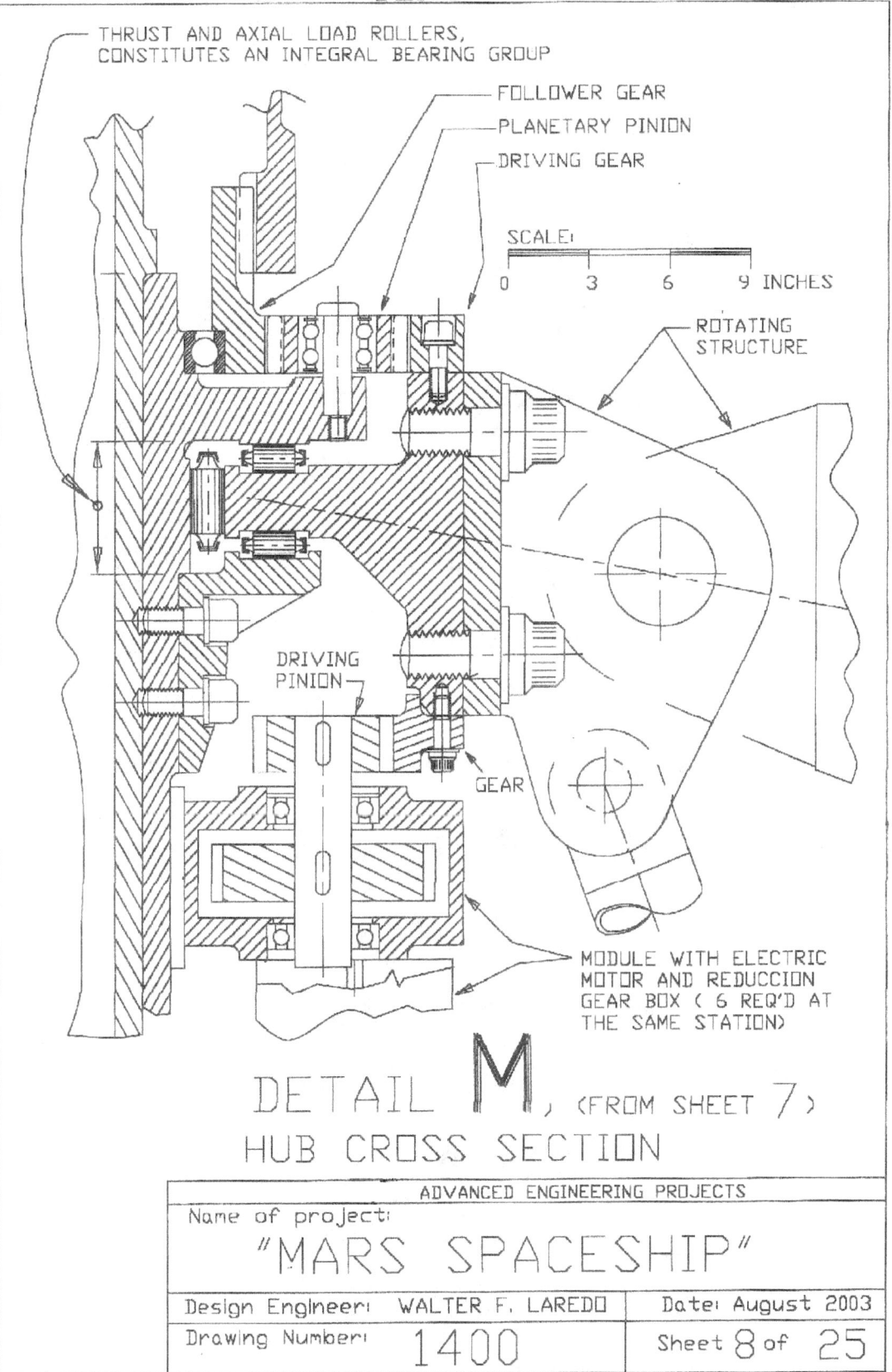

THRUST AND AXIAL LOAD ROLLERS,
CONSTITUTES AN INTEGRAL BEARING GROUP

FOLLOWER GEAR
PLANETARY PINION
DRIVING GEAR

SCALE:

0   3   6   9 INCHES

ROTATING
STRUCTURE

DRIVING
PINION

GEAR

MODULE WITH ELECTRIC
MOTOR AND REDUCCION
GEAR BOX (6 REQ'D AT
THE SAME STATION)

DETAIL M, (FROM SHEET 7)
HUB CROSS SECTION

ADVANCED ENGINEERING PROJECTS		
Name of project: "MARS SPACESHIP"		
Design Engineer: WALTER F. LAREDO		Date: August 2003
Drawing Number: 1400		Sheet 8 of 25

PLATE 144

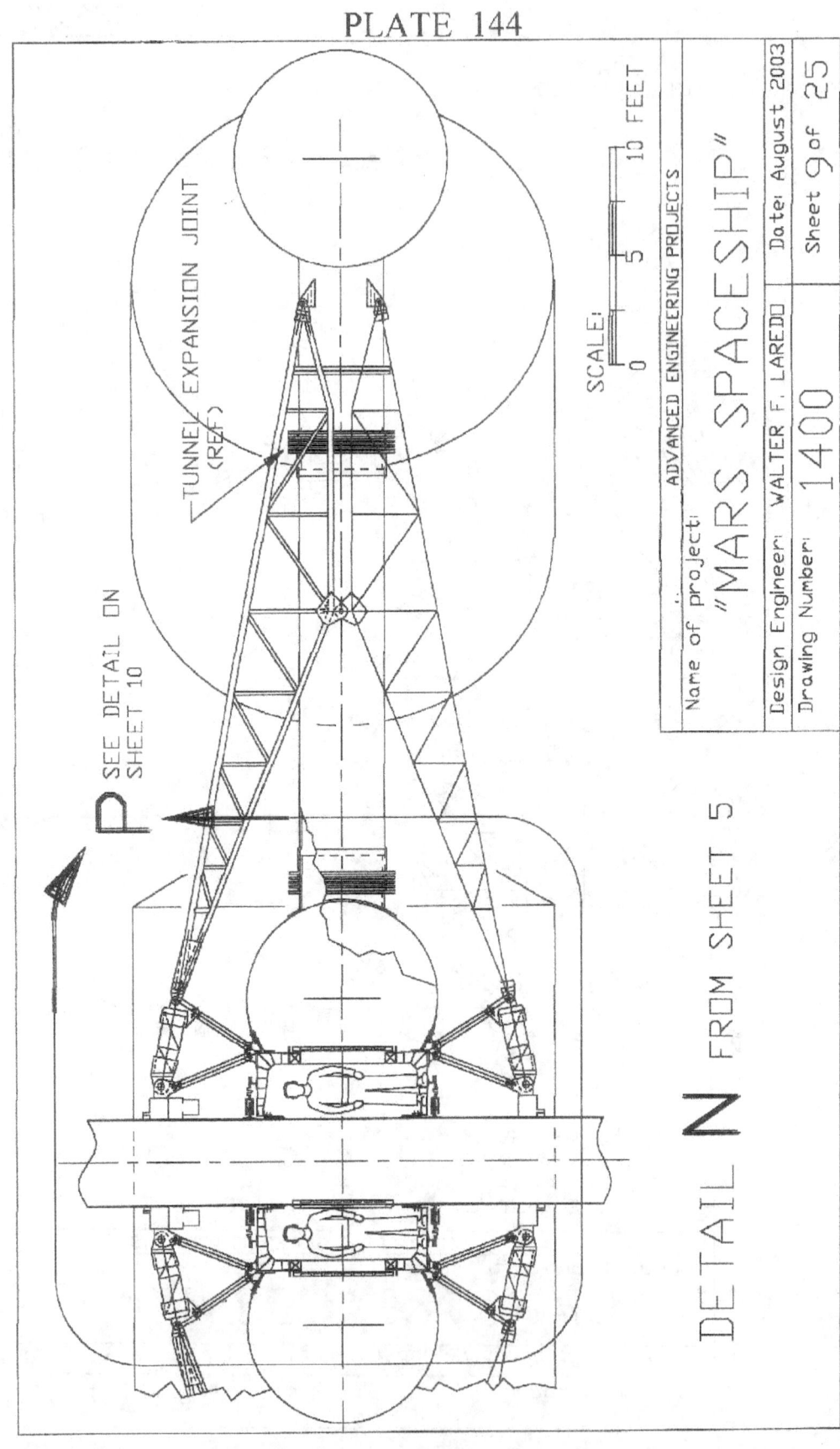

TUNNEL EXPANSION JOINT
(REF)

SEE DETAIL ON
SHEET 10

P

SCALE:

0   5   10 FEET

ADVANCED ENGINEERING PROJECTS

Name of project: "MARS SPACESHIP"

| Design Engineer: WALTER F. LAREDO | Date: August 2003 |
| Drawing Number: 1400 | Sheet 9 of 25 |

DETAIL N FROM SHEET 5

PLATE 145

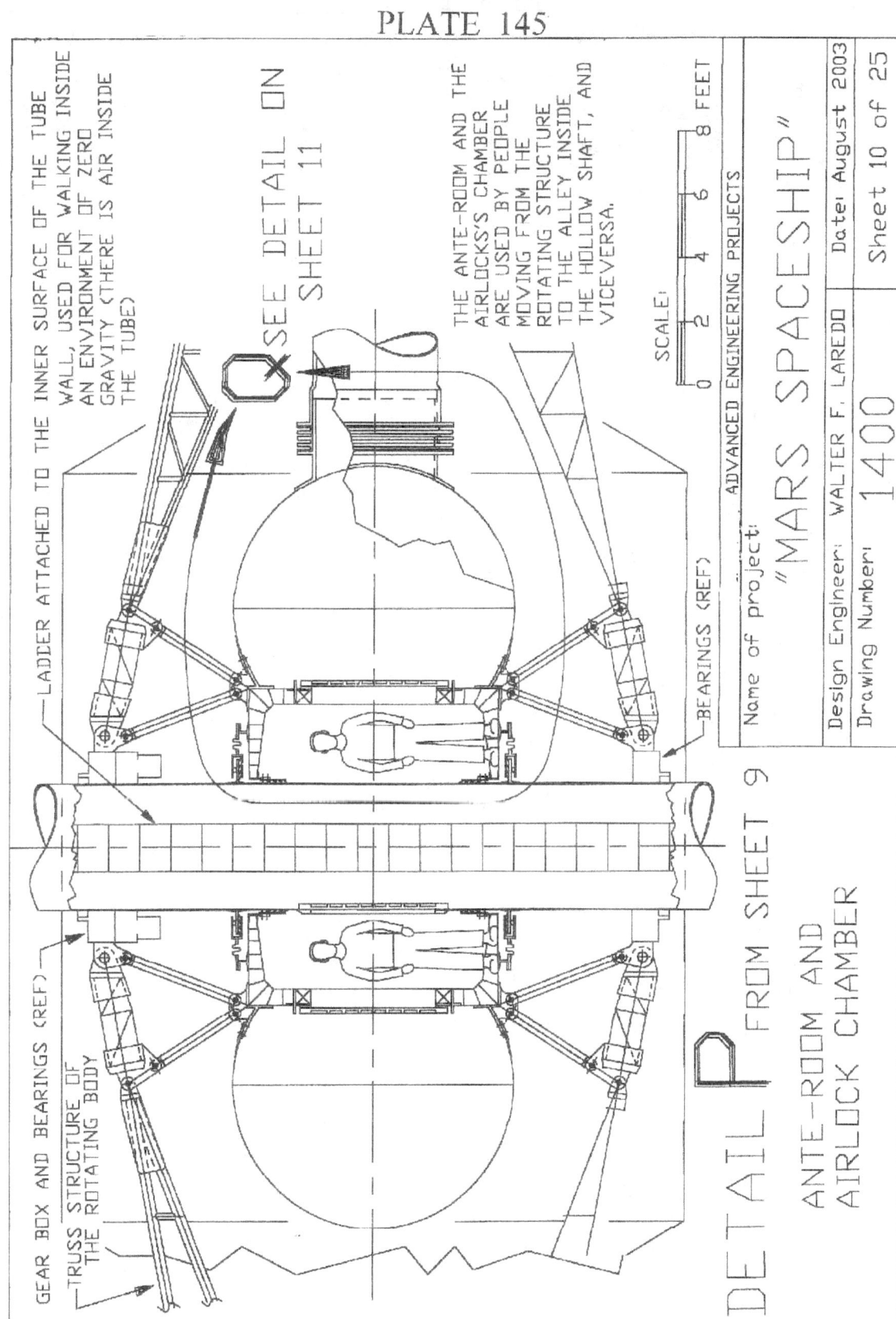

LADDER ATTACHED TO THE INNER SURFACE OF THE TUBE WALL, USED FOR WALKING INSIDE AN ENVIRONMENT OF ZERO GRAVITY (THERE IS AIR INSIDE THE TUBE)

SEE DETAIL ON SHEET 11

THE ANTE-ROOM AND THE AIRLOCKS'S CHAMBER ARE USED BY PEOPLE MOVING FROM THE ROTATING STRUCTURE TO THE ALLEY INSIDE THE HOLLOW SHAFT, AND VICEVERSA.

SCALE: 0 2 4 6 8 FEET

ADVANCED ENGINEERING PROJECTS

Name of project: "MARS SPACESHIP"

Design Engineer: WALTER F. LAREDO   Date: August 2003

Drawing Number: 1400   Sheet 10 of 25

BEARINGS (REF)

GEAR BOX AND BEARINGS (REF)

TRUSS STRUCTURE OF THE ROTATING BODY

DETAIL P FROM SHEET 9

ANTE-ROOM AND AIRLOCK CHAMBER

PLATE 146

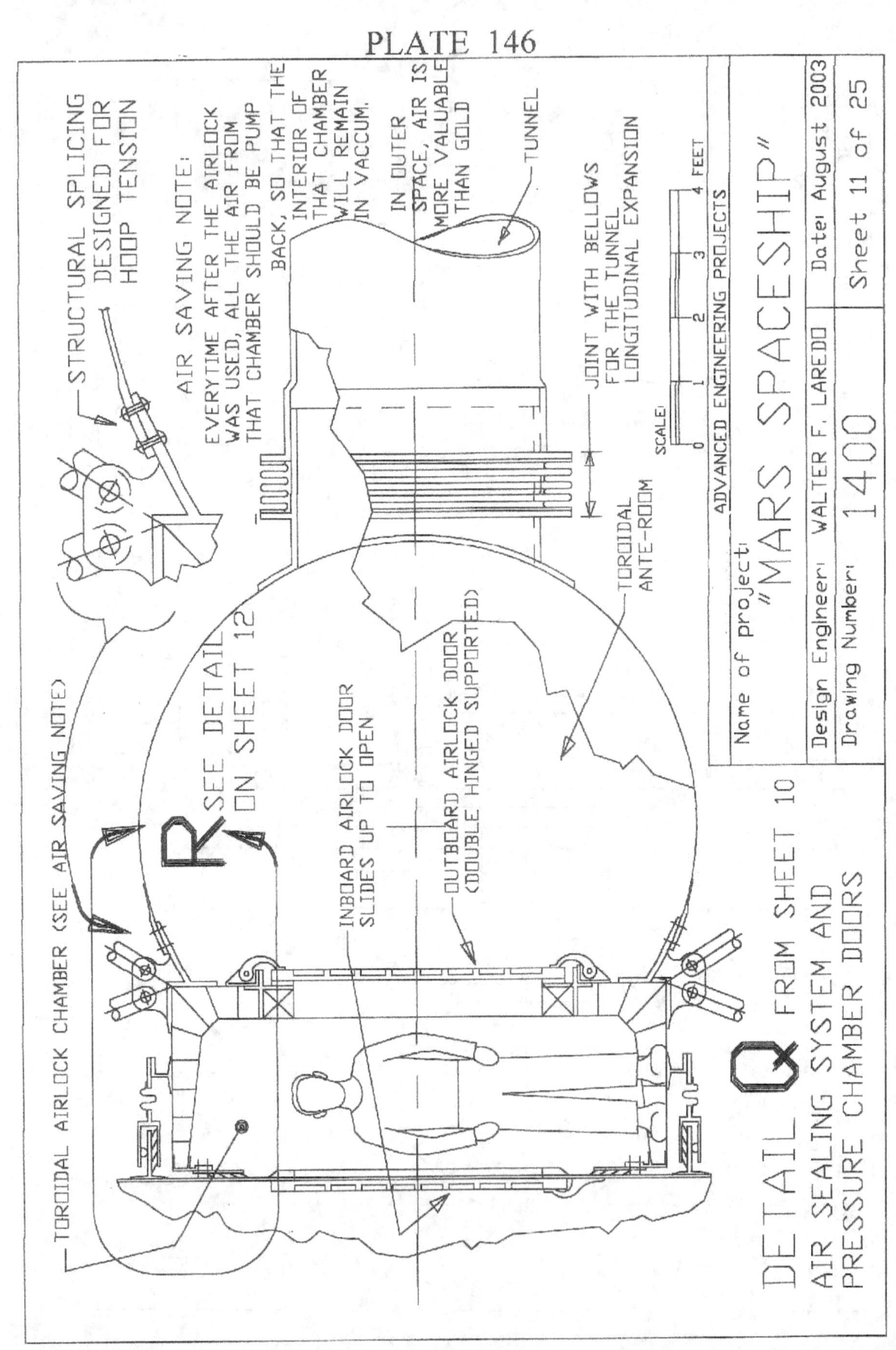

STRUCTURAL SPLICING DESIGNED FOR HOOP TENSION

AIR SAVING NOTE:
EVERYTIME AFTER THE AIRLOCK WAS USED, ALL THE AIR FROM THAT CHAMBER SHOULD BE PUMP BACK, SO THAT THE INTERIOR OF THAT CHAMBER WILL REMAIN IN VACCUM.

IN OUTER SPACE, AIR IS MORE VALUABLE THAN GOLD

TUNNEL

JOINT WITH BELLOWS FOR THE TUNNEL LONGITUDINAL EXPANSION

TOROIDAL ANTE-ROOM

SCALE:

0 1 2 3 4 FEET

ADVANCED ENGINEERING PROJECTS

Name of project: "MARS SPACESHIP"

Design Engineer: WALTER F. LAREDO   Date: August 2003

Drawing Number: 1400   Sheet 11 of 25

TOROIDAL AIRLOCK CHAMBER (SEE AIR SAVING NOTE)

SEE DETAIL R ON SHEET 12

INBOARD AIRLOCK DOOR SLIDES UP TO OPEN

OUTBOARD AIRLOCK DOOR (DOUBLE HINGED SUPPORTED)

DETAIL Q FROM SHEET 10

AIR SEALING SYSTEM AND PRESSURE CHAMBER DOORS

PLATE 147

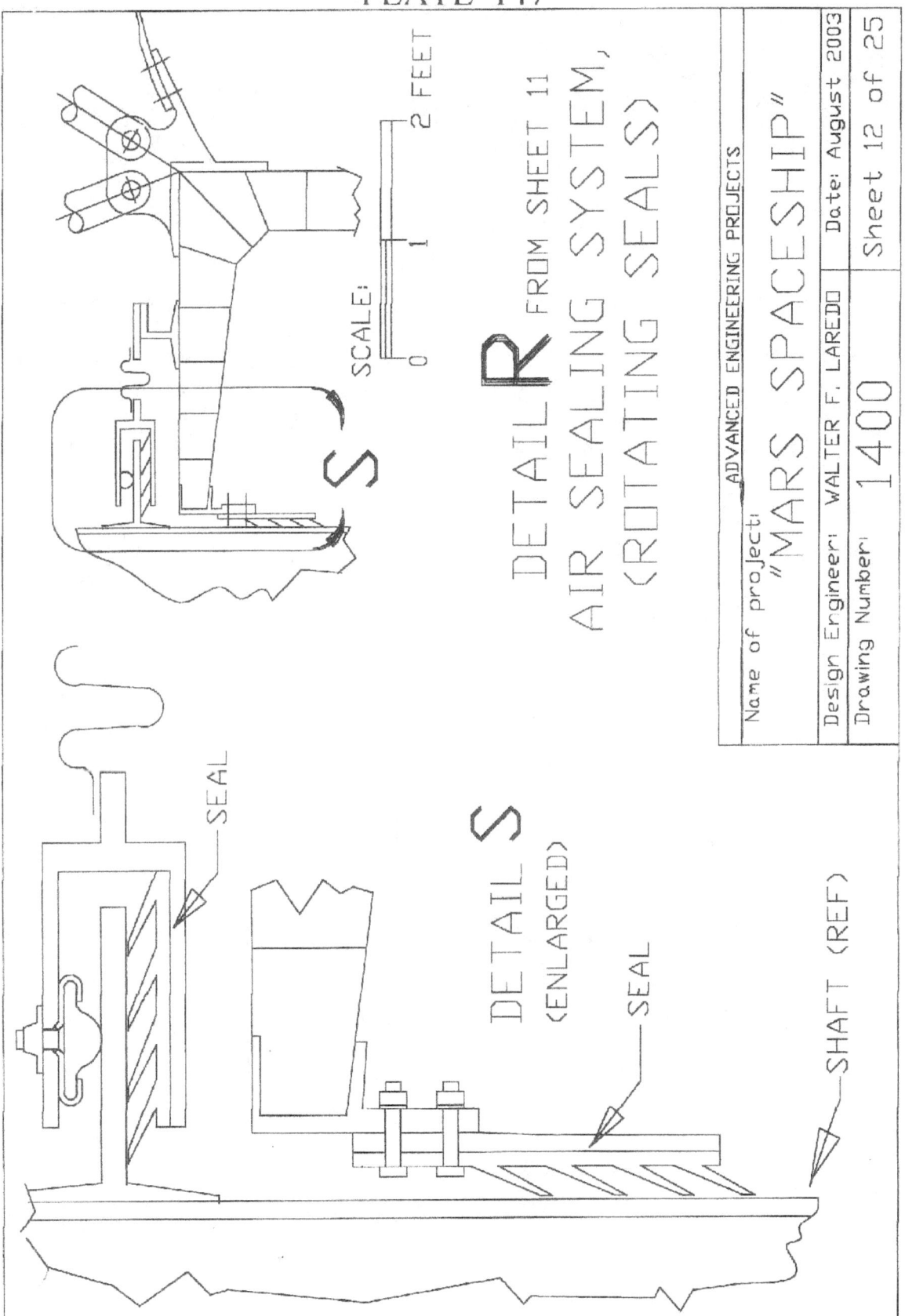

DETAIL R FROM SHEET 11
AIR SEALING SYSTEM,
(ROTATING SEALS)

DETAIL S (ENLARGED)

SEAL

SEAL

SHAFT (REF)

SCALE:

0   1   2 FEET

ADVANCED ENGINEERING PROJECTS	
Name of project: "MARS SPACESHIP"	
Design Engineer: WALTER F. LAREDO	Date: August 2003
Drawing Number: 1400	Sheet 12 of 25

PLATE 148

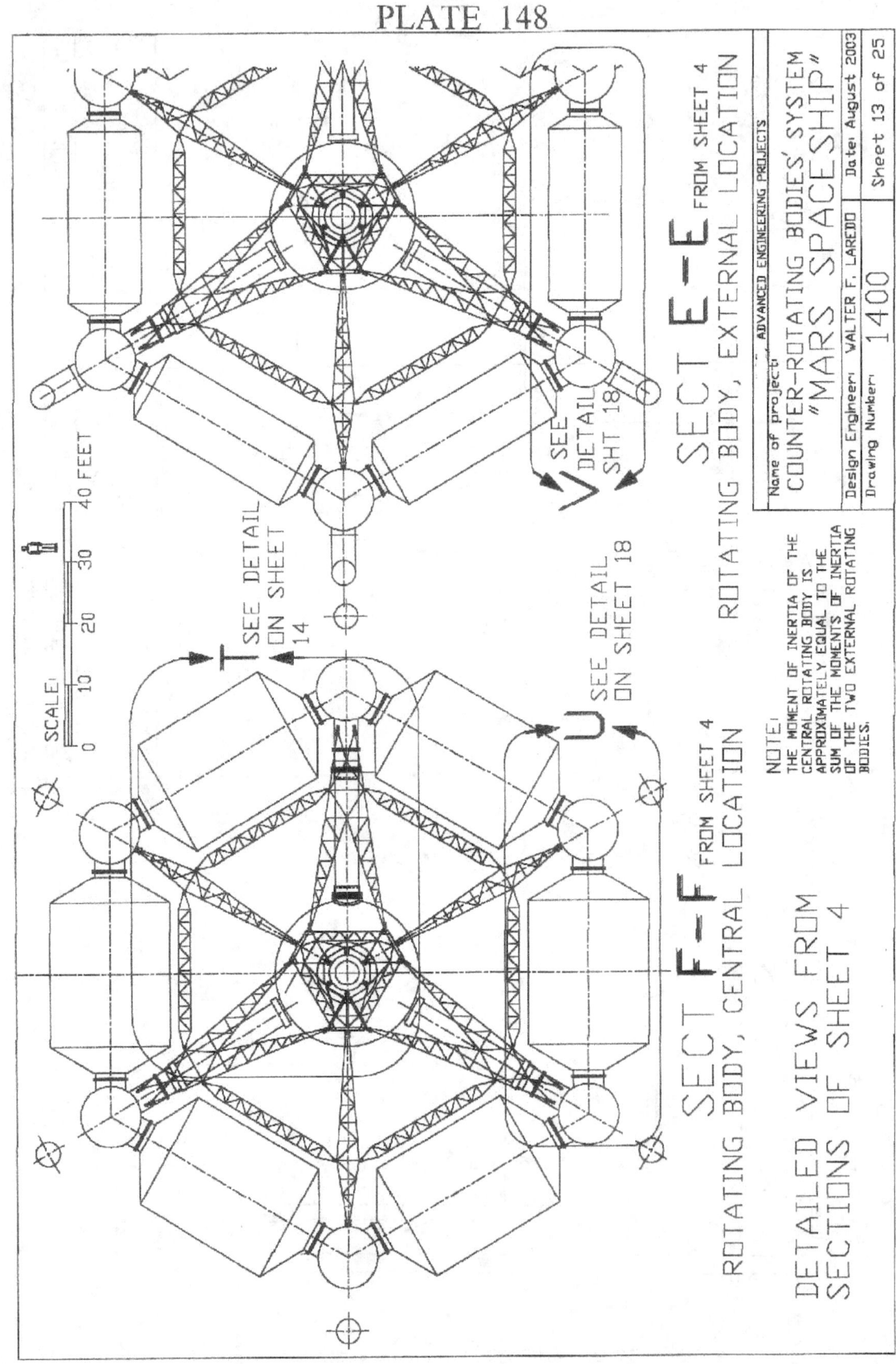

SECT E-E FROM SHEET 4

ROTATING BODY, EXTERNAL LOCATION

SEE DETAIL SHT 18

SEE DETAIL ON SHEET 14

SEE DETAIL ON SHEET 18

SCALE:
0   10   20   30   40 FEET

NOTE:
THE MOMENT OF INERTIA OF THE CENTRAL ROTATING BODY IS APPROXIMATELY EQUAL TO THE SUM OF THE MOMENTS OF INERTIA OF THE TWO EXTERNAL ROTATING BODIES.

SECT F-F FROM SHEET 4

ROTATING BODY, CENTRAL LOCATION

DETAILED VIEWS FROM SECTIONS OF SHEET 4

Name of project: ADVANCED ENGINEERING PROJECTS
COUNTER-ROTATING BODIES' SYSTEM "MARS SPACESHIP"

Design Engineer: WALTER F. LAREDO   Date: August 2003

Drawing Number: 1400   Sheet 13 of 25

PLATE 149

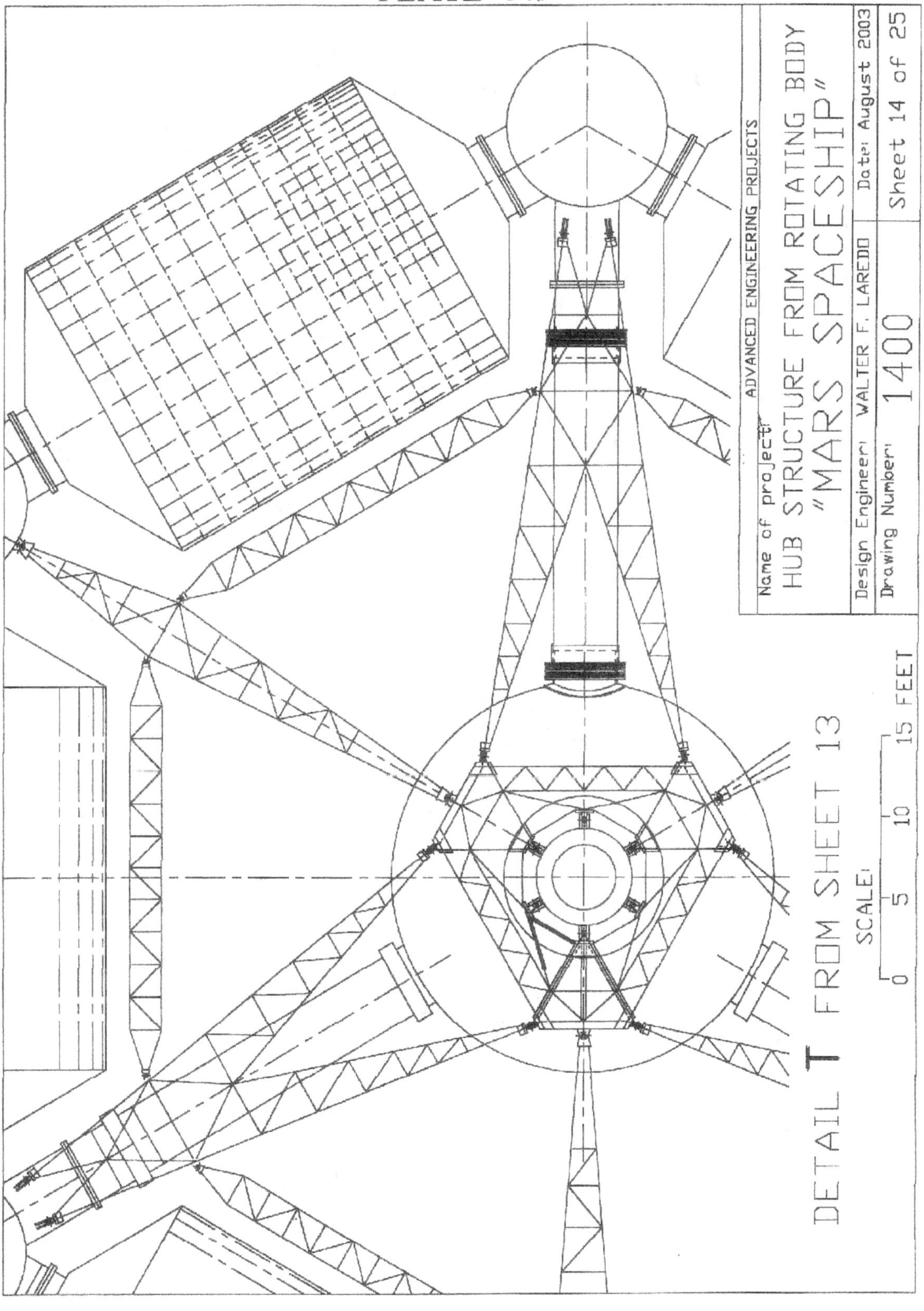

DETAIL T FROM SHEET 13

SCALE:

0   5   10   15 FEET

ADVANCED ENGINEERING PROJECTS

Name of project:
HUB STRUCTURE FROM ROTATING BODY
"MARS SPACESHIP"

Design Engineer: WALTER F. LAREDO

Date: August 2003

Drawing Number: 1400

Sheet 14 of 25

PLATE 150

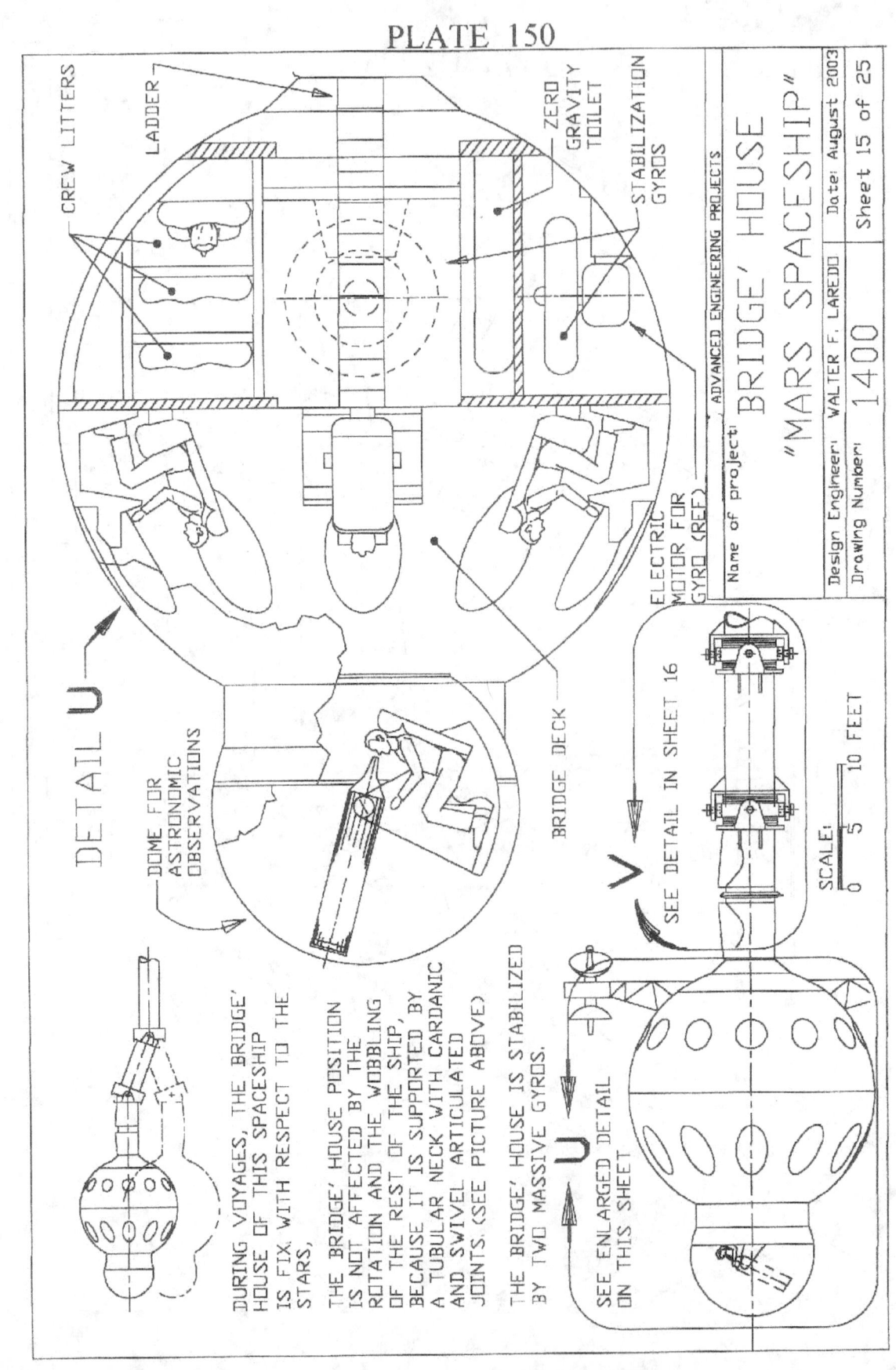

CREW LITTERS

LADDER

ZERO GRAVITY TOILET

STABILIZATION GYROS

DETAIL U

DOME FOR ASTRONOMIC OBSERVATIONS

ELECTRIC MOTOR FOR GYRO (REF.)

BRIDGE DECK

SEE DETAIL IN SHEET 16

V

U

SEE ENLARGED DETAIL ON THIS SHEET

DURING VOYAGES, THE BRIDGE' HOUSE OF THIS SPACESHIP IS FIX, WITH RESPECT TO THE STARS,

THE BRIDGE' HOUSE POSITION IS NOT AFFECTED BY THE ROTATION AND THE WOBBLING OF THE REST OF THE SHIP, BECAUSE IT IS SUPPORTED BY A TUBULAR NECK WITH CARDANIC AND SWIVEL ARTICULATED JOINTS. (SEE PICTURE ABOVE)

THE BRIDGE' HOUSE IS STABILIZED BY TWO MASSIVE GYROS.

SCALE:
0   5   10 FEET

Name of project:	ADVANCED ENGINEERING PROJECTS
	BRIDGE' HOUSE
	"MARS SPACESHIP"
Design Engineer: WALTER F. LAREDO	Date: August 2003
Drawing Number: 1400	Sheet 15 of 25

PLATE 151

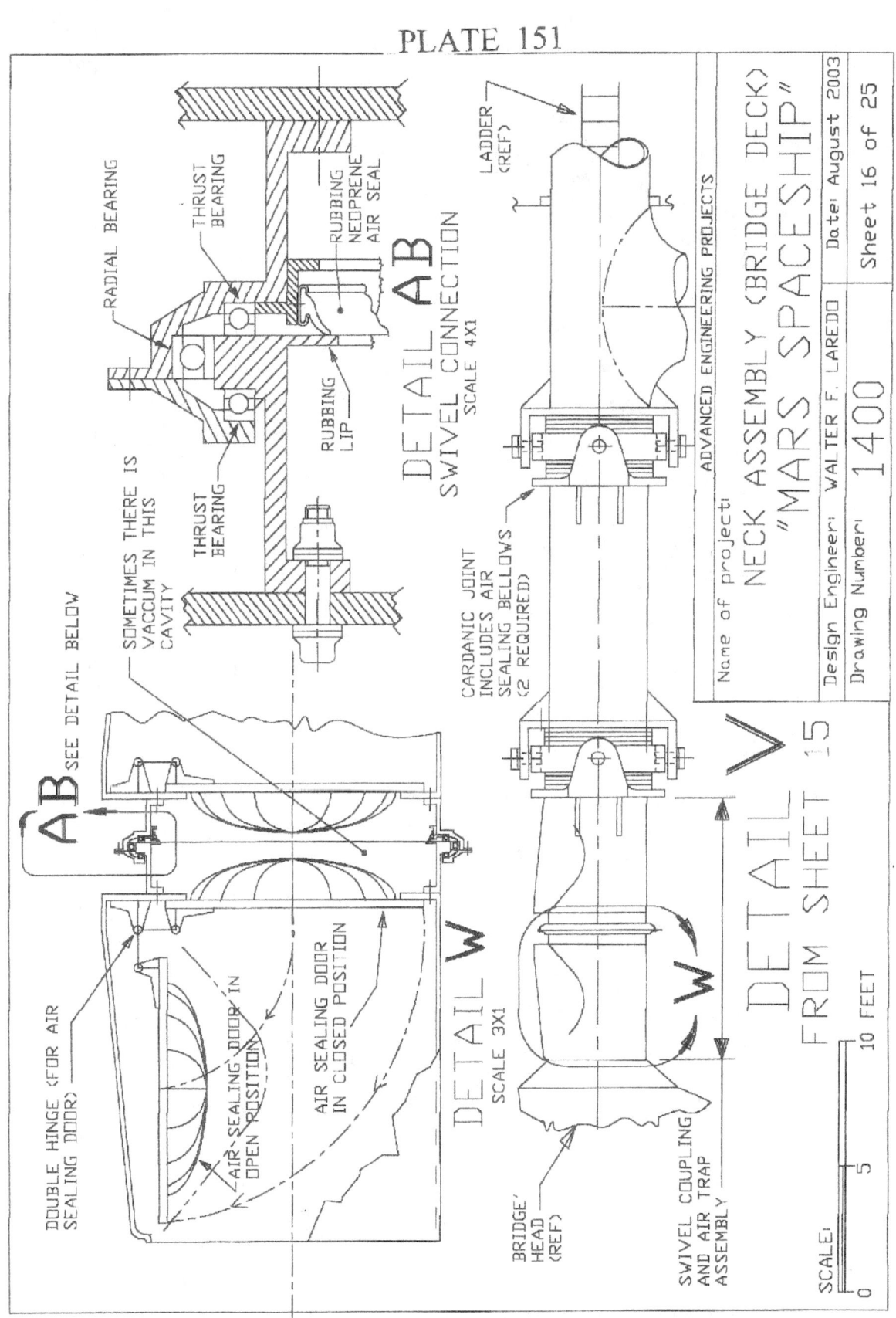

DETAIL AB
SWIVEL CONNECTION
SCALE 4X1

RADIAL BEARING

THRUST BEARING

RUBBING NEOPRENE AIR SEAL

RUBBING LIP

THRUST BEARING

SOMETIMES THERE IS VACUUM IN THIS CAVITY

AB SEE DETAIL BELOW

DOUBLE HINGE (FOR AIR SEALING DOOR)

AIR SEALING DOOR IN OPEN POSITION

AIR SEALING DOOR IN CLOSED POSITION

DETAIL W
SCALE 3X1

LADDER (REF)

CARDANIC JOINT INCLUDES AIR SEALING BELLOWS (2 REQUIRED)

DETAIL V
FROM SHEET 15

BRIDGE HEAD (REF)

SWIVEL COUPLING AND AIR TRAP ASSEMBLY

SCALE:
0   5   10 FEET

Name of project: ADVANCED ENGINEERING PROJECTS
NECK ASSEMBLY (BRIDGE DECK) "MARS SPACESHIP"

Design Engineer: WALTER F. LAREDO   Date: August 2003
Drawing Number: 1400   Sheet 16 of 25

PLATE 152

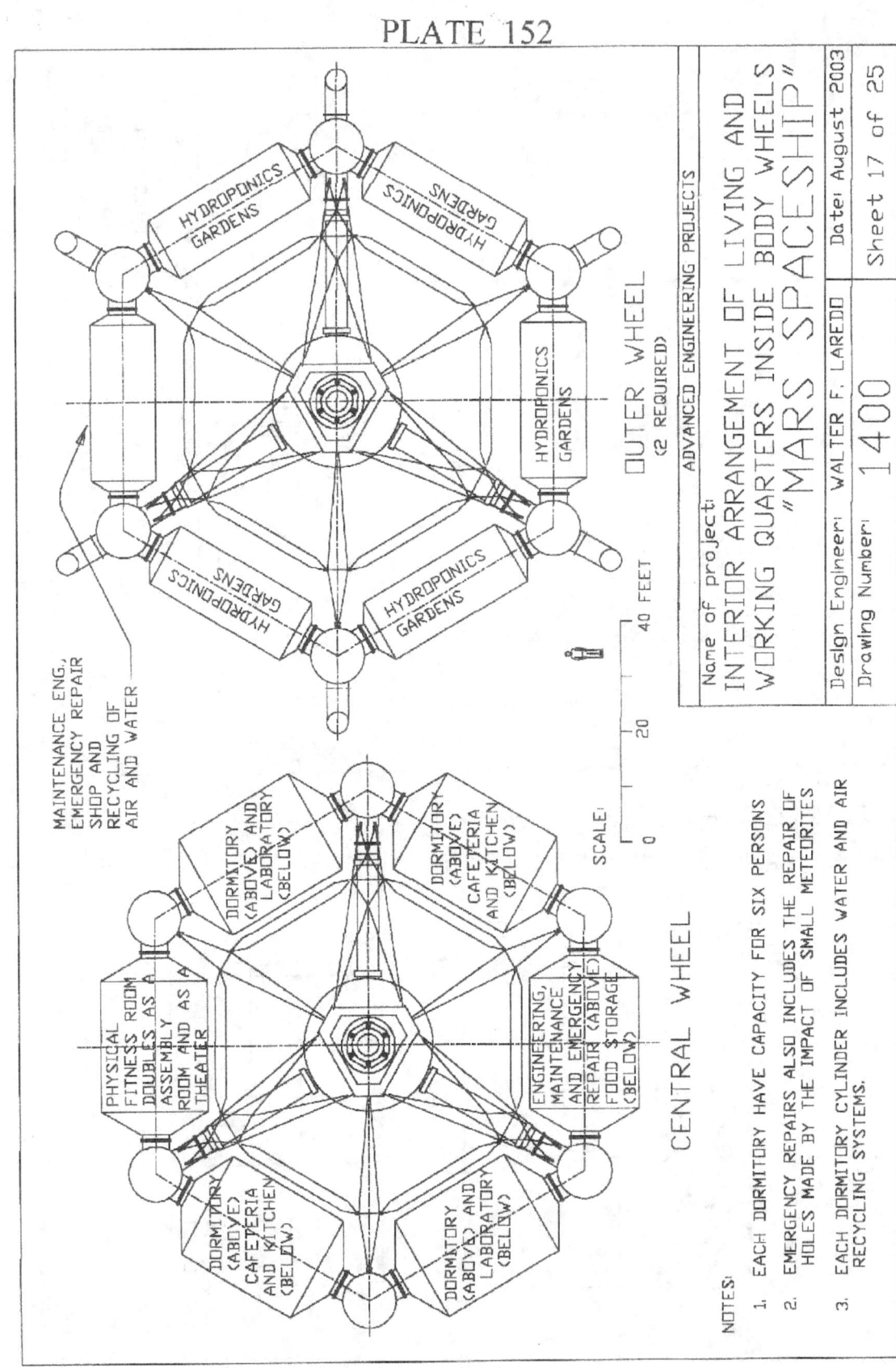

OUTER WHEEL
(2 REQUIRED)

MAINTENANCE ENG., EMERGENCY REPAIR SHOP AND RECYCLING OF AIR AND WATER

HYDROPONICS GARDENS

SCALE:

0   20   40 FEET

CENTRAL WHEEL

PHYSICAL FITNESS ROOM DOUBLES AS A ASSEMBLY ROOM AND AS A THEATER

DORMITORY (ABOVE) AND LABORATORY (BELOW)

DORMITORY (ABOVE) CAFETERIA AND KITCHEN (BELOW)

ENGINEERING, MAINTENANCE AND EMERGENCY REPAIR (ABOVE) FOOD STORAGE (BELOW)

DORMITORY (ABOVE) CAFETERIA AND KITCHEN (BELOW)

DORMITORY (ABOVE) AND LABORATORY (BELOW)

ADVANCED ENGINEERING PROJECTS

Name of project:
INTERIOR ARRANGEMENT OF LIVING AND WORKING QUARTERS INSIDE BODY WHEELS "MARS SPACESHIP"

| Design Engineer: WALTER F. LAREDO | Date: August 2003 |
| Drawing Number: 1400 | Sheet 17 of 25 |

NOTES:

1. EACH DORMITORY HAVE CAPACITY FOR SIX PERSONS

2. EMERGENCY REPAIRS ALSO INCLUDES THE REPAIR OF HOLES MADE BY THE IMPACT OF SMALL METEORITES

3. EACH DORMITORY CYLINDER INCLUDES WATER AND AIR RECYCLING SYSTEMS.

PLATE 153

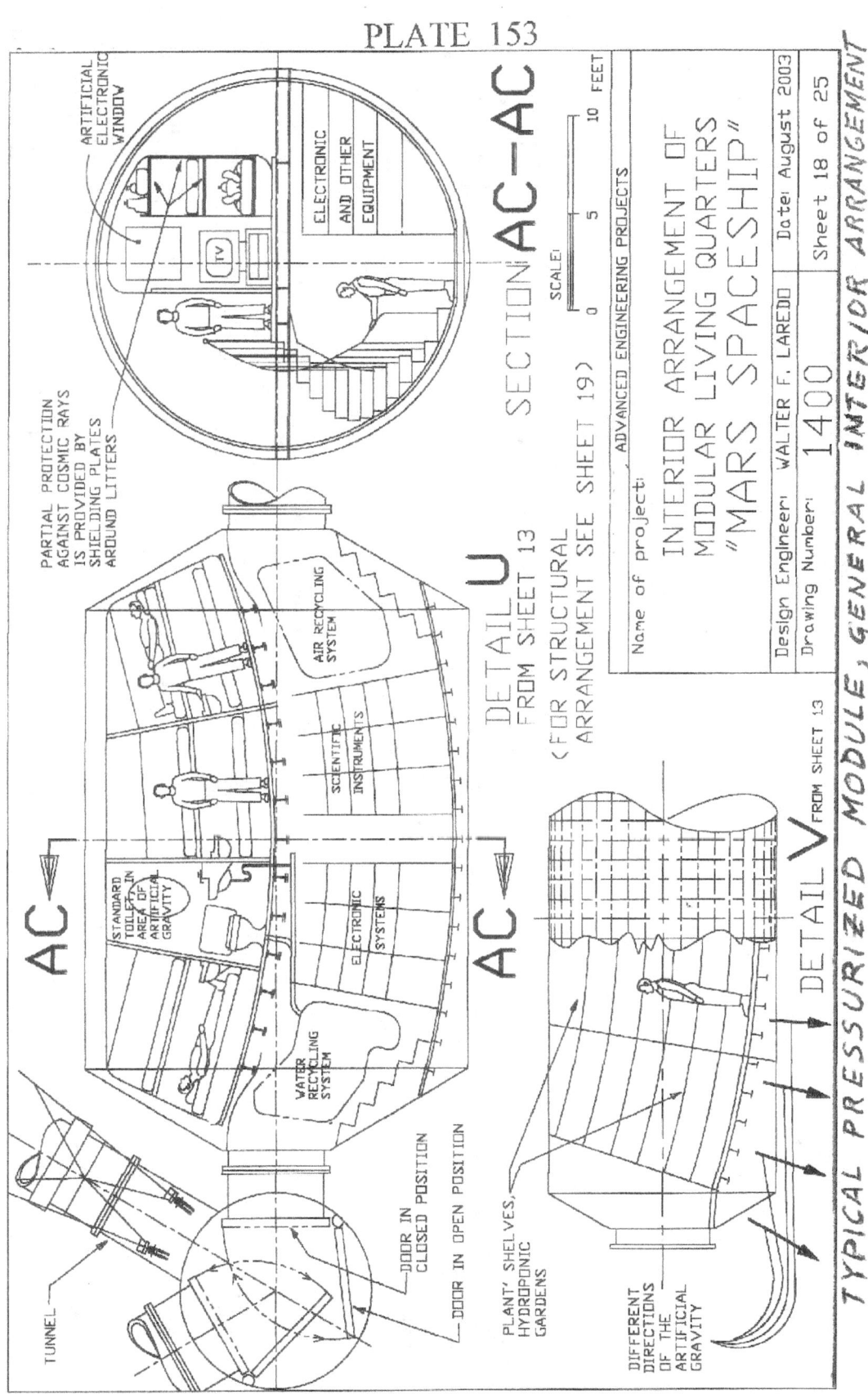

ARTIFICIAL ELECTRONIC WINDOW

ELECTRONIC AND OTHER EQUIPMENT

SECTION AC-AC

SCALE:

0   5   10 FEET

PARTIAL PROTECTION AGAINST COSMIC RAYS IS PROVIDED BY SHIELDING PLATES AROUND LITTERS

DETAIL U FROM SHEET 13

(FOR STRUCTURAL ARRANGEMENT SEE SHEET 19)

AIR RECYCLING SYSTEM

SCIENTIFIC INSTRUMENTS

ELECTRONIC SYSTEMS

STANDARD TOILET IN AREA OF ARTIFICIAL GRAVITY

WATER RECYCLING SYSTEM

AC

AC

DOOR IN CLOSED POSITION

DOOR IN OPEN POSITION

TUNNEL

PLANT SHELVES, HYDROPONIC GARDENS

DETAIL V FROM SHEET 13

DIFFERENT DIRECTIONS OF THE ARTIFICIAL GRAVITY

Name of project: ADVANCED ENGINEERING PROJECTS

INTERIOR ARRANGEMENT OF MODULAR LIVING QUARTERS "MARS SPACESHIP"

Design Engineer: WALTER F. LAREDO   Date: August 2003

Drawing Number: 1400   Sheet 18 of 25

TYPICAL PRESSURIZED MODULE, GENERAL INTERIOR ARRANGEMENT

PLATE 154

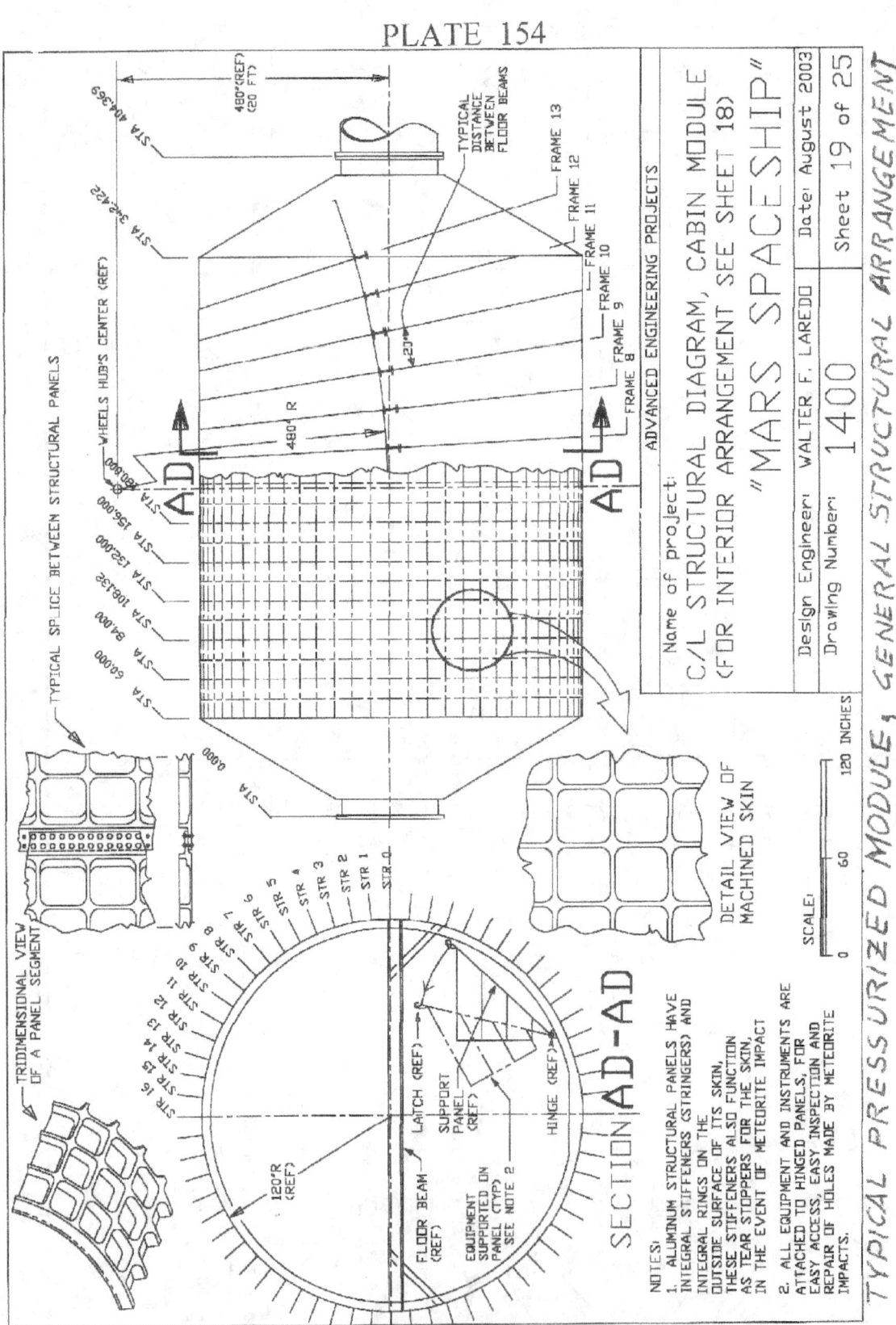

TRIDIMENSIONAL VIEW OF A PANEL SEGMENT

TYPICAL SPLICE BETWEEN STRUCTURAL PANELS

480"(REF) (20 FT)

STA 404.369

STA 344.422

WHEELS HUB'S CENTER (REF)

STA 180.000

STA 156.000

STA 132.000

STA 108.632

STA 84.000

STA 60.000

STA 0.000

480" R

20"

FRAME 13
FRAME 12
FRAME 11
FRAME 10
FRAME 9
FRAME 8

TYPICAL DISTANCE BETWEEN FLOOR BEAMS

A-D
A-D

DETAIL VIEW OF MACHINED SKIN

SECTION AD-AD

STR 0
STR 1
STR 2
STR 3
STR 4
STR 5
STR 6
STR 7
STR 8
STR 9
STR 10
STR 11
STR 12
STR 13
STR 14
STR 15
STR 16

120"R (REF)

FLOOR BEAM (REF)

LATCH (REF)

SUPPORT PANEL (REF)

HINGE (REF)

EQUIPMENT SUPPORTED ON PANEL (TYP) SEE NOTE 2

NOTES:
1. ALUMINUM STRUCTURAL PANELS HAVE INTEGRAL STIFFENERS (STRINGERS) AND INTEGRAL RINGS ON THE OUTSIDE SURFACE OF ITS SKIN. THESE STIFFENERS ALSO FUNCTION AS TEAR STOPPERS FOR THE SKIN, IN THE EVENT OF METEORITE IMPACT

2. ALL EQUIPMENT AND INSTRUMENTS ARE ATTACHED TO HINGED PANELS, FOR EASY ACCESS, EASY INSPECTION AND REPAIR OF HOLES MADE BY METEORITE IMPACTS.

SCALE:
0   60   120 INCHES

Name of project: ADVANCED ENGINEERING PROJECTS
C/L STRUCTURAL DIAGRAM, CABIN MODULE (FOR INTERIOR ARRANGEMENT SEE SHEET 18)
"MARS SPACESHIP"

| Design Engineer: WALTER F. LAREDO | Date: August 2003 |
| Drawing Number: 1400 | Sheet 19 of 25 |

TYPICAL PRESSURIZED MODULE; GENERAL STRUCTURAL ARRANGEMENT

PLATE 155

MARS SPACESHIP PROPULSION SYSTEMS

AT PRESENT TIME THERE IS NO PROPULSION SYSTEM WITH A SPECIFIC IMPULSE OF 4,000 SECONDS OR MORE. A SYSTEM IS NEEDED FOR THE LONG VOYAGES NECESSARY FOR EXPLORING PART OF THIS SOLAR SYSTEM, ONE THAT HAVE MUCH HIGHER SPECIFIC IMPULSE THAN CHEMICAL ROCKETS.

UNIVERSITIES, AEROSPACE INSTITUTIONS AND AEROSPACE AGENCIES OF MOST DEVELOPED COUNTRIES HAVE BEEN DOING RESEARCH ON SPACE PROPULSION CONCEPTS FOR MORE THAN TWENTY YEARS. PERHAPS IT WOULD NOT BE A BAD IDEA TO CREATE AN INTERNATIONAL ORGANIZATION SIMILAR TO THE ONE CREATED FOR THE INTERNATIONAL SPACE STATION TO DEVELOP AND DESIGN THE PROPULSION SYSTEM FOR THIS SPACESHIP, WHICH GROSS WEIGHT IS APPROXIMATELY 2,210,000 POUNDS (ONE MILLION KILOGRAMS).

INSTEAD OF A PROPULSION SYSTEM WITH ONE BIG SINGLE ENGINE, A SAFER ALTERNATIVE WOULD BE A SHIP WITH TWO OR THREE SMALLER PROPULSION SYSTEMS INSTALLED IN PARALLEL.

PROPULSION OPTIONS FOR THIS SPACESHIP ARE ELECTRIC ROCKETS AND NUCLEAR FUSION. THESE OPTIONS WILL BE CHOSEN NOT NECESSARILY FOR THE ENGINE WITH THE HIGHEST PERFORMANCE, BUT FOR THE ONE THAT WILL BECOME FIRST AVAILABLE, AFTER IT HAS BEEN DEVELOPED AND TESTED.

PROPULSION OPTIONS FOR THIS SPACESHIP ARE:

I. ELECTRICAL ROCKETS:
        ARCJET
        ION ENGINE
        MDP
        SUPERCONDUCTING PROYECTILE (MACROPARTICLE)

II. NUCLEAR FUSION:
        INERTIAL CONFINEMENT FUSION

FOR THE INERTIAL CONFINEMENT FUSION ENGINE IT WOULD BE NECESSARY FOR SCIENTISTS AND AEROSPACE INSTITUTIONS OF THE WORLD TO DEVELOP A COMPACT, RELATIVELY SMALL AND VERY POWERFULL LASER TRIGGER SYSTEM INCLUDING ITS LASER RODS COMPONENTS. THE ONLY INFORMATION THAT I KNOW, AND WOULD SUPPLY TO THE PROPULSION ENGINEERS ARE THE DIMENSIONS OF THE MOUNTING PLATE WHERE THOSE ENGINES SHOULD BE INSTALLED AND ITS BOLTS.

SUPPLEMENTARY INFORMATION, AS REFERENCE ONLY

WHEN I WAS A SMALL CHILD, I OVERHEARD A CONVERSATION BETWEEN MY FATHER AND A TOURIST, A SWISS ENGINEER. THE ENGINEER COMMENTED THAT IF THERE ARE ADVANCED CIVILIZATIONS IN THE UNIVERSE THAT HAVE ALREADY ATTAINED SPACE FLIGHT, THE CROSS SECTION OF THEIR ENGINES PROBABLY WOULD LOOK LIKE A CRESCENT MOON. IT WAS A CONCEPT THAT YEARS LATER APPEARED LOGICAL TO ME. THIS IS TRUE BECAUSE AN OPEN EXHAUST WILL NOT CONFINE THE ENERGY YIELD BY THE FUSION OF THE LITHIUM DEUTERIDE PELLETS, IN A CLOSED CHAMBER OR IN A CONVERGING-DIVERGING ROCKET NOZZLE, THE WALLS WOULD MELT. ALSO IT SEEMS LOGICAL THAT THE THICK HOLLOW WALLS FROM AN ENGINE, (WHICH CROSS SECTIONS APPEAR TO BE AS THE SILHOUETTE OF A CRESCENT MOON) WILL ALLOW MORE VOLUME FOR THE REFRIGERATING FLUID. THIS DESIGN WILL OPERATE AS A MODIFIED PUSHER PLATE OF MINIATURE HYDROGEN BOMBS.

I USED THESE CONCEPTS TO DO QUICK SKETCHES OF A FUSION INERTIAL CONFINEMENT REACTOR AS SHOWN IN THE DRAWINGS OF SHEETS 21, 22, 23 AND 24 OF THIS PROJECT, WHICH SHOWS A PARABOLIC FUSION CHAMBER INTEGRATED WITH A SHORT ALSO PARABOLIC EXHAUST.

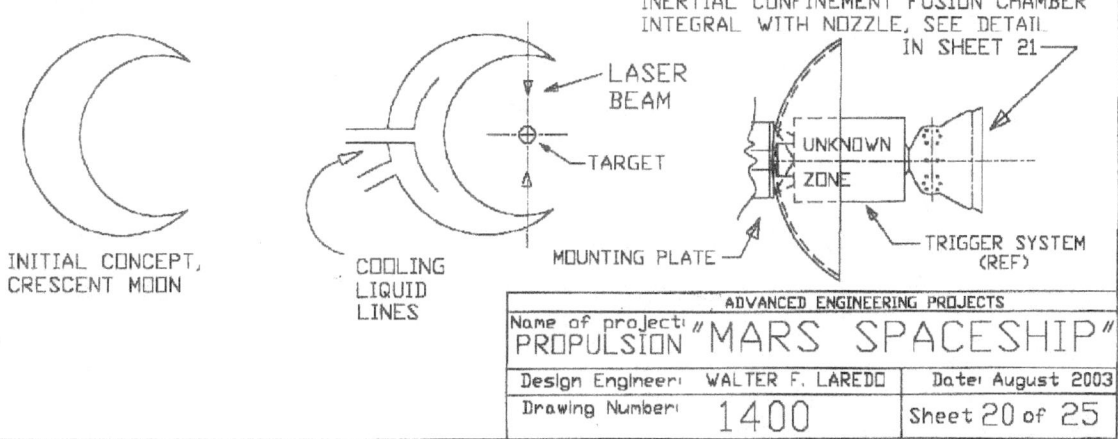

INITIAL CONCEPT, CRESCENT MOON

COOLING LIQUID LINES

LASER BEAM

TARGET

MOUNTING PLATE

INERTIAL CONFINEMENT FUSION CHAMBER INTEGRAL WITH NOZZLE, SEE DETAIL IN SHEET 21

UNKNOWN ZONE

TRIGGER SYSTEM (REF)

ADVANCED ENGINEERING PROJECTS		
Name of project: PROPULSION "MARS SPACESHIP"		
Design Engineer: WALTER F. LAREDO		Date: August 2003
Drawing Number: 1400		Sheet 20 of 25

PLATE 156

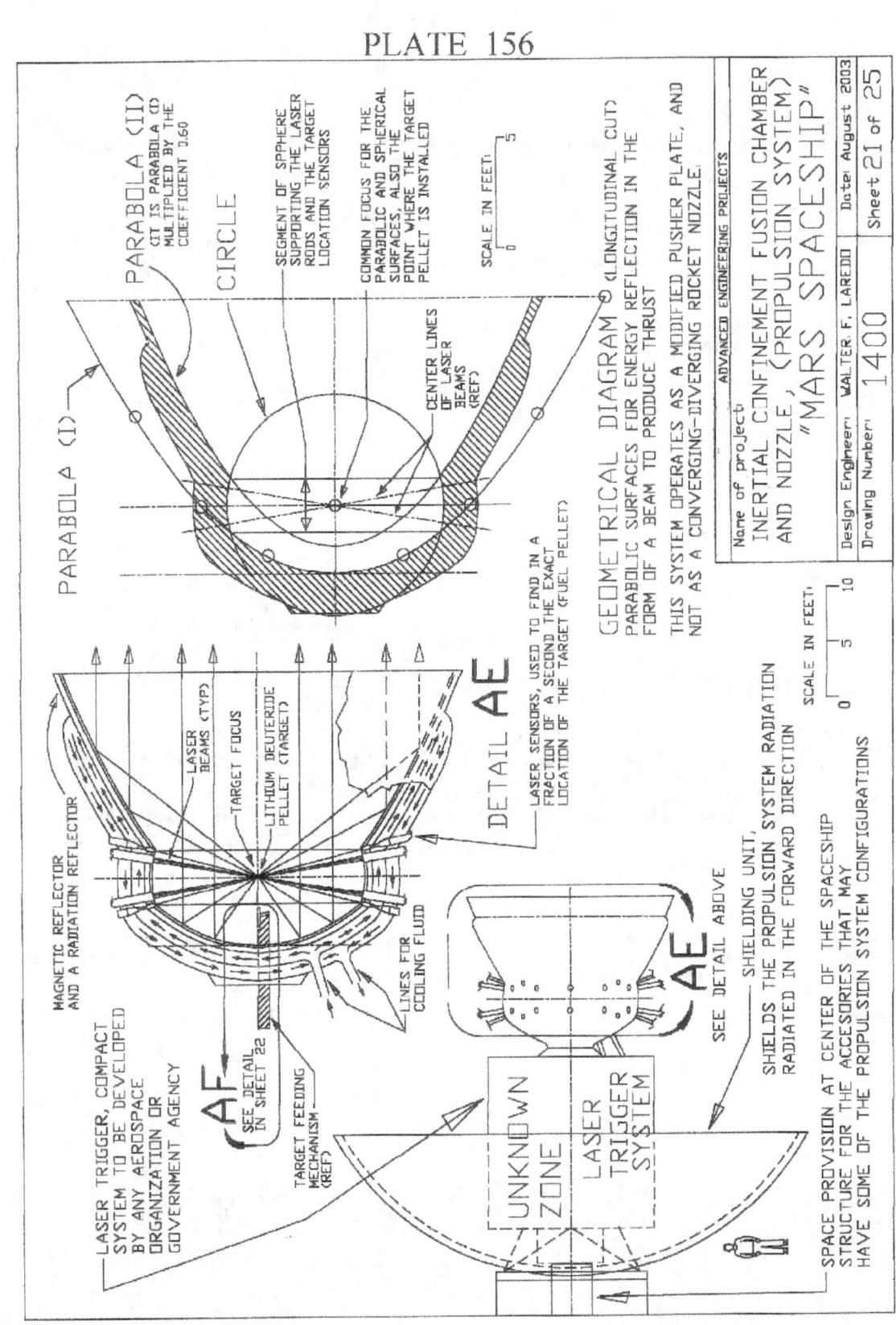

PLATE 157

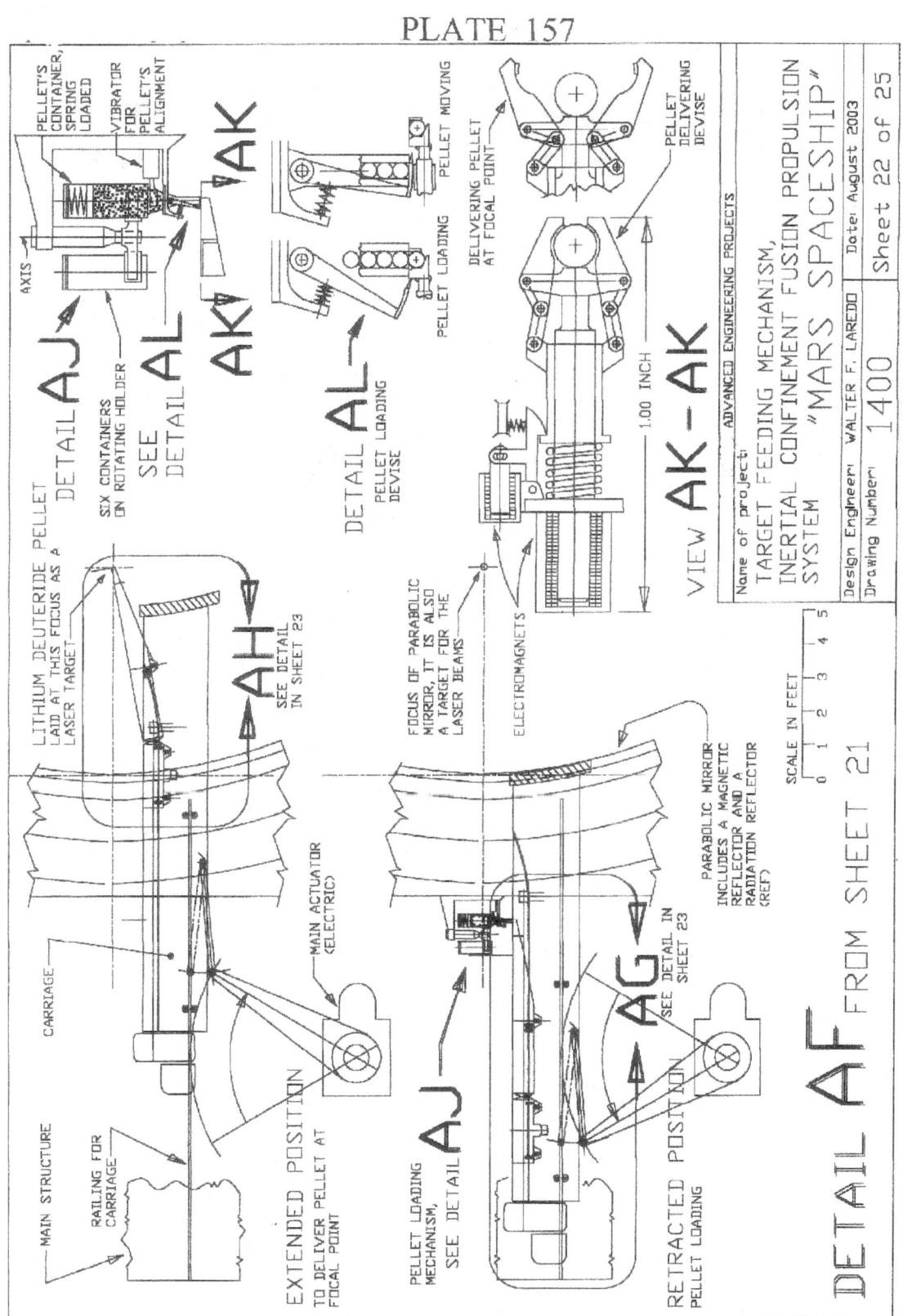

AXIS

PELLET'S CONTAINER, SPRING LOADED

VIBRATOR FOR PELLET'S ALIGNMENT

DETAIL AJ

SEE DETAIL AL

DETAIL AK

AK

SIX CONTAINERS ON ROTATING HOLDER

PELLET LOADING

PELLET MOVING

DETAIL AL

PELLET LOADING DEVISE

DELIVERING PELLET AT FOCAL POINT

PELLET DELIVERING DEVISE

1.00 INCH

VIEW AK-AK

LITHIUM DEUTERIDE PELLET LAID AT THIS FOCUS AS A LASER TARGET

SEE DETAIL IN SHEET 23

AH

FOCUS OF PARABOLIC MIRROR, IT IS ALSO A TARGET FOR THE LASER BEAMS

ELECTROMAGNETS

CARRIAGE

MAIN ACTUATOR (ELECTRIC)

EXTENDED POSITION

TO DELIVER PELLET AT FOCAL POINT

PELLET LOADING MECHANISM, SEE DETAIL AJ

AJ

MAIN STRUCTURE

RAILING FOR CARRIAGE

PARABOLIC MIRROR INCLUDES A MAGNETIC REFLECTOR AND A RADIATION REFLECTOR (REF)

SEE DETAIL IN SHEET 23

AG

RETRACTED POSITION

PELLET LOADING

DETAIL AF FROM SHEET 21

SCALE IN FEET

0 1 2 3 4 5

Name of project: ADVANCED ENGINEERING PROJECTS

TARGET FEEDING MECHANISM, INERTIAL CONFINEMENT FUSION PROPULSION SYSTEM "MARS SPACESHIP"

Design Engineer: WALTER F. LAREDO

Date: August 2003

Drawing Number: 1400

Sheet 22 of 25

PLATE 158

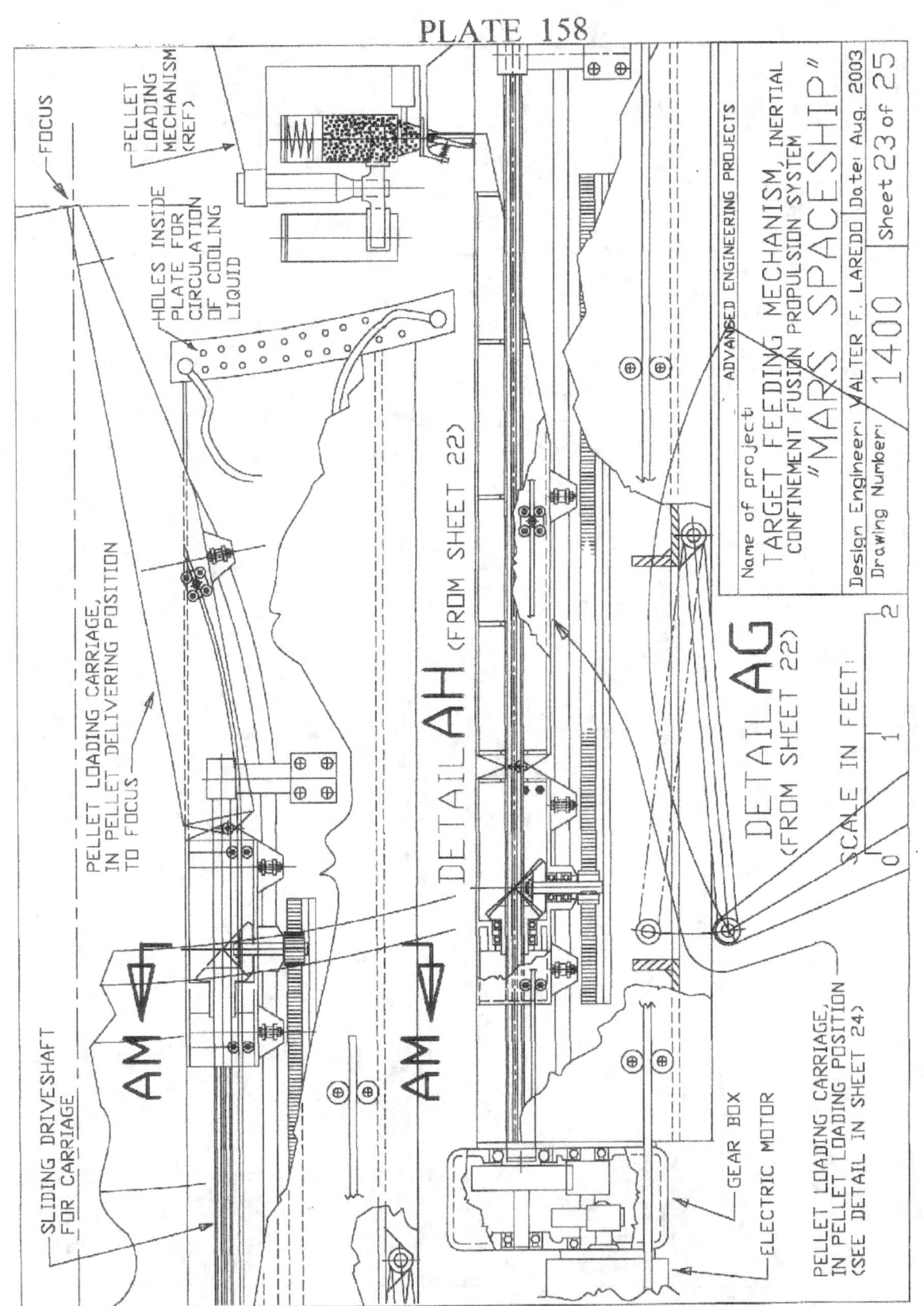

FOCUS

PELLET LOADING MECHANISM (REF)

SLIDING DRIVESHAFT FOR CARRIAGE

PELLET LOADING CARRIAGE, IN PELLET DELIVERING POSITION TO FOCUS

HOLES INSIDE PLATE FOR CIRCULATION OF COOLING LIQUID

AM

AM

DETAIL AH (FROM SHEET 22)

DETAIL AG (FROM SHEET 22)

SCALE IN FEET:

0   1   2

GEAR BOX

ELECTRIC MOTOR

PELLET LOADING CARRIAGE, IN PELLET LOADING POSITION (SEE DETAIL IN SHEET 24)

ADVANCED ENGINEERING PROJECTS

Name of project:
TARGET FEEDING MECHANISM, INERTIAL CONFINEMENT FUSION PROPULSION SYSTEM
"MARS SPACESHIP"

Design Engineer: WALTER F. LAREDO   Date: Aug. 2003

Drawing Number: 1400   Sheet 23 of 25

PLATE 159

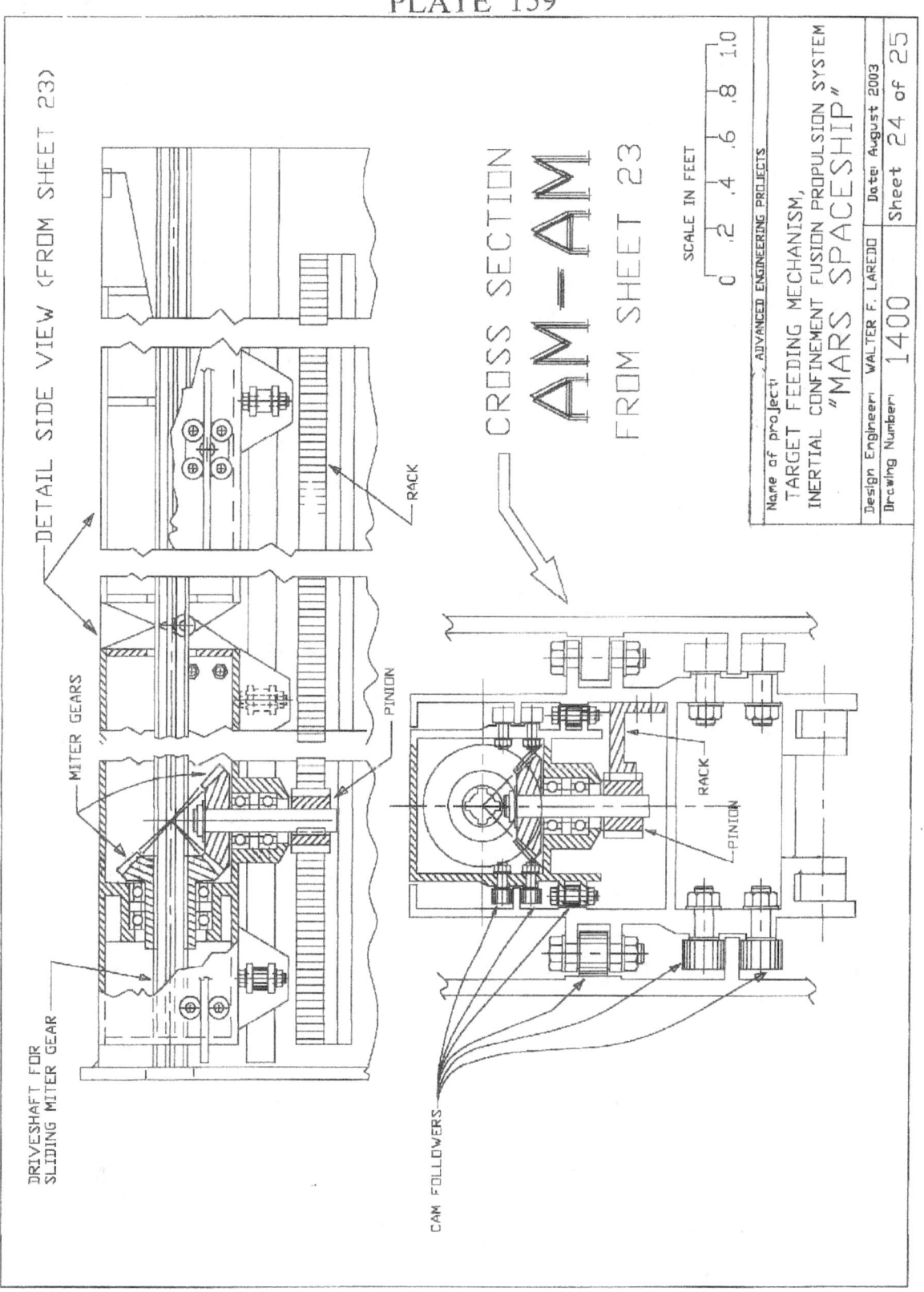

DETAIL SIDE VIEW (FROM SHEET 23)

DRIVESHAFT FOR
SLIDING MITER GEAR

MITER GEARS

PINION

RACK

CROSS SECTION

AM-AM

FROM SHEET 23

CAM FOLLOWERS

RACK

PINION

SCALE IN FEET

0   .2   .4   .6   .8   1.0

ADVANCED ENGINEERING PROJECTS

Name of project:
TARGET FEEDING MECHANISM,
INERTIAL CONFINEMENT FUSION PROPULSION SYSTEM
"MARS SPACESHIP"

Design Engineer: WALTER F. LAREDO   Date: August 2003
Drawing Number: 1400   Sheet 24 of 25

PLATE 160

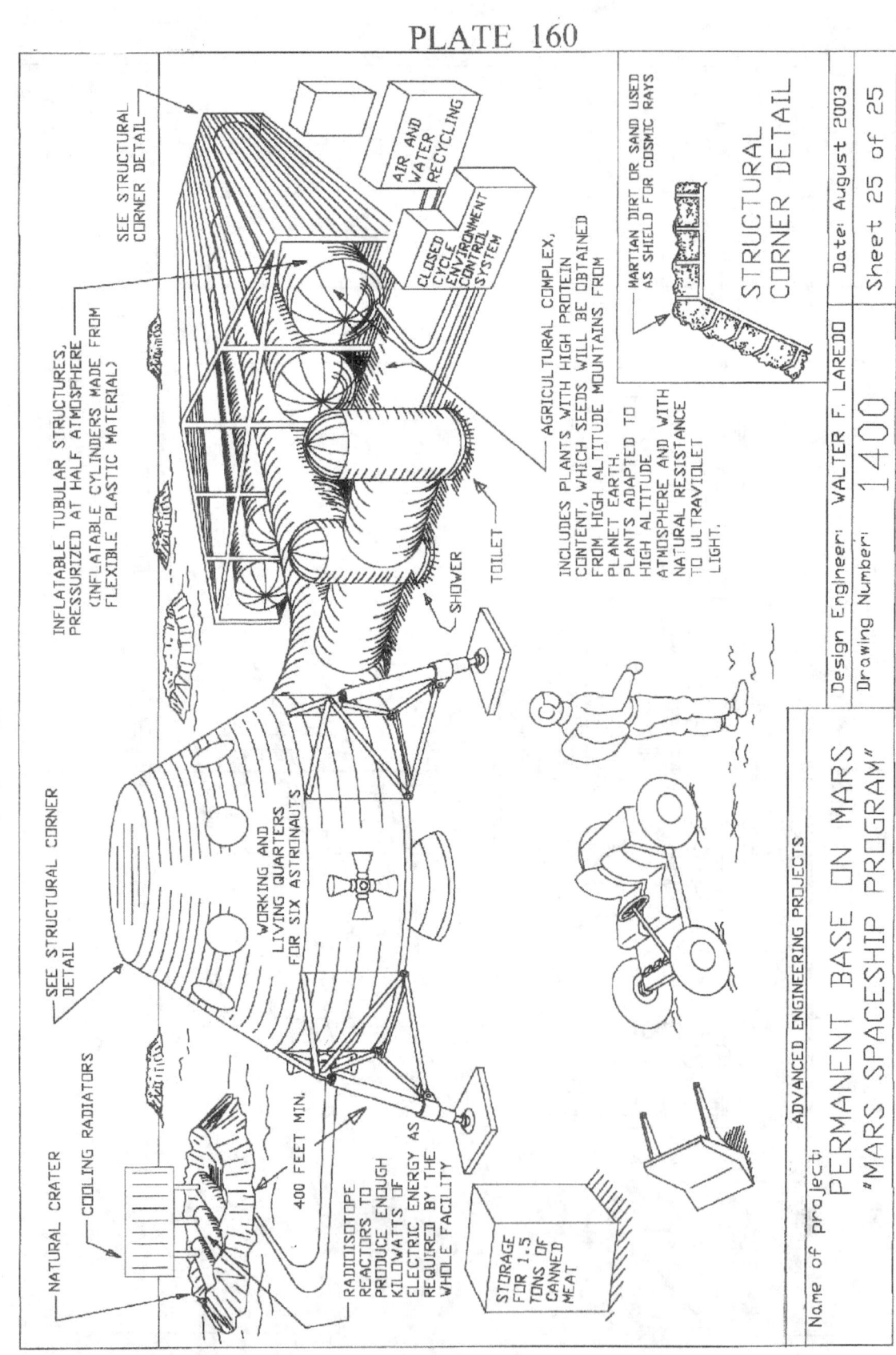

STRUCTURAL CORNER DETAIL

MARTIAN DIRT OR SAND USED AS SHIELD FOR COSMIC RAYS

Design Engineer: WALTER F. LAREDO
Date: August 2003

Drawing Number: 1400
Sheet 25 of 25

Name of project:
PERMANENT BASE ON MARS
"MARS SPACESHIP PROGRAM"

ADVANCED ENGINEERING PROJECTS

SEE STRUCTURAL CORNER DETAIL

INFLATABLE TUBULAR STRUCTURES, PRESSURIZED AT HALF ATMOSPHERE (INFLATABLE CYLINDERS MADE FROM FLEXIBLE PLASTIC MATERIAL)

AIR AND WATER RECYCLING

CLOSED CYCLE ENVIRONMENT CONTROL SYSTEM

AGRICULTURAL COMPLEX, INCLUDES PLANTS WITH HIGH PROTEIN CONTENT, WHICH SEEDS WILL BE OBTAINED FROM HIGH ALTITUDE MOUNTAINS FROM PLANET EARTH. PLANTS ADAPTED TO HIGH ALTITUDE ATMOSPHERE AND WITH NATURAL RESISTANCE TO ULTRAVIOLET LIGHT.

TOILET

SHOWER

WORKING AND LIVING QUARTERS FOR SIX ASTRONAUTS

SEE STRUCTURAL CORNER DETAIL

NATURAL CRATER

COOLING RADIATORS

400 FEET MIN.

RADIOISOTOPE REACTORS TO PRODUCE ENOUGH KILOWATTS OF ELECTRIC ENERGY AS REQUIRED BY THE WHOLE FACILITY

STORAGE FOR 1.5 TONS OF CANNED MEAT

PLATE 161

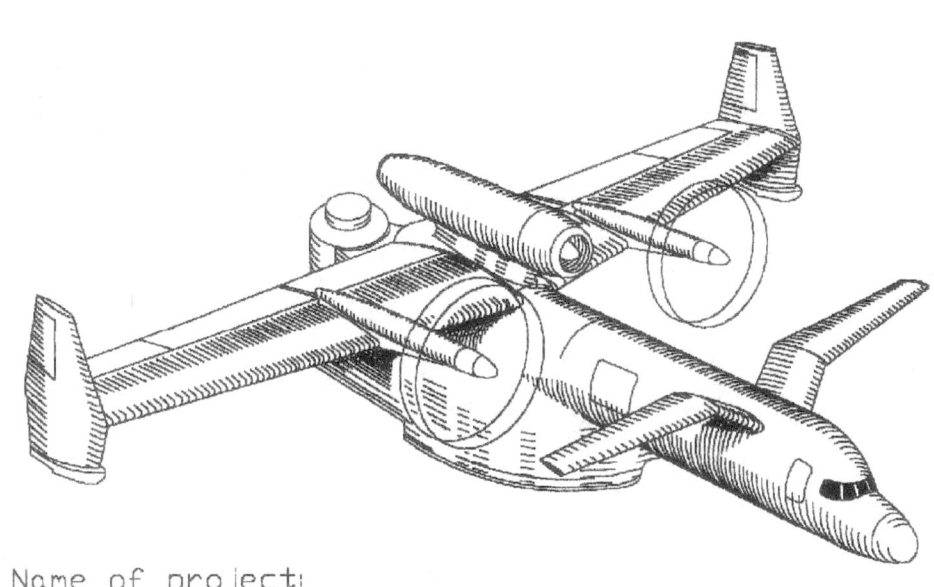

Name of project:

PRELIMINARY DESIGN OF
CARGO AIRCRAFT WITH
NUCLEAR PROPULSION

CRUISING SPEED   292 MPH
SERVICE CEILING   FROM 15 TO 5,000 FT
RANGE   LIMITLESS, REACTOR HAVE FUEL FOR 10 YEARS
                             OF CONTINUOUS USE

OVERALL LENGHT   108 FEET
WING SPAN   120 FEET
WING AREA   3,180 SQ FT
GROSS WEIGHT   254,400 LB
WING LOADING   80 LB/SQ FT
POWER LOADING   8.48 PER SHP
TURBINE OUTPUT POWER   30,000 SHP = 22,371 KW
REACTOR THERMAL POWER   67,000 KW
EFFICIENCY   33 %
REACTOR SHIELDING WEIGHT 70,000 LB

FLIGHT
PROFILE:

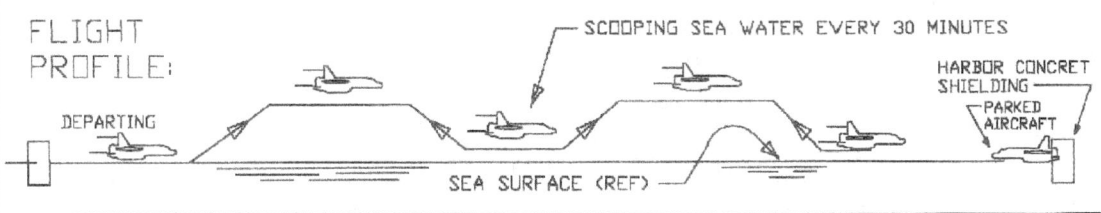

ADVANCED ENGINEERING PROJECTS		
Design Engineer: Walter F. Laredo		Date: August 2003
Drawing Number: 1500		Sheet 1 of 9

PLATE 162

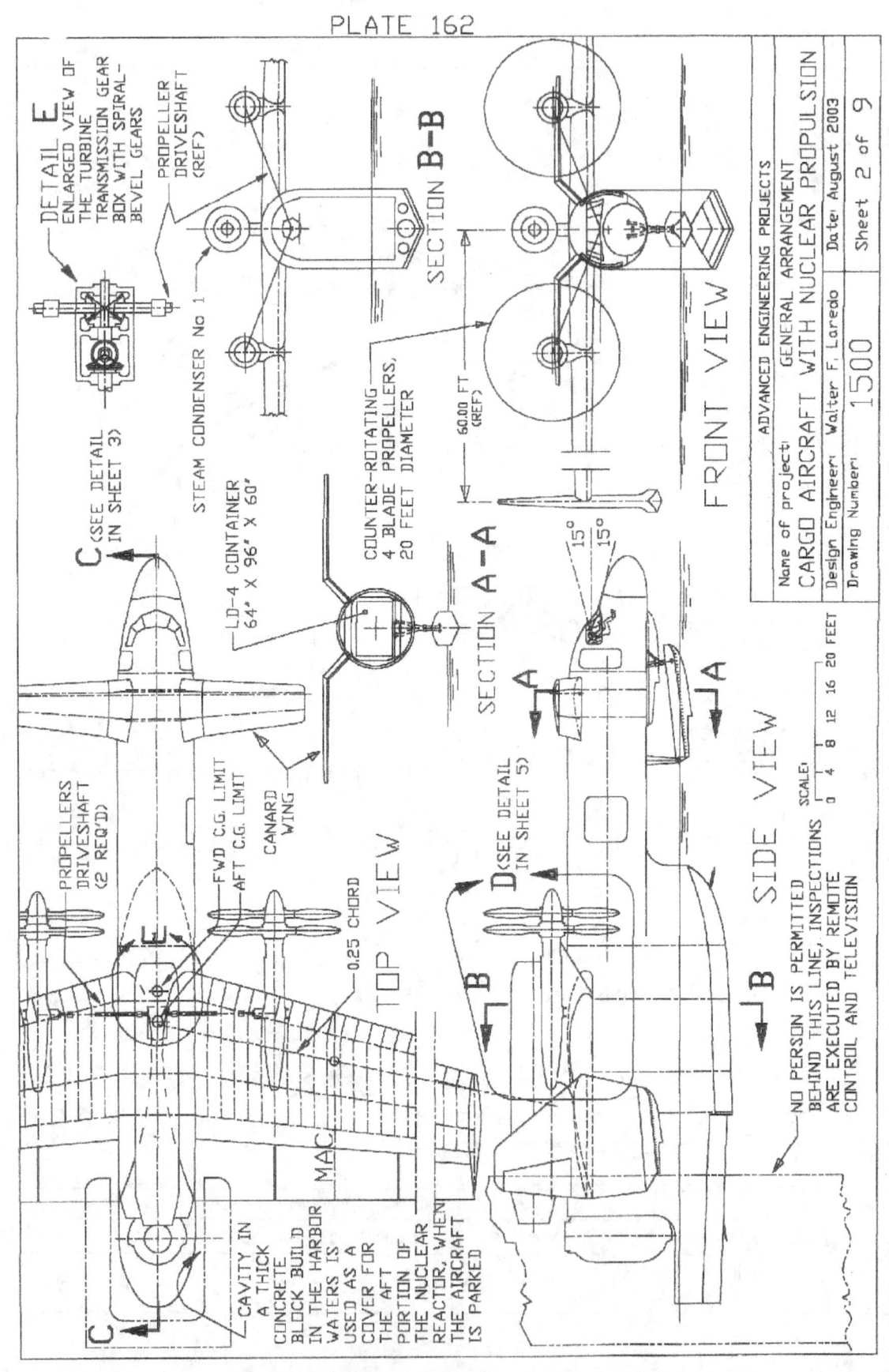

DETAIL E
ENLARGED VIEW OF THE TURBINE TRANSMISSION GEAR BOX WITH SPIRAL-BEVEL GEARS

PROPELLER DRIVESHAFT (REF)

SECTION B-B

STEAM CONDENSER No 1

LD-4 CONTAINER 64" X 96" X 60"

COUNTER-ROTATING 4 BLADE PROPELLERS, 20 FEET DIAMETER

60.00 FT (REF)

FRONT VIEW

C (SEE DETAIL IN SHEET 3)

SECTION A-A

15°
15°

0 4 8 12 16 20 FEET
SCALE:

SIDE VIEW

D (SEE DETAIL IN SHEET 5)

B

B

PROPELLERS DRIVESHAFT (2 REQ'D)

FWD C.G. LIMIT
AFT C.G. LIMIT

CANARD WING

0.25 CHORD

MAC

TOP VIEW

E

C

CAVITY IN A THICK CONCRETE BLOCK BUILD IN THE HARBOR WATERS IS USED AS A COVER FOR THE AFT PORTION OF THE NUCLEAR REACTOR, WHEN THE AIRCRAFT IS PARKED

NO PERSON IS PERMITTED BEHIND THIS LINE, INSPECTIONS ARE EXECUTED BY REMOTE CONTROL AND TELEVISION

Name of project: GENERAL ARRANGEMENT
CARGO AIRCRAFT WITH NUCLEAR PROPULSION

ADVANCED ENGINEERING PROJECTS

Design Engineer: Walter F. Laredo   Date: August 2003

Drawing Number: 1500   Sheet 2 of 9

PLATE 163

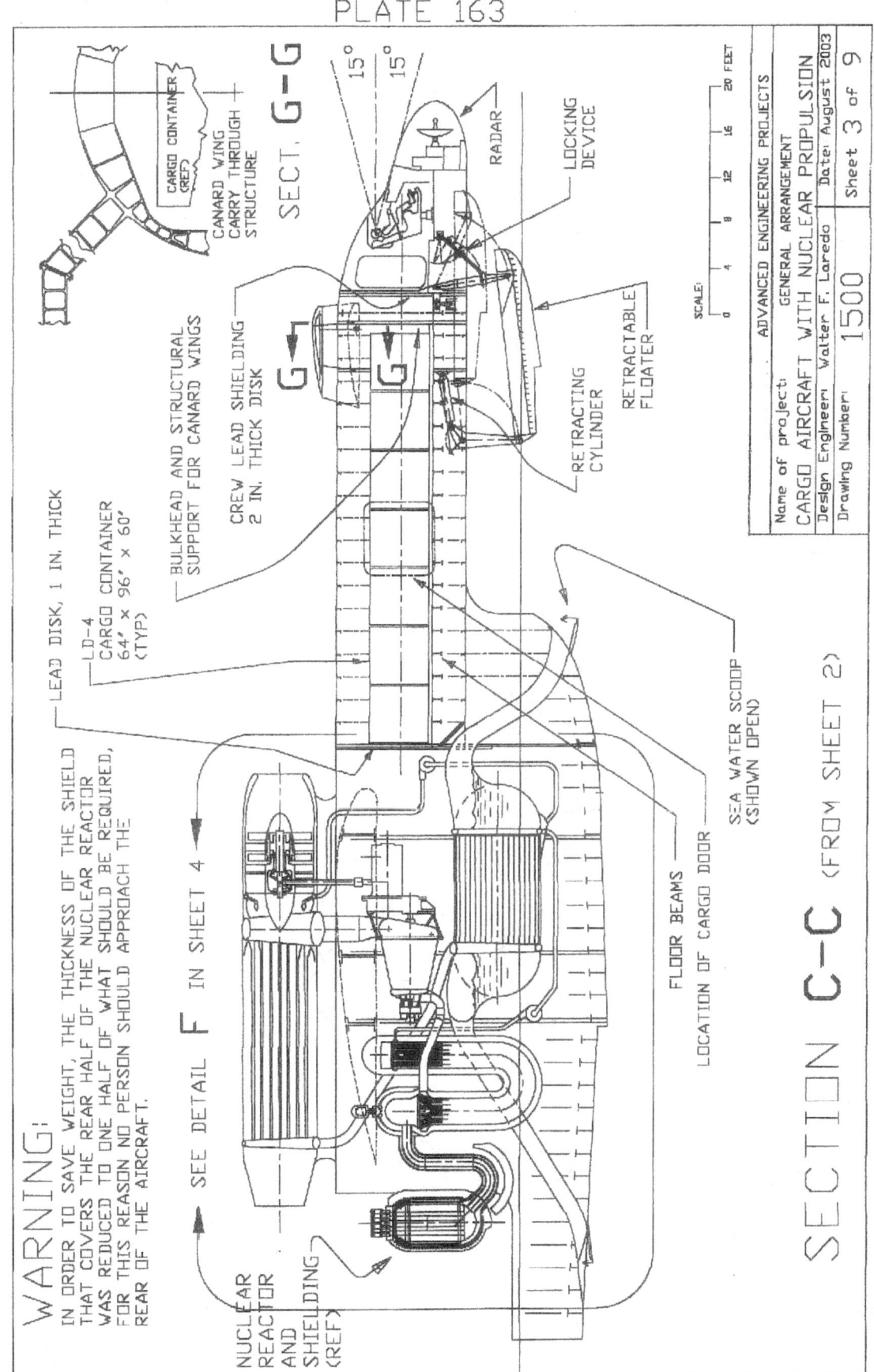

WARNING:
IN ORDER TO SAVE WEIGHT, THE THICKNESS OF THE SHIELD THAT COVERS THE REAR HALF OF THE NUCLEAR REACTOR WAS REDUCED TO ONE HALF OF WHAT SHOULD BE REQUIRED, FOR THIS REASON NO PERSON SHOULD APPROACH THE REAR OF THE AIRCRAFT.

CARGO CONTAINER (REF)

CANARD WING CARRY THROUGH STRUCTURE

SECT. G-G

15°
15°

RADAR

LOCKING DEVICE

RETRACTING CYLINDER

RETRACTABLE FLOATER

LEAD DISK, 1 IN. THICK

LD-4 CARGO CONTAINER 64' × 96' × 60' (TYP)

BULKHEAD AND STRUCTURAL SUPPORT FOR CANARD WINGS

CREW LEAD SHIELDING 2 IN. THICK DISK

G

G

SEE DETAIL F IN SHEET 4

NUCLEAR REACTOR AND SHIELDING (REF)

FLOOR BEAMS

LOCATION OF CARGO DOOR

SEA WATER SCOOP (SHOWN OPEN)

SECTION C-C (FROM SHEET 2)

SCALE:
0 4 8 12 16 20 FEET

ADVANCED ENGINEERING PROJECTS		
Name of project:	GENERAL ARRANGEMENT	
CARGO AIRCRAFT WITH NUCLEAR PROPULSION		
Design Engineer: Walter F. Laredo		Date: August 2003
Drawing Number:	1500	Sheet 3 of 9

PLATE 164

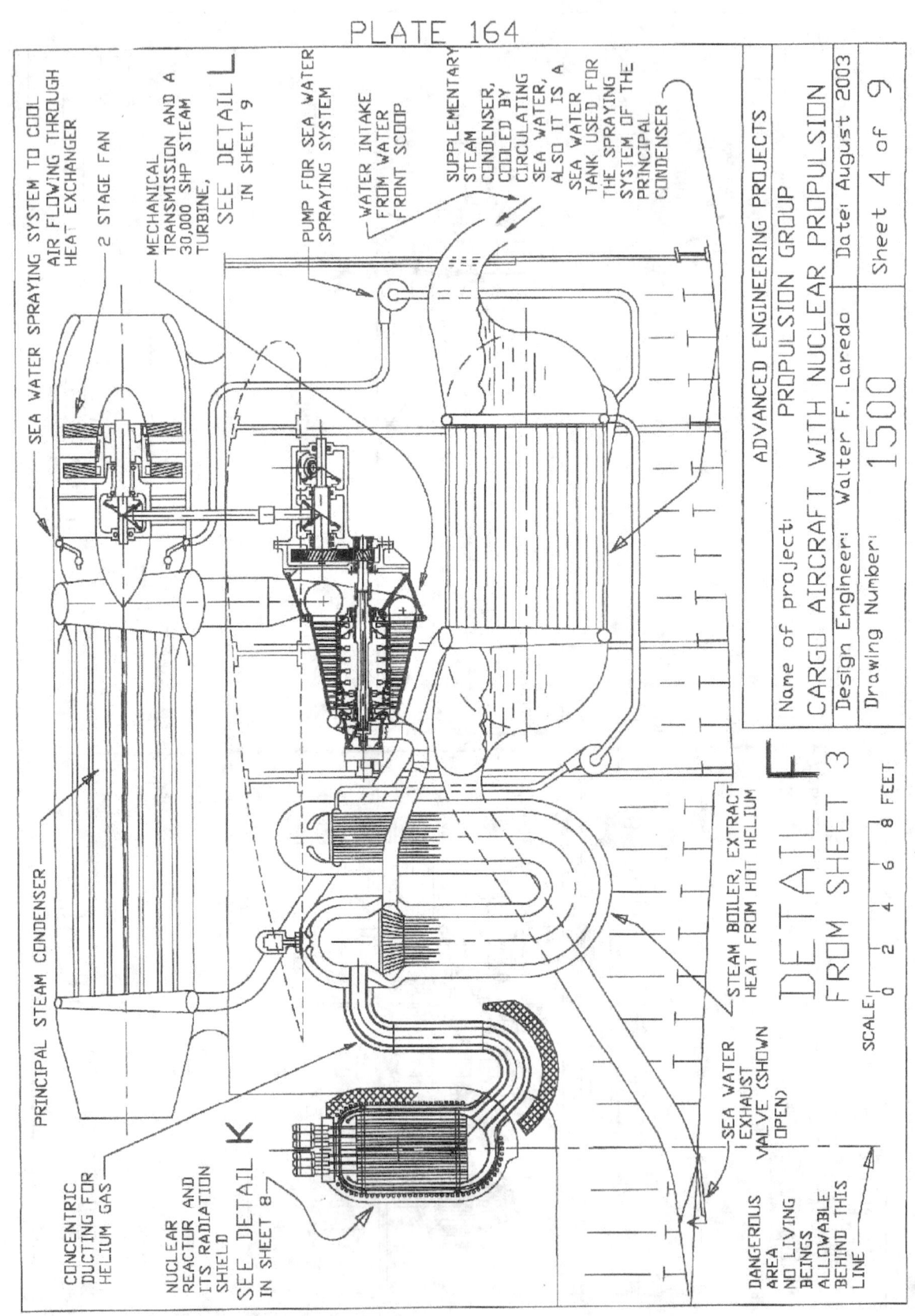

SEA WATER SPRAYING SYSTEM TO COOL AIR FLOWING THROUGH HEAT EXCHANGER

2 STAGE FAN

MECHANICAL TRANSMISSION AND A 30,000 SHP STEAM TURBINE.

SEE DETAIL L IN SHEET 9

PUMP FOR SEA WATER SPRAYING SYSTEM

WATER INTAKE FROM WATER FRONT SCOOP

SUPPLEMENTARY STEAM CONDENSER, COOLED BY CIRCULATING SEA WATER, ALSO IT IS A SEA WATER TANK USED FOR THE SPRAYING SYSTEM OF THE PRINCIPAL CONDENSER

PRINCIPAL STEAM CONDENSER

CONCENTRIC DUCTING FOR HELIUM GAS

NUCLEAR REACTOR AND ITS RADIATION SHIELD

SEE DETAIL K IN SHEET 8

STEAM BOILER, EXTRACT HEAT FROM HOT HELIUM

SEA WATER EXHAUST VALVE (SHOWN OPEN)

DANGEROUS AREA NO LIVING BEINGS ALLOWABLE BEHIND THIS LINE

DETAIL F FROM SHEET 3

SCALE: 0 2 4 6 8 FEET

Name of project: ADVANCED ENGINEERING PROJECTS PROPULSION GROUP
CARGO AIRCRAFT WITH NUCLEAR PROPULSION

Design Engineer: Walter F. Laredo   Date: August 2003

Drawing Number: 1500   Sheet 4 of 9

PLATE 165

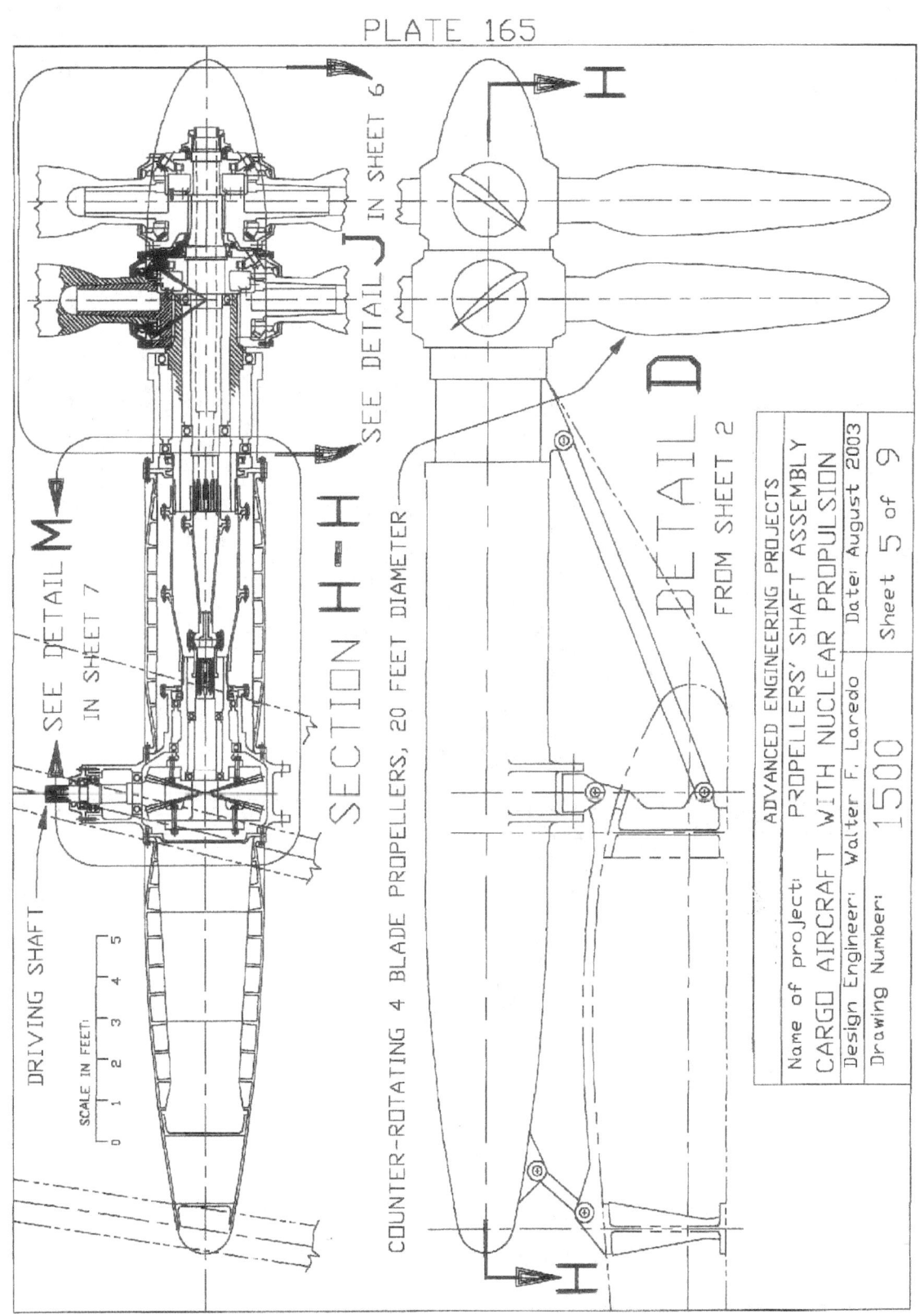

DRIVING SHAFT

SCALE IN FEET:
0 1 2 3 4 5

SECTION H-H

SEE DETAIL M IN SHEET 7

SEE DETAIL J IN SHEET 6

COUNTER-ROTATING 4 BLADE PROPELLERS, 20 FEET DIAMETER

DETAIL D

FROM SHEET 2

ADVANCED ENGINEERING PROJECTS	
Name of project: PROPELLERS' SHAFT ASSEMBLY CARGO AIRCRAFT WITH NUCLEAR PROPULSION	
Design Engineer: Walter F. Laredo	Date: August 2003
Drawing Number: 1500	Sheet 5 of 9

PLATE 166

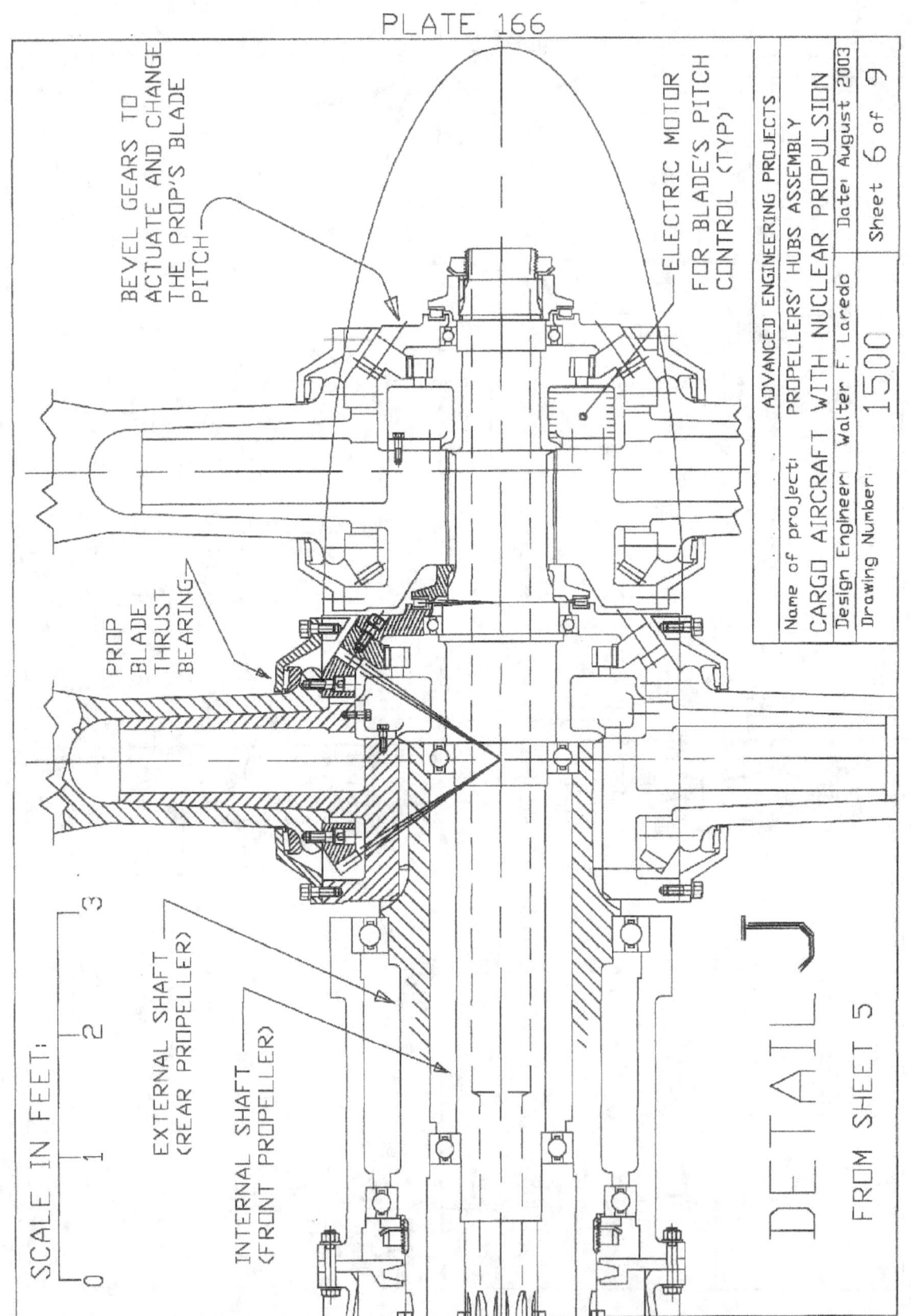

SCALE IN FEET:

0   1   2   3

BEVEL GEARS TO ACTUATE AND CHANGE THE PROP'S BLADE PITCH

ELECTRIC MOTOR FOR BLADE'S PITCH CONTROL (TYP)

PROP BLADE THRUST BEARING

EXTERNAL SHAFT (REAR PROPELLER)

INTERNAL SHAFT (FRONT PROPELLER)

DETAIL J

FROM SHEET 5

Name of project:	ADVANCED ENGINEERING PROJECTS
PROPELLERS' HUBS ASSEMBLY CARGO AIRCRAFT WITH NUCLEAR PROPULSION	
Design Engineer: Walter F. Laredo	Date: August 2003
Drawing Number:	1500
	Sheet 6 of 9

PLATE 167

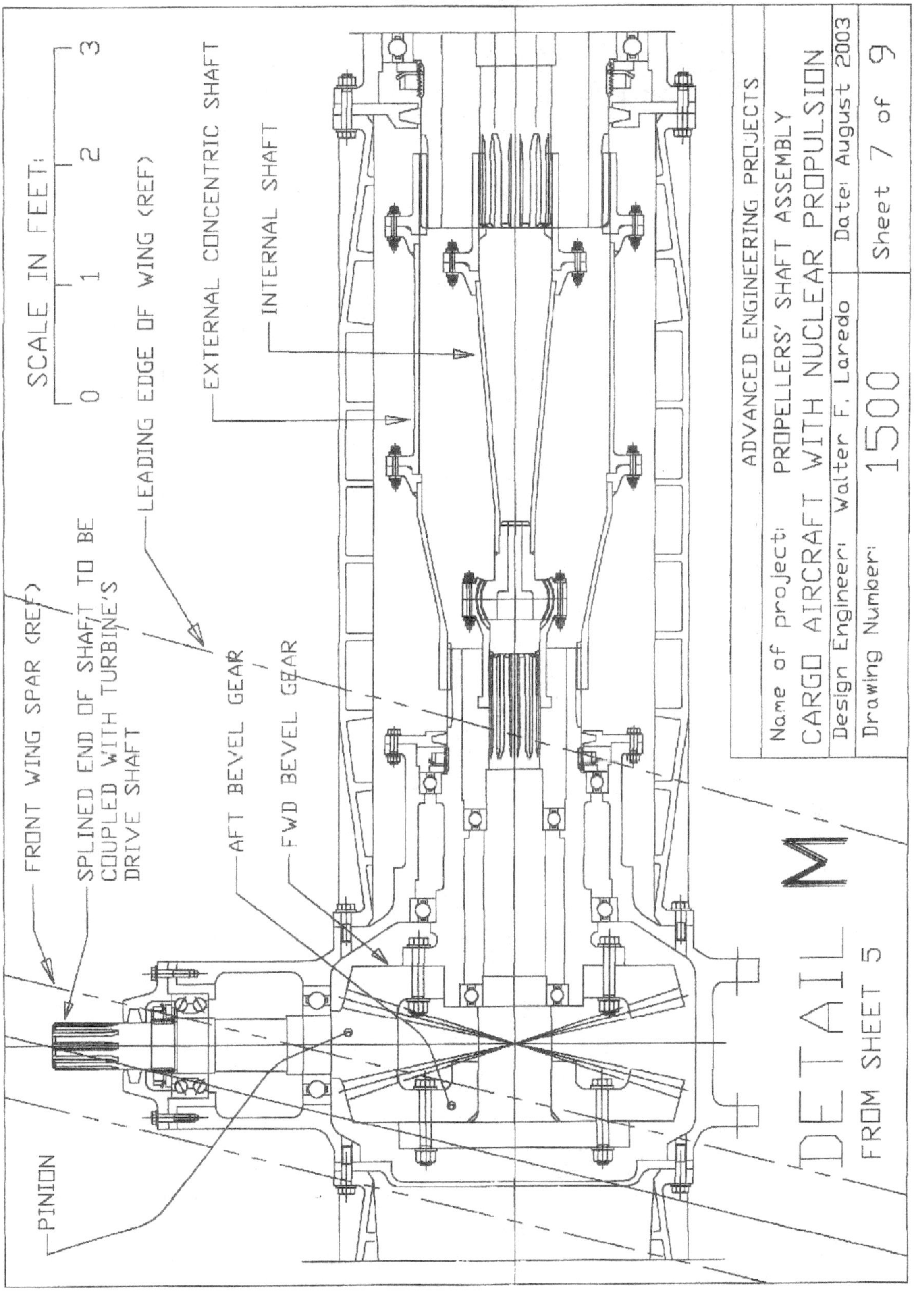

SCALE IN FEET:

0 1 2 3

LEADING EDGE OF WING (REF)

EXTERNAL CONCENTRIC SHAFT

INTERNAL SHAFT

FRONT WING SPAR (REF)

SPLINED END OF SHAFT TO BE COUPLED WITH TURBINE'S DRIVE SHAFT

AFT BEVEL GEAR

FWD BEVEL GEAR

PINION

DETAIL M

FROM SHEET 5

Name of project:	PROPELLERS' SHAFT ASSEMBLY	
CARGO AIRCRAFT WITH NUCLEAR PROPULSION		
Design Engineer: Walter F. Laredo		Date: August 2003
Drawing Number:	1500	Sheet 7 of 9

ADVANCED ENGINEERING PROJECTS

PLATE 168

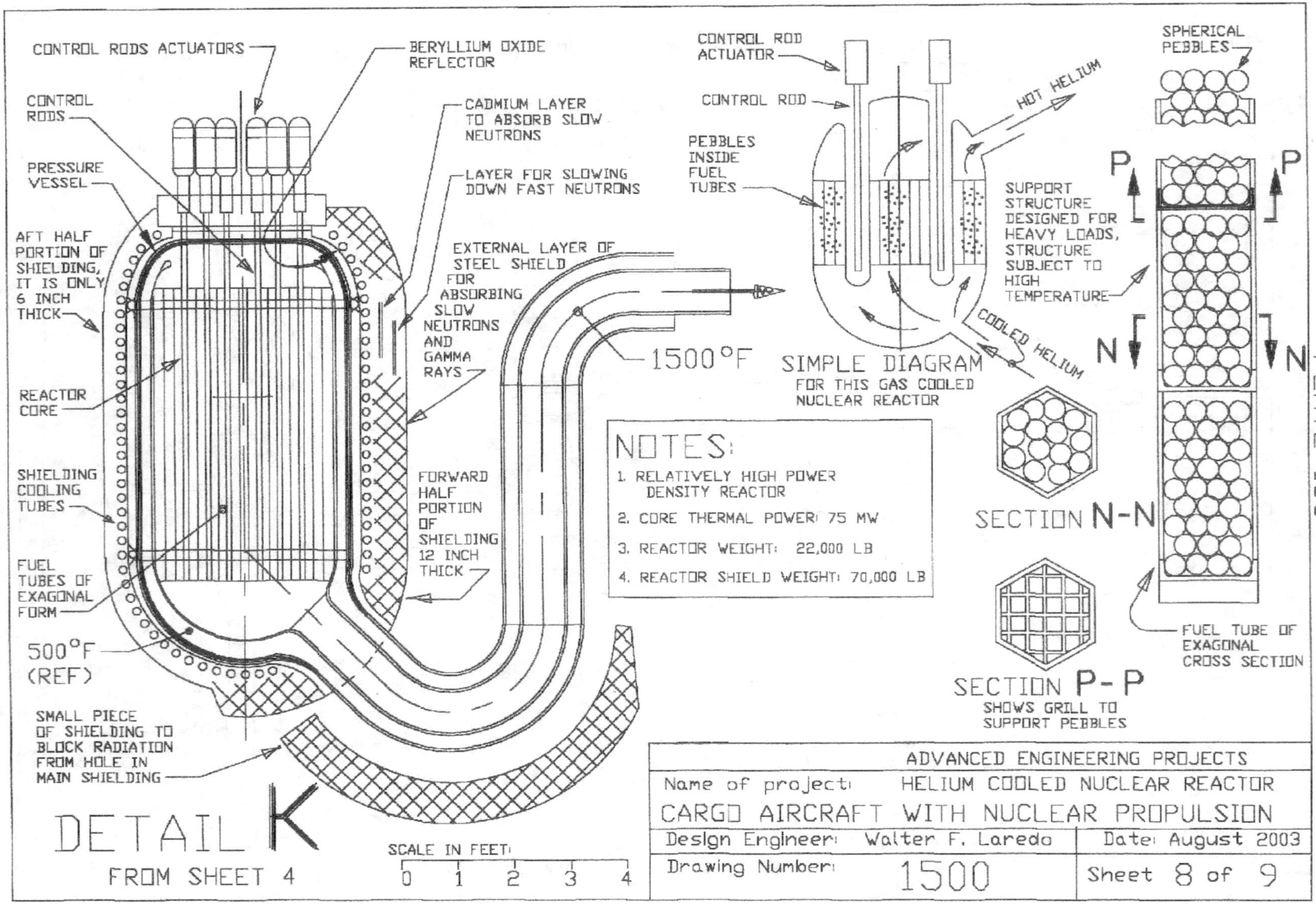
CONTROL RODS ACTUATORS
BERYLLIUM OXIDE REFLECTOR
CONTROL RODS
CADMIUM LAYER TO ABSORB SLOW NEUTRONS
PRESSURE VESSEL
LAYER FOR SLOWING DOWN FAST NEUTRONS
AFT HALF PORTION OF SHIELDING, IT IS ONLY 6 INCH THICK
EXTERNAL LAYER OF STEEL SHIELD FOR ABSORBING SLOW NEUTRONS AND GAMMA RAYS
REACTOR CORE
SHIELDING COOLING TUBES
FORWARD HALF PORTION OF SHIELDING 12 INCH THICK
FUEL TUBES OF EXAGONAL FORM
500°F (REF)
1500°F
SMALL PIECE OF SHIELDING TO BLOCK RADIATION FROM HOLE IN MAIN SHIELDING
DETAIL K
FROM SHEET 4
SCALE IN FEET:
0 1 2 3 4
CONTROL ROD ACTUATOR
CONTROL ROD
PEBBLES INSIDE FUEL TUBES
HOT HELIUM
COOLED HELIUM
SIMPLE DIAGRAM
FOR THIS GAS COOLED NUCLEAR REACTOR
SPHERICAL PEBBLES
P P
SUPPORT STRUCTURE DESIGNED FOR HEAVY LOADS, STRUCTURE SUBJECT TO HIGH TEMPERATURE
N N
FUEL TUBE OF EXAGONAL CROSS SECTION
SECTION N-N
SECTION P-P
SHOWS GRILL TO SUPPORT PEBBLES
NOTES:
1. RELATIVELY HIGH POWER DENSITY REACTOR
2. CORE THERMAL POWER: 75 MW
3. REACTOR WEIGHT: 22,000 LB
4. REACTOR SHIELD WEIGHT: 70,000 LB
ADVANCED ENGINEERING PROJECTS
Name of project: HELIUM COOLED NUCLEAR REACTOR
CARGO AIRCRAFT WITH NUCLEAR PROPULSION
Design Engineer: Walter F. Laredo
Date: August 2003
Drawing Number: 1500
Sheet 8 of 9

PLATE 169

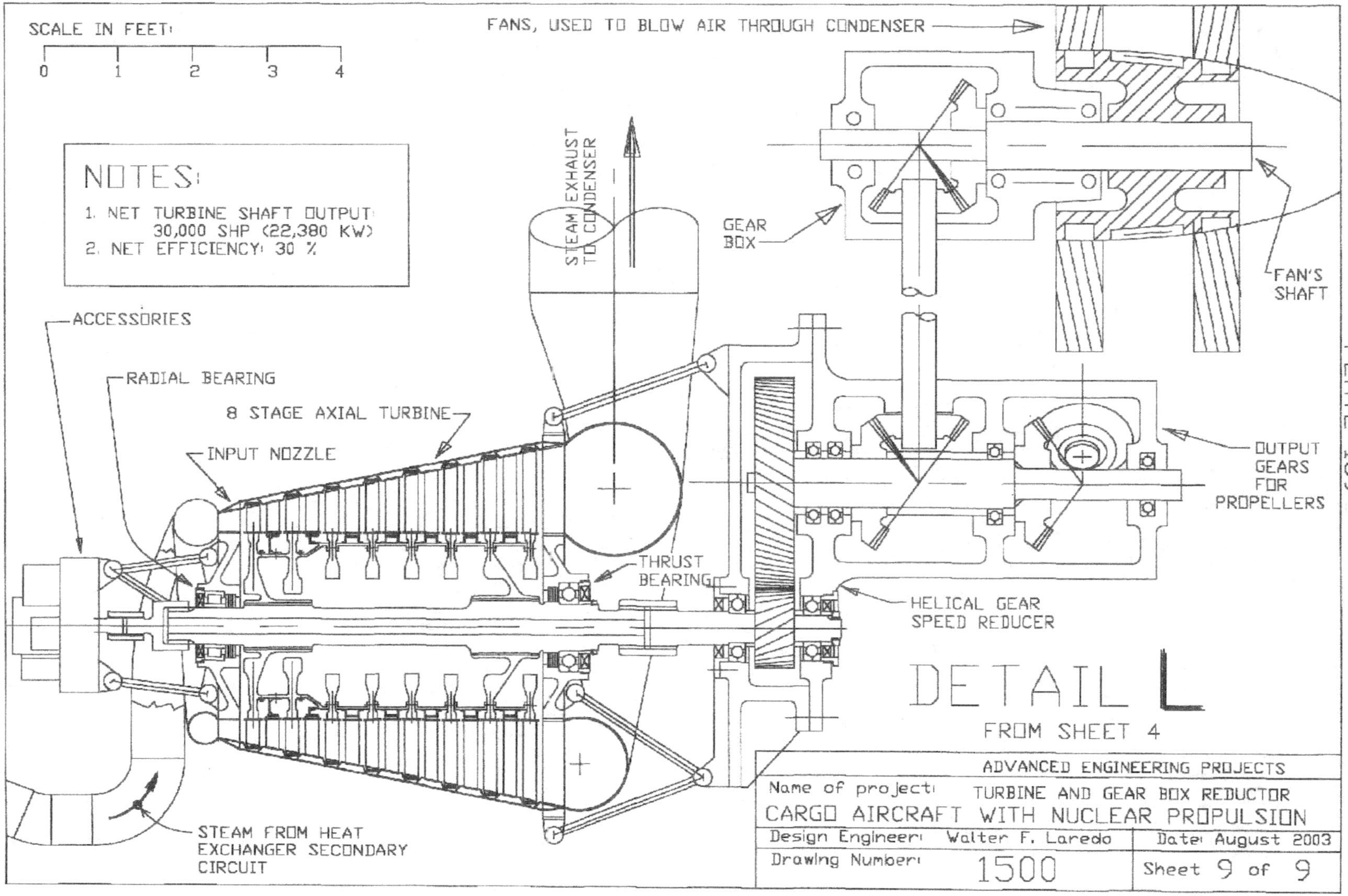

FANS, USED TO BLOW AIR THROUGH CONDENSER

FAN'S SHAFT

OUTPUT GEARS FOR PROPELLERS

GEAR BOX

HELICAL GEAR SPEED REDUCER

STEAM EXHAUST TO CONDENSER

DETAIL L

FROM SHEET 4

THRUST BEARING

SCALE IN FEET:
0 1 2 3 4

NOTES:
1. NET TURBINE SHAFT OUTPUT:
   30,000 SHP (22,380 Kw)
2. NET EFFICIENCY: 30 %

ACCESSORIES

RADIAL BEARING

8 STAGE AXIAL TURBINE

INPUT NOZZLE

STEAM FROM HEAT EXCHANGER SECONDARY CIRCUIT

ADVANCED ENGINEERING PROJECTS
Name of project: TURBINE AND GEAR BOX REDUCTOR
CARGO AIRCRAFT WITH NUCLEAR PROPULSION
Design Engineer: Walter F. Laredo   Date: August 2003
Drawing Number: 1500   Sheet 9 of 9

PLATE 170

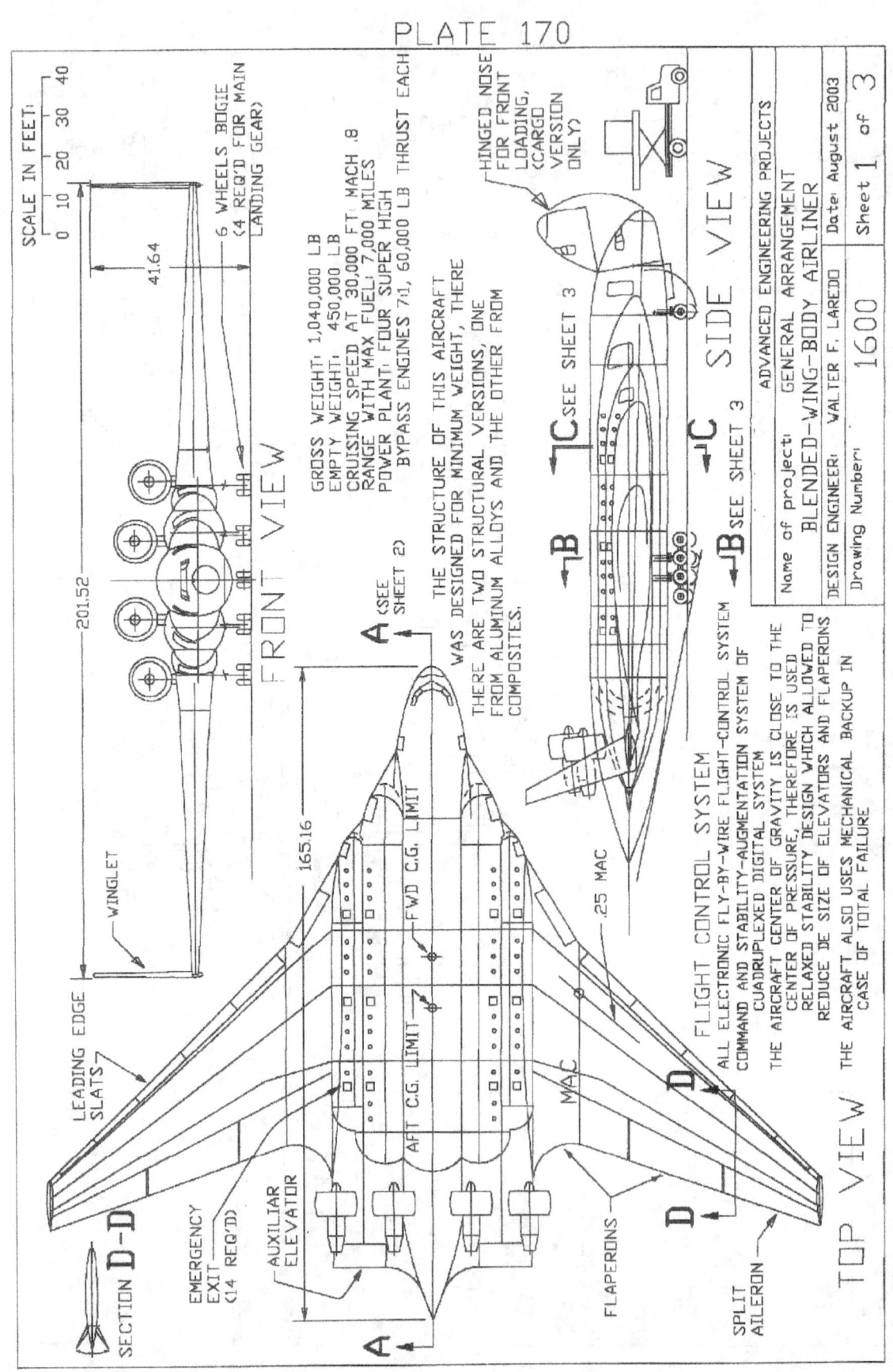

SCALE IN FEET:
0 10 20 30 40

201.52

41.64

6 WHEELS BOGIE
(4 REQ'D FOR MAIN
LANDING GEAR)

WINGLET

LEADING EDGE
SLATS

EMERGENCY
EXIT
(14 REQ'D)

SECTION D-D

AUXILIAR
ELEVATOR

FLAPERONS

SPLIT
AILERON

FRONT VIEW

GROSS WEIGHT: 1,040,000 LB
EMPTY WEIGHT: 450,000 LB
CRUISING SPEED AT 30,000 FT: MACH .8
RANGE WITH MAX FUEL: 7,000 MILES
POWER PLANT: FOUR SUPER HIGH
BYPASS ENGINES 71, 60,000 LB THRUST EACH

THE STRUCTURE OF THIS AIRCRAFT
WAS DESIGNED FOR MINIMUM WEIGHT, THERE
ARE TWO STRUCTURAL VERSIONS, ONE
FROM ALUMINUM ALLOYS AND THE OTHER FROM
COMPOSITES.

A (SEE SHEET 2)

HINGED NOSE
FOR FRONT
LOADING,
(CARGO
VERSION
ONLY?)

B
C SEE SHEET 3
C
B SEE SHEET 3

SIDE VIEW

165.16

FWD C.G. LIMIT

AFT C.G. LIMIT

.25 MAC

MAC

D

D

A

FLIGHT CONTROL SYSTEM

ALL ELECTRONIC FLY-BY-WIRE FLIGHT-CONTROL SYSTEM
COMMAND AND STABILITY-AUGMENTATION SYSTEM OF
CUADRUPLEXED DIGITAL SYSTEM

THE AIRCRAFT CENTER OF GRAVITY IS CLOSE TO THE
CENTER OF PRESSURE, THEREFORE IS USED
RELAXED STABILITY DESIGN WHICH ALLOWED TO
REDUCE DE SIZE OF ELEVATORS AND FLAPERONS

THE AIRCRAFT ALSO USES MECHANICAL BACKUP IN
CASE OF TOTAL FAILURE

TOP VIEW

Name of project: GENERAL ARRANGEMENT
ADVANCED ENGINEERING PROJECTS
BLENDED-WING-BODY AIRLINER
Date: August 2003

DESIGN ENGINEER: WALTER F. LAREDO
Sheet 1 of 3

Drawing Number: 1600

PLATE 171

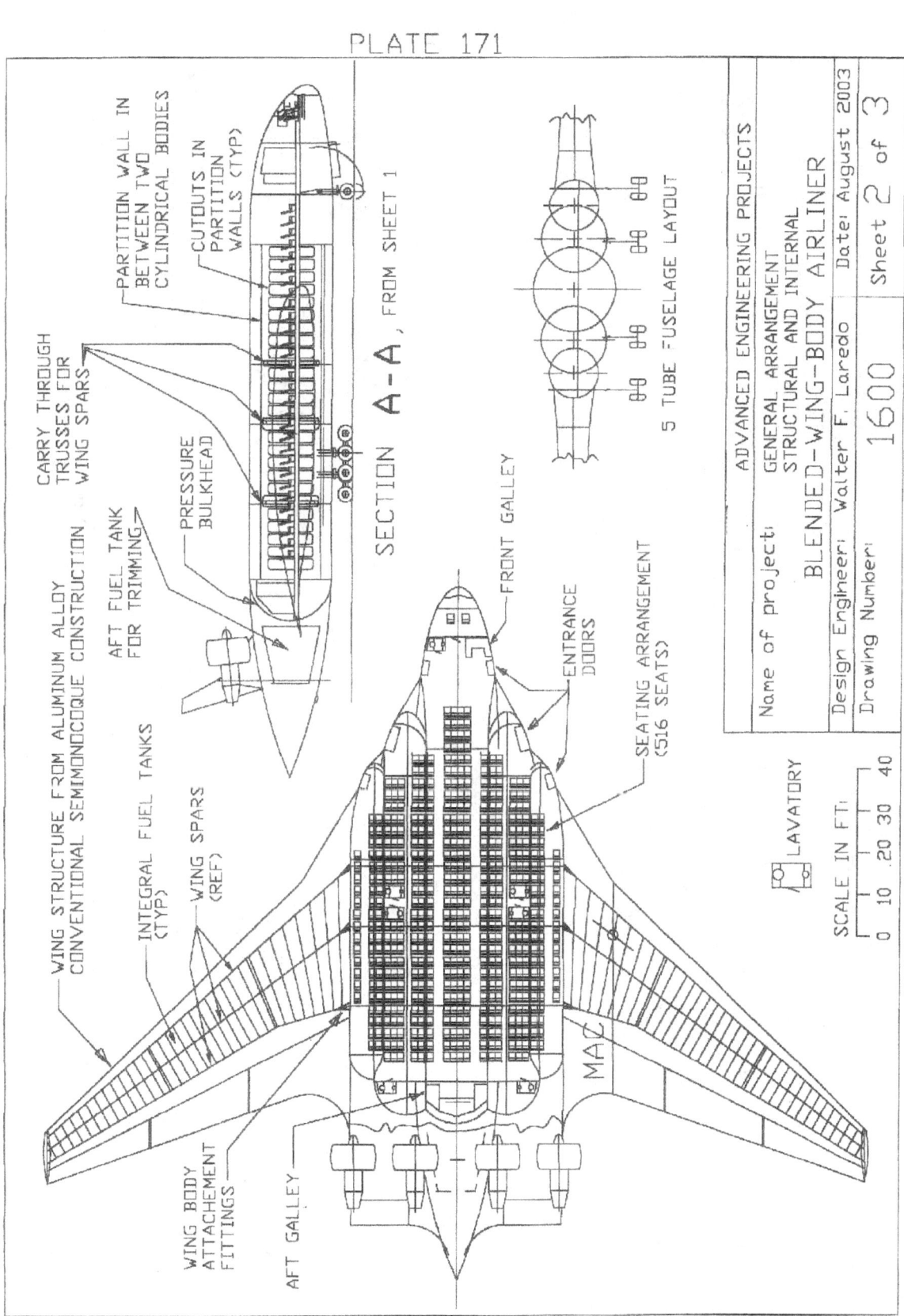

PARTITION WALL IN BETWEEN TWO CYLINDRICAL BODIES

CUTOUTS IN PARTITION WALLS (TYP)

CARRY THROUGH TRUSSES FOR WING SPARS

AFT FUEL TANK FOR TRIMMING

PRESSURE BULKHEAD

WING STRUCTURE FROM ALUMINUM ALLOY CONVENTIONAL SEMIMONOCOQUE CONSTRUCTION

INTEGRAL FUEL TANKS (TYP)

WING SPARS (REF)

WING BODY ATTACHEMENT FITTINGS

AFT GALLEY

SECTION A-A, FROM SHEET 1

5 TUBE FUSELAGE LAYOUT

FRONT GALLEY

ENTRANCE DOORS

SEATING ARRANGEMENT (516 SEATS)

MAC

LAVATORY

SCALE IN FT:

0   10   20   30   40

Name of project:	ADVANCED ENGINEERING PROJECTS		
	GENERAL ARRANGEMENT STRUCTURAL AND INTERNAL		
	BLENDED-WING-BODY AIRLINER		
Design Engineer:	Walter F. Laredo	Date: August 2003	
Drawing Number:	1600	Sheet 2 of 3	

PLATE 172

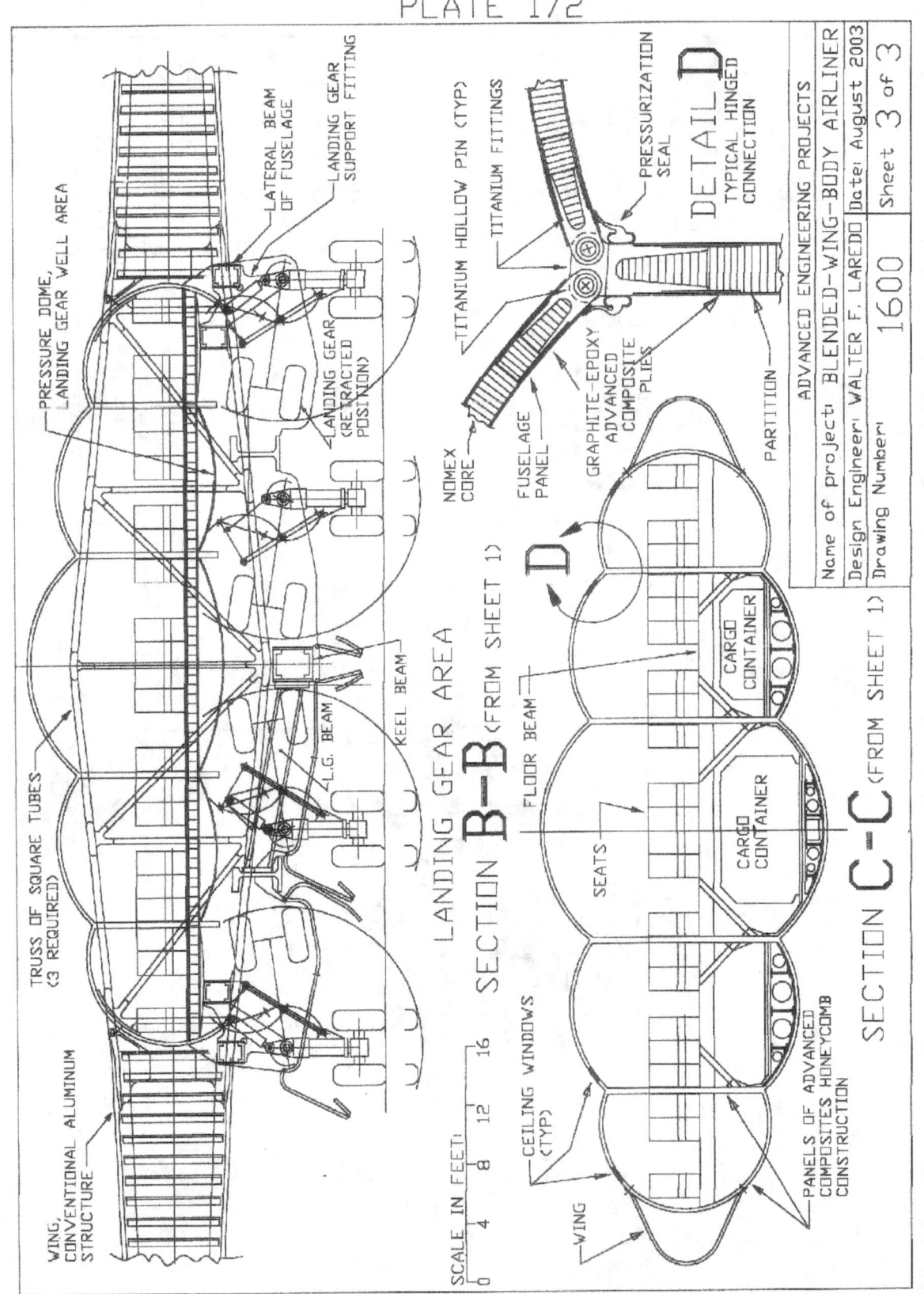

PRESSURE DOME, LANDING GEAR WELL AREA

LATERAL BEAM OF FUSELAGE

LANDING GEAR SUPPORT FITTING

LANDING GEAR (RETRACTED POSITION)

TRUSS OF SQUARE TUBES (3 REQUIRED)

WING, CONVENTIONAL ALUMINUM STRUCTURE

KEEL BEAM

L.G. BEAM

NOMEX CORE

LANDING GEAR AREA
SECTION B-B (FROM SHEET 1)

SCALE IN FEET:
0 4 8 12 16

TITANIUM HOLLOW PIN (TYP)

TITANIUM FITTINGS

PRESSURIZATION SEAL

DETAIL D
TYPICAL HINGED CONNECTION

FUSELAGE PANEL

GRAPHITE-EPOXY ADVANCED COMPOSITE PLIES

PARTITION

CEILING WINDOWS (TYP)

FLOOR BEAM

SEATS

CARGO CONTAINER

CARGO CONTAINER

WING

PANELS OF ADVANCED COMPOSITES HONEYCOMB CONSTRUCTION

SECTION C-C (FROM SHEET 1)

ADVANCED ENGINEERING PROJECTS

Name of project: BLENDED-WING-BODY AIRLINER

Design Engineer: WALTER F. LAREDO   Date: August 2003

Drawing Number: 1600   Sheet 3 of 3

PLATE 173

AEROSPACE PROJECT
COMMUTING SYSTEM BETWEEN EARTH AND THE INTERNATIONAL SPACE STATION

IN THE 21 CENTURY, SPACE AIRPORTS WILL BE BUSY PLACES WHERE COMMUTING FLIGHTS BETWEEN EARTH AND THE SPACE STATION WILL BECOME ROUTINE WITH THE USE OF EFFICIENT SPACE SHUTTLES AND SCRAM-JET SPACE PLANES. THIS COMMUTING SYSTEM WILL BE USED BY THE SPACE AGENCIES, GOVERNMENTS, SPACE TOURISTS, EXPLORERS, SCIENTISTS AND SPACE INDUSTRIALISTS.

IN THIS PROJECT ARE INCLUDED A SERIES OF DESIGNS, THE FIRST AND SECOND SHEETS SHOWS A SPACE AIRPORT DESIGNED FOR CONTINUOUS OPERATION, WHERE SPACECRAFT COULD BE LAUNCH ONE EVERY TWO HOURS.

A LARGE SPACE STATION WOULD BE NECESSARY TO USE AS AN OUTER-SPACE PORT, USED FOR THE DEPARTURE AND ARRIVAL OF INTERPLANETARY SPACESHIPS, TO OR FROM DIFFERENT DESTINATIONS OF OUR SOLAR SYSTEM, SPACESHIPS THAT ALWAYS WILL REMAIN IN OUTER-SPACE, EITHER, WHEN TRAVELING IN DEEP SPACE OR WHEN ACHORED TO A SPACE STATION. SHIPS THAT NEVER DESCENDS TO THE SURFACE FROM EARTH OR ANY OTHER CELESTIAL BODY.

OUR CIVILIZATION SHOULD CONTINUE WITH EXPLORATION, MOSTLY TO OUR MOON AND MARS, IN ORDER TO LEARN MORE ABOUT OUR SOLAR SYSTEM.

A WORLD CATACLYSM IS DUE, BUT NOBODY KNOWS WHEN IT WILL HAPPENS, MOST LIKELY IT WILL BE LINK WITH THOSE EARTH PERIODIC EXTINCTIONS CAUSED BY COMETS, METEORS AND ASTEROIDS, EVENTS THAT IN THE PAST HAD OBLITERATED MOST EARTH SPECIES, NEXT TIME WHEN A FOREIGN OBJECT WILL APPROACH TO STRIKE EARTH AGAIN, WILL BE THE FIRST TIME WITNESSED BY HUMANS BEINGS,

AND PERHAPS THE LAST TIME BEFORE OUR SPECIES DESAPPEARS.

SOME DAY, THE CONFIRMATION OF AN APPROACHING OBJECT OR COMET MAY DRIVE THE DESPERATE HUMANITY TO ORGANIZE AN INTERPLANETARY EXODUS IN ORDER TO PERPETUATE OUR SPECIES, SENDING THOUSANDS OF HUMAN COUPLES AND WITH THEM SURVIVAL EQUIPMENT, SEEDS OF EDIBLE PLANTS AND FRUITS TO BE PLANTED LATER ON ALIEN LAND, AND ALSO TO SEND ANIMALS THAT COULD MULTIPLY TO BE USED AS SOURCE OF FOOD.

THOSE VOYAGES MAY LAST FROM FEW MONTHS TO MANY YEARS, FINALLY ARRIVING INTO AUTONOMOUS SELF SUPPORTING BASES, LOCATED IN MARS AND IN OTHER MOONS FROM OUR SOLAR SYSTEM, WITH BASES BUILT IN ADVANCED BY ASTRONAUTS SENT IN PREVIOUS MISSIONS.

THE POWERS OF THE WORLD SHOULD JOIN, IN ORDER TO BUILD TWO INTERNATIONAL SPACE-PORTS (ANTE-ROOMS OF THE UNIVERSE) LOCATED IN THE TWO HIGHEST PLATEAUS OF THE WORLD, ONE ON THE TIBET REGION OF THE HIMALAYAS AND THE OTHER IN THE ALTIPLANO PLATEAU OF THE ANDES MOUNTAINS OF SOUTH AMERICA, SPACE-PORTS WHICH SHOULD INCLUDE TILTTED UP LAUNCHING RAMPS WITH EXTRA LONG 2 STAGES CATAPULTS, STEAM GENERATION PLANTS, AND IF GEOTHERMAL SOURCES ARE AVAILABLE, TAPPING HIGH PRESSURE STEAM, THE SPACE BASE SHOULD ALSO INCLUDE LONG LANDING STRIPS LOCATED ON THE SALTED DRY LAKES OF THE PLATEAUS.

HIGH SPEED HANGING MONORAIL TRAINS SHOULD BE USE TO CONNECT THE HIGH ALTITUDE SPACE-PORT WITH THE NEAREST LOW LANDS AND OCEAN PORT WHERE THE LIQUID OXYGEN AND LIQUID HYDROGEN PROCESS AND STORAGE PLANTS ARE LOCATED.

ADVANCED ENGINEERING PROJECTS		
Name of project: LAUNCHING FACILITY COMMUTING SYSTEM BETWEEN EARTH AND THE SPACE STATION		
Design Engineer: WALTER F. LAREDO	Date: Feb. 1997	
Drawing Number: 1700	Sheet 1 of 11	

PLATE 174

MAP OF SOUTH AMERICA

ATLANTIC OCEAN

PACIFIC OCEAN

BRAZIL

BOLIVIA

ARGENTINA

CHILE

PERU

EASTERN ANDES RANGE

HIGH ALTITUDE PLATEAU WITH THIN ATMOSPHERE

COLD DESERT

HIGHWAY

TALLEST PEAK IN THE REGION

LANDING STRIP FOR RETURNING SPACECRAFT

WAITING LINE

PROPELLANT FILLING STA.

TITICACA LAKE, HIGHEST OF THE WORLD

2 MILE LONG, LAUNCHING CATAPULT 2 REQUIRED IN PARALLEL

LIQUEFACTION PLANT FOR GASES AS OXYGEN AND HYDROGEN

SUSPENSION MONORAIL FOR A PASSENGER TRAIN BUT IN SPECIAL FOR A TRAIN TO TRANSPORT LIQUEFY GASES OF OXYGEN, HYDROGEN AND HELIUM.

ADVANCED ENGINEERING PROJECTS

Name of project: AEROSPACE BASE TO LAUNCH SPACE-PLANES AND PIGGY-BACK SPACE SHUTTLES, PROPOSED TO BE BUILD IN THE ALTIPLANO PLATEAU OF BOLIVIA

| Design Engineer: WALTER F. LAREDO | Date: Feb. 1997 |
| Drawing Number: 1700 | Sheet 2 of 11 |

PLATE 175

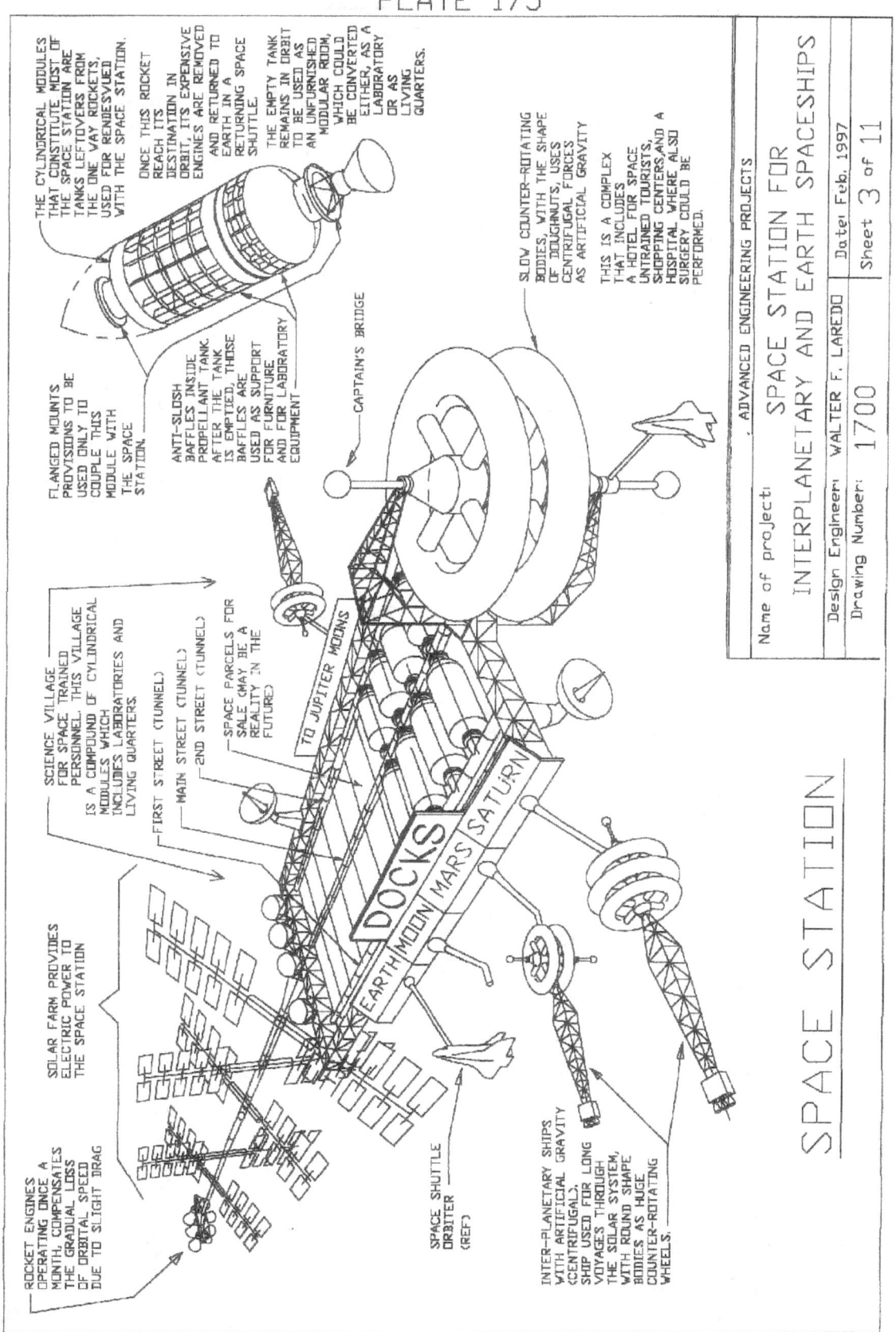

THE CYLINDRICAL MODULES THAT CONSTITUTE MOST OF THE SPACE STATION ARE TANKS LEFTOVERS FROM THE ONE WAY ROCKETS, USED FOR RENDESVUED WITH THE SPACE STATION.

ONCE THIS ROCKET REACH ITS DESTINATION IN ORBIT, ITS EXPENSIVE ENGINES ARE REMOVED AND RETURNED TO EARTH IN A RETURNING SPACE SHUTTLE.

THE EMPTY TANK REMAINS IN ORBIT TO BE USED AS AN UNFURNISHED MODULAR ROOM, WHICH COULD BE CONVERTED EITHER, AS A LABORATORY OR AS A LIVING QUARTERS.

FLANGED MOUNTS PROVISIONS TO BE USED ONLY TO COUPLE THIS MODULE WITH THE SPACE STATION.

ANTI-SLOSH BAFFLES INSIDE PROPELLANT TANK. AFTER THE TANK IS EMPTIED, THOSE BAFFLES ARE USED AS SUPPORT FOR FURNITURE, AND FOR LABORATORY EQUIPMENT.

SLOW COUNTER-ROTATING BODIES, WITH THE SHAPE OF DOUGHNUTS, USES CENTRIFUGAL FORCES AS ARTIFICIAL GRAVITY

THIS IS A COMPLEX THAT INCLUDES A HOTEL FOR SPACE UNTRAINED TOURISTS, AND SHOPPING CENTERS, AND A HOSPITAL WHERE ALSO SURGERY COULD BE PERFORMED.

CAPTAIN'S BRIDGE

SCIENCE VILLAGE FOR SPACE TRAINED PERSONNEL. THIS VILLAGE IS A COMPOUND OF CYLINDRICAL MODULES WHICH INCLUDES LABORATORIES AND LIVING QUARTERS.

FIRST STREET (TUNNEL)

MAIN STREET (TUNNEL)

2ND STREET (TUNNEL)

SPACE PARCELS FOR SALE (MAY BE A REALITY IN THE FUTURE)

TO JUPITER MOONS

DOCKS

EARTH MOON MARS SATURN

ROCKET ENGINES OPERATING ONCE A MONTH, COMPENSATES THE GRADUAL LOSS OF ORBITAL SPEED DUE TO SLIGHT DRAG

SOLAR FARM PROVIDES ELECTRIC POWER TO THE SPACE STATION

SPACE SHUTTLE ORBITER (REF)

INTER-PLANETARY SHIPS WITH ARTIFICIAL GRAVITY (CENTRIFUGAL). SHIP USED FOR LONG VOYAGES THROUGH THE SOLAR SYSTEM, WITH ROUND SHAPE BODIES AS HUGE COUNTER-ROTATING WHEELS.

SPACE STATION

ADVANCED ENGINEERING PROJECTS		
Name of project:	SPACE STATION FOR INTERPLANETARY AND EARTH SPACESHIPS	
Design Engineer: WALTER F. LAREDO		Date: Feb. 1997
Drawing Number: 1700		Sheet 3 of 11

PLATE 176

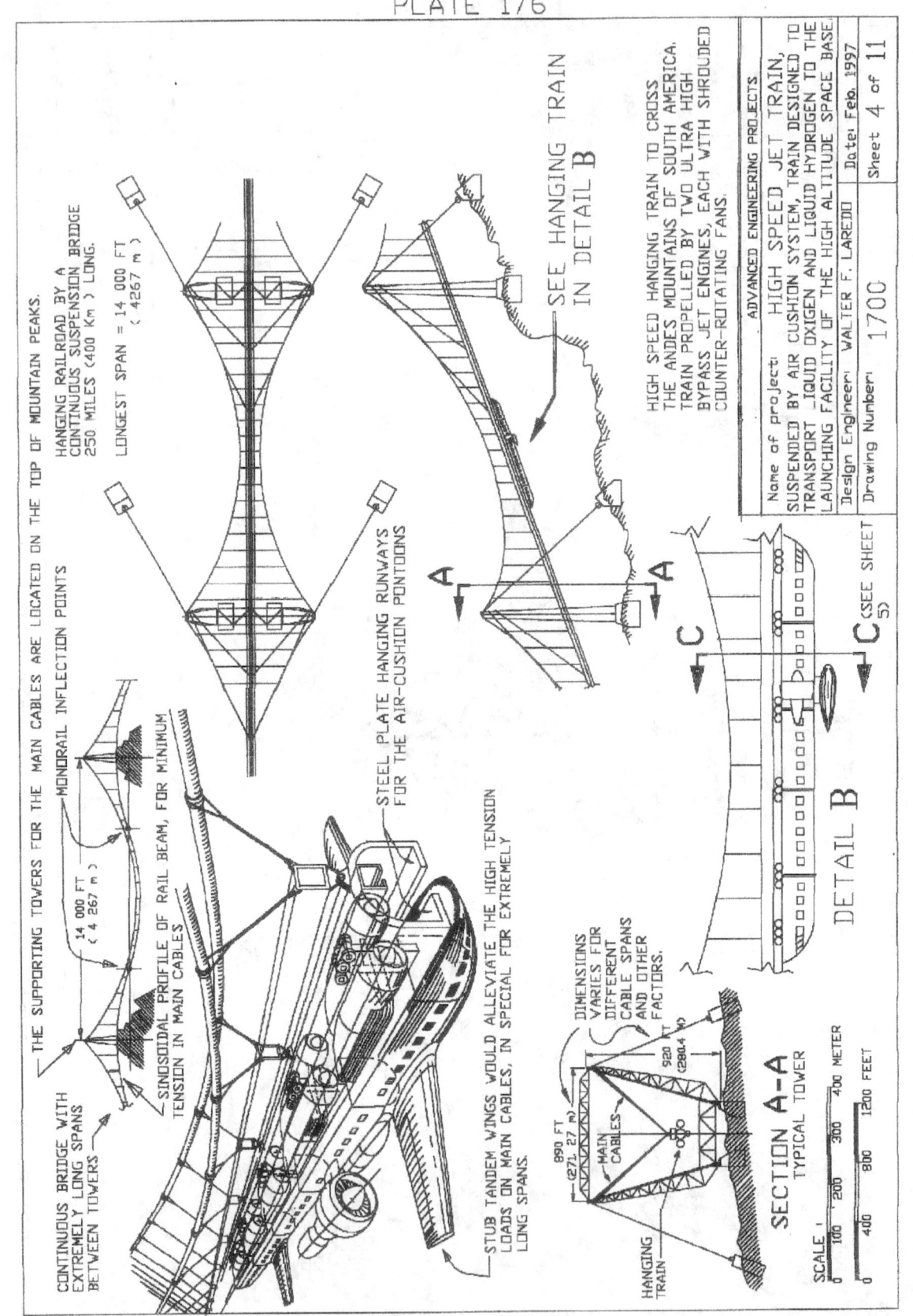

THE SUPPORTING TOWERS FOR THE MAIN CABLES ARE LOCATED ON THE TOP OF MOUNTAIN PEAKS.

MONORAIL INFLECTION POINTS

HANGING RAILROAD BY A
CONTINUOUS SUSPENSION BRIDGE
250 MILES (400 Km) LONG.

LONGEST SPAN = 14 000 FT
(4267 m)

SEE HANGING TRAIN
IN DETAIL B

HIGH SPEED HANGING TRAIN TO CROSS
THE ANDES MOUNTAINS OF SOUTH AMERICA.
TRAIN PROPELLED BY TWO ULTRA HIGH
BYPASS JET ENGINES, EACH WITH SHROUDED
COUNTER-ROTATING FANS.

CONTINUOUS BRIDGE WITH
EXTREMELY LONG SPANS
BETWEEN TOWERS

14 000 FT
(4 267 m)

SINUSOIDAL PROFILE OF RAIL BEAM, FOR MINIMUM
TENSION IN MAIN CABLES

STEEL PLATE HANGING RUNWAYS
FOR THE AIR-CUSHION PONTOONS

STUB TANDEM WINGS WOULD ALLEVIATE THE HIGH TENSION
LOADS ON MAIN CABLES, IN SPECIAL FOR EXTREMELY
LONG SPANS.

A

A

C

C (SEE SHEET
5)

DETAIL B

DIMENSIONS
VARIES FOR
DIFFERENT
CABLE SPANS
AND OTHER
FACTORS.

920 FT
(280.4 m)

890 FT
(271.27 m)

MAIN
CABLES

HANGING
TRAIN

SECTION A-A
TYPICAL TOWER

SCALE :

0   100   200   300   400 METER

0   400   800   1200 FEET

ADVANCED ENGINEERING PROJECTS

Name of project:	HIGH SPEED JET TRAIN,

SUSPENDED BY AIR CUSHION SYSTEM, TRAIN DESIGNED TO
TRANSPORT LIQUID OXIGEN AND LIQUID HYDROGEN TO THE
LAUNCHING FACILITY OF THE HIGH ALTITUDE SPACE BASE.

Design Engineer: WALTER F. LAREDO	Date: Feb. 1997
Drawing Number: 1700	Sheet 4 of 11

PLATE 177

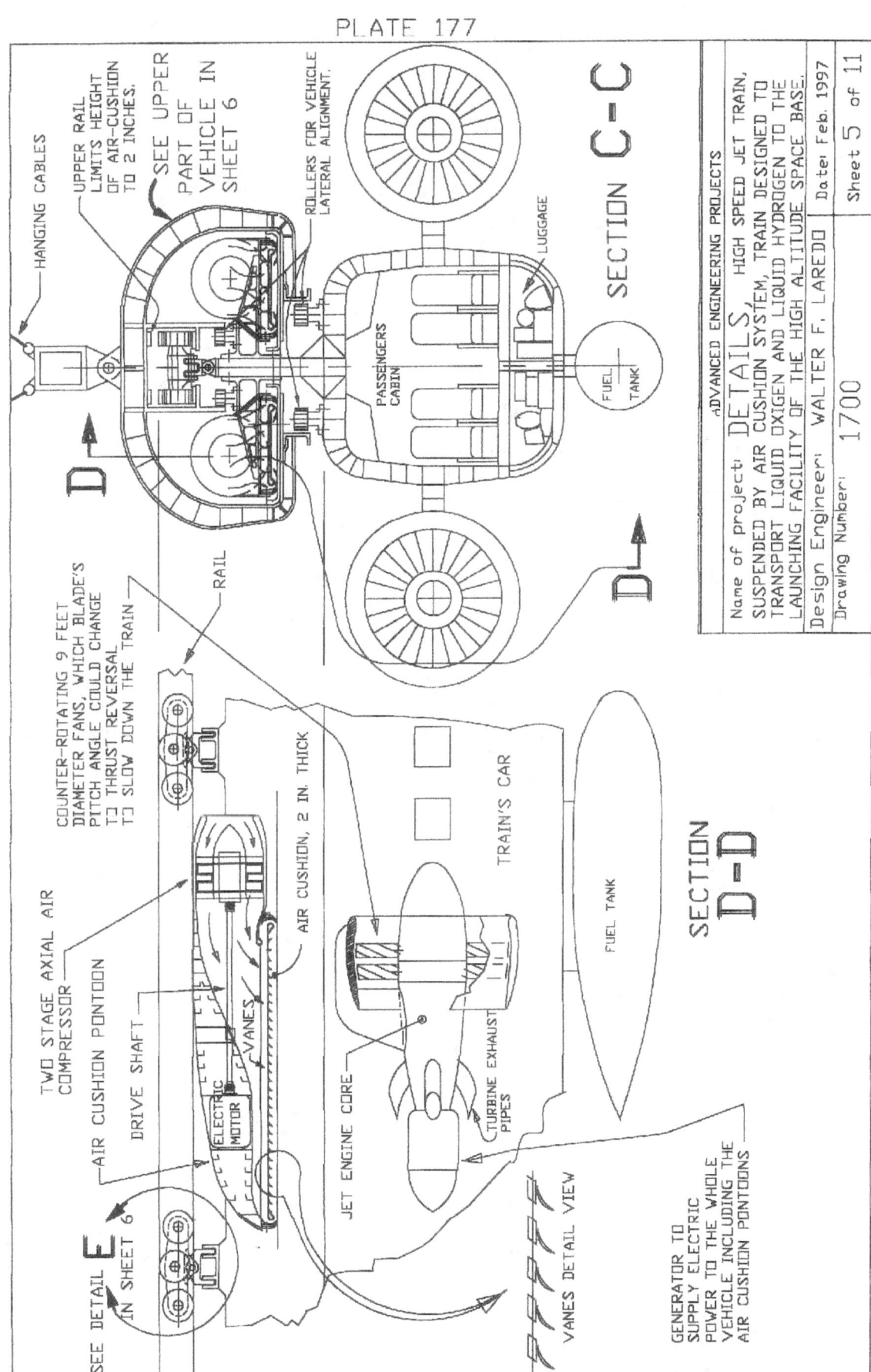

HANGING CABLES

UPPER RAIL LIMITS HEIGHT OF AIR-CUSHION TO 2 INCHES.

SEE UPPER PART OF VEHICLE IN SHEET 6

ROLLERS FOR VEHICLE LATERAL ALIGNMENT.

LUGGAGE

PASSENGERS CABIN

FUEL TANK

SECTION C-C

COUNTER-ROTATING 9 FEET DIAMETER FANS, WHICH BLADE'S PITCH ANGLE COULD CHANGE TO THRUST REVERSAL TO SLOW DOWN THE TRAIN

RAIL

TWO STAGE AXIAL AIR COMPRESSOR

AIR CUSHION PONTOON

DRIVE SHAFT

AIR CUSHION, 2 IN. THICK

VANES

ELECTRIC MOTOR

TRAIN'S CAR

FUEL TANK

JET ENGINE CORE

TURBINE EXHAUST PIPES

SECTION D-D

SEE DETAIL E IN SHEET 6

VANES DETAIL VIEW

GENERATOR TO SUPPLY ELECTRIC POWER TO THE WHOLE VEHICLE INCLUDING THE AIR CUSHION PONTOONS

Name of project: DETAILS, HIGH SPEED JET TRAIN, SUSPENDED BY AIR CUSHION SYSTEM, TRAIN DESIGNED TO TRANSPORT LIQUID OXIGEN AND LIQUID HYDROGEN TO THE LAUNCHING FACILITY OF THE HIGH ALTITUDE SPACE BASE.

ADVANCED ENGINEERING PROJECTS

Design Engineer: WALTER F. LAREDO

Date: Feb. 1997

Drawing Number: 1700

Sheet 5 of 11

PLATE 178

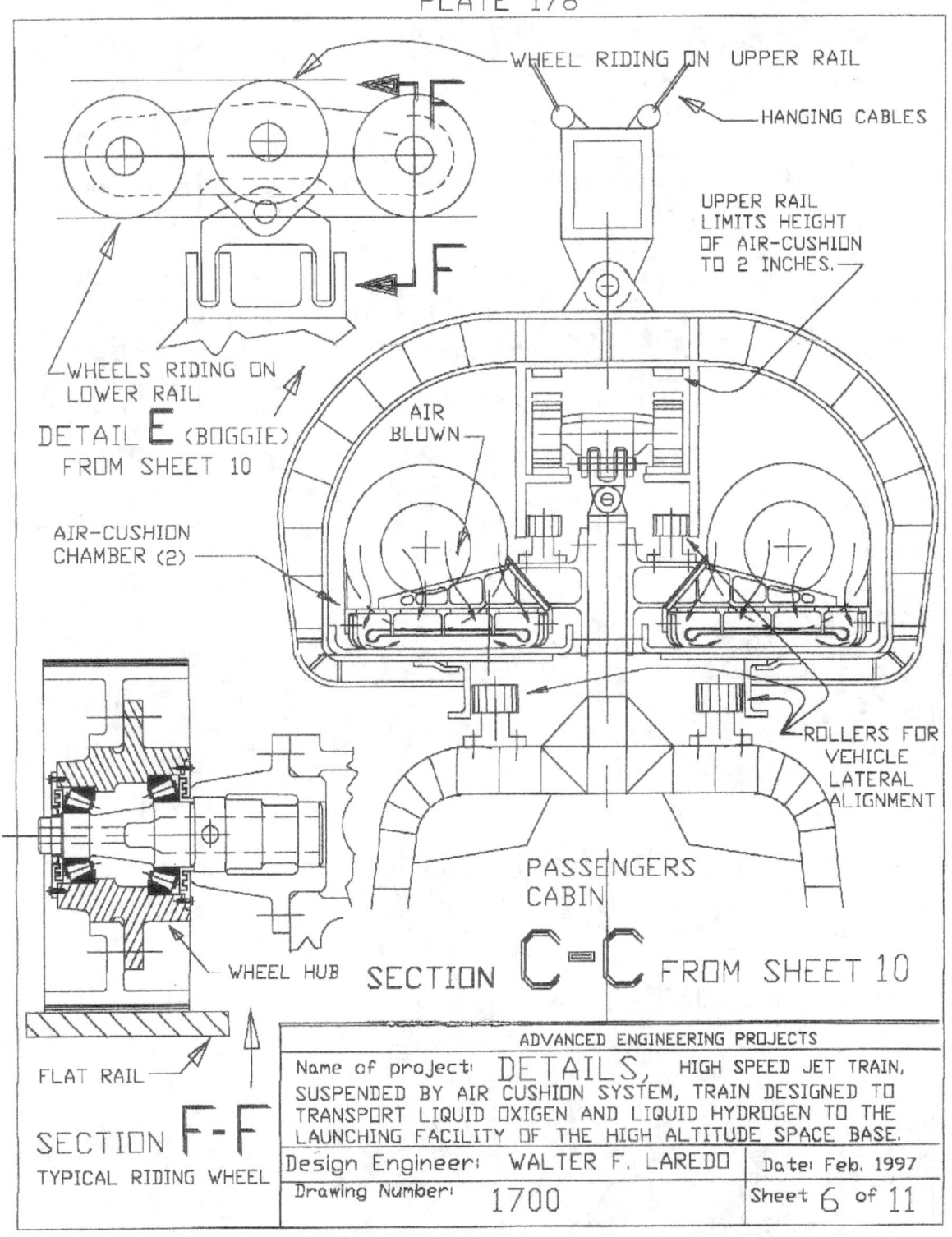

WHEEL RIDING ON UPPER RAIL

HANGING CABLES

UPPER RAIL LIMITS HEIGHT OF AIR-CUSHION TO 2 INCHES.

WHEELS RIDING ON LOWER RAIL

DETAIL E (BOGGIE) FROM SHEET 10

AIR BLOWN

AIR-CUSHION CHAMBER (2)

ROLLERS FOR VEHICLE LATERAL ALIGNMENT

PASSENGERS CABIN

SECTION C-C FROM SHEET 10

WHEEL HUB

FLAT RAIL

SECTION F-F
TYPICAL RIDING WHEEL

ADVANCED ENGINEERING PROJECTS		
Name of project: DETAILS, HIGH SPEED JET TRAIN, SUSPENDED BY AIR CUSHION SYSTEM, TRAIN DESIGNED TO TRANSPORT LIQUID OXIGEN AND LIQUID HYDROGEN TO THE LAUNCHING FACILITY OF THE HIGH ALTITUDE SPACE BASE.		
Design Engineer: WALTER F. LAREDO		Date: Feb. 1997
Drawing Number: 1700		Sheet 6 of 11

PLATE 179

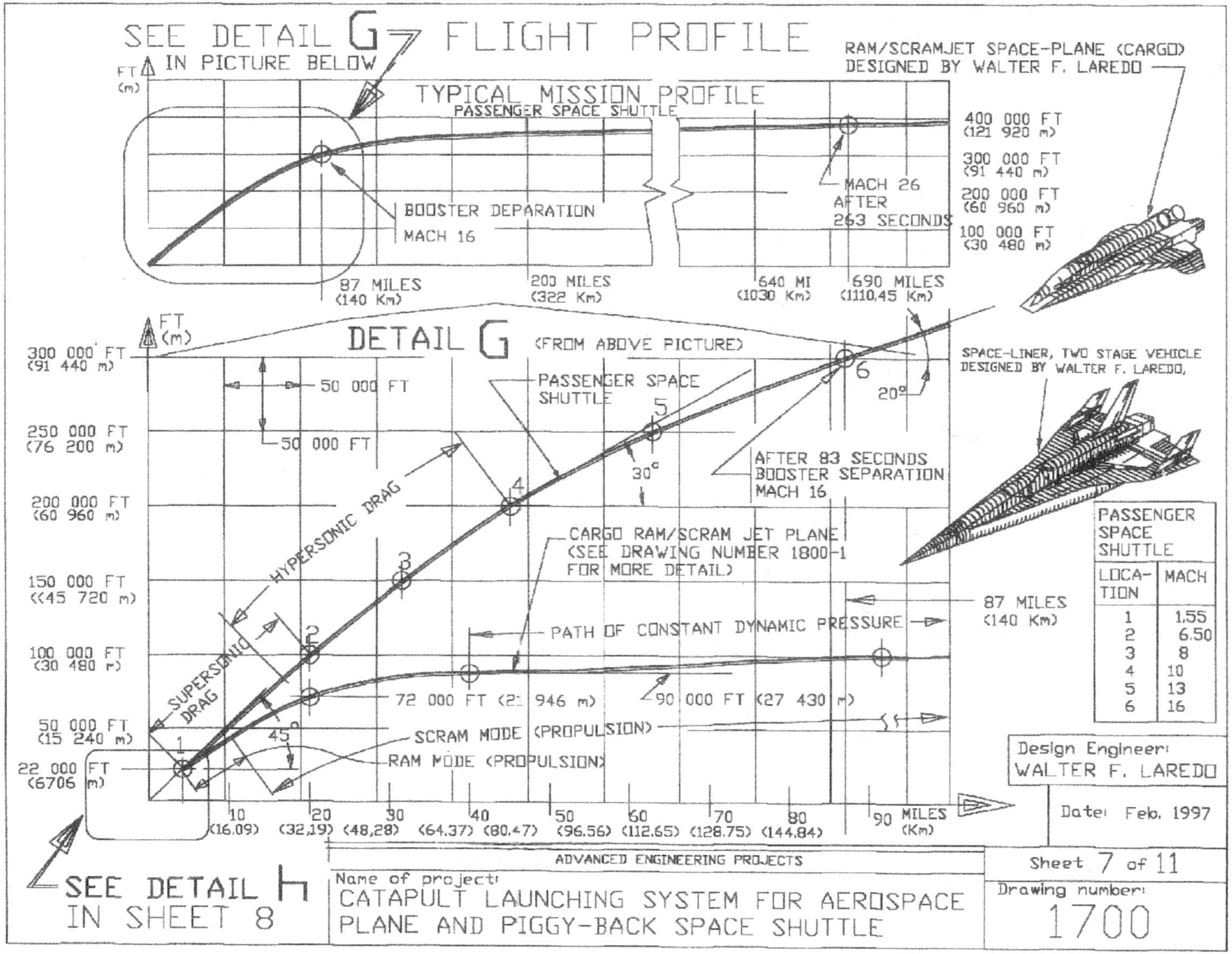

FLIGHT PROFILE

SEE DETAIL G IN PICTURE BELOW

RAM/SCRAMJET SPACE-PLANE (CARGO)
DESIGNED BY WALTER F. LAREDO

SPACE-LINER, TWO STAGE VEHICLE
DESIGNED BY WALTER F. LAREDO,

TYPICAL MISSION PROFILE
PASSENGER SPACE SHUTTLE

BOOSTER DEPARATION
MACH 16

87 MILES
(140 Km)

200 MILES
(322 Km)

640 MI
(1030 Km)

690 MILES
(1110.45 Km)

MACH 26
AFTER
263 SECONDS

400 000 FT
(121 920 m)

300 000 FT
(91 440 m)

200 000 FT
(60 960 m)

100 000 FT
(30 480 m)

PASSENGER SPACE SHUTTLE	
LOCA-TION	MACH
1	1.55
2	6.50
3	8
4	10
5	13
6	16

87 MILES
(140 Km)

Design Engineer:
WALTER F. LAREDO

Date: Feb. 1997

Sheet 7 of 11

Drawing number:
1700

DETAIL G

(FROM ABOVE PICTURE)

PASSENGER SPACE SHUTTLE

20°

30°

45°

AFTER 83 SECONDS
BOOSTER SEPARATION
MACH 16

CARGO RAM/SCRAM JET PLANE
(SEE DRAWING NUMBER 1800-1
FOR MORE DETAIL)

PATH OF CONSTANT DYNAMIC PRESSURE

90 000 FT (27 430 m)

72 000 FT (2: 946 m)

SCRAM MODE (PROPULSION)

RAM MODE (PROPULSION)

HYPERSONIC DRAG

SUPERSONIC DRAG

SEE DETAIL H IN SHEET 8

50 000 FT

50 000 FT

300 000 FT
(91 440 m)

250 000 FT
(76 200 m)

200 000 FT
(60 960 m)

150 000 FT
(45 720 m)

100 000 FT
(30 480 m)

50 000 FT
(15 240 m)

22 000 FT
(6706 m)

FT
(m)

FT
(m)

10	20	30	40	50	60	70	80	90 MILES
(16.09)	(32.19)	(48.28)	(64.37)	(80.47)	(96.56)	(112.65)	(128.75)	(144.84) (Km)

ADVANCED ENGINEERING PROJECTS

Name of project:
CATAPULT LAUNCHING SYSTEM FOR AEROSPACE
PLANE AND PIGGY-BACK SPACE SHUTTLE

PLATE 180

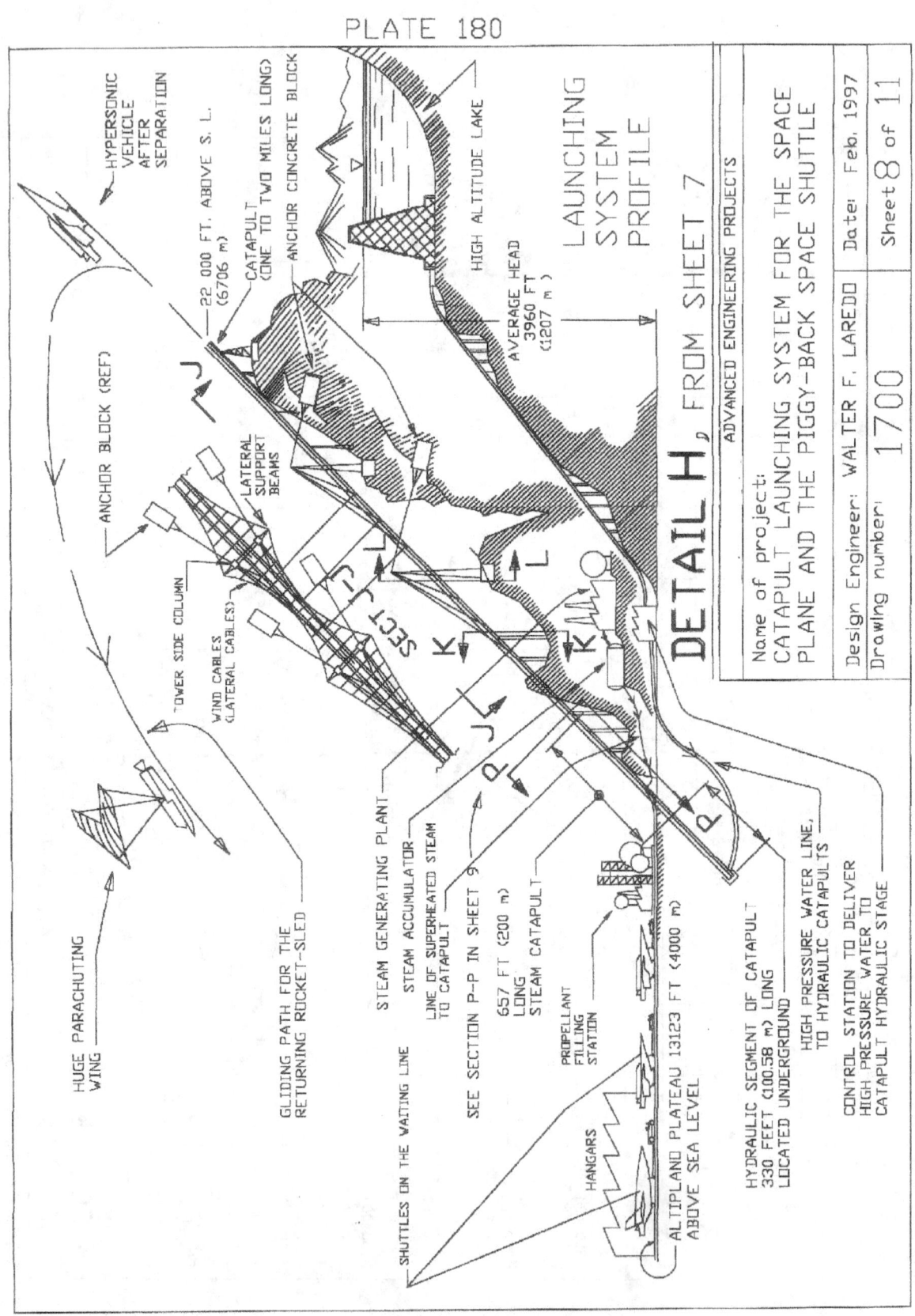

HYPERSONIC VEHICLE AFTER SEPARATION

22 000 FT. ABOVE S. L. (6706 m)

CATAPULT (ONE TO TWO MILES LONG)

ANCHOR CONCRETE BLOCK

HIGH ALTITUDE LAKE

AVERAGE HEAD 3960 FT (1207 m)

ANCHOR BLOCK (REF)

TOWER SIDE COLUMN

LATERAL SUPPORT BEAMS

WIND CABLES (LATERAL CABLES)

SECT J-J

L

L

K

K

J

J

P

P

HUGE PARACHUTING WING

GLIDING PATH FOR THE RETURNING ROCKET-SLED

STEAM GENERATING PLANT

STEAM ACCUMULATOR

LINE OF SUPERHEATED STEAM TO CATAPULT

SEE SECTION P-P IN SHEET 9

657 FT (200 m) LONG STEAM CATAPULT

PROPELLANT FILLING STATION

SHUTTLES ON THE WAITING LINE

HANGARS

ALTIPLANO PLATEAU 13123 FT (4000 m) ABOVE SEA LEVEL

HYDRAULIC SEGMENT OF CATAPULT 330 FEET (100.58 m) LONG LOCATED UNDERGROUND

HIGH PRESSURE WATER LINE, TO HYDRAULIC CATAPULTS

CONTROL STATION TO DELIVER HIGH PRESSURE WATER TO CATAPULT HYDRAULIC STAGE

DETAIL H, FROM SHEET 7

LAUNCHING SYSTEM PROFILE

ADVANCED ENGINEERING PROJECTS

Name of project: CATAPULT LAUNCHING SYSTEM FOR THE SPACE PLANE AND THE PIGGY-BACK SPACE SHUTTLE

Date: Feb. 1997

Sheet 8 of 11

Design Engineer: WALTER F. LAREDO

Drawing number: 1700

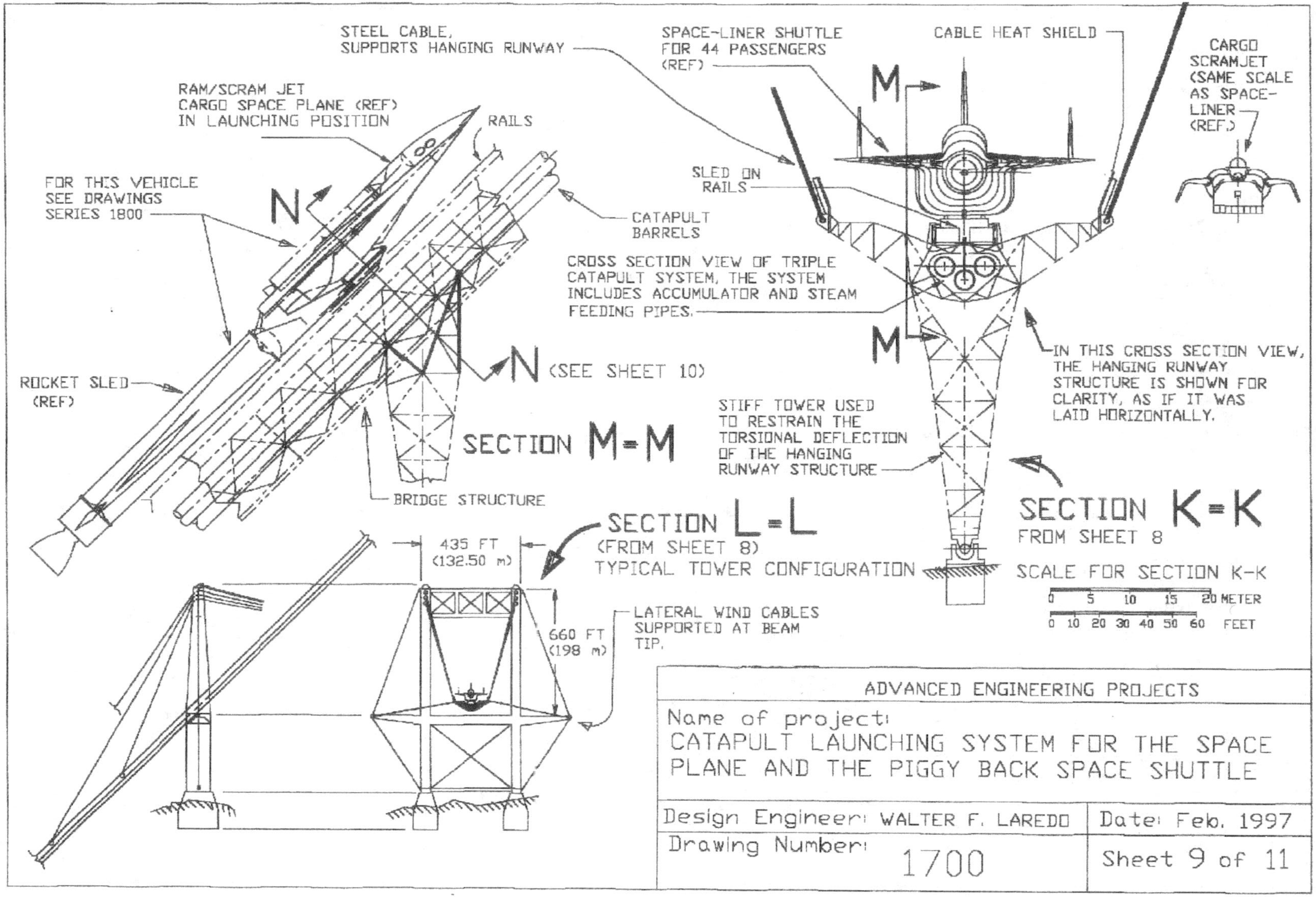

PLATE 181

CARGO SCRAMJET (SAME SCALE AS SPACE-LINER (REF.)

CABLE HEAT SHIELD

SPACE-LINER SHUTTLE FOR 44 PASSENGERS (REF)

SLED ON RAILS

CATAPULT BARRELS

CROSS SECTION VIEW OF TRIPLE CATAPULT SYSTEM, THE SYSTEM INCLUDES ACCUMULATOR AND STEAM FEEDING PIPES.

N (SEE SHEET 10)

SECTION M=M

IN THIS CROSS SECTION VIEW, THE HANGING RUNWAY STRUCTURE IS SHOWN FOR CLARITY, AS IF IT WAS LAID HORIZONTALLY.

SECTION K=K
FROM SHEET 8

SCALE FOR SECTION K-K

0 5 10 15 20 METER
0 10 20 30 40 50 60 FEET

STIFF TOWER USED TO RESTRAIN THE TORSIONAL DEFLECTION OF THE HANGING RUNWAY STRUCTURE

SECTION L=L
(FROM SHEET 8)
TYPICAL TOWER CONFIGURATION

LATERAL WIND CABLES SUPPORTED AT BEAM TIP.

STEEL CABLE, SUPPORTS HANGING RUNWAY

RAM/SCRAM JET CARGO SPACE PLANE (REF) IN LAUNCHING POSITION

RAILS

FOR THIS VEHICLE SEE DRAWINGS SERIES 1800

ROCKET SLED (REF)

N

BRIDGE STRUCTURE

660 FT (198 m)

435 FT (132.50 m)

ADVANCED ENGINEERING PROJECTS

Name of project:
CATAPULT LAUNCHING SYSTEM FOR THE SPACE PLANE AND THE PIGGY BACK SPACE SHUTTLE

| Design Engineer: WALTER F. LAREDO | Date: Feb. 1997 |
| Drawing Number: 1700 | Sheet 9 of 11 |

PLATE 182

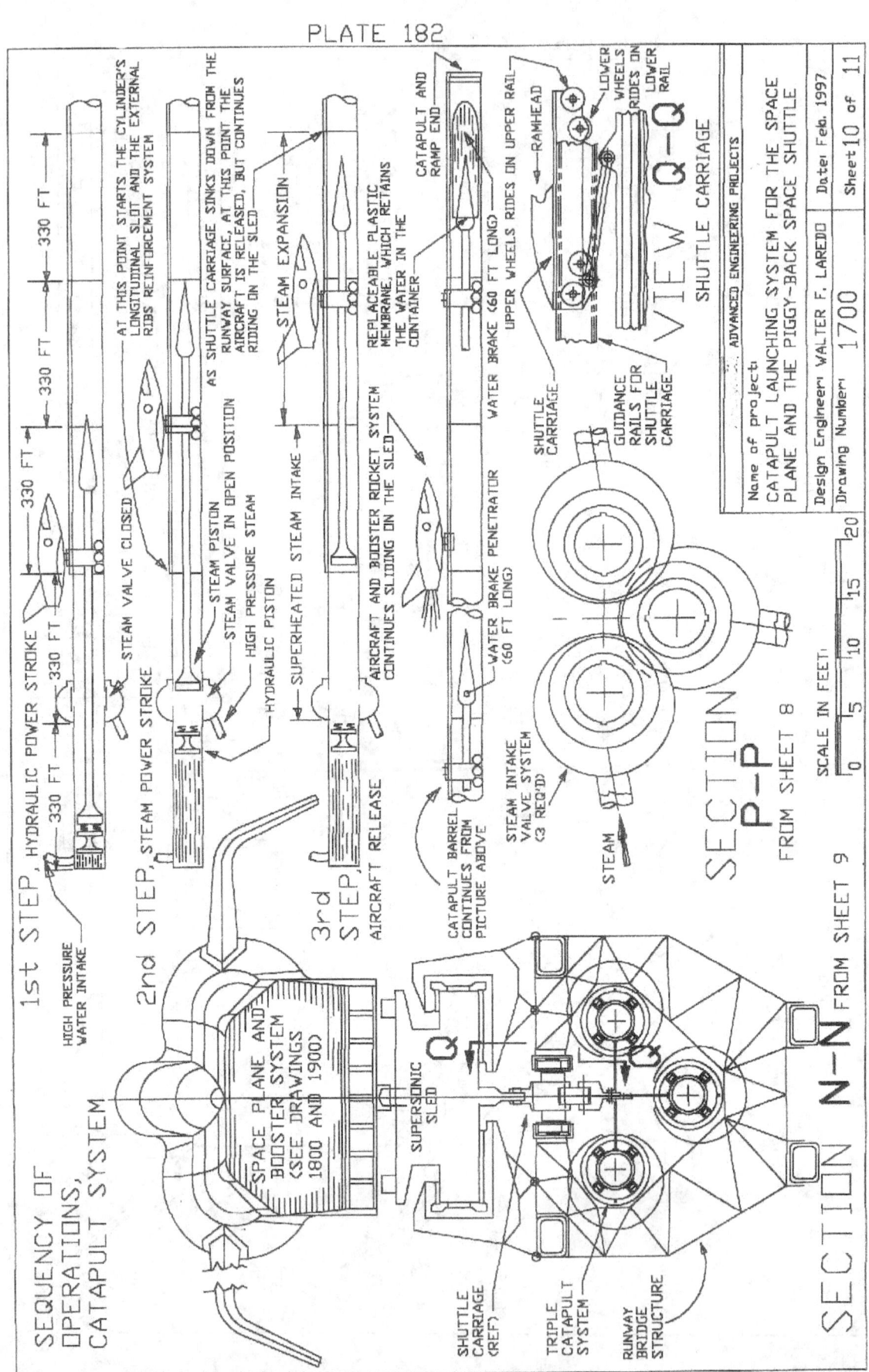

SEQUENCE OF OPERATIONS, CATAPULT SYSTEM

1st STEP, HYDRAULIC POWER STROKE

HIGH PRESSURE WATER INTAKE

AT THIS POINT STARTS THE CYLINDER'S LONGITUDINAL SLOT AND THE EXTERNAL RIBS REINFORCEMENT SYSTEM

STEAM VALVE CLOSED

2nd STEP, STEAM POWER STROKE

STEAM PISTON
STEAM VALVE IN OPEN POSITION

HIGH PRESSURE STEAM

HYDRAULIC PISTON

AS SHUTTLE CARRIAGE SINKS DOWN FROM THE RUNWAY SURFACE, AT THIS POINT THE AIRCRAFT IS RELEASED, BUT CONTINUES RIDING ON THE SLED

STEAM EXPANSION

3rd STEP,

SUPERHEATED STEAM INTAKE

AIRCRAFT RELEASE

REPLACEABLE PLASTIC MEMBRANE, WHICH RETAINS THE WATER IN THE CONTAINER

CATAPULT AND RAMP END

AIRCRAFT AND BOOSTER ROCKET SYSTEM CONTINUES SLIDING ON THE SLED

WATER BRAKE (60 FT LONG)

UPPER WHEELS RIDES ON UPPER RAIL

RAMHEAD

LOWER WHEELS RIDES ON LOWER RAIL

VIEW Q-Q

SHUTTLE CARRIAGE

CATAPULT BARREL CONTINUES FROM PICTURE ABOVE

WATER BRAKE PENETRATOR (60 FT LONG)

STEAM INTAKE VALVE SYSTEM (3 REQ'D)

STEAM

SHUTTLE CARRIAGE

GUIDANCE RAILS FOR SHUTTLE CARRIAGE

SECTION P-P
FROM SHEET 8

SCALE IN FEET:
0 5 10 15 20

SPACE PLANE AND BOOSTER SYSTEM (SEE DRAWINGS 1800 AND 1900)

SUPERSONIC SLED

SHUTTLE CARRIAGE (REF)

TRIPLE CATAPULT SYSTEM

RUNWAY BRIDGE STRUCTURE

SECTION N-N FROM SHEET 9

ADVANCED ENGINEERING PROJECTS

Name of project:
CATAPULT LAUNCHING SYSTEM FOR THE SPACE PLANE AND THE PIGGY-BACK SPACE SHUTTLE

Design Engineer: WALTER F. LAREDO

Date: Feb. 1997

Sheet 10 of 11

Drawing Number: 1700

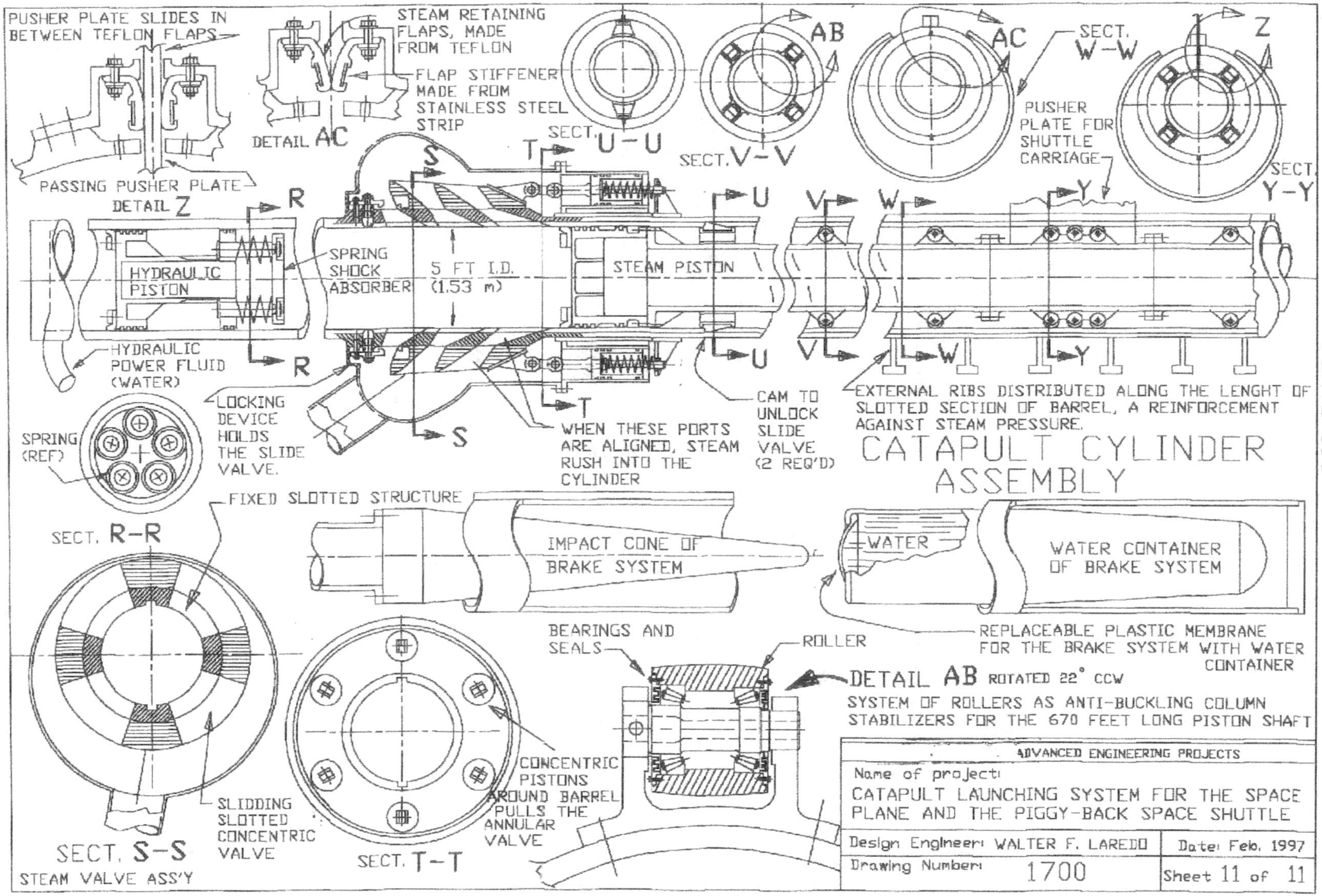

PLATE 183

CATAPULT CYLINDER ASSEMBLY

SECT. Y-Y

SECT. W-W

PUSHER PLATE FOR SHUTTLE CARRIAGE

AC

AB

SECT. V-V

SECT. U-U

EXTERNAL RIBS DISTRIBUTED ALONG THE LENGHT OF SLOTTED SECTION OF BARREL, A REINFORCEMENT AGAINST STEAM PRESSURE.

WATER CONTAINER OF BRAKE SYSTEM

REPLACEABLE PLASTIC MEMBRANE FOR THE BRAKE SYSTEM WITH WATER CONTAINER

WATER

DETAIL AB ROTATED 22° CCW

SYSTEM OF ROLLERS AS ANTI-BUCKLING COLUMN STABILIZERS FOR THE 670 FEET LONG PISTON SHAFT

ROLLER

STEAM PISTON

CAM TO UNLOCK SLIDE VALVE (2 REG'D)

WHEN THESE PORTS ARE ALIGNED, STEAM RUSH INTO THE CYLINDER

STEAM RETAINING FLAPS, MADE FROM TEFLON

FLAP STIFFENER MADE FROM STAINLESS STEEL STRIP

5 FT I.D. (1.53 m)

SPRING SHOCK ABSORBER

IMPACT CONE OF BRAKE SYSTEM

BEARINGS AND SEALS

CONCENTRIC PISTONS AROUND BARREL PULLS THE ANNULAR VALVE

SECT. T-T

DETAIL AC

PUSHER PLATE SLIDES IN BETWEEN TEFLON FLAPS

PASSING PUSHER PLATE

DETAIL Z

LOCKING DEVICE HOLDS THE SLIDE VALVE.

HYDRAULIC PISTON

HYDRAULIC POWER FLUID (WATER)

SPRING (REF)

SECT. R-R

FIXED SLOTTED STRUCTURE

SLIDING SLOTTED CONCENTRIC VALVE

SECT. S-S

STEAM VALVE ASS'Y

Name of project:
CATAPULT LAUNCHING SYSTEM FOR THE SPACE PLANE AND THE PIGGY-BACK SPACE SHUTTLE

ADVANCED ENGINEERING PROJECTS

| Design Engineer: WALTER F. LAREDO | Date: Feb. 1997 |
| Drawing Number: 1700 | Sheet 11 of 11 |

PLATE 184

AEROSPACE PROJECT

DUAL-MODE RAMJET/SCRAMJET VEHICLE

PRELIMINARY DESIGN

IN ORDER TO IMPLEMENT SPACE EXPLORATION IN REALISTIC
WAY, FIRST WOULD BE NECESSARY TO BUILD TWO
INTERNATIONAL SPACE BASES ON THE TWO HIGHEST PLATEAUS
OF THE WORLD, EACH BASE SHOULD INCLUDE PLANTS TO SUPPLY
STEAM FOR THE CATAPULTS, ALSO SHOULD HAVE LONG LANDING
STRIPS ON THE PLATEAUS' DRIED SALTED LAKES, ALSO EACH
BASE SHOULD INCLUDE A TILTED LAUNCHING RAMP WITH AN
EXTRALONG TWO STAGED CATAPULT, A HYDRAULIC STAGE
FOLLOWED BY A STEAM ONE.

NOTE: FOR CLARITY IN THE FOLLOWING DRAWINGS WAS NOT SHOWN THE
INTERNAL INSULATION OF THE EXTERNAL STRUCTURE AND THE FOAM INSULATION
THAT COVERS THE TANKS AND OTHER CRYOGENIC COMPONENTS.

ADVANCED ENGINEERING PROJECTS	
Name of project:	
HYPERSONIC SPACE PLANE (SCRAMJET)	
FOR CARGO TRANSPORTATION SYSTEM IN BETWEEN EARTH AND THE SPACE STATION	
Design Engineer: WALTER F. LAREDO	Date: Feb. 1997
Drawing Number: 1800	Sheet 1 of 21

PLATE 185

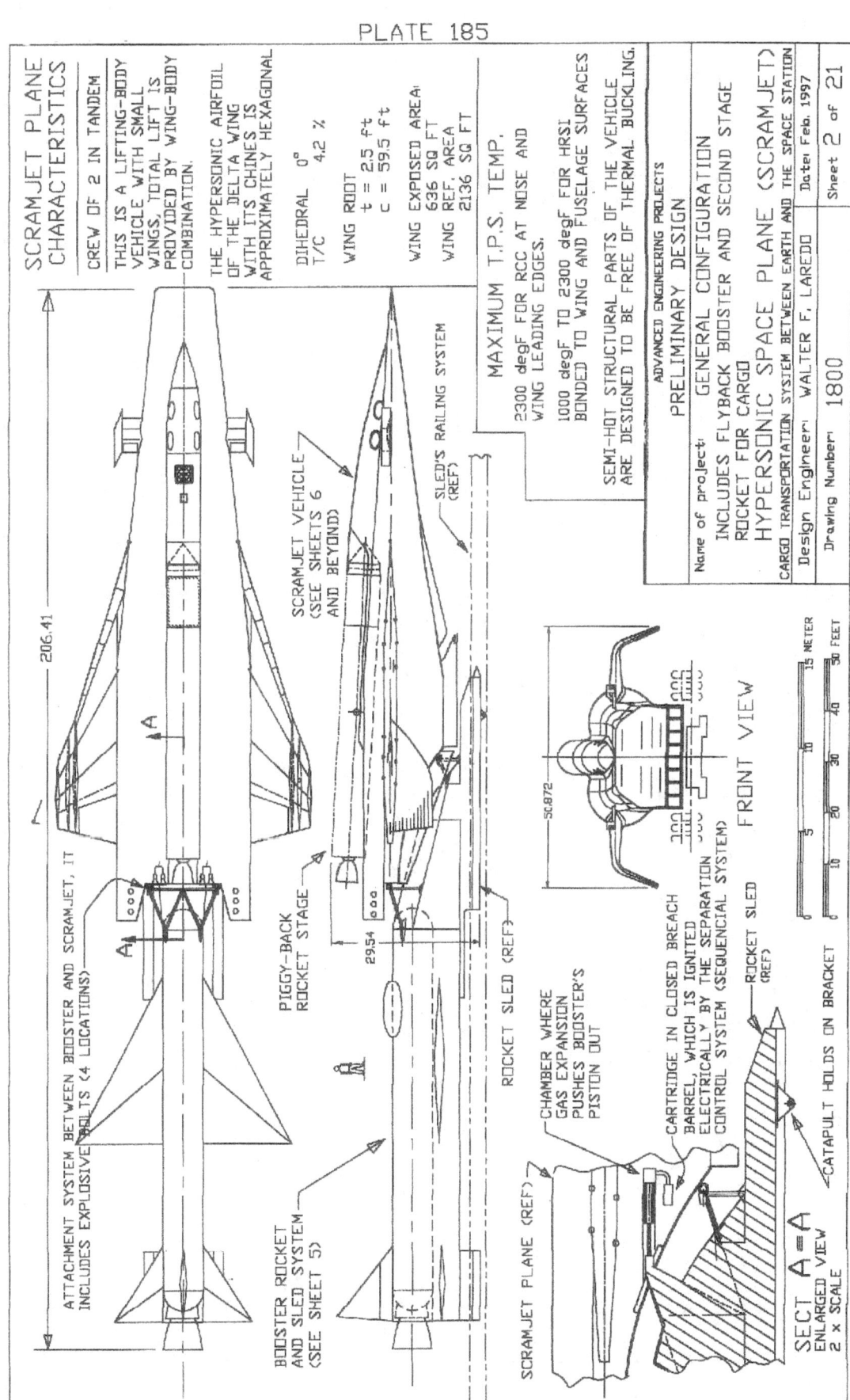

SCRAMJET PLANE CHARACTERISTICS

CREW OF 2 IN TANDEM

THIS IS A LIFTING-BODY VEHICLE WITH SMALL WINGS, TOTAL LIFT IS PROVIDED BY WING-BODY COMBINATION.

THE HYPERSONIC AIRFOIL OF THE DELTA WING WITH ITS CHINES IS APPROXIMATELY HEXAGONAL

DIHEDRAL   0°
T/C   4.2 %

WING ROOT
t = 2.5 ft
c = 59.5 ft

WING EXPOSED AREA: 636 SQ FT
WING REF. AREA   2136 SQ FT

MAXIMUM T.P.S. TEMP.

2300 degF FOR RCC AT NOSE AND WING LEADING EDGES.

1000 degF TO 2300 degF FOR HRSI BONDED TO WING AND FUSELAGE SURFACES.

SEMI-HOT STRUCTURAL PARTS OF THE VEHICLE ARE DESIGNED TO BE FREE OF THERMAL BUCKLING.

ADVANCED ENGINEERING PROJECTS
PRELIMINARY DESIGN
GENERAL CONFIGURATION

Name of project:
INCLUDES FLYBACK BOOSTER AND SECOND STAGE ROCKET FOR CARGO
HYPERSONIC SPACE PLANE (SCRAMJET)
CARGO TRANSPORTATION SYSTEM BETWEEN EARTH AND THE SPACE STATION

Design Engineer: WALTER F. LAREDO   Date: Feb. 1997

Drawing Number: 1800   Sheet 2 of 21

206.41

SCRAMJET VEHICLE (SEE SHEETS 6 AND BEYOND)

SLED'S RAILING SYSTEM (REF)

A

PIGGY-BACK ROCKET STAGE

29.54

ROCKET SLED (REF)

ATTACHMENT SYSTEM BETWEEN BOOSTER AND SCRAMJET, IT INCLUDES EXPLOSIVE BOLTS (4 LOCATIONS)

A

BOOSTER ROCKET AND SLED SYSTEM (SEE SHEET 5)

SCRAMJET PLANE (REF)

50.872

FRONT VIEW

15 METER
50 FEET

CHAMBER WHERE GAS EXPANSION PUSHES BOOSTER'S PISTON OUT

CARTRIDGE IN CLOSED BREACH BARREL, WHICH IS IGNITED ELECTRICALLY BY THE SEPARATION CONTROL SYSTEM (SEQUENCIAL SYSTEM)

ROCKET SLED (REF)

CATAPULT HOLDS ON BRACKET

SECT A=A
ENLARGED VIEW
2 x SCALE

PLATE 186

VERTICAL STABILIZERS

TWIN TYPE WITH DOUBLE WEDGE CROSS SECTIONS.

AREA, EACH =131.25 SQ FT
TOTAL AREA = 262.50 SQ FT
MOMENT ARM = lvs = 25 FT

$\overline{Vs}$ = VOLUME COEFFICIENT
(VERT. STABILIZER)

$$\overline{Vs} = \frac{lvs \times Svs}{b \times Sw} = 0.06$$

DESIGN C.G. RANGE
FROM ___ TO ___ % M.A.C.

DESIGN SPEEDS

AT LAUNCHING FROM CATAPULT (A SPECIAL CATAPULT 1650 FT LONG)   300 MPH

AT THE POINT WHEN SCRAMJET PLANE SEPARATES FROM
BOOSTER   MACH 3.3

AT THE POINT WHEN PIGGY-BACK ROCKET SEPARATES FROM SCRAMJET PLANE   mACH 12

ROCKET MAX. MACH   MACH 26

MINIMUM LANDING SPEED OF SCRAMJET PLANE (0.0 FUEL)   175 MPH

RANGE FACTOR

(M x (L/D)) S.F.C.

25 TO 35

FLIGHT CONTROL SYSTEM

ALL ELECTRONIC FLY-BY-WIRE FLIGHT-CONTROL SYSTEM.
COMMAND AND STABILITY-AUGMENTATION SYSTEM OF CUADRUPLEXED DIGITAL SYSTEM.
THE AIRCRAFT CENTER OF GRAVITY IS CLOSE TO THE CENTER OF PRESSURE, THEREFORE IS USED RELAXED STABILITY DESIGN WHICH WILL ALLOW THE USE OF RELATIVELY SMALL SIZE ELEVATORS AND FLAPERONS.

THE FLIGHT CONTROL SYSTEM ALSO CONTROLS THE REACTION CONTROL SYSTEM AT HIGH ALTITUDE WHERE AERODYNAMIC CONTROL SURFACES ARE NOT EFFICIENT.

THIS SCRAMJET PLANE ALSO USES SOME MECHANICAL BACKUP IN CASE OF TOTAL FAILURE.
MOST MECHANISMS AND AERODYNAMIC SURFACES ARE OPERATED BY ROTARY ACTUATORS.

PROPULSION SYSTEM AND SPECIFIC IMPULSES

AT TAKE-OFF THE VEHICLE SLIDES ON A SLED, ON A TILT 1.5 TO 2 MILES LONG RAMP, THE VEHICLE IS INITIALLY LAUNCH BY A HYDRAULIC CATAPULT, THEN FOLLOWED BY THE STEAM CATAPULT, THE LAUNCHING STARTS AFTER THE SRB WAS IGNITED.

THE BOOSTER HAVE A THRUST OF 650 000 LB AND A SPECIFIC IMPULSE OF 260 SECONDS.
THE MAIN PROPULSION SYSTEM OF THE HYPERSONIC PLANE IS PROVIDED BY A CLUSTER OF 8 DUAL-MODE RAMJET/SCRAMJET ENGINE MODULES, WITH A TOTAL THRUST OF 460 000 LB AT LIFT-OFF FROM A HIGH ALTITUDE BASE AT THE ALTIPLAND PLATEAU IN THE ANDES MOUNTAINS.

THEORETICAL SPECIFIC IMPULSE OF THE AEROSPACE PLANE.

RAMJET MODE AT 40 000 FT AND M3, Isp = 3300 SEC.
SCRAMJET MODE:
  AT 100 000 FT AND MACH 5,   Isp = 3000 SEC
  AT 100 000 FT AND MACH 12,   Isp = 1900 SEC

HOWEVER IN PRACTICE WITH THE REAL VEHICLE
  Isp = 1500 to 2000 SEC.

SECOND STAGE ROCKET
  THRUST IN VACCUM : 55 000 LB.
  SPECIFIC IMPULSE : 450 SECONDS

WEIGHTS

PAYLOAD	4 000 LB
PIGGY-BACK ROCKET WITH PAYLOAD	31 000 LB
EMPTY WEIGHT OF SCRAMJET PLANE AT LANDING	170 511 LB
LAUNCHING WEIGHT OF SCRAMJET PLANE WITH PIGGY-BACK ROCKET AND PAYLOAD	246 000 LB
WEIGHT OF THE SRB	270 000 LB
WEIGHT OF SLED INCLUDING SRB	290 000 LB
TOTAL LAUNCHING WEIGHT OF COMPOUND VEHICLE	536 000 LB

PROPELLANT WEIGHT

SCRAMJET PLANE	LH2	38 489 LB
	LO2	6 000 LB
AUX.TANK		44 489 L3
ROCKET (SECOND STAGE)	LH2	3 090 LB
	LO2	18 435 LB
		21 525 LB

L/W AT LAUNCHING IS APPROXIMATELY 1.2

ADVANCED ENGINEERING PROJECTS

PRELIMINARY DESIGN

Name of project:
GENERAL ARRANGEMENT, INCLUDES FLYBACK BOOSTER AND SECOND STAGE CARGO ROCKET

HYPERSONIC SPACE PLANE (SCRAMJET)

CARGO TRANSPORTATION SYSTEM BETWEEN EARTH AND THE SPACE STATION

Design Engineer: WALTER F. LAREDO

Date: Feb. 1997

Drawing Number: 1800   Sheet 3 of 21

PLATE 187

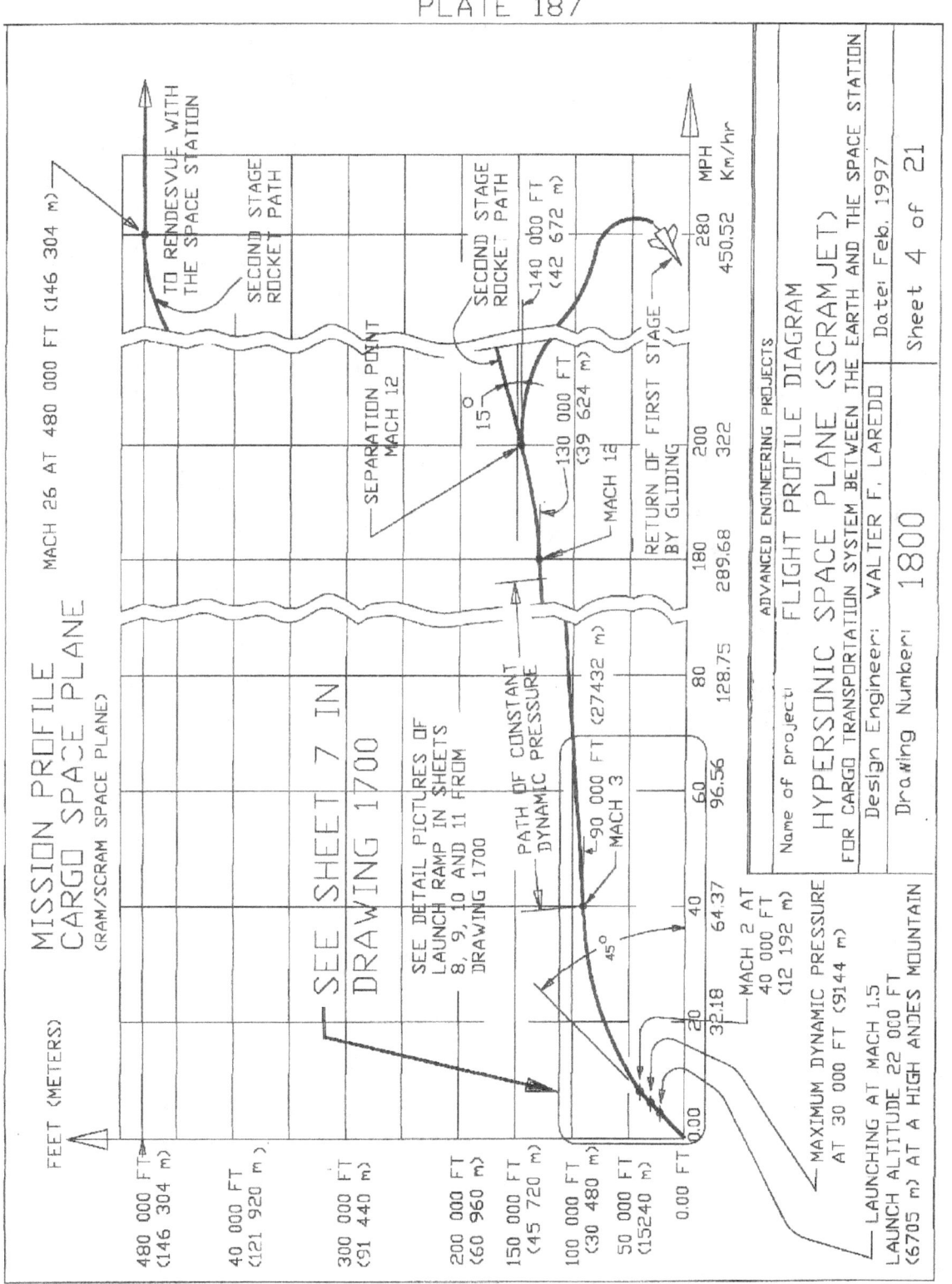

MISSION PROFILE
CARGO SPACE PLANE
(RAM/SCRAM SPACE PLANE)

FEET (METERS)

MACH 26 AT 480 000 FT (146 304 m)

TO RENDESVUE WITH
THE SPACE STATION

SECOND STAGE
ROCKET PATH

SEPARATION POINT
MACH 12

SECOND STAGE
ROCKET PATH

140 000 FT
(42 672 m)

15°

130 000 FT
(39 624 m)

MACH 18

RETURN OF FIRST STAGE
BY GLIDING

SEE SHEET 7 IN
DRAWING 1700

SEE DETAIL PICTURES OF
LAUNCH RAMP IN SHEETS
8, 9, 10 AND 11 FROM
DRAWING 1700

PATH OF CONSTANT
DYNAMIC PRESSURE

90 000 FT (27432 m)
MACH 3

45°

MACH 2 AT
40 000 FT
(12 192 m)

MAXIMUM DYNAMIC PRESSURE
AT 30 000 FT (9144 m)

LAUNCHING AT MACH 1.5
LAUNCH ALTITUDE 22 000 FT
(6705 m) AT A HIGH ANDES MOUNTAIN

480 000 FT
(146 304 m)

40 000 FT
(121 920 m)

300 000 FT
(91 440 m)

200 000 FT
(60 960 m)

150 000 FT
(45 720 m)

100 000 FT
(30 480 m)

50 000 FT
(15240 m)

0.00 FT

0.00   20   40   60   80   180   200   280   MPH
0.00   32.18   64.37   96.56 128.75 289.68 322 450.52 Km/hr

ADVANCED ENGINEERING PROJECTS

Name of project: FLIGHT PROFILE DIAGRAM

HYPERSONIC SPACE PLANE (SCRAM JET)
FOR CARGO TRANSPORTATION SYSTEM BETWEEN THE EARTH AND THE SPACE STATION

Design Engineer: WALTER F. LAREDO   Date: Feb. 1997

Drawing Number: 1800   Sheet 4 of 21

PLATE 188

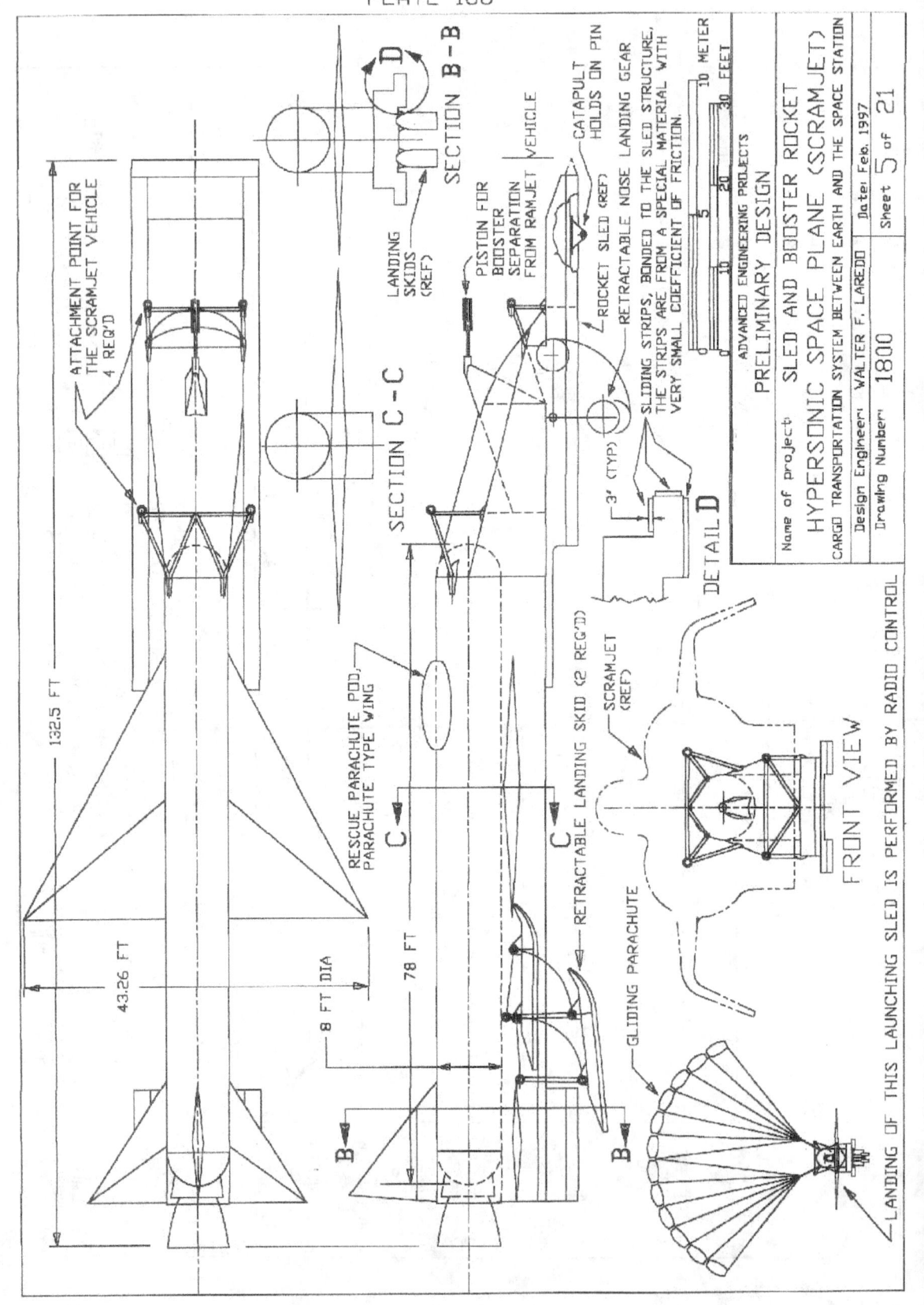

SECTION B-B

SECTION C-C

DETAIL D

FRONT VIEW

ATTACHMENT POINT FOR THE SCRAMJET VEHICLE 4 REQ'D

LANDING SKIDS (REF)

PISTON FOR BOOSTER SEPARATION FROM RAMJET VEHICLE

CATAPULT HOLDS ON PIN

ROCKET SLED (REF)

RETRACTABLE NOSE LANDING GEAR

SLIDING STRIPS, BONDED TO THE SLED STRUCTURE, THE STRIPS ARE FRIM A SPECIAL MATERIAL WITH VERY SMALL COEFFICIENT OF FRICTION.

3" (TYP)

10 METER
5
30 FEET
20
10

132.5 FT

43.26 FT

8 FT DIA

78 FT

RESCUE PARACHUTE POD, PARACHUTE TYPE WING

RETRACTABLE LANDING SKID (2 REQ'D)

SCRAMJET (REF)

GLIDING PARACHUTE

LANDING OF THIS LAUNCHING SLED IS PERFORMED BY RADIO CONTROL

ADVANCED ENGINEERING PROJECTS

PRELIMINARY DESIGN

Name of project: SLED AND BOOSTER ROCKET HYPERSONIC SPACE PLANE (SCRAMJET)

CARGO TRANSPORTATION SYSTEM BETWEEN EARTH AND THE SPACE STATION

Design Engineer: WALTER F. LAREDO   Date: Feb. 1997

Drawing Number: 1800   Sheet 5 of 21

PLATE 189

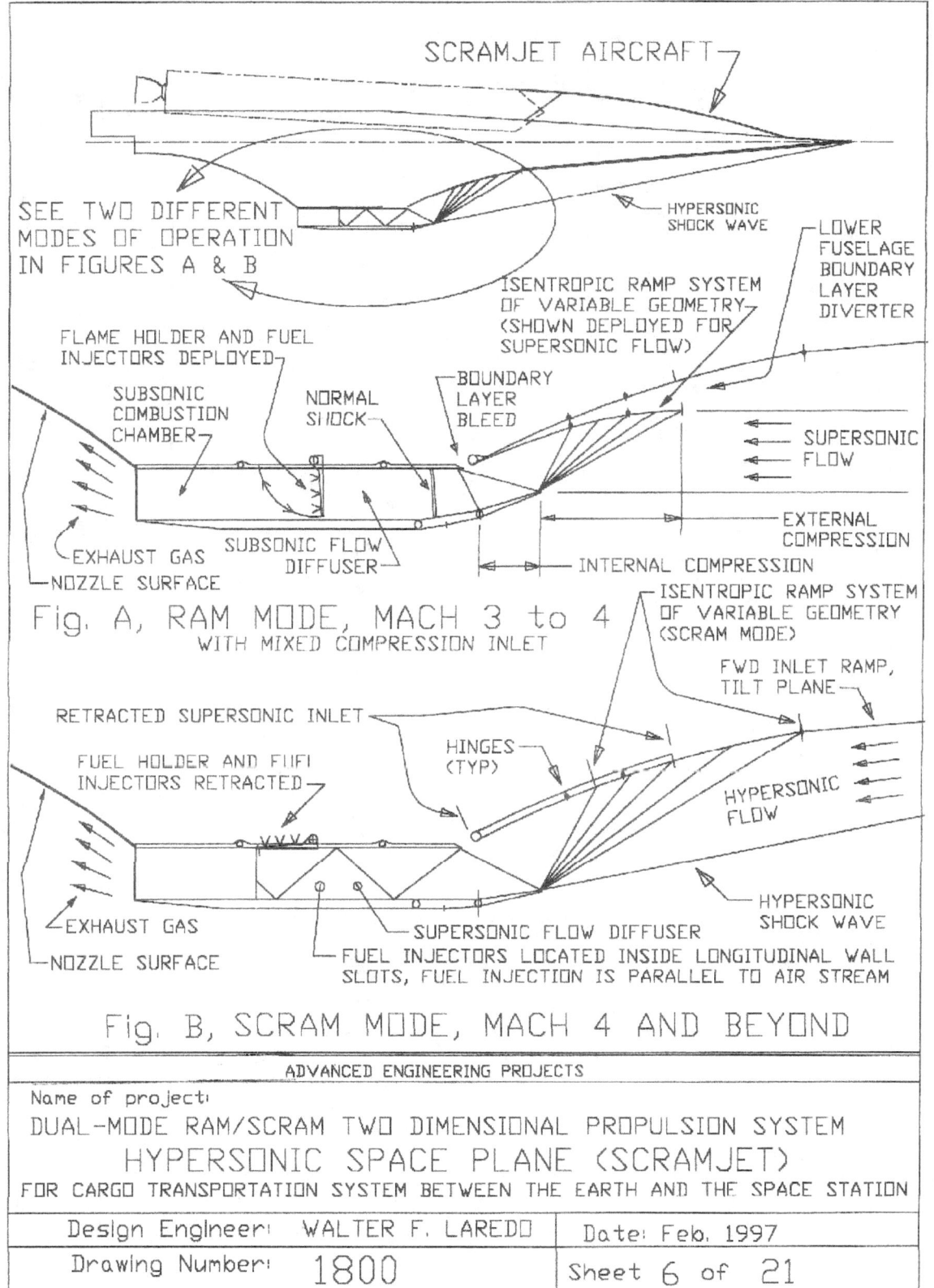

SCRAMJET AIRCRAFT

SEE TWO DIFFERENT
MODES OF OPERATION
IN FIGURES A & B

HYPERSONIC
SHOCK WAVE

ISENTROPIC RAMP SYSTEM
OF VARIABLE GEOMETRY
(SHOWN DEPLOYED FOR
SUPERSONIC FLOW)

LOWER
FUSELAGE
BOUNDARY
LAYER
DIVERTER

FLAME HOLDER AND FUEL
INJECTORS DEPLOYED

SUBSONIC
COMBUSTION
CHAMBER

NORMAL
SHOCK

BOUNDARY
LAYER
BLEED

SUPERSONIC
FLOW

EXHAUST GAS
NOZZLE SURFACE

SUBSONIC FLOW
DIFFUSER

INTERNAL COMPRESSION

EXTERNAL
COMPRESSION

Fig. A, RAM MODE, MACH 3 to 4
WITH MIXED COMPRESSION INLET

ISENTROPIC RAMP SYSTEM
OF VARIABLE GEOMETRY
(SCRAM MODE)

FWD INLET RAMP,
TILT PLANE

RETRACTED SUPERSONIC INLET

FUEL HOLDER AND FUEL
INJECTORS RETRACTED

HINGES
(TYP)

HYPERSONIC
FLOW

EXHAUST GAS
NOZZLE SURFACE

SUPERSONIC FLOW DIFFUSER

FUEL INJECTORS LOCATED INSIDE LONGITUDINAL WALL
SLOTS, FUEL INJECTION IS PARALLEL TO AIR STREAM

HYPERSONIC
SHOCK WAVE

Fig. B, SCRAM MODE, MACH 4 AND BEYOND

ADVANCED ENGINEERING PROJECTS

Name of project:
DUAL-MODE RAM/SCRAM TWO DIMENSIONAL PROPULSION SYSTEM
HYPERSONIC SPACE PLANE (SCRAMJET)
FOR CARGO TRANSPORTATION SYSTEM BETWEEN THE EARTH AND THE SPACE STATION

Design Engineer: WALTER F. LAREDO	Date: Feb. 1997
Drawing Number: 1800	Sheet 6 of 21

PLATE 190

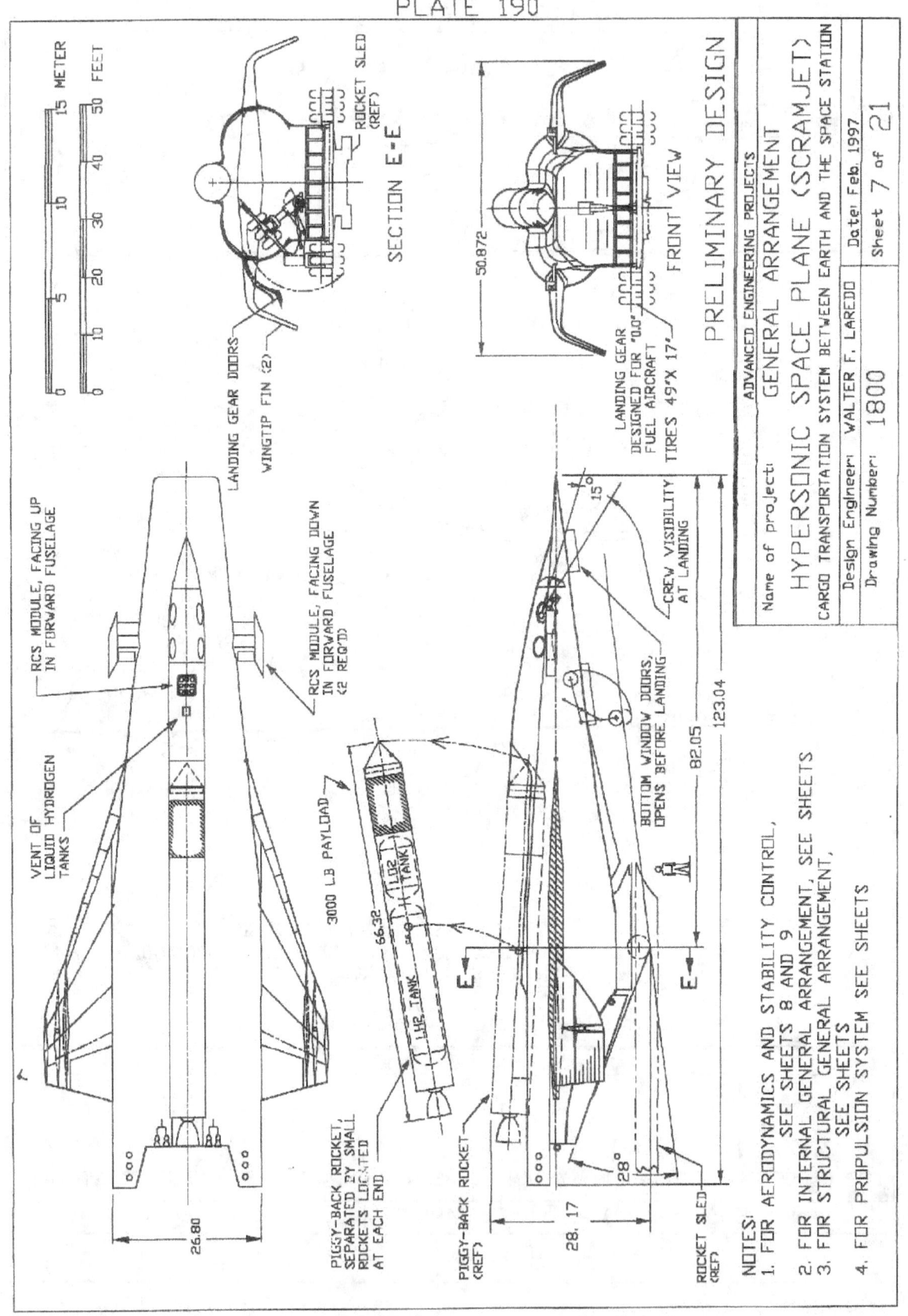

SECTION E-E

METER
FEET

LANDING GEAR DOORS
WINGTIP FIN (2)
ROCKET SLED (REF)

50.872

FRONT VIEW

LANDING GEAR DESIGNED FOR "DC" FUEL AIRCRAFT TIRES 49"X 17'

PRELIMINARY DESIGN

ADVANCED ENGINEERING PROJECTS		
Name of project:	GENERAL ARRANGEMENT	
	HYPERSONIC SPACE PLANE (SCRAMJET)	
	CARGO TRANSPORTATION SYSTEM BETWEEN EARTH AND THE SPACE STATION	
Design Engineer: WALTER F. LAREDO		Date: Feb. 1997
Drawing Number: 1800		Sheet 7 of 21

RCS MODULE, FACING UP IN FORWARD FUSELAGE

RCS MODULE, FACING DOWN IN FORWARD FUSELAGE (2 REQ'D)

VENT OF LIQUID HYDROGEN TANKS

PIGGY-BACK ROCKET, SEPARATED BY SMALL ROCKETS LOCATED AT EACH END

26.80

3000 LB PAYLOAD

LH2 TANK
LH2 TANK

66.32

PIGGY-BACK ROCKET (REF)

15°

CREW VISIBILITY AT LANDING

BOTTOM WINDOW DOORS, OPENS BEFORE LANDING

123.04

82.05

28.17

28°

ROCKET SLED (REF)

E

E

NOTES:
1. FOR AERODYNAMICS AND STABILITY CONTROL, SEE SHEETS 8 AND 9
2. FOR INTERNAL GENERAL ARRANGEMENT, SEE SHEETS
3. FOR STRUCTURAL GENERAL ARRANGEMENT, SEE SHEETS
4. FOR PROPULSION SYSTEM SEE SHEETS

PLATE 191

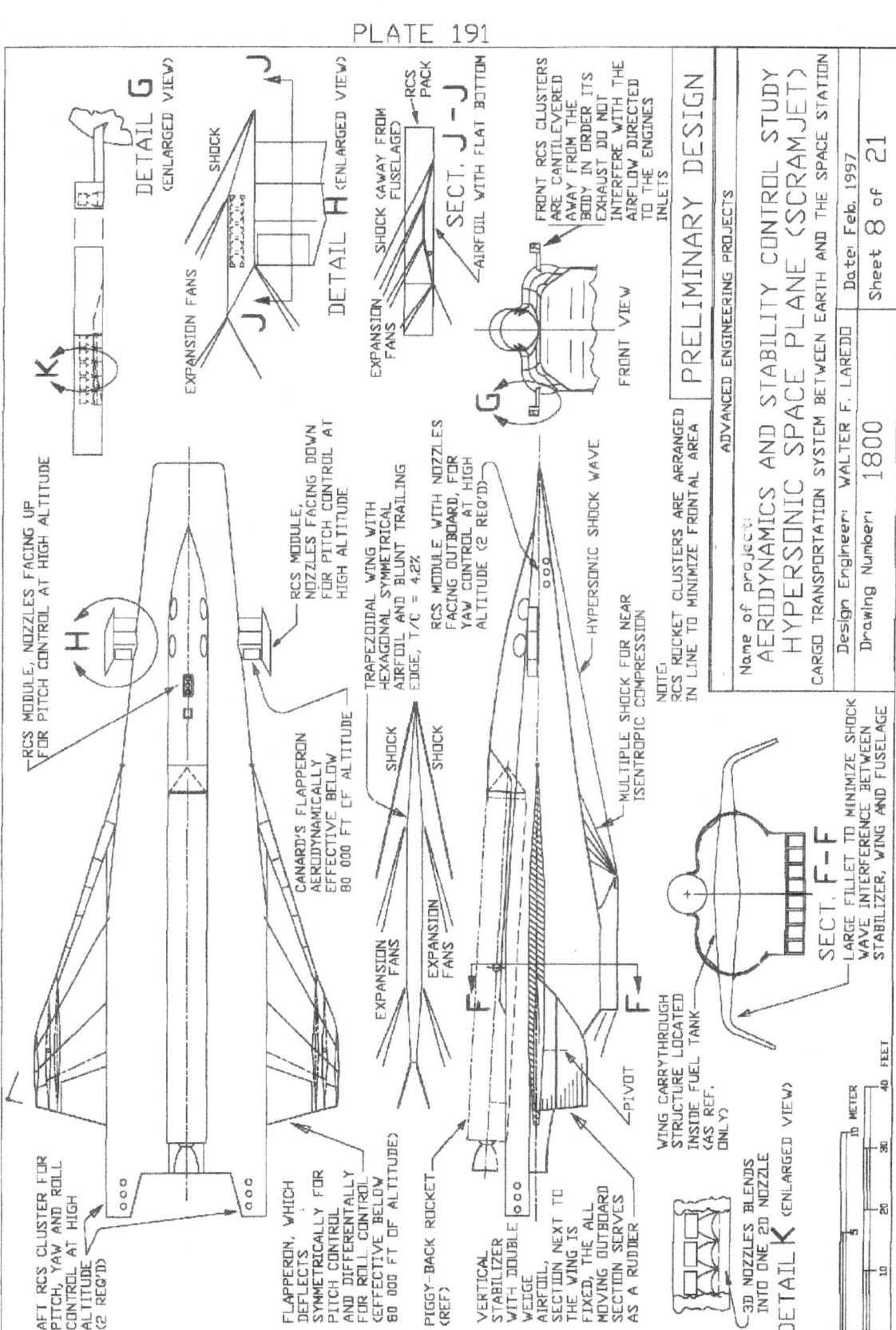

DETAIL G (ENLARGED VIEW)

SHOCK

DETAIL H (ENLARGED VIEW)

EXPANSION FANS

SHOCK (AWAY FROM FUSELAGE)

RCS PACK

SECT. J-J

AIRFOIL WITH FLAT BOTTOM

EXPANSION FANS

FRONT RCS CLUSTERS ARE CANTILEVERED AWAY FROM THE BODY IN ORDER ITS EXHAUST DO NOT INTERFERE WITH THE AIRFLOW DIRECTED TO THE ENGINES INLETS

FRONT VIEW

RCS ROCKET CLUSTERS ARE ARRANGED IN LINE TO MINIMIZE FRONTAL AREA

ADVANCED ENGINEERING PROJECTS

PRELIMINARY DESIGN

Name of project: AERODYNAMICS AND STABILITY CONTROL STUDY HYPERSONIC SPACE PLANE (SCRAMJET)

CARGO TRANSPORTATION SYSTEM BETWEEN EARTH AND THE SPACE STATION

| Design Engineer: WALTER F. LAREDO | Date: Feb. 1997 |
| Drawing Number: 1800 | Sheet 8 of 21 |

RCS MODULE, NOZZLES FACING UP FOR PITCH CONTROL AT HIGH ALTITUDE

RCS MODULE, NOZZLES FACING DOWN FOR PITCH CONTROL AT HIGH ALTITUDE

TRAPEZOIDAL WING WITH HEXAGONAL SYMMETRICAL AIRFOIL AND BLUNT TRAILING EDGE, T/C = 4.2%

RCS MODULE WITH NOZZLES FACING OUTBOARD, FOR YAW CONTROL AT HIGH ALTITUDE (2 REQ'D)

HYPERSONIC SHOCK WAVE

MULTIPLE SHOCK FOR NEAR ISENTROPIC COMPRESSION

NOTE: RCS ROCKET CLUSTERS ARE ARRANGED IN LINE TO MINIMIZE FRONTAL AREA

AFT RCS CLUSTER FOR PITCH, YAW AND ROLL CONTROL AT HIGH ALTITUDE (2 REQ'D)

FLAPPERON, WHICH DEFLECTS SYMMETRICALLY FOR PITCH CONTROL AND DIFFERENTIALLY FOR ROLL CONTROL (EFFECTIVE BELOW 80 000 FT OF ALTITUDE)

CANARD'S FLAPPERON AERODYNAMICALLY EFFECTIVE BELOW 80 000 FT OF ALTITUDE)

PIGGY-BACK ROCKET (REF)

SHOCK

SHOCK

EXPANSION FANS

EXPANSION FANS

VERTICAL STABILIZER WITH DOUBLE WEDGE AIRFOIL, SECTION NEXT TO THE WING IS FIXED, THE ALL MOVING OUTBOARD SECTION SERVES AS A RUDDER

PIVOT

WING CARRYTHROUGH STRUCTURE LOCATED INSIDE FUEL TANK (AS REF. ONLY)

SECT. F-F

LARGE FILLET TO MINIMIZE SHOCK WAVE INTERFERENCE BETWEEN STABILIZER, WING AND FUSELAGE

3D NOZZLES BLENDS INTO ONE 2D NOZZLE

DETAIL K (ENLARGED VIEW)

METER

FEET

0   10   20   30   40

0   5   10   15

PLATE 192

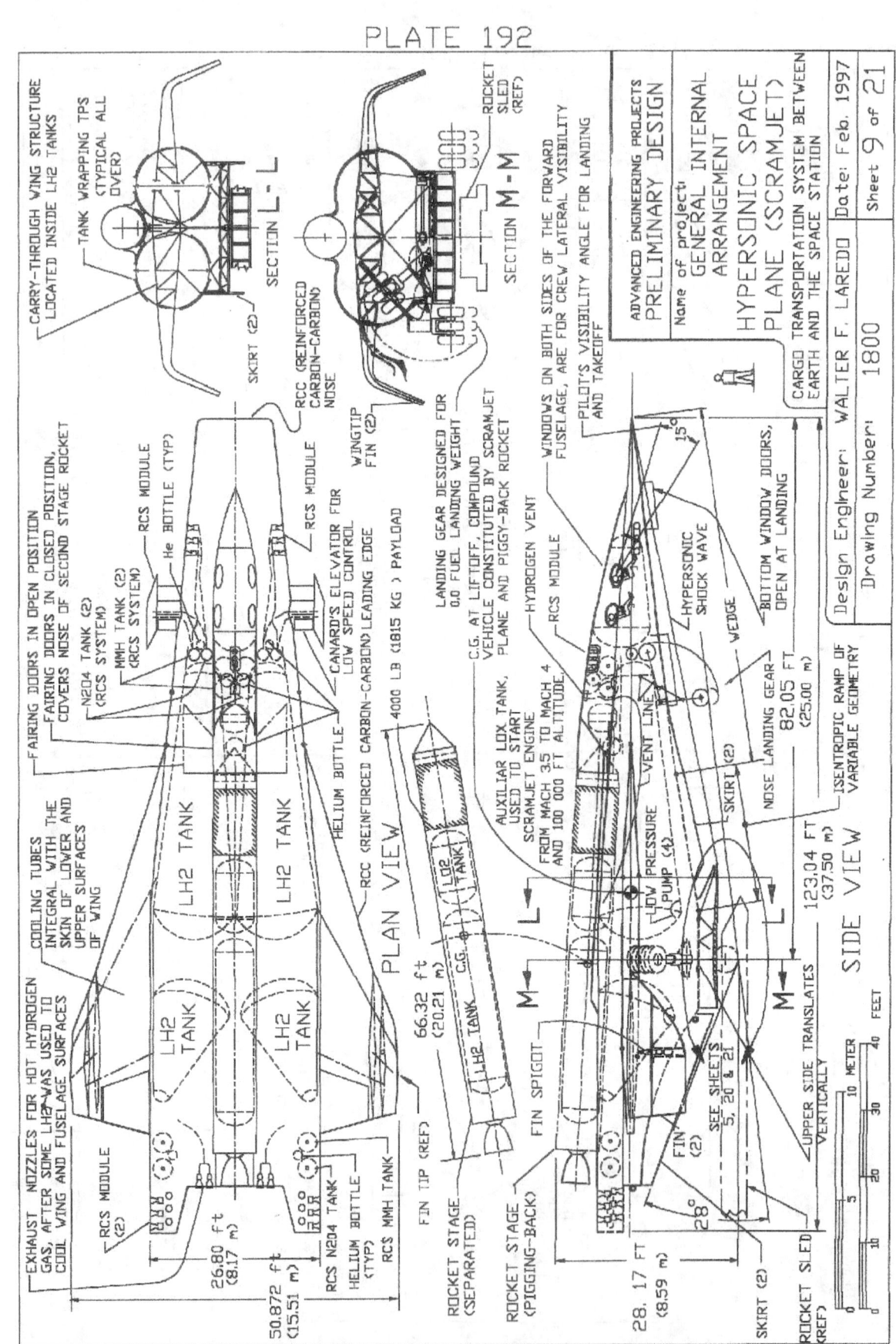

CARRY-THROUGH WING STRUCTURE LOCATED INSIDE LH2 TANKS

TANK WRAPPING TPS (TYPICAL ALL OVER)

SECTION L-L

SKIRT (2)

ROCKET SLED (REF)

RCC (REINFORCED CARBON-CARBON) NOSE

SECTION M-M

WINDOWS ON BOTH SIDES OF THE FORWARD FUSELAGE, ARE FOR CREW LATERAL VISIBILITY

PILOT'S VISIBILITY ANGLE FOR LANDING AND TAKEOFF

15°

HYPERSONIC SHOCK WAVE

WEDGE

BOTTOM WINDOW DOORS, OPEN AT LANDING

NOSE LANDING GEAR

82.05 FT (25.00 m)

ISENTROPIC RAMP OF VARIABLE GEOMETRY

123.04 FT (37.50 m)

SIDE VIEW

HYDROGEN VENT

RCS MODULE

VENT LINE

LOW PRESSURE LH2 PUMP (4)

SKIRT (2)

28°

ROCKET SLED (REF)

SKIRT (2)

UPPER SIDE TRANSLATES VERTICALLY

SEE SHEETS 5, 20 & 21

FIN (2)

FIN SPIGOT

28.17 FT (8.59 m)

LH2 TANK

AUXILIAR LOX TANK, USED TO START SCRAMJET ENGINE

FROM MACH 3.5 TO MACH 4 AND 100 000 FT ALTITUDE.

C.G. AT LIFTOFF, COMPOUND VEHICLE CONSTITUTED BY SCRAMJET PLANE AND PIGGY-BACK ROCKET

LANDING GEAR DESIGNED FOR 0.0 FUEL LANDING WEIGHT

4000 LB (1815 KG) PAYLOAD

PLAN VIEW

LH2 TANK

LOX TANK

LH2 TANK

FIN TIP (REF)

C.G.

66.32 ft (20.21 m)

ROCKET STAGE (SEPARATED)

ROCKET STAGE (PIGGY-BACK)

HELIUM BOTTLE

RCC (REINFORCED CARBON-CARBON) LEADING EDGE

CANARD'S ELEVATOR FOR LOW SPEED CONTROL

WINGTIP FIN (2)

He BOTTLE (TYP)

RCS MODULE

RCS MODULE

N2O4 TANK (2) (RCS SYSTEM)

MMH TANK (2) (RCS SYSTEM)

LH2 TANK

LH2 TANK

LH2 TANK

EXHAUST NOZZLES FOR HOT HYDROGEN GAS, AFTER SOME LH2 WAS USED TO COOL WING AND FUSELAGE SURFACES

COOLING TUBES INTEGRAL WITH THE SKIN OF LOWER AND UPPER SURFACES OF WING

FAIRING DOORS IN OPEN POSITION

FAIRING DOORS IN CLOSED POSITION, COVERS NOSE OF SECOND STAGE ROCKET

RCS MODULE (2)

RCS N2O4 TANK

HELIUM BOTTLE (TYP)

RCS MMH TANK

RCS MODULE (2)

50.872 ft (15.51 m)

26.80 ft (8.17 m)

FEET

METER

ADVANCED ENGINEERING PROJECTS

PRELIMINARY DESIGN

Name of project: GENERAL INTERNAL ARRANGEMENT

HYPERSONIC SPACE PLANE (SCRAMJET)

CARGO TRANSPORTATION SYSTEM BETWEEN EARTH AND THE SPACE STATION

Design Engineer: WALTER F. LAREDO

Date: Feb. 1997

Sheet 9 of 21

Drawing Number: 1800

PLATE 193

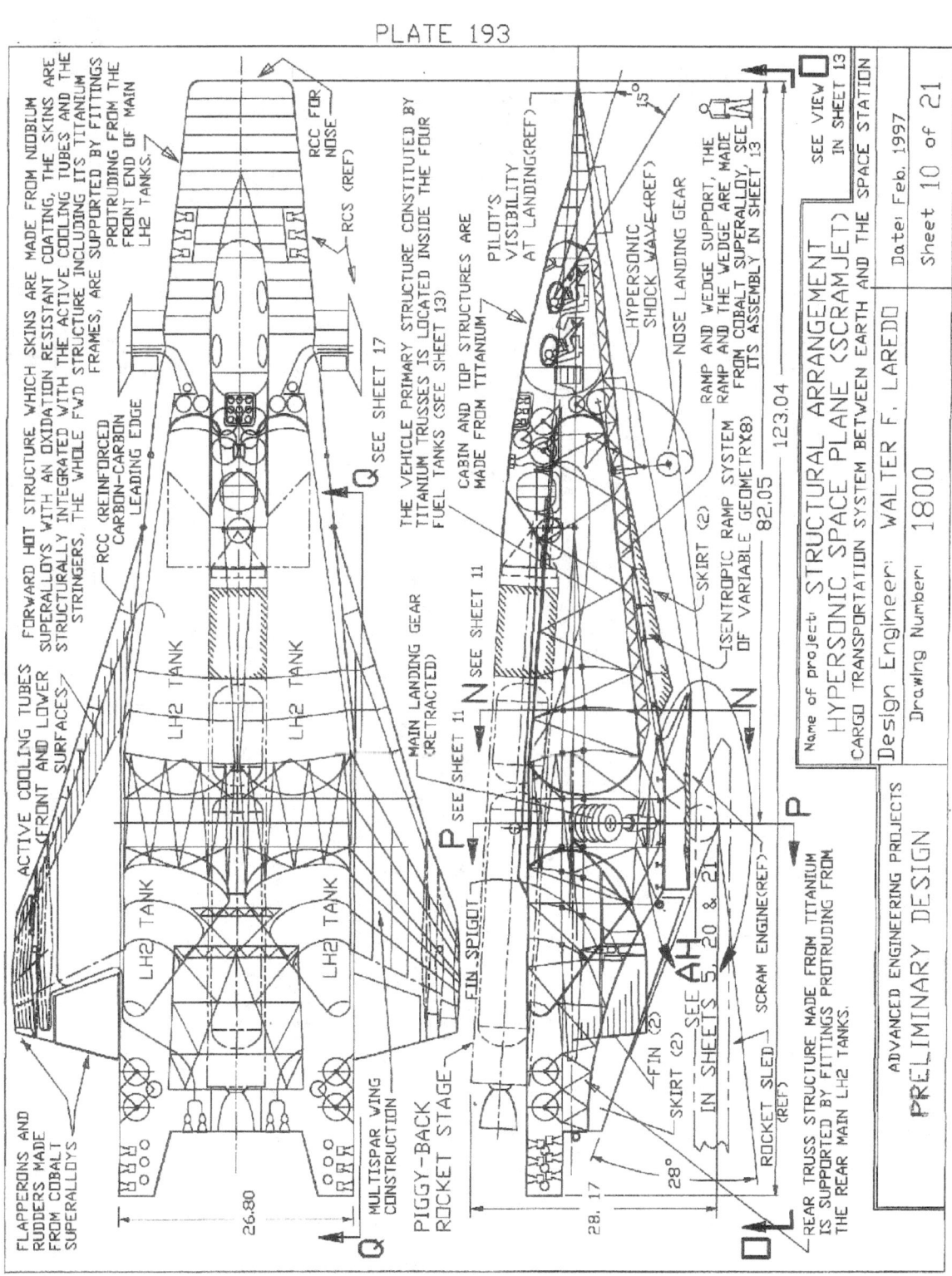

FLAPPERONS AND RUDDERS MADE FROM COBALT SUPERALLOYS

ACTIVE COOLING TUBES (FRONT AND LOWER SURFACES)

FORWARD HOT STRUCTURE WHICH SKINS ARE MADE FROM NIOBIUM SUPERALLOYS WITH AN OXIDATION RESISTANT COATING. THE SKINS ARE STRUCTURALLY INTEGRATED WITH THE ACTIVE COOLING TUBES AND THE STRINGERS. THE WHOLE FWD STRUCTURE INCLUDING ITS TITANIUM FRAMES, ARE SUPPORTED BY FITTINGS PROTRUDING FROM THE FRONT END OF MAIN LH2 TANKS.

RCC (REINFORCED CARBON-CARBON) LEADING EDGE

RCC FOR NOSE (REF)

RCS (REF)

SEE SHEET 17

THE VEHICLE PRIMARY STRUCTURE CONSTITUTED BY TITANIUM TRUSSES IS LOCATED INSIDE THE FOUR FUEL TANKS (SEE SHEET 13)

CABIN AND TOP STRUCTURES ARE MADE FROM TITANIUM

PILOT'S VISIBILITY AT LANDING(REF)

HYPERSONIC SHOCK WAVE(REF)

15°

NOSE LANDING GEAR

RAMP AND WEDGE SUPPORT. THE RAMP AND THE WEDGE ARE MADE FROM COBALT SUPERALLOY, SEE ITS ASSEMBLY IN SHEET 13

SEE VIEW □ IN SHEET 13

ISENTROPIC RAMP SYSTEM OF VARIABLE GEOMETRY(8)

SKIRT (2)

123.04

82.05

LH2 TANK

LH2 TANK

MAIN LANDING GEAR (RETRACTED)

P SEE/SHEET 11

N SEE SHEET 11

P

N

N

LH2 TANK

LH2 TANK

26.80

MULTISPAR WING CONSTRUCTION

PIGGY-BACK ROCKET STAGE

FIN SPIGOT

FIN (2)

SKIRT (2)

SEE AH IN SHEETS 5, 20 & 21

SCRAM ENGINE(REF)

ROCKET SLED (REF)

28°

28.17

REAR TRUSS STRUCTURE MADE FROM TITANIUM IS SUPPORTED BY FITTINGS PROTRUDING FROM THE REAR MAIN LH2 TANKS.

Q

Q

T

L

P

Name of project: STRUCTURAL ARRANGEMENT
HYPERSONIC SPACE PLANE (SCRAMJET)
CARGO TRANSPORTATION SYSTEM BETWEEN EARTH AND THE SPACE STATION

Design Engineer: WALTER F. LAREDO   Date: Feb. 1997

Drawing Number: 1800   Sheet 10 of 21

ADVANCED ENGINEERING PROJECTS

PRELIMINARY DESIGN

PLATE 194

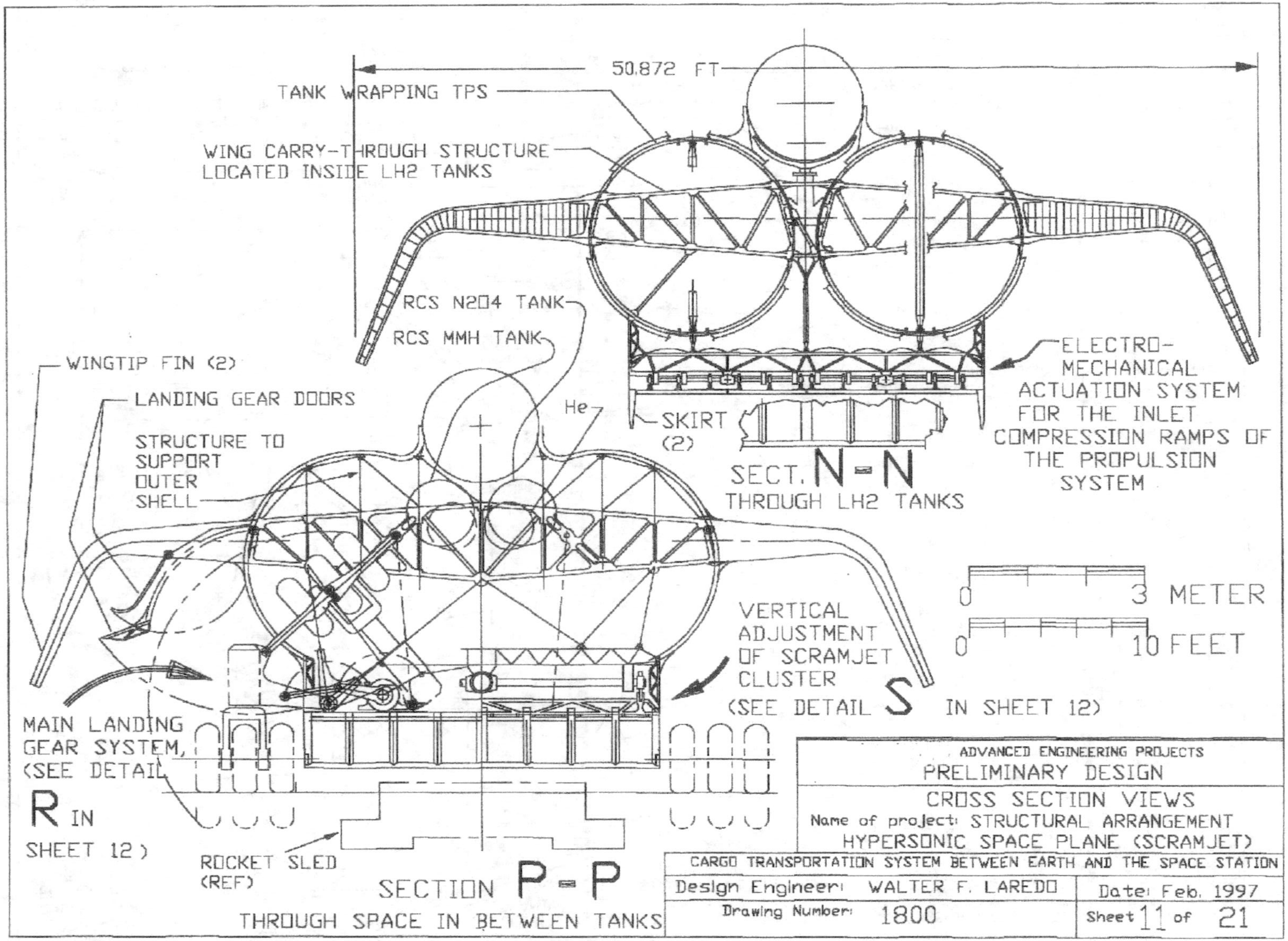

50.872 FT

TANK WRAPPING TPS

WING CARRY-THROUGH STRUCTURE
LOCATED INSIDE LH2 TANKS

RCS N2O4 TANK

RCS MMH TANK

WINGTIP FIN (2)

LANDING GEAR DOORS

STRUCTURE TO
SUPPORT
OUTER
SHELL

MAIN LANDING
GEAR SYSTEM,
(SEE DETAIL R IN
SHEET 12)

ROCKET SLED
(REF)

SECTION P-P
THROUGH SPACE IN BETWEEN TANKS

He

VERTICAL
ADJUSTMENT
OF SCRAMJET
CLUSTER
(SEE DETAIL S
IN SHEET 12)

ELECTRO-
MECHANICAL
ACTUATION SYSTEM
FOR THE INLET
COMPRESSION RAMPS OF
THE PROPULSION
SYSTEM

SKIRT
(2)

SECT. N-N
THROUGH LH2 TANKS

3 METER

10 FEET

ADVANCED ENGINEERING PROJECTS
PRELIMINARY DESIGN
CROSS SECTION VIEWS

Name of project: STRUCTURAL ARRANGEMENT
HYPERSONIC SPACE PLANE (SCRAMJET)

CARGO TRANSPORTATION SYSTEM BETWEEN EARTH AND THE SPACE STATION

Design Engineer: WALTER F. LAREDO   Date: Feb. 1997

Drawing Number: 1800   Sheet 11 of 21

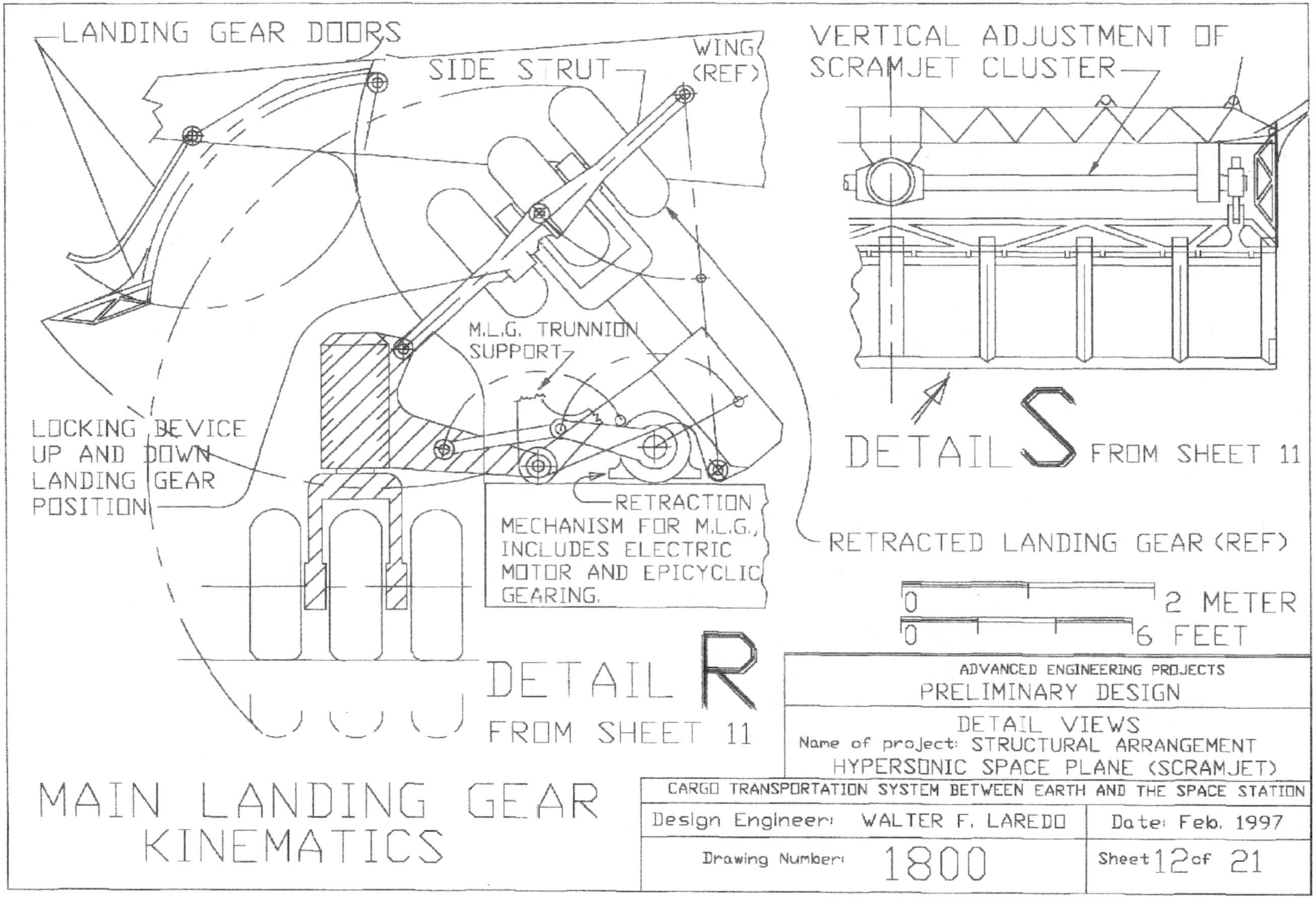

PLATE 195

VERTICAL ADJUSTMENT OF SCRAMJET CLUSTER

DETAIL S FROM SHEET 11

RETRACTED LANDING GEAR (REF)

2 METER

6 FEET

LANDING GEAR DOORS

WING (REF)

SIDE STRUT

M.L.G. TRUNNION SUPPORT

RETRACTION MECHANISM FOR M.L.G., INCLUDES ELECTRIC MOTOR AND EPICYCLIC GEARING.

DETAIL R FROM SHEET 11

LOCKING DEVICE
UP AND DOWN
LANDING GEAR
POSITION

MAIN LANDING GEAR KINEMATICS

ADVANCED ENGINEERING PROJECTS
PRELIMINARY DESIGN

DETAIL VIEWS

Name of project: STRUCTURAL ARRANGEMENT
HYPERSONIC SPACE PLANE (SCRAMJET)

CARGO TRANSPORTATION SYSTEM BETWEEN EARTH AND THE SPACE STATION	
Design Engineer: WALTER F. LAREDO	Date: Feb. 1997
Drawing Number: 1800	Sheet 12 of 21

PLATE 196

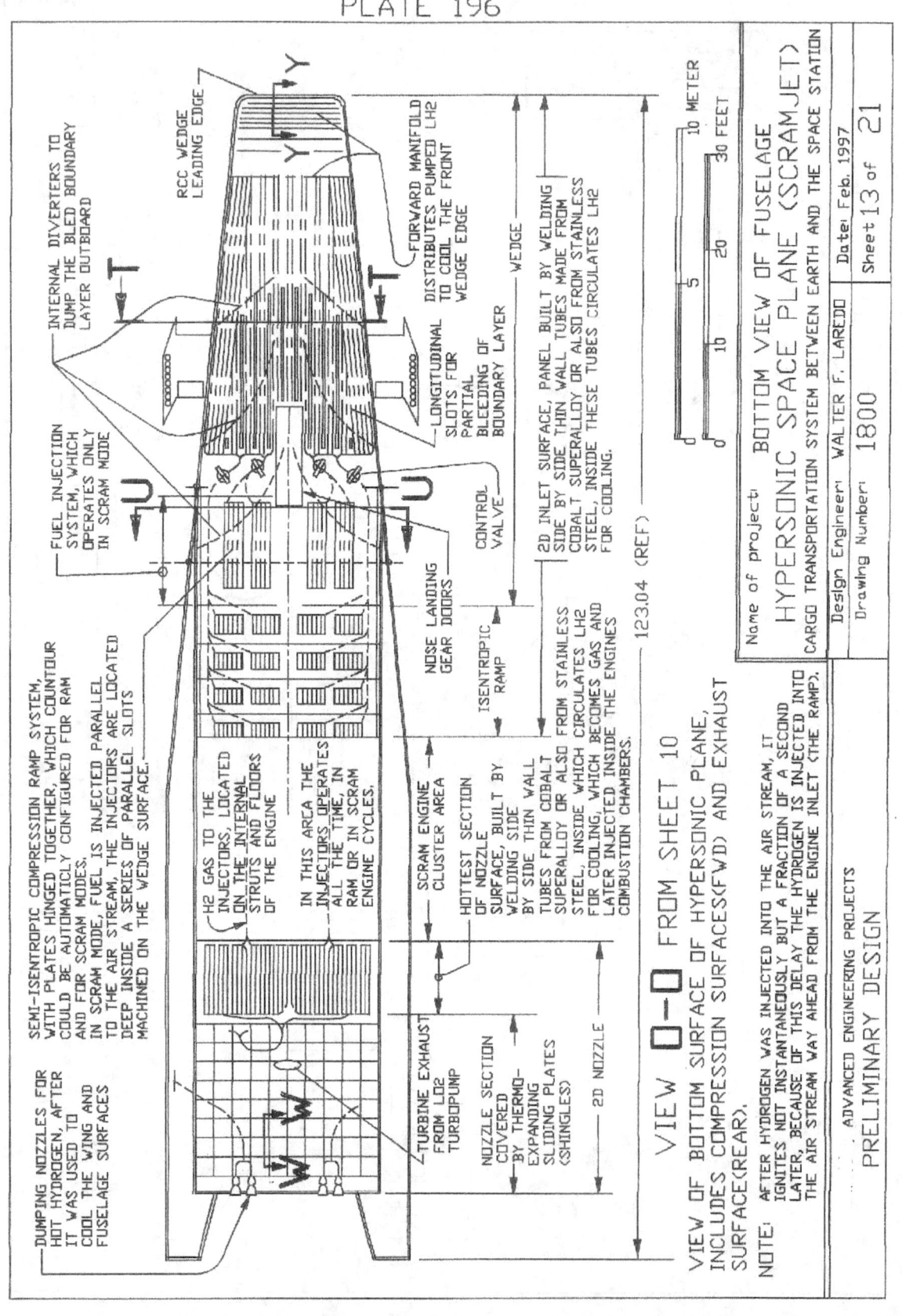

DUMPING NOZZLES FOR HOT HYDROGEN, AFTER IT WAS USED TO COOL THE WING AND FUSELAGE SURFACES

SEMI-ISENTROPIC COMPRESSION RAMP SYSTEM, WITH PLATES HINGED TOGETHER, WHICH COUNTOUR COULD BE AUTOMATICLY CONFIGURED FOR RAM AND FOR SCRAM MODES. IN SCRAM MODE, FUEL IS INJECTED PARALLEL TO THE AIR STREAM, THE INJECTORS ARE LOCATED DEEP INSIDE A SERIES OF PARALLEL SLOTS MACHINED ON THE WEDGE SURFACE.

RCC WEDGE LEADING EDGE

INTERNAL DIVERTERS TO DUMP THE BLED BOUNDARY LAYER OUTBOARD

FORWARD MANIFOLD DISTRIBUTES PUMPED LH2 TO COOL THE FRONT WEDGE EDGE

FUEL INJECTION SYSTEM, WHICH OPERATES ONLY IN SCRAM MODE

LONGITUDINAL SLOTS FOR PARTIAL BLEEDING OF BOUNDARY LAYER

WEDGE

H2 GAS TO THE INJECTORS, LOCATED ON THE INTERNAL STRUTS AND FLOORS OF THE ENGINE

IN THIS AREA THE INJECTORS OPERATES ALL THE TIME IN RAM OR IN SCRAM ENGINE CYCLES.

CONTROL VALVE

2D INLET SURFACE, PANEL BUILT BY WELDING SIDE BY SIDE THIN WALL TUBES MADE FROM COBALT SUPERALLOY OR ALSO FROM STAINLESS STEEL, INSIDE THESE TUBES CIRCULATES LH2 FOR COOLING.

NOSE LANDING GEAR DOORS

ISENTROPIC RAMP

123.04 (REF)

SCRAM ENGINE CLUSTER AREA

HOTTEST SECTION OF NOZZLE SURFACE, BUILT BY WELDING SIDE BY SIDE THIN WALL TUBES FROM COBALT SUPERALLOY OR ALSO FROM STAINLESS STEEL, INSIDE WHICH CIRCULATES LH2 FOR COOLING, WHICH BECOMES GAS AND LATER INJECTED INSIDE THE ENGINES COMBUSTION CHAMBERS.

TURBINE EXHAUST FROM LO2 TURBOPUMP

NOZZLE SECTION COVERED BY THERMO-EXPANDING SLIDING PLATES (SHINGLES)

2D NOZZLE

VIEW O-O FROM SHEET 10

VIEW OF BOTTOM SURFACE OF HYPERSONIC PLANE, INCLUDES COMPRESSION SURFACES(FWD) AND EXHAUST SURFACE(REAR).

NOTE: AFTER HYDROGEN WAS INJECTED INTO THE AIR STREAM, IT IGNITES NOT INSTANTANEOUSLY BUT A FRACTION OF A SECOND LATER, BECAUSE OF THIS DELAY THE HYDROGEN IS INJECTED INTO THE AIR STREAM WAY AHEAD FROM THE ENGINE INLET (THE RAMP).

10 METER

5

30 FEET

10   20

Name of project: BOTTOM VIEW OF FUSELAGE
HYPERSONIC SPACE PLANE (SCRAMJET)
CARGO TRANSPORTATION SYSTEM BETWEEN EARTH AND THE SPACE STATION

| Design Engineer: WALTER F. LAREDO | Date: Feb. 1997 |
| Drawing Number: 1800 | Sheet 13 of 21 |

ADVANCED ENGINEERING PROJECTS
PRELIMINARY DESIGN

PLATE 197

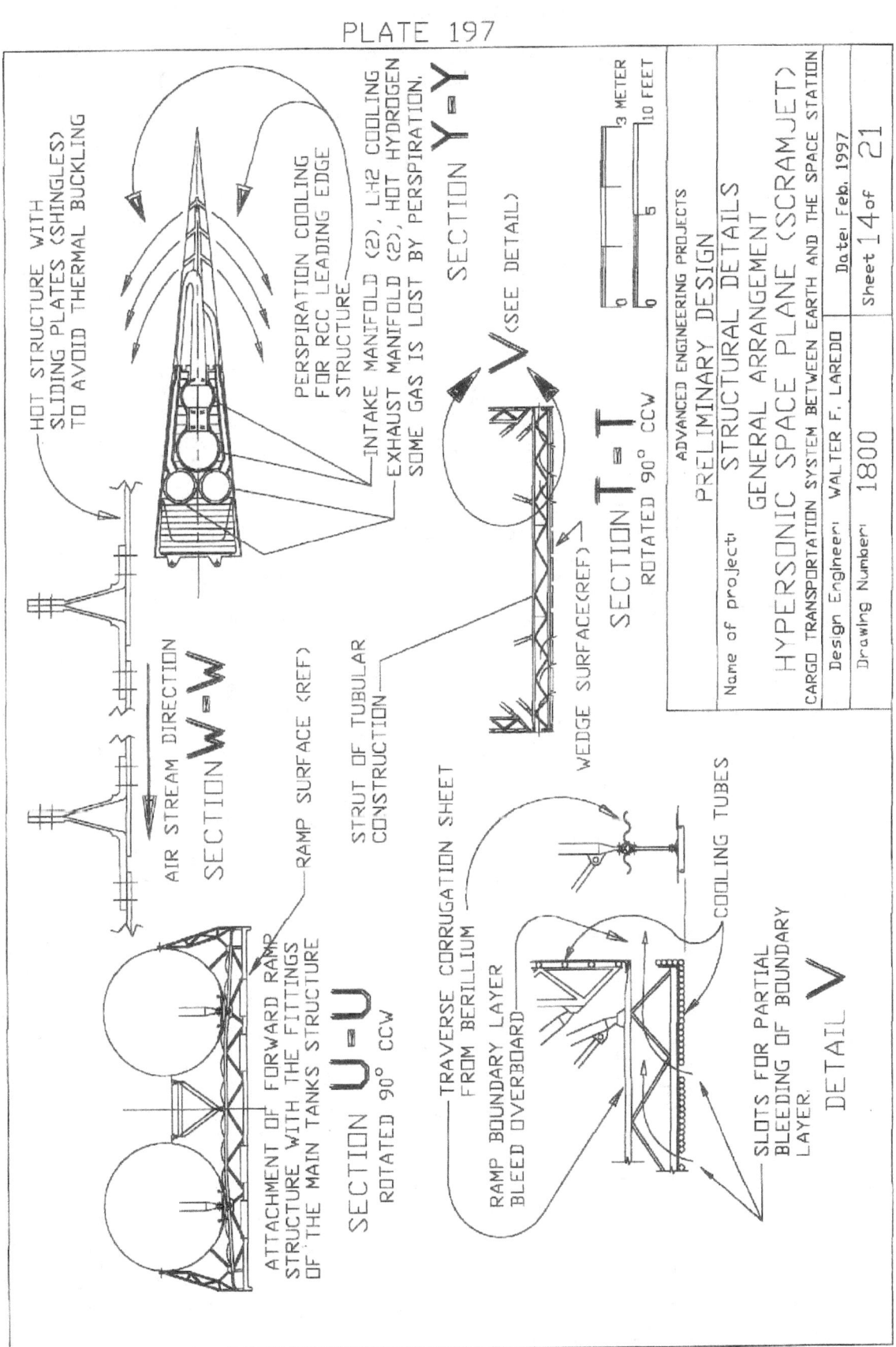

SECTION **W—W**

AIR STREAM DIRECTION

HOT STRUCTURE WITH SLIDING PLATES (SHINGLES) TO AVOID THERMAL BUCKLING

PERSPIRATION COOLING FOR RCC LEADING EDGE STRUCTURE

INTAKE MANIFOLD (2), LH2 COOLING
EXHAUST MANIFOLD (2), HOT HYDROGEN
SOME GAS IS LOST BY PERSPIRATION.

RAMP SURFACE (REF)

STRUT OF TUBULAR CONSTRUCTION

SECTION **U—U**
ROTATED 90° CCW

ATTACHMENT OF FORWARD RAMP STRUCTURE WITH THE FITTINGS OF THE MAIN TANKS STRUCTURE

TRAVERSE CORRUGATION SHEET FROM BERILLIUM

RAMP BOUNDARY LAYER BLEED OVERBOARD

SECTION **Y—Y**

(SEE DETAIL)

SECTION **T—T**
ROTATED 90° CCW

WEDGE SURFACE (REF)

COOLING TUBES

SLOTS FOR PARTIAL BLEEDING OF BOUNDARY LAYER.

DETAIL **V**

3 METER

10 FEET

ADVANCED ENGINEERING PROJECTS

PRELIMINARY DESIGN

STRUCTURAL DETAILS

GENERAL ARRANGEMENT

Name of project: HYPERSONIC SPACE PLANE (SCRAMJET)

CARGO TRANSPORTATION SYSTEM BETWEEN EARTH AND THE SPACE STATION

| Design Engineer: WALTER F. LAREDO | Date: Feb. 1997 |
| Drawing Number: 1800 | Sheet 14 of 21 |

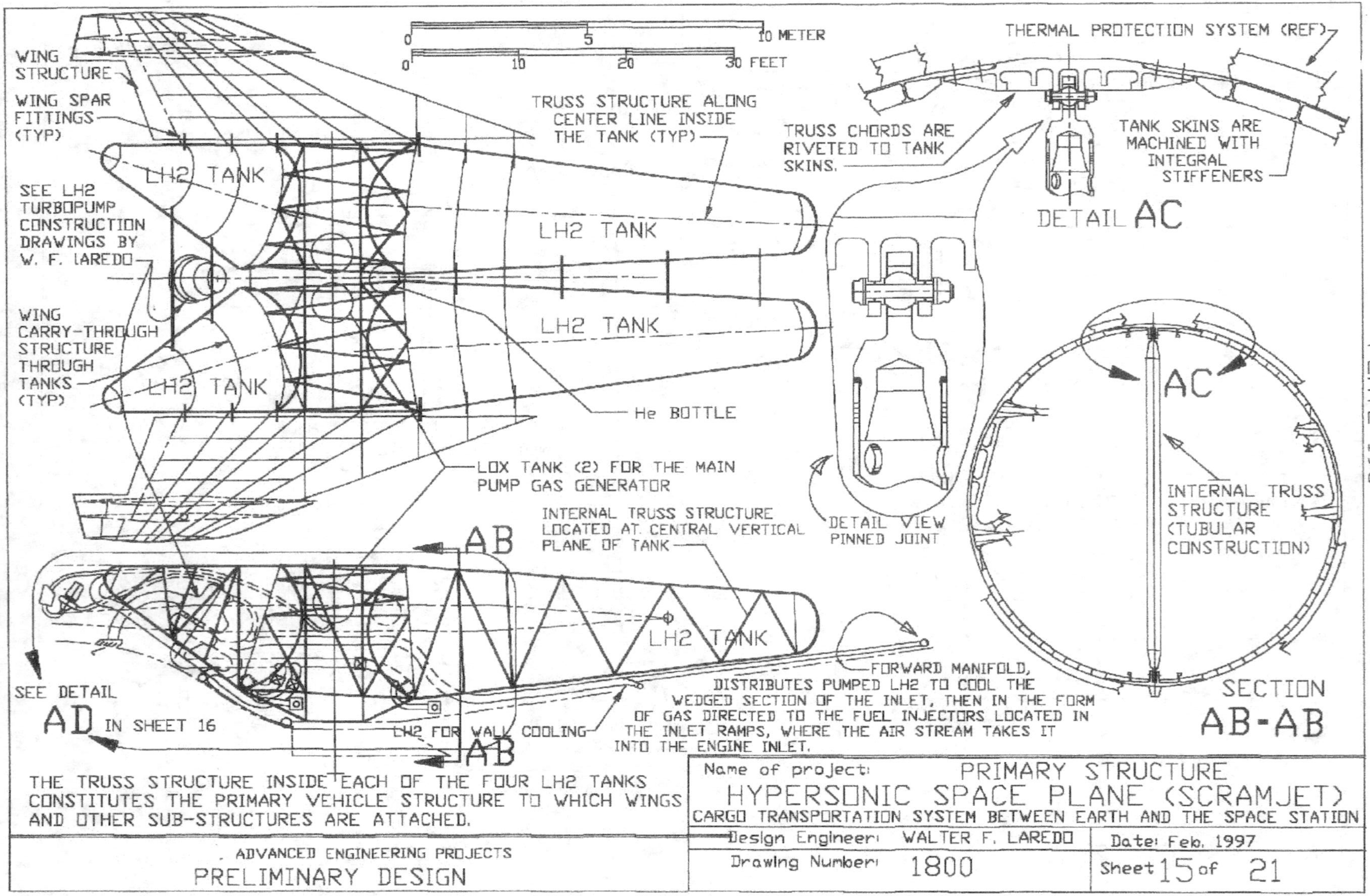

PLATE 198

THERMAL PROTECTION SYSTEM (REF)

TANK SKINS ARE MACHINED WITH INTEGRAL STIFFENERS

DETAIL AC

AC

INTERNAL TRUSS STRUCTURE (TUBULAR CONSTRUCTION)

SECTION AB-AB

TRUSS CHORDS ARE RIVETED TO TANK SKINS.

DETAIL VIEW PINNED JOINT

FORWARD MANIFOLD, DISTRIBUTES PUMPED LH2 TO COOL THE WEDGED SECTION OF THE INLET, THEN IN THE FORM OF GAS DIRECTED TO THE FUEL INJECTORS LOCATED IN THE INLET RAMPS, WHERE THE AIR STREAM TAKES IT INTO THE ENGINE INLET.

10 METER
30 FEET
5
20
15
10
0
0

TRUSS STRUCTURE ALONG CENTER LINE INSIDE THE TANK (TYP)

LH2 TANK

LH2 TANK

He BOTTLE

LOX TANK (2) FOR THE MAIN PUMP GAS GENERATOR

INTERNAL TRUSS STRUCTURE LOCATED AT CENTRAL VERTICAL PLANE OF TANK

AB

AB

LH2 TANK

LH2 FOR WALL COOLING

WING STRUCTURE

WING SPAR FITTINGS (TYP)

SEE LH2 TURBOPUMP CONSTRUCTION DRAWINGS BY W. F. LAREDO

WING CARRY-THROUGH STRUCTURE THROUGH TANKS (TYP)

LH2 TANK

LH2 TANK

SEE DETAIL AD IN SHEET 16

THE TRUSS STRUCTURE INSIDE EACH OF THE FOUR LH2 TANKS CONSTITUTES THE PRIMARY VEHICLE STRUCTURE TO WHICH WINGS AND OTHER SUB-STRUCTURES ARE ATTACHED.

ADVANCED ENGINEERING PROJECTS

PRELIMINARY DESIGN

Name of project: PRIMARY STRUCTURE		
HYPERSONIC SPACE PLANE (SCRAMJET)		
CARGO TRANSPORTATION SYSTEM BETWEEN EARTH AND THE SPACE STATION		
Design Engineer: WALTER F. LAREDO	Date: Feb. 1997	
Drawing Number: 1800	Sheet 15 of 21	

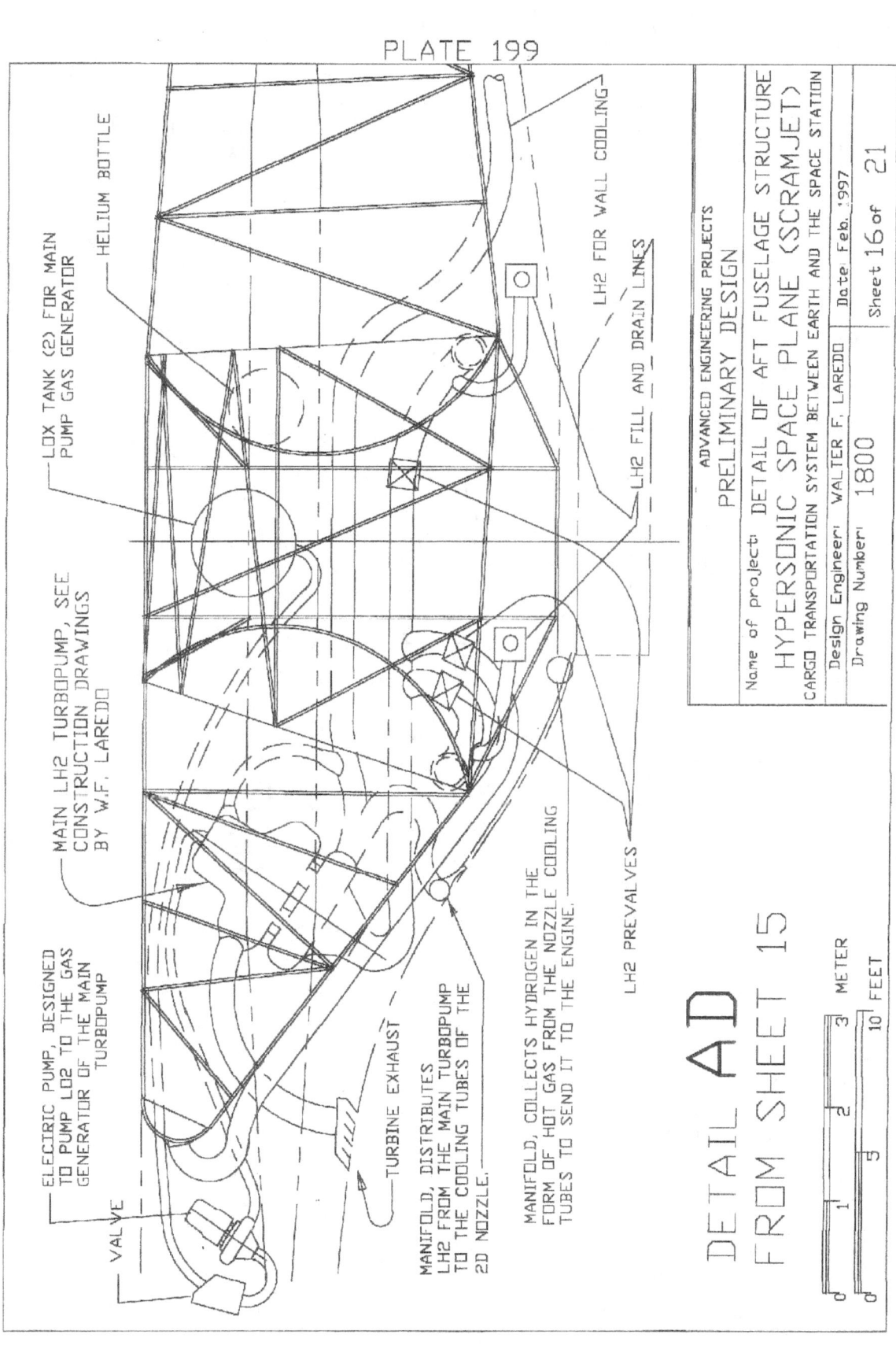

PLATE 199

HELIUM BOTTLE

LOX TANK (2) FOR MAIN
PUMP GAS GENERATOR

LH2 FOR WALL COOLING

LH2 FILL AND DRAIN LINES

MAIN LH2 TURBOPUMP, SEE
CONSTRUCTION DRAWINGS
BY W.F. LAREDO

ELECTRIC PUMP, DESIGNED
TO PUMP LO2 TO THE GAS
GENERATOR OF THE MAIN
TURBOPUMP

VALVE

TURBINE EXHAUST

MANIFOLD, DISTRIBUTES
LH2 FROM THE MAIN TURBOPUMP
TO THE COOLING TUBES OF THE
2D NOZZLE.

MANIFOLD, COLLECTS HYDROGEN IN THE
FORM OF HOT GAS FROM THE NOZZLE COOLING
TUBES TO SEND IT TO THE ENGINE.

LH2 PREVALVES

DETAIL AD
FROM SHEET 15

METER

FEET

ADVANCED ENGINEERING PROJECTS

PRELIMINARY DESIGN

Name of project: DETAIL OF AFT FUSELAGE STRUCTURE

HYPERSONIC SPACE PLANE (SCRAMJET)

CARGO TRANSPORTATION SYSTEM BETWEEN EARTH AND THE SPACE STATION

| Design Engineer: WALTER F. LAREDO | Date: Feb. 1997 |
| Drawing Number: 1800 | Sheet 16 of 21 |

PLATE 200

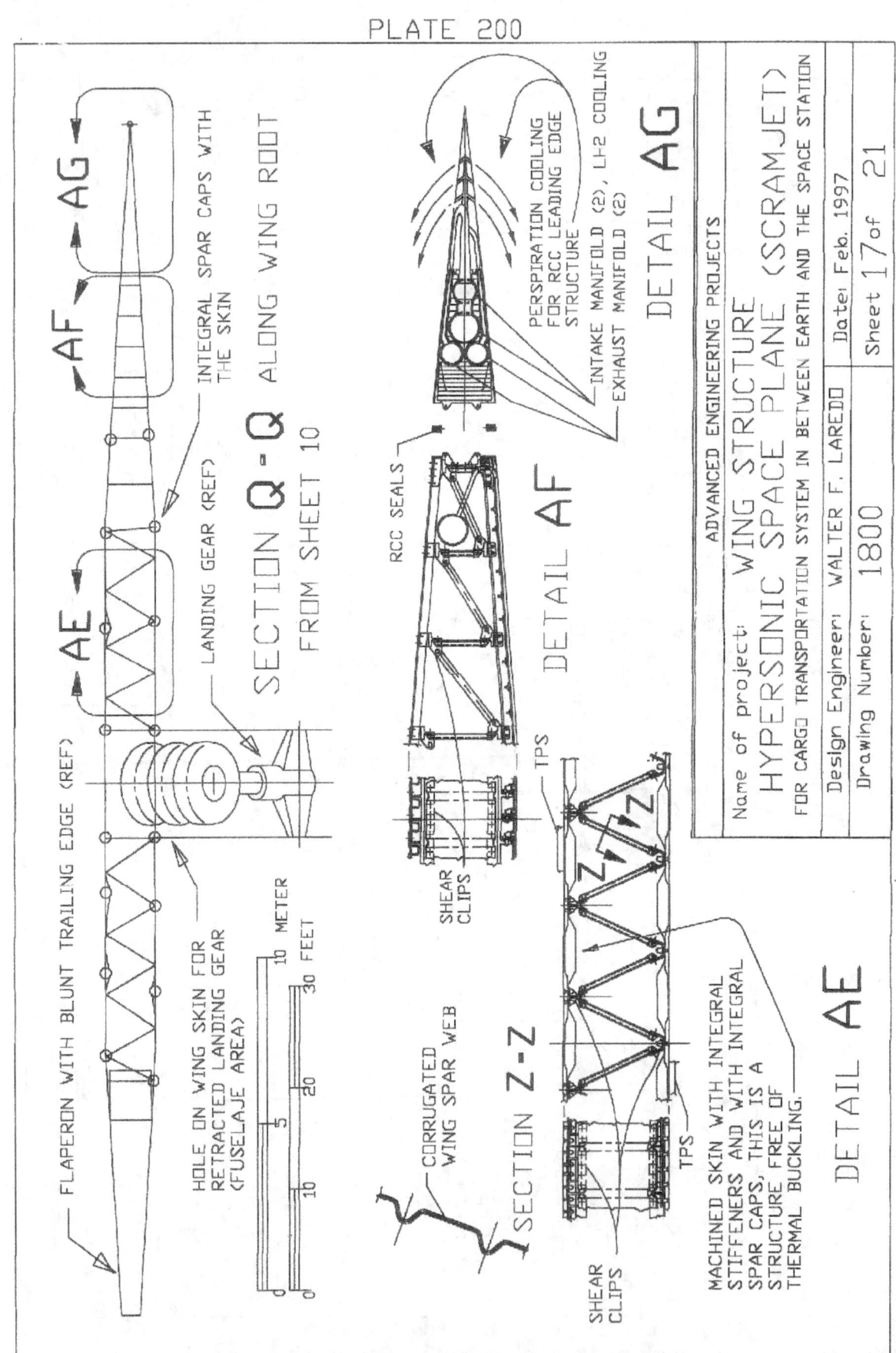

FLAPERON WITH BLUNT TRAILING EDGE (REF)

HOLE ON WING SKIN FOR RETRACTED LANDING GEAR (FUSELAJE AREA)

INTEGRAL SPAR CAPS WITH THE SKIN

LANDING GEAR (REF)

SECTION Q-Q ALONG WING ROOT
FROM SHEET 10

PERSPIRATION COOLING FOR RCC LEADING EDGE STRUCTURE

INTAKE MANIFOLD (2), LH2 COOLING
EXHAUST MANIFOLD (2)

DETAIL AG

RCC SEALS

DETAIL AF

TPS

SHEAR CLIPS

CORRUGATED WING SPAR WEB

SECTION Z-Z

TPS

SHEAR CLIPS

MACHINED SKIN WITH INTEGRAL STIFFENERS AND WITH INTEGRAL SPAR CAPS, THIS IS A STRUCTURE FREE OF THERMAL BUCKLING.

DETAIL AE

10 METER
30 FEET

ADVANCED ENGINEERING PROJECTS

Name of project: WING STRUCTURE
HYPERSONIC SPACE PLANE (SCRAMJET)
FOR CARGO TRANSPORTATION SYSTEM IN BETWEEN EARTH AND THE SPACE STATION

Design Engineer: WALTER F. LAREDO   Date: Feb. 1997

Drawing Number: 1800   Sheet 17 of 21

PLATE 201

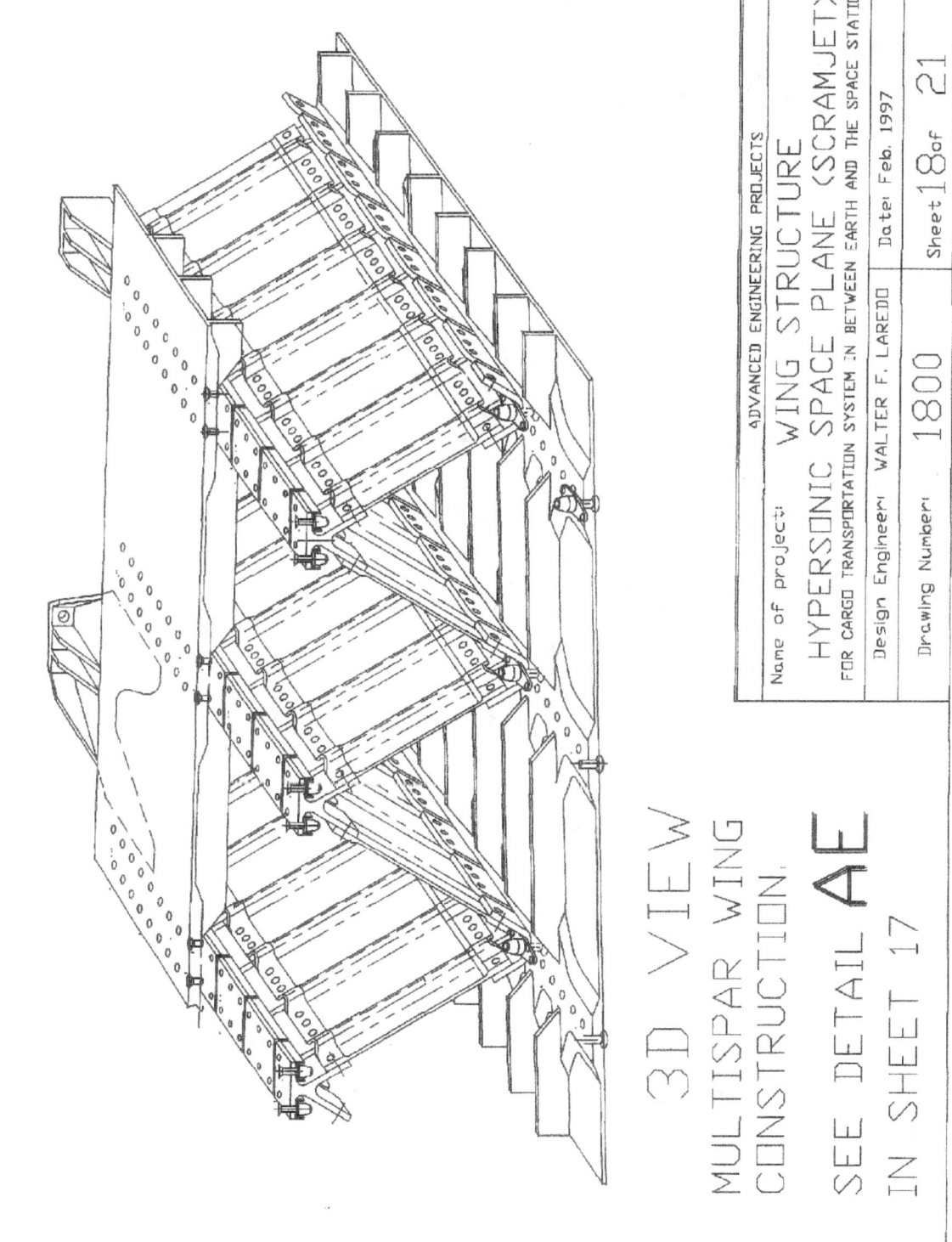

3D VIEW
MULTISPAR WING
CONSTRUCTION.

SEE DETAIL AE
IN SHEET 17

ADVANCED ENGINEERING PROJECTS

Name of project: WING STRUCTURE

HYPERSONIC SPACE PLANE (SCRAMJET)

FOR CARGO TRANSPORTATION SYSTEM IN BETWEEN EARTH AND THE SPACE STATION

Design Engineer: WALTER F. LAREDO	Date: Feb. 1997
Drawing Number: 1800	Sheet 18 of 21

PLATE 202

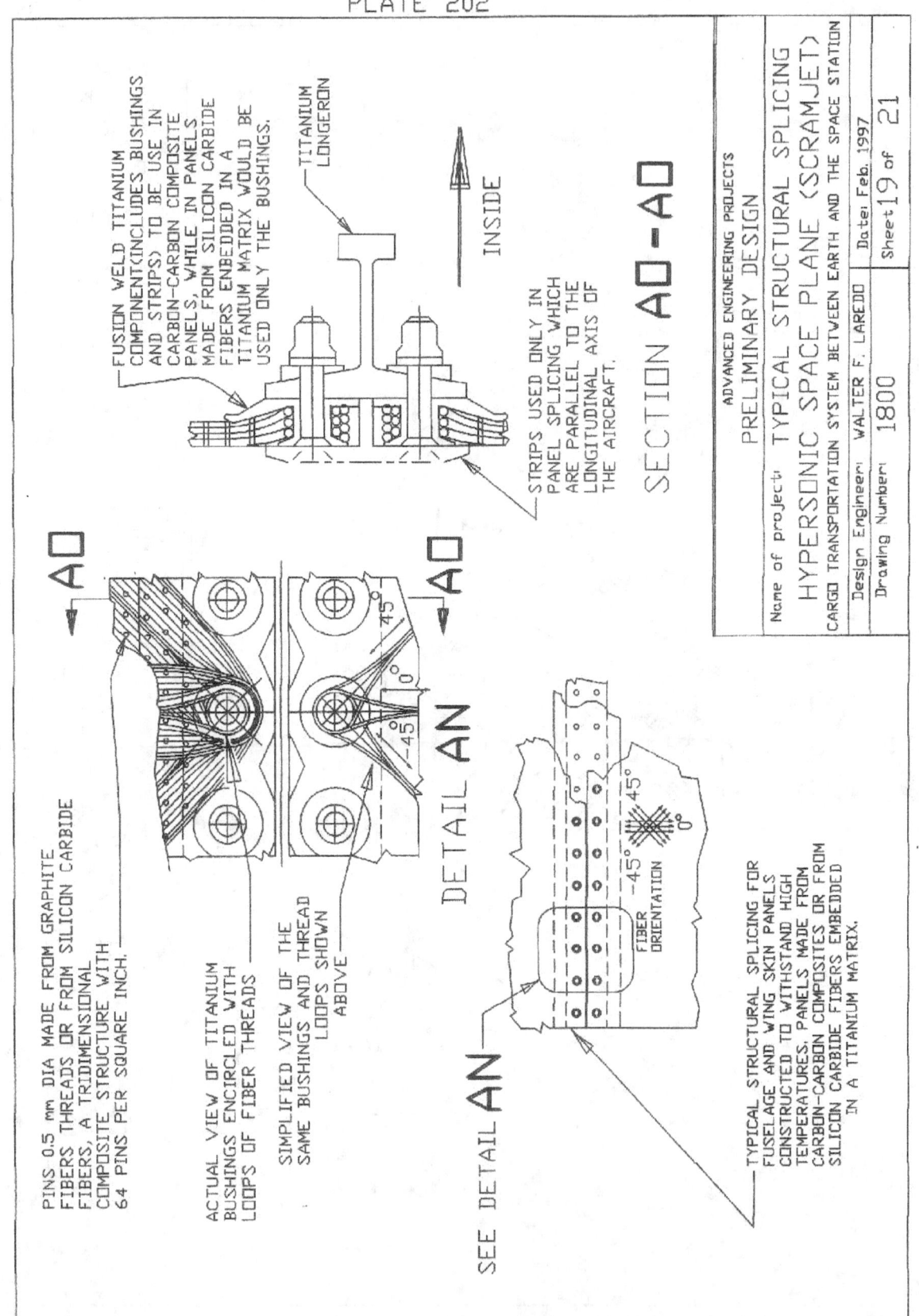

FUSION WELD TITANIUM COMPONENT(INCLUDES BUSHINGS AND STRIPS) TO BE USE IN CARBON-CARBON COMPOSITE PANELS, WHILE IN PANELS MADE FROM SILICON CARBIDE FIBERS EMBEDDED IN A TITANIUM MATRIX WOULD BE USED ONLY THE BUSHINGS.

TITANIUM LONGERON

INSIDE

STRIPS USED ONLY IN PANEL SPLICING WHICH ARE PARALLEL TO THE LONGITUDINAL AXIS OF THE AIRCRAFT.

SECTION AO-AO

PINS 0.5 mm DIA MADE FROM GRAPHITE FIBERS THREADS OR FROM SILICON CARBIDE FIBERS, A TRIDIMENSIONAL COMPOSITE STRUCTURE WITH 64 PINS PER SQUARE INCH.

ACTUAL VIEW OF TITANIUM BUSHINGS ENCIRCLED WITH LOOPS OF FIBER THREADS

SIMPLIFIED VIEW OF THE SAME BUSHINGS AND THREAD LOOPS SHOWN ABOVE

AO

AO

45°
0°
-45°

DETAIL AN

SEE DETAIL AN

45°
-45°
0°

FIBER ORIENTATION

TYPICAL STRUCTURAL SPLICING FOR FUSELAGE AND WING SKIN PANELS CONSTRUCTED TO WITHSTAND HIGH TEMPERATURES. PANELS MADE FROM CARBON-CARBON COMPOSITES OR FROM SILICON CARBIDE FIBERS EMBEDDED IN A TITANIUM MATRIX.

ADVANCED ENGINEERING PROJECTS
PRELIMINARY DESIGN

Name of project: TYPICAL STRUCTURAL SPLICING
HYPERSONIC SPACE PLANE (SCRAMJET)
CARGO TRANSPORTATION SYSTEM BETWEEN EARTH AND THE SPACE STATION

| Design Engineer: WALTER F. LAREDO | Date: Feb. 1997 |
| Drawing Number: 1800 | Sheet 19 of 21 |

PLATE 203

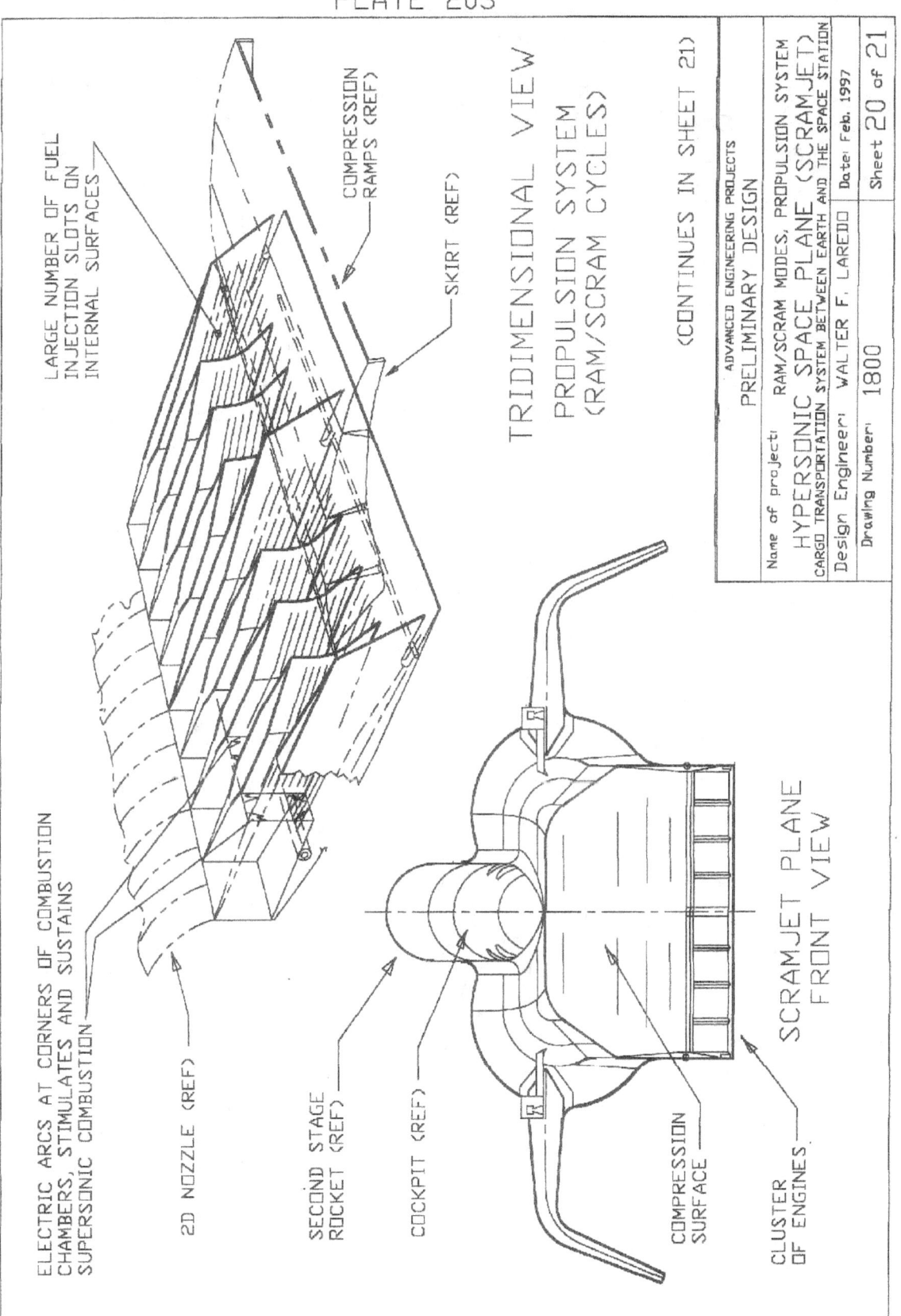

LARGE NUMBER OF FUEL
INJECTION SLOTS ON
INTERNAL SURFACES

COMPRESSION
RAMPS (REF)

SKIRT (REF)

TRIDIMENSIONAL VIEW

PROPULSION SYSTEM
(RAM/SCRAM CYCLES)

(CONTINUES IN SHEET 21)

ELECTRIC ARCS AT CORNERS OF COMBUSTION
CHAMBERS, STIMULATES AND SUSTAINS
SUPERSONIC COMBUSTION

2D NOZZLE (REF)

SECOND STAGE
ROCKET (REF)

COCKPIT (REF)

COMPRESSION
SURFACE

CLUSTER
OF ENGINES

SCRAMJET PLANE
FRONT VIEW

ADVANCED ENGINEERING PROJECTS		
PRELIMINARY DESIGN		
Name of project: RAM/SCRAM MODES, PROPULSION SYSTEM		
HYPERSONIC SPACE PLANE (SCRAMJET)		
CARGO TRANSPORTATION SYSTEM BETWEEN EARTH AND THE SPACE STATION		
Design Engineer: WALTER F. LAREDO	Date: Feb. 1997	
Drawing Number: 1800	Sheet 20 of 21	

PLATE 204

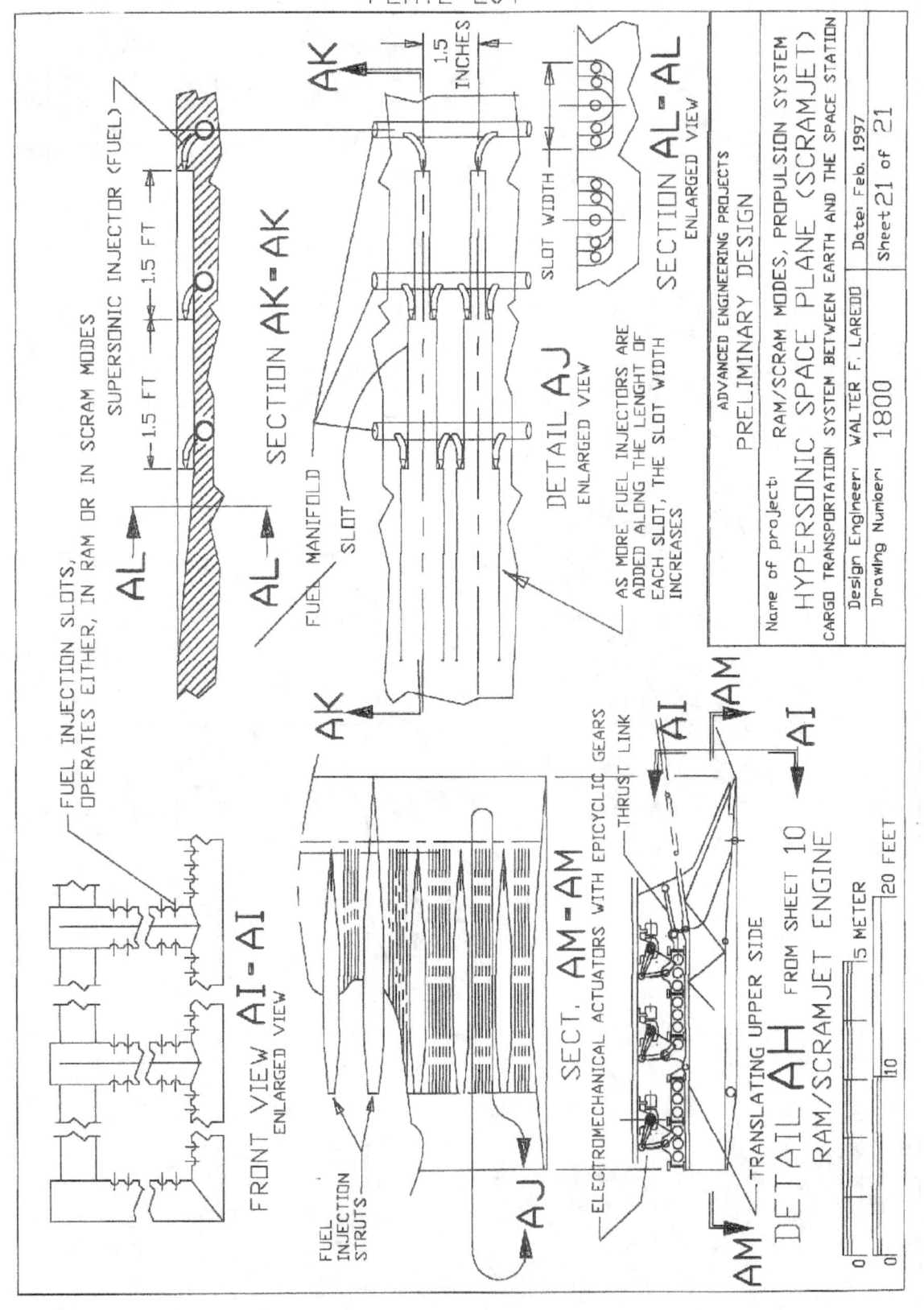

FUEL INJECTION SLOTS, OPERATES EITHER, IN RAM OR IN SCRAM MODES

SUPERSONIC INJECTOR (FUEL)

1.5 FT   1.5 FT   1.5 FT

SECTION AK-AK

FUEL MANIFOLD

SLOT

DETAIL AJ
ENLARGED VIEW

AS MORE FUEL INJECTORS ARE ADDED ALONG THE LENGHT OF EACH SLOT, THE SLOT WIDTH INCREASES

SLOT WIDTH

1.5 INCHES

SECTION AL-AL
ENLARGED VIEW

FRONT VIEW AI-AI
ENLARGED VIEW

FUEL INJECTION STRUTS

SECT. AM-AM

ELECTROMECHANICAL ACTUATORS WITH EPICYCLIC GEARS

THRUST LINK

TRANSLATING UPPER SIDE

DETAIL AH FROM SHEET 10
RAM/SCRAMJET ENGINE

5 METER
10
0   20 FEET

ADVANCED ENGINEERING PROJECTS		
RAM/SCRAM MODES, PROPULSION SYSTEM		
PRELIMINARY DESIGN		
Name of project:		
HYPERSONIC SPACE PLANE (SCRAMJET)		
CARGO TRANSPORTATION SYSTEM BETWEEN EARTH AND THE SPACE STATION		
Design Engineer: WALTER F. LAREDO	Date: Feb. 1997	
Drawing Number: 1800	Sheet 21 of 21	

PLATE 205

AEROSPACE PROJECT
PRELIMINARY DESIGN

PIGGY-BACK
SPACE SHUTTLE
(REF)

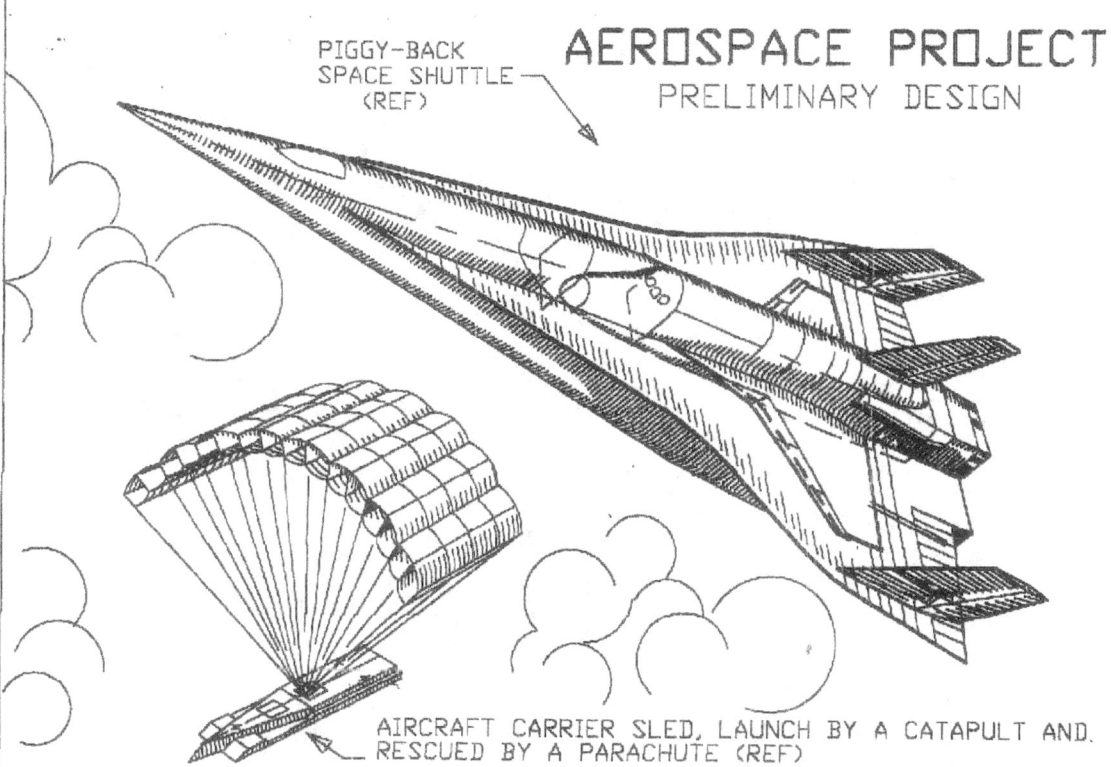

AIRCRAFT CARRIER SLED, LAUNCH BY A CATAPULT AND.
RESCUED BY A PARACHUTE (REF)

INTERNATIONAL SPACE BASES ON HIGH PLATEAUS OF THE WORLD SHOULD
INCLUDE TILTED LAUNCHING RAMPS WITH EXTRALONG TWO STAGGED CATAPULTS,
THE SAME BASES SHOULD HAVE ALSO LONG LANDING STRIPS ON THE DRIED
SALTED LAKES OF THE PLATEAU FOR THE RETURNING GLIDDING BOOSTER PLANE
AND LATER OF THE SPACE SHUTTLE.

NOTE: FOR CLARITY IN THE FOLLOWING DRAWINGS WAS NOT SHOWN THE
EXTERNAL THERMOPROTECCION SYSTEM OF THE SPACE VEHICLES, ALSO WAS
NOT SHOWN THE FOAM INSULATION THAT COVERS THE TANKS AND OTHER
CRYOGENIC COMPONENTS.

ADVANCED ENGINEERING PROJECTS	
Name of project: GENERAL ARRANGEMENT	
HYPERSONIC PIGGYBACK PLANE	
TRANSPORTATION SYSTEM BETWEEN EARTH AND THE SPACE STATION	
Design Engineer: WALTER F. LAREDO	Date: Feb. 1997
Drawing Number: 1900 SERIES	Sheet 1 of 12

PLATE 206

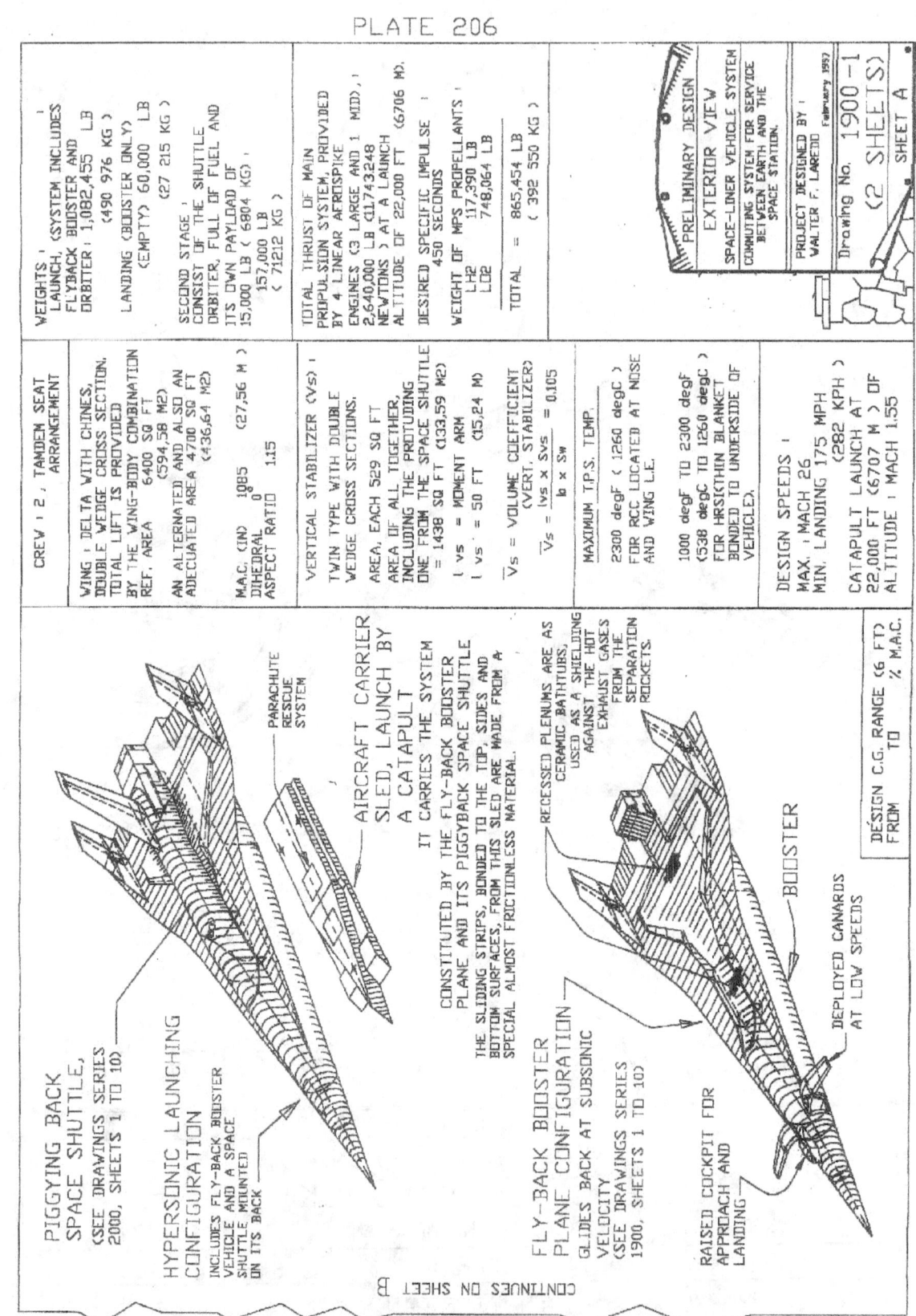

WEIGHTS : LAUNCH, (SYSTEM INCLUDES FLYBACK BOOSTER AND ORBITER : 1,082,455 LB
(490 976 KG)

LANDING (BOOSTER ONLY)
(EMPTY) 60,000 LB
(27 215 KG)

SECOND STAGE :
CONSIST OF THE SHUTTLE ORBITER, FULL OF FUEL AND ITS OWN PAYLOAD OF
15,000 LB (6804 KG) :
157,000 LB
(71212 KG)

TOTAL THRUST OF MAIN PROPULSION SYSTEM, PROVIDED BY 4 LINEAR AEROSPIKE ENGINES (3 LARGE AND 1 MID.) :
2,640,000 LB (11743248 NEWTONS) AT A LAUNCH ALTITUDE OF 22,000 FT (6706 M).

DESIRED SPECIFIC IMPULSE :
450 SECONDS

WEIGHT OF MPS PROPELLANTS :
LH2   117,390 LB
LO2   748,064 LB

TOTAL = 865,454 LB
(392 550 KG)

PRELIMINARY DESIGN
EXTERIOR VIEW
SPACE-LINER VEHICLE SYSTEM
COMMUTING SYSTEM FOR SERVICE BETWEEN EARTH AND THE SPACE STATION.

PROJECT DESIGNED BY :
WALTER F. LAREDO

Drawing No. 1900-1
February 1997

(2 SHEETS)
SHEET A

CREW : 2 , TAMDEN SEAT ARRANGEMENT

WING : DELTA WITH CHINES, DOUBLE WEDGE CROSS SECTION.
TOTAL LIFT IS PROVIDED BY THE WING-BODY COMBINATION
REF. AREA   6400 SQ FT
(594.58 M2)

AN ALTERNATED AND ALSO AN ADEQUATED AREA 4700 SQ FT
(436.64 M2)

M.A.C. (IN)   1085   (27.56 M)
DIHEDRAL   0°
ASPECT RATIO   1.15

VERTICAL STABILIZER (Vs) :

TWIN TYPE WITH DOUBLE WEDGE CROSS SECTIONS.

AREA, EACH 529 SQ FT
AREA OF ALL TOGETHER, INCLUDING THE PROTUDING ONE FROM THE SPACE SHUTTLE
= 1438 SQ FT (133.59 M2)

l vs = MOMENT ARM
l vs = 50 FT   (15.24 M)

$\overline{V}s$ = VOLUME COEFFICIENT
(VERT. STABILIZER)

$\overline{V}s = \dfrac{lvs \times Svs}{lo \times Sw} = 0.105$

MAXIMUM T.P.S. TEMP.

2300 degF (1260 degC)
FOR RCC LOCATED AT NOSE AND WING L.E.

1000 degF TO 2300 degF
(538 degC TO 1260 degC)
FOR HRSI(THIN BLANKET BONDED TO UNDERSIDE OF VEHICLE).

DESIGN SPEEDS :
MAX. : MACH 26
MIN. LANDING 175 MPH
(282 KPH)

CATAPULT LAUNCH AT 22,000 FT (6707 M) OF ALTITUDE : MACH 1.55

PIGGYING BACK SPACE SHUTTLE,
(SEE DRAWINGS SERIES 2000, SHEETS 1 TO 10)

HYPERSONIC LAUNCHING CONFIGURATION
INCLUDES FLY-BACK BOOSTER VEHICLE AND A SPACE SHUTTLE MOUNTED ON ITS BACK

PARACHUTE RESCUE SYSTEM

AIRCRAFT CARRIER SLED, LAUNCH BY A CATAPULT
IT CARRIES THE SYSTEM

CONSTITUTED BY THE FLY-BACK BOOSTER PLANE AND ITS PIGGYBACK SPACE SHUTTLE
THE SLIDING STRIPS, BONDED TO THE TOP, SIDES AND BOTTOM SURFACES, ARE MADE FROM A SPECIAL ALMOST FRICTIONLESS MATERIAL.

RECESSED PLENUMS ARE AS CERAMIC BATHTUBS, USED AS A SHIELDING AGAINST THE HOT EXHAUST GASES FROM THE SEPARATION ROCKETS.

BOOSTER

FLY-BACK BOOSTER PLANE CONFIGURATION
GLIDES BACK AT SUBSONIC VELOCITY
(SEE DRAWINGS SERIES 1900, SHEETS 1 TO 10)

RAISED COCKPIT FOR APPROACH AND LANDING

DEPLOYED CANARDS AT LOW SPEEDS

DESIGN C.G. RANGE (6 FT) FROM   TO   % M.A.C.

CONTINUES ON SHEET B

PLATE 207

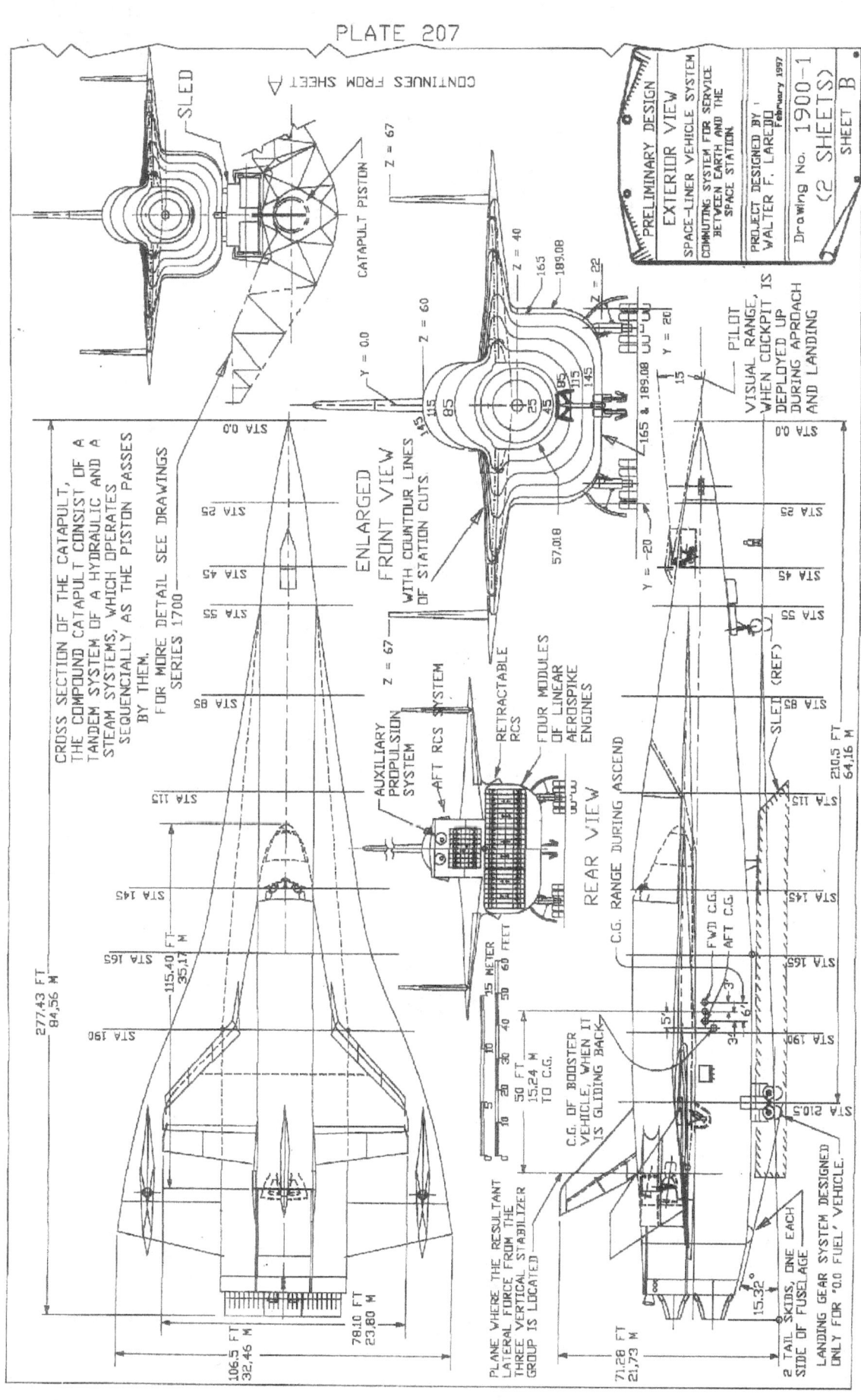

CONTINUES FROM SHEET A

SLED

CATAPULT PISTON

Z = 67

ENLARGED FRONT VIEW
WITH CONTOUR LINES OF STATION CUTS

PRELIMINARY DESIGN
EXTERIOR VIEW
SPACE-LINER VEHICLE SYSTEM
COMMUTING SYSTEM FOR SERVICE
BETWEEN EARTH AND THE
SPACE STATION

PROJECT DESIGNED BY
WALTER F. LAREDO
February 1997

Drawing No. 1900-1
(2 SHEETS)
SHEET B

PILOT VISUAL RANGE, WHEN COCKPIT IS DEPLOYED UP DURING APPROACH AND LANDING

CROSS SECTION OF THE CATAPULT, THE COMPOUND CATAPULT CONSIST OF A TANDEM SYSTEM OF A HYDRAULIC AND A STEAM SYSTEMS, WHICH OPERATES SEQUENCIALLY AS THE PISTON PASSES BY THEM.
FOR MORE DETAIL SEE DRAWINGS SERIES 1700

AUXILIARY PROPULSION SYSTEM
AFT RCS SYSTEM
RETRACTABLE RCS
FOUR MODULES OF LINEAR AEROSPIKE ENGINES

REAR VIEW

C.G. RANGE DURING ASCEND

FWD C.G.
AFT C.G.

SLED (REF)

210.5 FT
64.16 M

PLANE WHERE THE RESULTANT LATERAL FORCE FROM THE THREE VERTICAL STABILIZER GROUP IS LOCATED

C.G. OF BOOSTER VEHICLE, WHEN IT IS GLIDING BACK

277.43 FT
84.56 M

115.40 FT
35.17 M

106.5 FT
32.46 M

78.10 FT
23.80 M

71.28 FT
21.73 M

15.32°

2 TAIL SKIDS, ONE EACH SIDE OF FUSELAGE

LANDING GEAR SYSTEM DESIGNED ONLY FOR "0.0 FUEL" VEHICLE.

C.G. OF BOOSTER VEHICLE, WHEN IT IS GLIDING BACK
50 FT
15.24 M
TO C.G.

PLATE 208

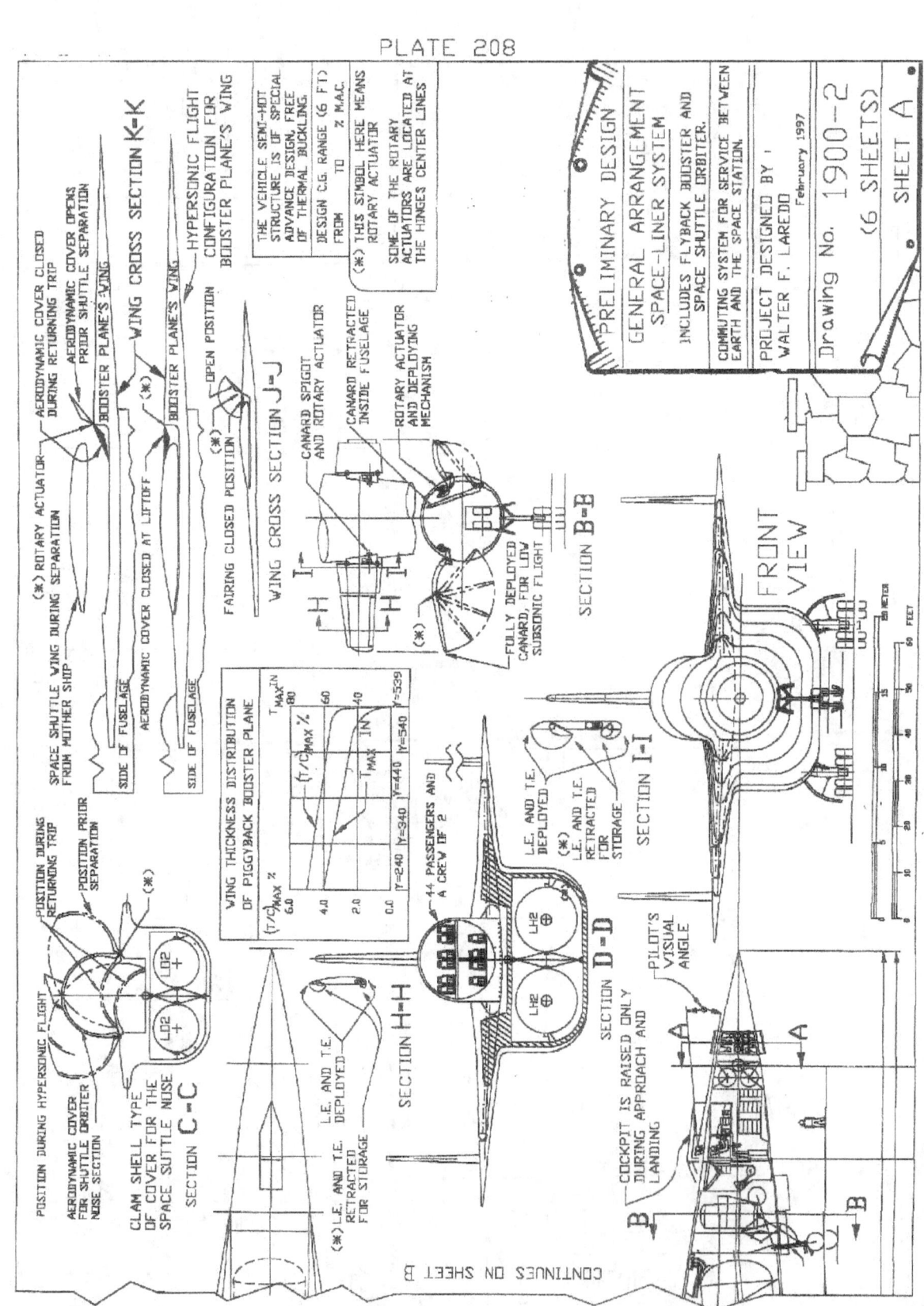

CONTINUES ON SHEET B

PLATE 209

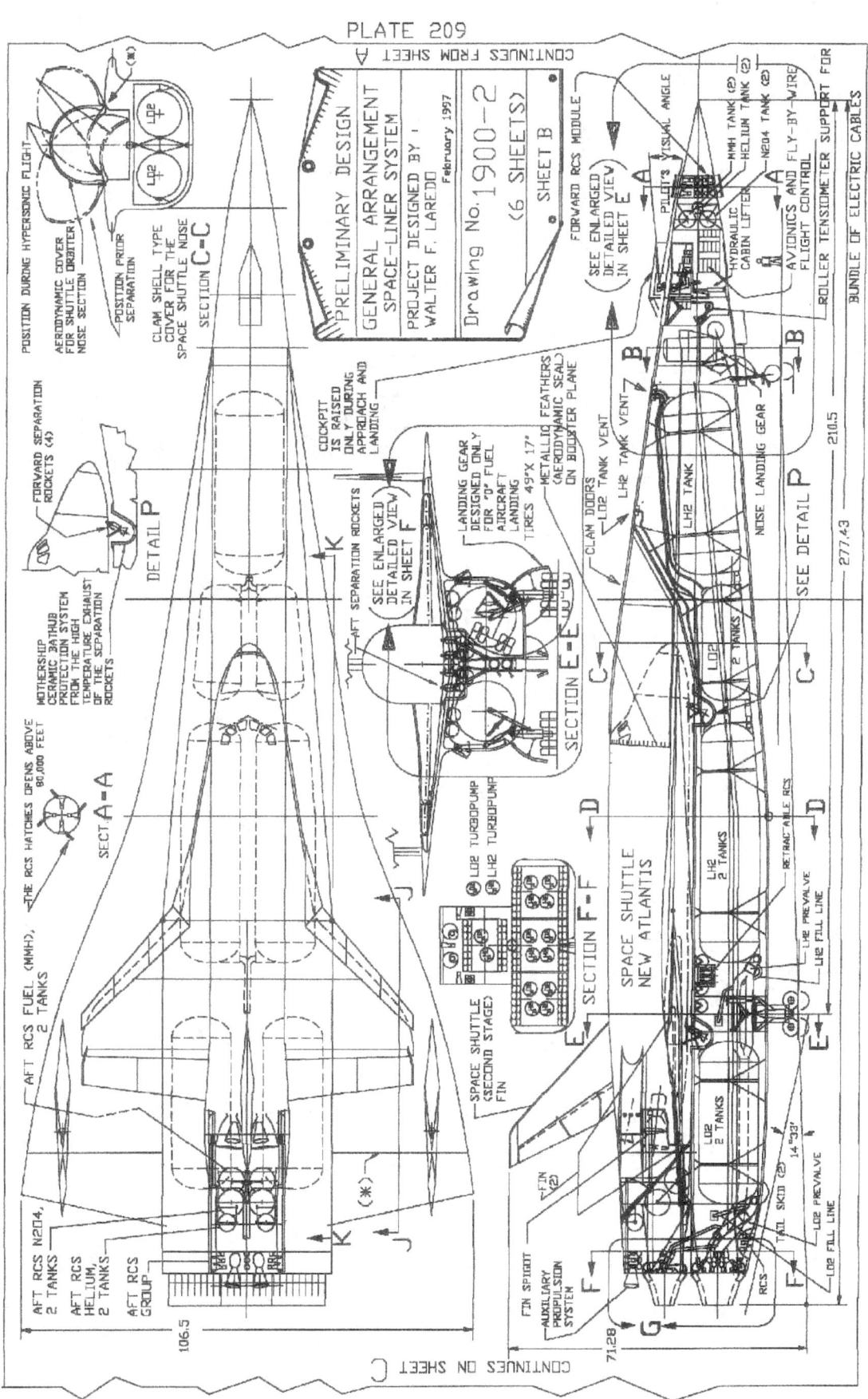

PRELIMINARY DESIGN
GENERAL ARRANGEMENT
SPACE-LINER SYSTEM

PROJECT DESIGNED BY :
WALTER F. LAREDO   February 1997

Drawing No. 1900-2
(6 SHEETS)   SHEET B

CONTINUES FROM SHEET A

CONTINUES ON SHEET C

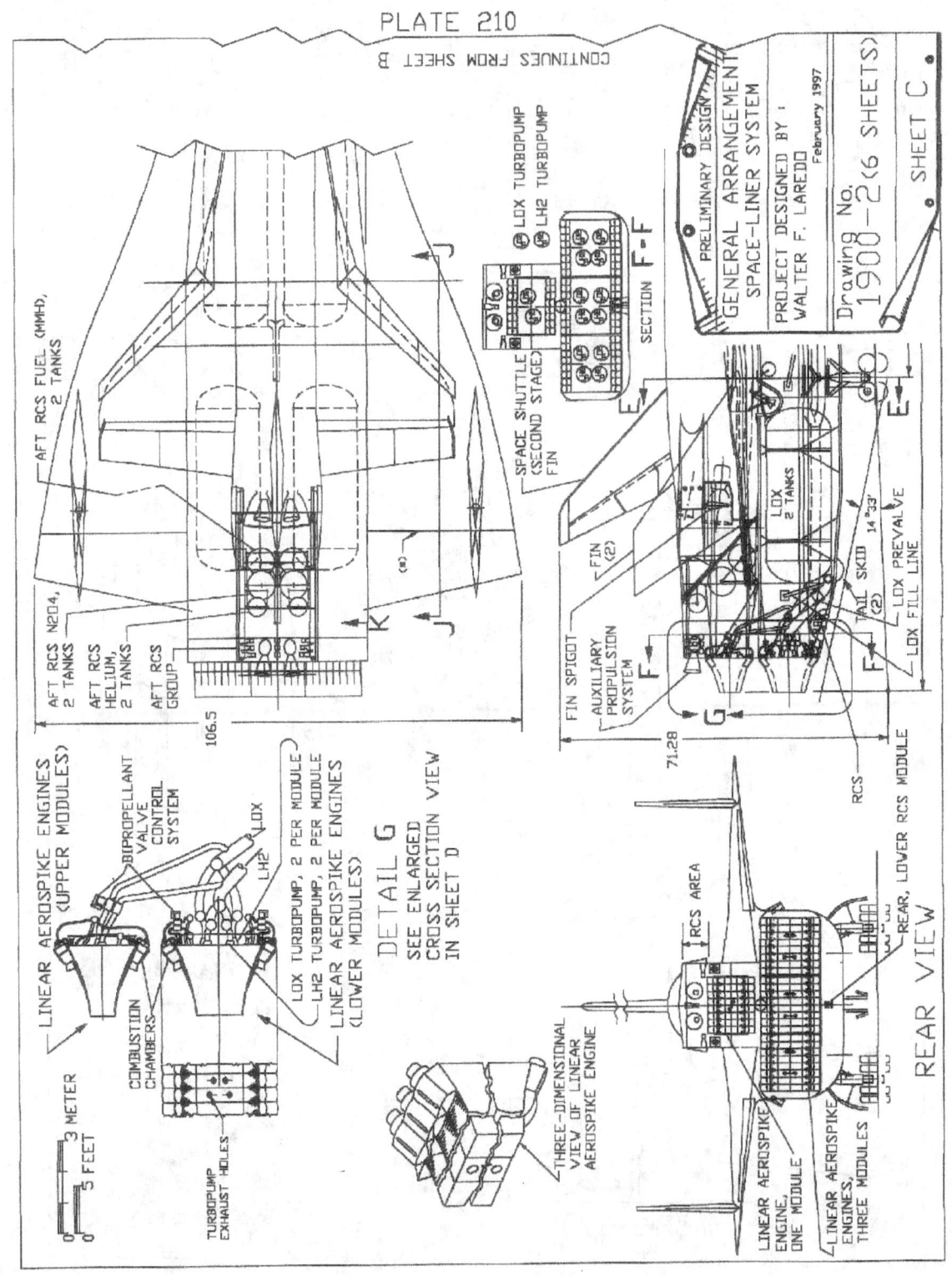

PLATE 211

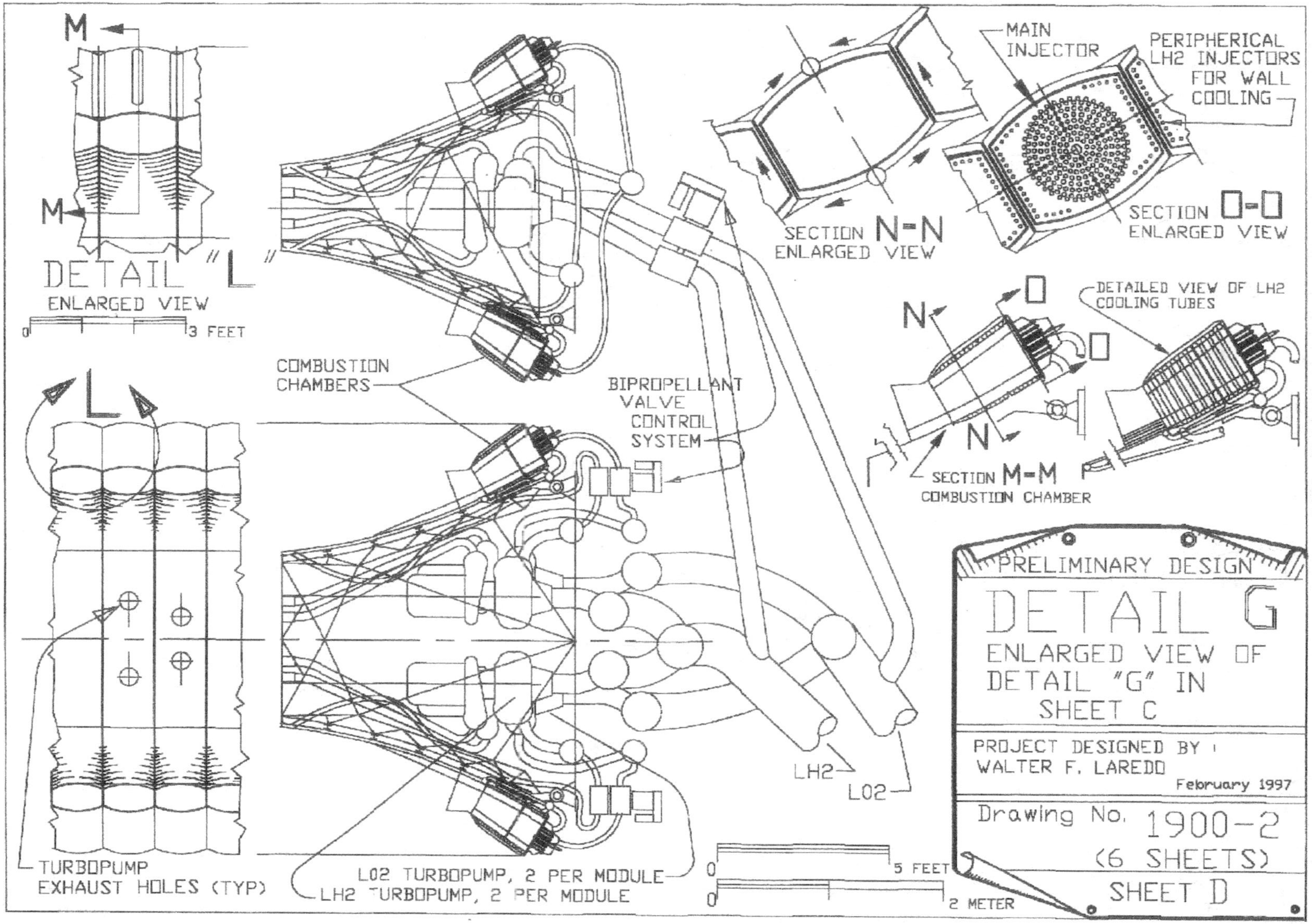

PERIPHERICAL
LH2 INJECTORS
FOR WALL
COOLING

MAIN
INJECTOR

SECTION O-O
ENLARGED VIEW

DETAILED VIEW OF LH2
COOLING TUBES

SECTION M-M
COMBUSTION CHAMBER

SECTION N-N
ENLARGED VIEW

PRELIMINARY DESIGN

DETAIL G
ENLARGED VIEW OF
DETAIL "G" IN
SHEET C

PROJECT DESIGNED BY :
WALTER F. LAREDO

February 1997

Drawing No. 1900-2
(6 SHEETS)
SHEET D

2 METER

5 FEET

LH2

L02

BIPROPELLANT
VALVE
CONTROL
SYSTEM

COMBUSTION
CHAMBERS

L02 TURBOPUMP, 2 PER MODULE
LH2 TURBOPUMP, 2 PER MODULE

DETAIL "L"
ENLARGED VIEW

3 FEET

TURBOPUMP
EXHAUST HOLES (TYP)

PLATE 212

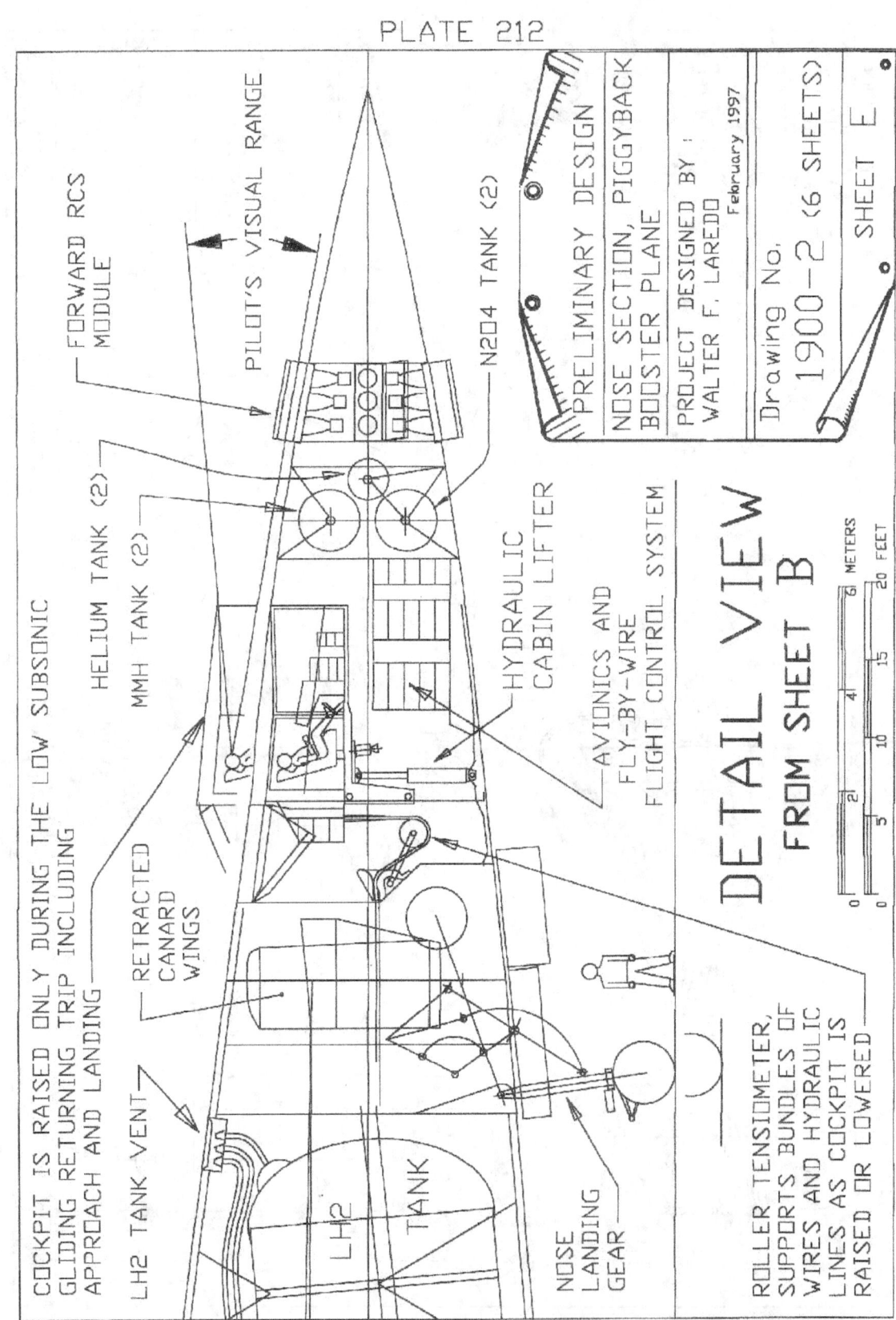

COCKPIT IS RAISED ONLY DURING THE LOW SUBSONIC
GLIDING RETURNING TRIP INCLUDING
APPROACH AND LANDING

LH2 TANK VENT

RETRACTED
CANARD
WINGS

FORWARD RCS
MODULE

HELIUM TANK (2)

MMH TANK (2)

PILOT'S VISUAL RANGE

N2O4 TANK (2)

HYDRAULIC
CABIN LIFTER

AVIONICS AND
FLY-BY-WIRE
FLIGHT CONTROL SYSTEM

LH2
TANK

NOSE
LANDING
GEAR

ROLLER TENSIOMETER,
SUPPORTS BUNDLES OF
WIRES AND HYDRAULIC
LINES AS COCKPIT IS
RAISED OR LOWERED

PRELIMINARY DESIGN

NOSE SECTION, PIGGYBACK
BOOSTER PLANE

PROJECT DESIGNED BY :
WALTER F. LAREDO
February 1997

Drawing No.
1900-2 (6 SHEETS)

SHEET E

DETAIL VIEW
FROM SHEET B

METERS
20 FEET

PLATE 213

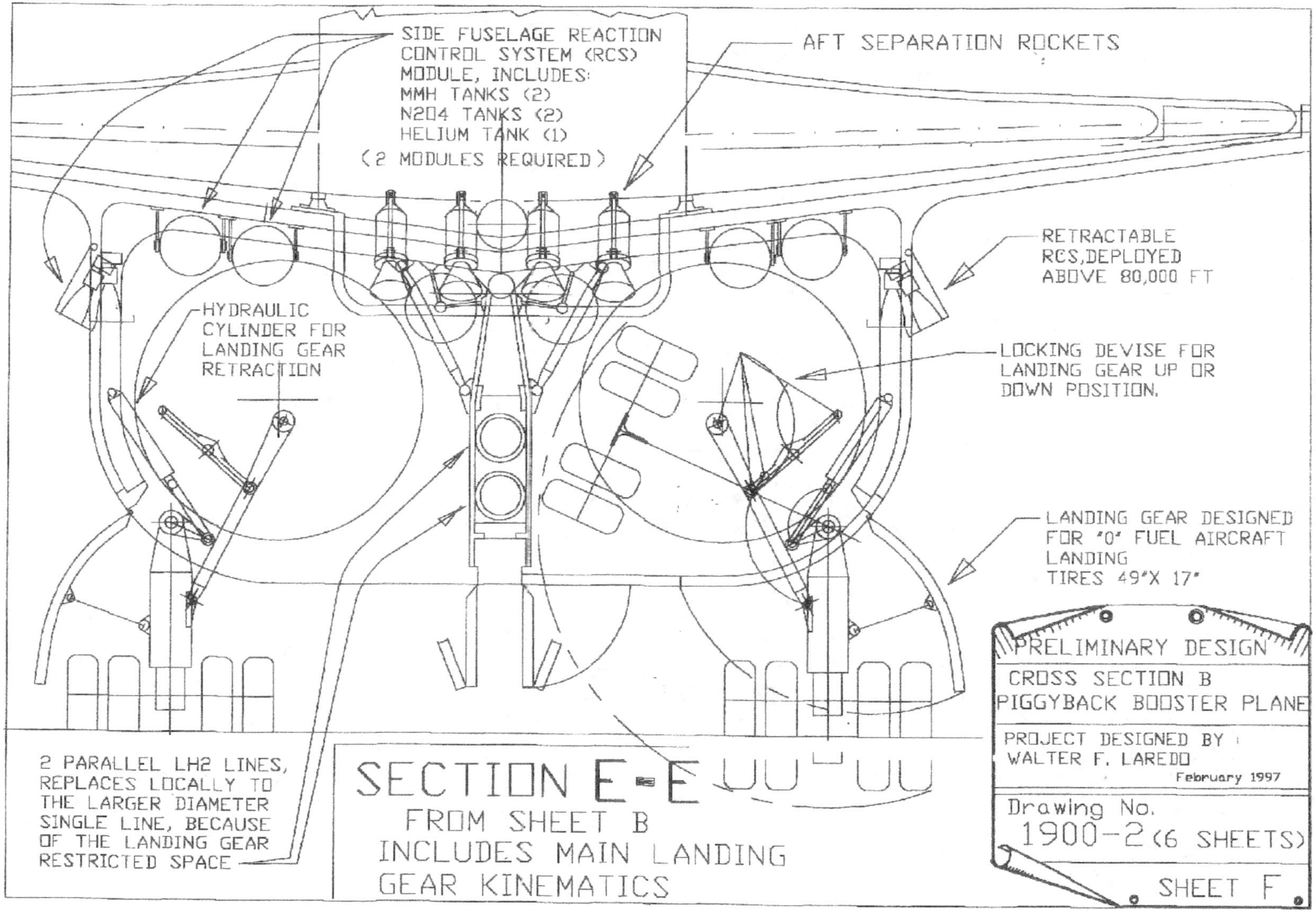

AFT SEPARATION ROCKETS

RETRACTABLE
RCS,DEPLOYED
ABOVE 80,000 FT

LOCKING DEVISE FOR
LANDING GEAR UP OR
DOWN POSITION.

LANDING GEAR DESIGNED
FOR '0' FUEL AIRCRAFT
LANDING
TIRES 49"X 17"

PRELIMINARY DESIGN

CROSS SECTION B
PIGGYBACK BOOSTER PLANE

PROJECT DESIGNED BY :
WALTER F. LAREDO
February 1997

Drawing No.
1900-2 (6 SHEETS)

SHEET F

SIDE FUSELAGE REACTION
CONTROL SYSTEM (RCS)
MODULE, INCLUDES:
MMH TANKS (2)
N2O4 TANKS (2)
HELIUM TANK (1)
(2 MODULES REQUIRED)

HYDRAULIC
CYLINDER FOR
LANDING GEAR
RETRACTION

SECTION E-E

FROM SHEET B
INCLUDES MAIN LANDING
GEAR KINEMATICS

2 PARALLEL LH2 LINES,
REPLACES LOCALLY TO
THE LARGER DIAMETER
SINGLE LINE, BECAUSE
OF THE LANDING GEAR
RESTRICTED SPACE

PLATE 214

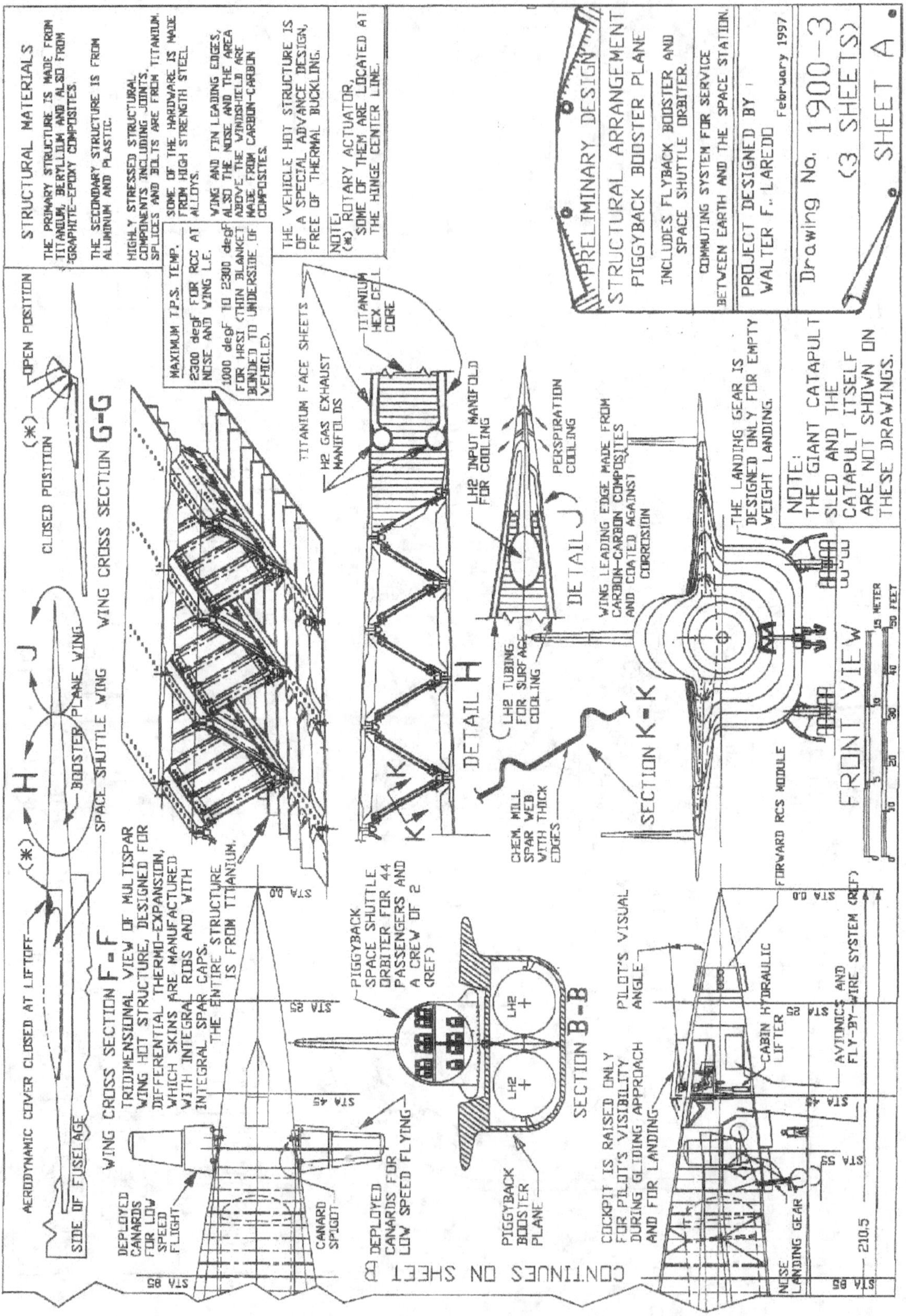

STRUCTURAL MATERIALS

THE PRIMARY STRUCTURE IS MADE FROM TITANIUM, BERYLLIUM AND ALSO FROM GRAPHITE-EPOXY COMPOSITES.

THE SECONDARY STRUCTURE IS FROM ALUMINUM AND PLASTIC.

HIGHLY STRESSED STRUCTURAL COMPONENTS INCLUDING JOINTS, SPLICES AND BOLTS ARE FROM TITANIUM.

SOME OF THE HARDWARE IS MADE FROM HIGH STRENGTH STEEL ALLOYS.

WING AND FIN LEADING EDGES, ALSO THE NOSE AND THE AREA ABOVE THE WINDSHIELD ARE MADE FROM CARBON-CARBON COMPOSITES.

THE VEHICLE HOT STRUCTURE IS OF A SPECIAL ADVANCE DESIGN, FREE OF THERMAL BUCKLING.

NOTE:
(*) ROTARY ACTUATOR, SOME OF THEM ARE LOCATED AT THE HINGE CENTER LINE.

PRELIMINARY DESIGN

STRUCTURAL ARRANGEMENT
PIGGYBACK BOOSTER PLANE

INCLUDES FLYBACK BOOSTER AND SPACE SHUTTLE ORBITER.

COMMUTING SYSTEM FOR SERVICE BETWEEN EARTH AND THE SPACE STATION.

PROJECT DESIGNED BY :
WALTER F. LAREDO   February 1997

Drawing No. 1900-3
(3 SHEETS)

SHEET A

AERODYNAMIC COVER CLOSED AT LIFTOFF

(*) OPEN POSITION

CLOSED POSITION

SIDE OF FUSELAGE

BOOSTER PLANE WING

SPACE SHUTTLE WING

WING CROSS SECTION G-G

WING CROSS SECTION F-F

TRIDIMENSIONAL VIEW OF MULTISPAR WING HOT STRUCTURE, DESIGNED FOR DIFFERENTIAL THERMO-EXPANSION, WHICH SKINS ARE MANUFACTURED WITH INTEGRAL RIBS AND WITH INTEGRAL SPAR CAPS.
THE ENTIRE STRUCTURE IS FROM TITANIUM.

MAXIMUM T.P.S. TEMP.
2300 degF FOR RCC AT NOSE AND WING L.E.
1000 degF TO 2300 degF FOR HRSI (THIN BLANKET BONDED TO UNDERSIDE OF VEHICLE).

TITANIUM FACE SHEETS

TITANIUM HEX CELL CORE

H2 GAS EXHAUST MANIFOLDS

LH2 INPUT MANIFOLD FOR COOLING

PERSPIRATION COOLING

DETAIL J

WING LEADING EDGE MADE FROM CARBON-CARBON COMPOSITES AND COATED AGAINST CORROSION

LH2 TUBING FOR SURFACE COOLING

DETAIL H

CHEM. MILL SPAR WEB WITH THICK EDGES

SECTION K-K

THE LANDING GEAR IS DESIGNED ONLY FOR EMPTY WEIGHT LANDING.

NOTE:
THE GIANT CATAPULT SLED AND THE CATAPULT ITSELF ARE NOT SHOWN ON THESE DRAWINGS.

FORWARD RCS MODULE

FRONT VIEW

DEPLOYED CANARDS FOR LOW SPEED FLIGHT

CANARD SPIGOT

STA 00

STA 25

STA 45

DEPLOYED CANARDS FOR LOW SPEED FLYING

PIGGYBACK SPACE SHUTTLE ORBITER FOR 44 PASSENGERS AND A CREW OF 2 (REF)

PIGGYBACK BOOSTER PLANE

SECTION B-B

LH2

LH2

PILOT'S VISUAL ANGLE

COCKPIT IS RAISED ONLY FOR PILOT'S VISIBILITY DURING GLIDING APPROACH AND FOR LANDING

CABIN HYDRAULIC LIFTER

AVIONICS AND FLY-BY-WIRE SYSTEM (REF)

NOSE LANDING GEAR

210.5

STA 00

STA 45

STA 55

STA 85

STA 85

STA 85

CONTINUES ON SHEET B

PLATE 215

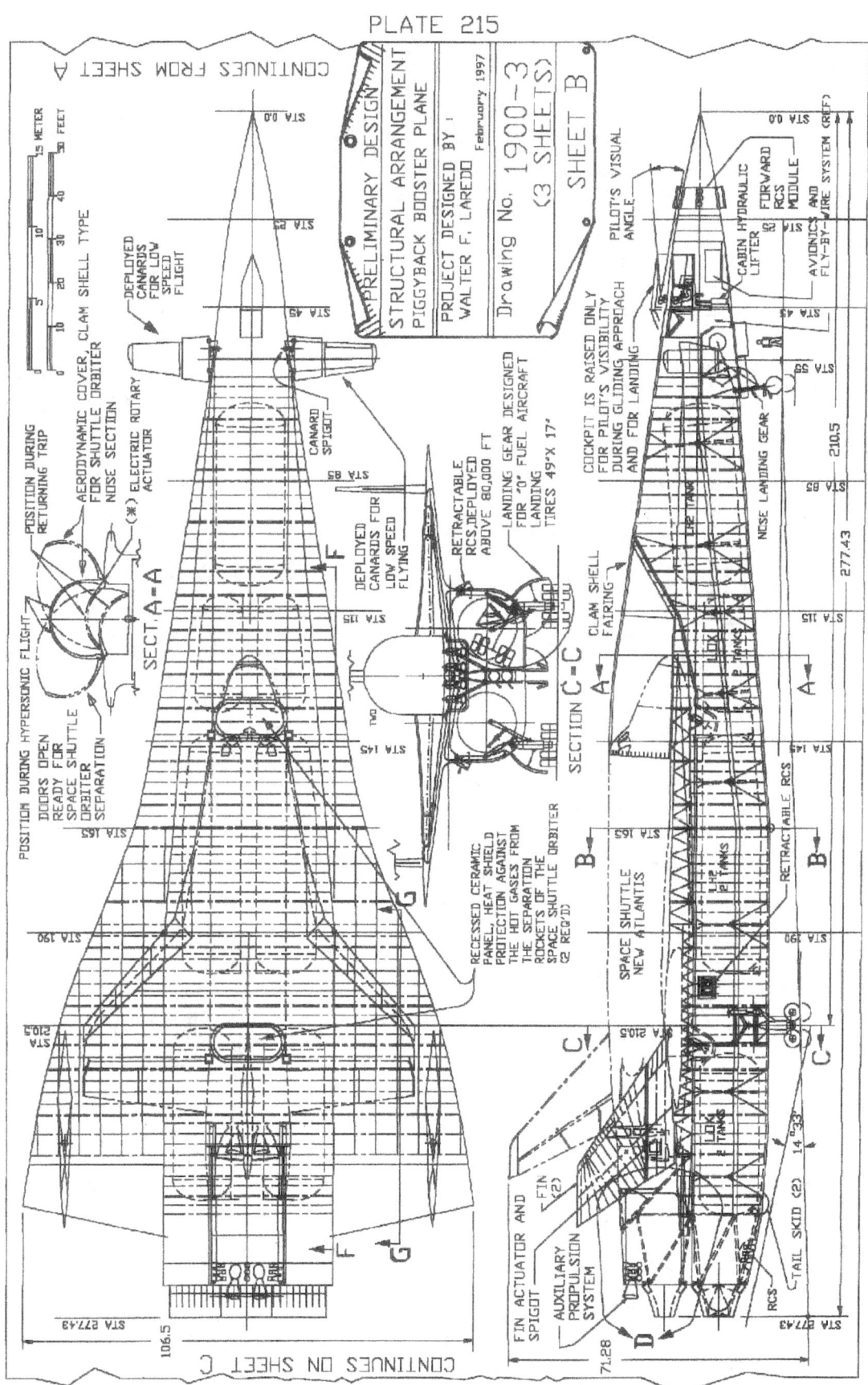

CONTINUES FROM SHEET A

PRELIMINARY DESIGN
STRUCTURAL ARRANGEMENT
PIGGYBACK BOOSTER PLANE
PROJECT DESIGNED BY:
WALTER F. LAREDO   February 1997
Drawing No. 1900-3
(3 SHEETS)
SHEET B

15 METER
50 FEET

POSITION DURING RETURNING TRIP

AERODYNAMIC COVER, CLAM SHELL TYPE
FOR SHUTTLE ORBITER NOSE SECTION

DEPLOYED CANARDS FOR LOW SPEED FLIGHT

(*) ELECTRIC ROTARY ACTUATOR

CANARD SPIGOT

POSITION DURING HYPERSONIC FLIGHT
DOORS OPEN READY FOR SPACE SHUTTLE ORBITER SEPARATION

SECT. A-A

STA 0.0
STA 25
STA 45
STA 85
STA 115
STA 145
STA 165
STA 190
STA 210.5
STA 277.43

DEPLOYED CANARDS FOR LOW SPEED FLYING

RETRACTABLE RCS.DEPLOYED ABOVE 80,000 FT

LANDING GEAR DESIGNED FOR "0" FUEL AIRCRAFT LANDING
TIRES 49"X 17"

SECTION C-C

RECESSED CERAMIC PANEL, HEAT SHIELD PROTECTION AGAINST THE HOT GASES FROM THE SEPARATION ROCKETS OF THE SPACE SHUTTLE ORBITER
(2 REQ'D)

106.5

CONTINUES ON SHEET C

PILOT'S VISUAL ANGLE

CABIN HYDRAULIC LIFTER

FORWARD RCS MODULE

AVIONICS AND FLY-BY-WIRE SYSTEM (REF)

STA 0.0
STA 25
STA 45
STA 55
STA 85
STA 115
STA 165
STA 190
STA 210.5
STA 277.43

COCKPIT IS RAISED ONLY FOR PILOT'S VISIBILITY DURING GLIDING APPROACH AND FOR LANDING

NOSE LANDING GEAR

210.5

277.43

CLAM SHELL FAIRING

LOX TANK

LH2 TANK

LO2 TANKS

LH2 TANKS

RETRACTABLE RCS

SPACE SHUTTLE NEW ATLANTIS

LOX TANKS

14.33

FIN (2)

AUXILIARY PROPULSION SYSTEM

FIN ACTUATOR AND SPIGOT

RCS

TAIL SKID (2)

71.28

CONTINUES ON SHEET C

PLATE 216

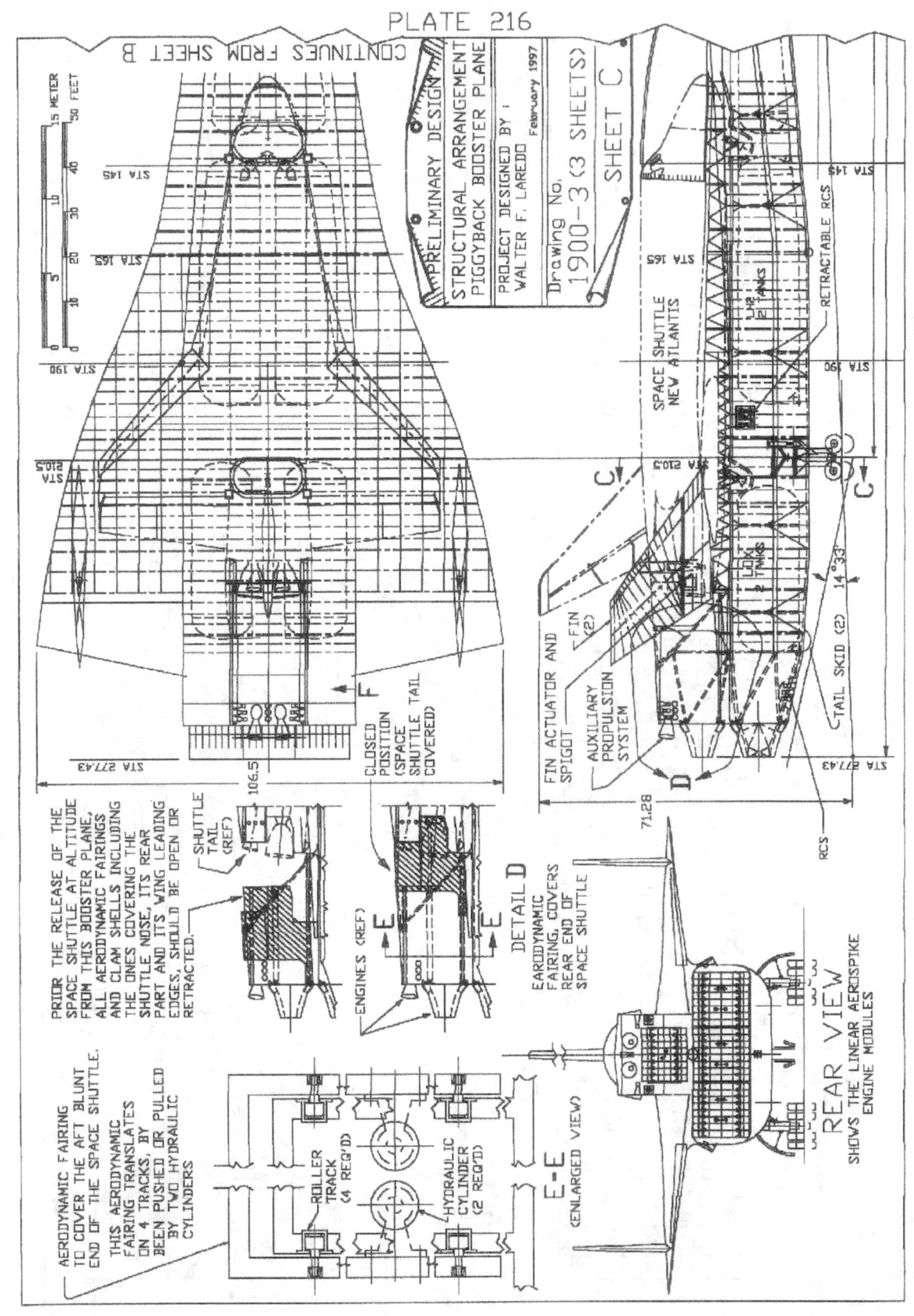

CONTINUES FROM SHEET B

PRELIMINARY DESIGN
STRUCTURAL ARRANGEMENT
PIGGYBACK BOOSTER PLANE

PROJECT DESIGNED BY :
WALTER F. LAREDO   February 1997

Drawing No.
1900-3 (3 SHEETS)

SHEET C

15 METER
50 FEET

STA 145
STA 165
STA 190
STA 210.5

SPACE SHUTTLE
NEW ATLANTIS

RETRACTABLE RCS

LH2 TANKS

LOX TANKS

FIN (2)

14°33'

FIN ACTUATOR AND SPIGOT

AUXILIARY PROPULSION SYSTEM

TAIL SKID (2)

RCS

STA 277.43

71.28

106.5

STA 277.43

CLOSED POSITION (SPACE SHUTTLE TAIL COVERED)

SHUTTLE TAIL (REF)

ENGINES (REF)

DETAIL D
AERODYNAMIC FAIRING, COVERS REAR END OF SPACE SHUTTLE

ROLLER TRACK (4 REQ'D)

HYDRAULIC CYLINDER (2 REQ'D)

E-E (ENLARGED VIEW)

REAR VIEW
SHOWS THE LINEAR AEROSPIKE ENGINE MODULES

AERODYNAMIC FAIRING TO COVER THE AFT BLUNT END OF THE SPACE SHUTTLE.

THIS AERODYNAMIC FAIRING TRANSLATES ON 4 TRACKS, BY BEEN PUSHED OR PULLED BY TWO HYDRAULIC CYLINDERS

PRIOR THE RELEASE OF THE SPACE SHUTTLE AT ALTITUDE FROM THIS BOOSTER PLANE, ALL AERODYNAMIC FAIRINGS AND CLAM SHELLS INCLUDING THE ONES COVERING THE SHUTTLE NOSE, ITS REAR PART AND ITS WING LEADING EDGES, SHOULD BE OPEN OR RETRACTED.

PLATE 217

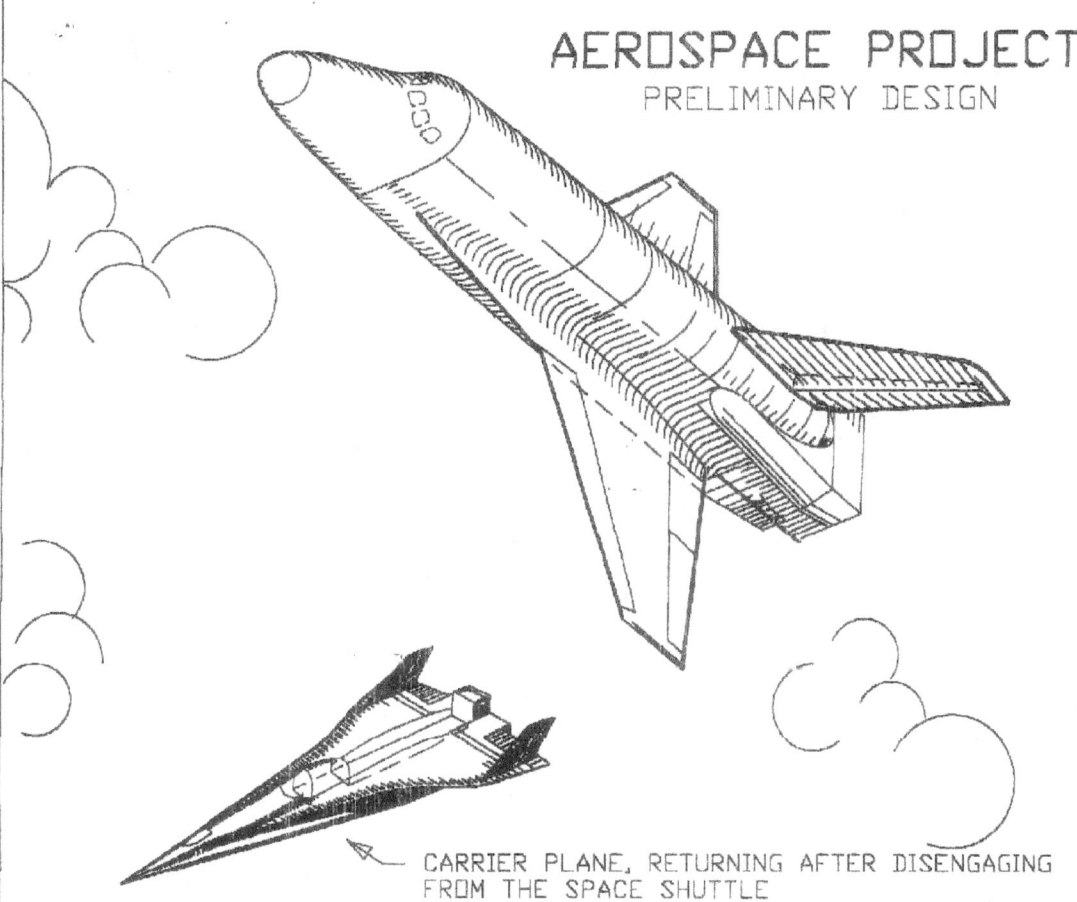

AEROSPACE PROJECT
PRELIMINARY DESIGN

CARRIER PLANE, RETURNING AFTER DISENGAGING
FROM THE SPACE SHUTTLE

NOTE: FOR CLARITY IN ALL FOLLOWING DRAWINGS WAS NOT
SHOWN THE VEHICLE'S EXTERNAL THERMOPROTECTION SYSTEM, ALSO
NOT SHOWN THE FOAM INSULATION THAT COVERS THE TANKS AND OTHER
CRYOGENIC COMPONENTS.

ADVANCED ENGINEERING PROJECTS

Name of project:
SPACE SHUTTLE, SPECIAL CONFIGURATION
TRANSPORTATION SYSTEM BETWEEN EARTH AND THE SPACE STATION

Design Engineer: WALTER F. LAREDO	Date: Feb. 1997
Drawing Number: 2000	Sheet 1 (TOTAL OF 11 SHEETS)

PLATE 218

MAIN PROPULSION SYSTEM
(2) ENGINES RATED AT :
115,080 LB THRUST IN THE VACUUM,
91,823 LB THRUST S.S.L.
DESIRED SPECIFIC IMPULSE = 450 SECONDS

PROPELLANT WEIGHT
LH2   13,138 LB
LO2   78,650 LB
TOTAL = 91,787 LB

RATIO OF TOTAL LIFTOFF WEIGHT TO 0.0 FUEL WEIGHT = 1.72

FOR FINAL DESIGN, REDUCE WING AREA AS REQUIRED

PRELIMINARY DESIGN
GENERAL ARRANGEMENT
SHUTTLE ORBITER
"NEW ATLANTIS"

VEHICLE DESIGNED TO COMMUTE INTERPLANETARY TRAVELERS BETWEEN EARTH AND A SPACE STATION USED AS SPACEPORT FOR SPACESHIP

PROJECT DESIGNED BY :
WALTER F. LAREDO
February 1997

Drawing No. 2000   TOTAL OF 11 SHEETS

SHEET 2A

TOTAL NUMBER OF TRAVELLERS= 46
(2 CREW + 44 PASSENGERS)

VERTICAL STABILIZER (Vs) :
AREA   380 SQ FT
lvs   20 FT

VOL. COEFFICIENT = $\bar{V}s$

$\bar{V}s = \dfrac{lvs \times Svs}{b \times Sw} = 0.036$

DESIGN C.G. RANGE
WITH 0 FUEL :
FROM 22 TO 28 % M.A.C.
BEFORE IGNITION :
FROM 33 TO 37 % M.A.C.

DESIGN SPEEDS :
MAX. MACH 26
MIN. LANDING 175 MPH

WING : COULD BE USE SAME WING AS NASA SPACE SHUTTLE.
ASPECT RATIO   2.26
AREA   2690 SQ FT
SWEEP   45 DEGREES
M.A.C. (IN.)   474.8
DIHEDRAL (T.E.) 3 90'

ORBITER WEIGHTS :
LAUNCH   157,000 LB
LANDING   65,000 LB
DESIGN PAYLOAD   15,000 LB

MAXIMUM T.P.S. TEMPERATURES
2300 degF FOR RCC , USED AT NOSE AND WING L.E.
1200 degF TO 2300 degF FOR HRST TILES LOCATED AT UNDERSIZE OF VEHICLE

(COMPONENT'S LIST, CONTINUES FROM SHEET 2B)

ORBITAL MANEUVERING SYSTEM (OMS)
18. OMS ENGINE (2)
19. OMS FUEL(MMH), 2 TANKS
20. OMS HELIUM, 2 TANKS
21. OMS N2O4, 2 TANKS

REACTION CONTROL SYSTEM (RCS)
22. AFT RCS MODULE (2)
23. AFT RCS HELIUM BOTTLE
24. AFT RCS N2O4, 2 TANKS
25. AFT RCS MMH, 2 TANKS
26. NOSE RCS MODULE

CABIN AREA
27. PRESSURIZED CHAMBER
28. AVIONIC BAY AND VARIOUS KINDS OF EQUIPMENT, INCLUDING THE ENVIRONMENT CONTROL SYSTEM.
29. AIR LOCK
30. INTERNAL AIR LOCK HATCH
31. EXTERNAL ACCESS, AIRLOCK HATCH
32. TOILET OF SPECIAL DESIGN
33. WARDROBE
34. AVIONIC BAY
35. COCKPIT
36. OUTER ACCESS HATCH
37. LADDER
38. FLOOR BEAM SUPPORT COLUMNS
39. MAIN LANDING GEAR
40. BODY FLAP
41. NOSE LANDING GEAR
42. BALLAST BLOCK FROM DEPLETED URANIUM, 2 REQ'D
43. NOSE CAP
44. EXPENDABLE AFT SEPARATION MOTORS, 4 REQ'D
45. EXPENDABLE FORWARD SEPARATION MOTORS, 4 REQUIRED
46. WING
47. FUSELAGE
48. TAIL SKID

CONTINUES ON SHEET 2B

PLATE 219

CONTINUES FROM SHEET 2A

LIST OF COMPONENTS

MAIN PROPULSION SYSTEM

1. MAIN ROCKET ENGINE (2)
2. LO2 FILL AND DRAIN AREA
3. LH2 MANIFOLD
4. LO2 MANIFOLD
5. LH2 LOW PRESSURE PUMP
6. LH2 FILL AND DRAIN LINE
7. LO2 PRE-VALVE
8. LO2 LOW PRESSURE PUMP
9. LH2 PRE-VALVE
10. LO2 LINE
11. LH2 TANK VENT
12. LO2 TANK VENT
13. MAIN LH2 TANK
14. MAIN LO2 TANK
15. 2 HELIUM TANKS
16. PITCH ACTUATOR
17. YAW ACTUATOR

(CONTINUES IN SHEET 2A)

PRELIMINARY DESIGN

GENERAL ARRANGEMENT
SHUTTLE ORBITER
"NEW ATLANTIS"

VEHICLE DESIGNED TO COMMUTE
INTERPLANETARY TRAVELERS
BETWEEN EARTH AND A
SPACE STATION (SPACEPORT)

PROJECT DESIGNED BY :
WALTER F. LAREDO

February 1997

Drawing No.   TOTAL OF
2000   11 SHEETS

SHEET 2B

NOTES:

1. ENLARGED VIEWS FROM DETAIL DRAWINGS
   IN SHEET 1C ARE SHOWN IN FOLLOWING
   SHEETS.

2. FOR CLARITY WAS NOT SHOWN
   IN THESE DRAWINGS THE
   VEHICLE'S EXTERNAL THERMO
   PROTECTION SYSTEM AND THE
   INTERNAL HEAT INSULATION

3. TO AVOID DUPLICATION OF DEVELOPMENTAL WORK, FOR SOME
   COMPONENTS, COULD BE USE THE ONES FROM THE EXISTING
   NASA'S SPACE SHUTTLE ORBITERS, THOSE COMPONENTS ARE
   THE FOLLOWING:

   NOSE CAP, MAIN AND NOSE LANDING GEARS,
   NOSE REACTION CONTROL SYSTEM (RCS) AND THE
   OUTBOARD WING STRUCTURE, WITHOUT INCLUDING
   THE WING CARRY-THROUGH STRUCTURE.

4. THERE IS SOME RESEMBLANCE BETWEEN THE GENERAL
   ARRANGEMENT OF THIS VEHICLE AND THE GENERAL
   ARRANGEMENTS OF NASA'S SPACE SHUTTLE ORBITER,
   BOTH VEHICLES ARE ABOUT THE SAME SIZE.
   IN THIS VEHICLE THE WING AND THE MAIN LANDING
   GEAR WAS SHIFT TWO FEET FORWARD, IN ORDER TO
   IMPROVE ITS AERODYNAMIC STABILITY FOR SOME FLIGHT
   CONDITIONS, INCLUDING DURING THE RE-ENTRY, SUPERSONIC
   AND SUBSONIC GLIDING, FOLLOWED BY LOW SPEED
   APPROACH AND LANDING.

5. THE EXTERNAL COVER OF THE AFT RCS MODULE WAS DESIGNED
   FOR MINIMUM HYPERSONIC DRAG DURING ASCEND.

6. DEPLETED URANIUM IS USED AS A BALLAST TO BALANCE
   THE VEHICLE.

CONTINUES ON SHEET 2C

PLATE 220

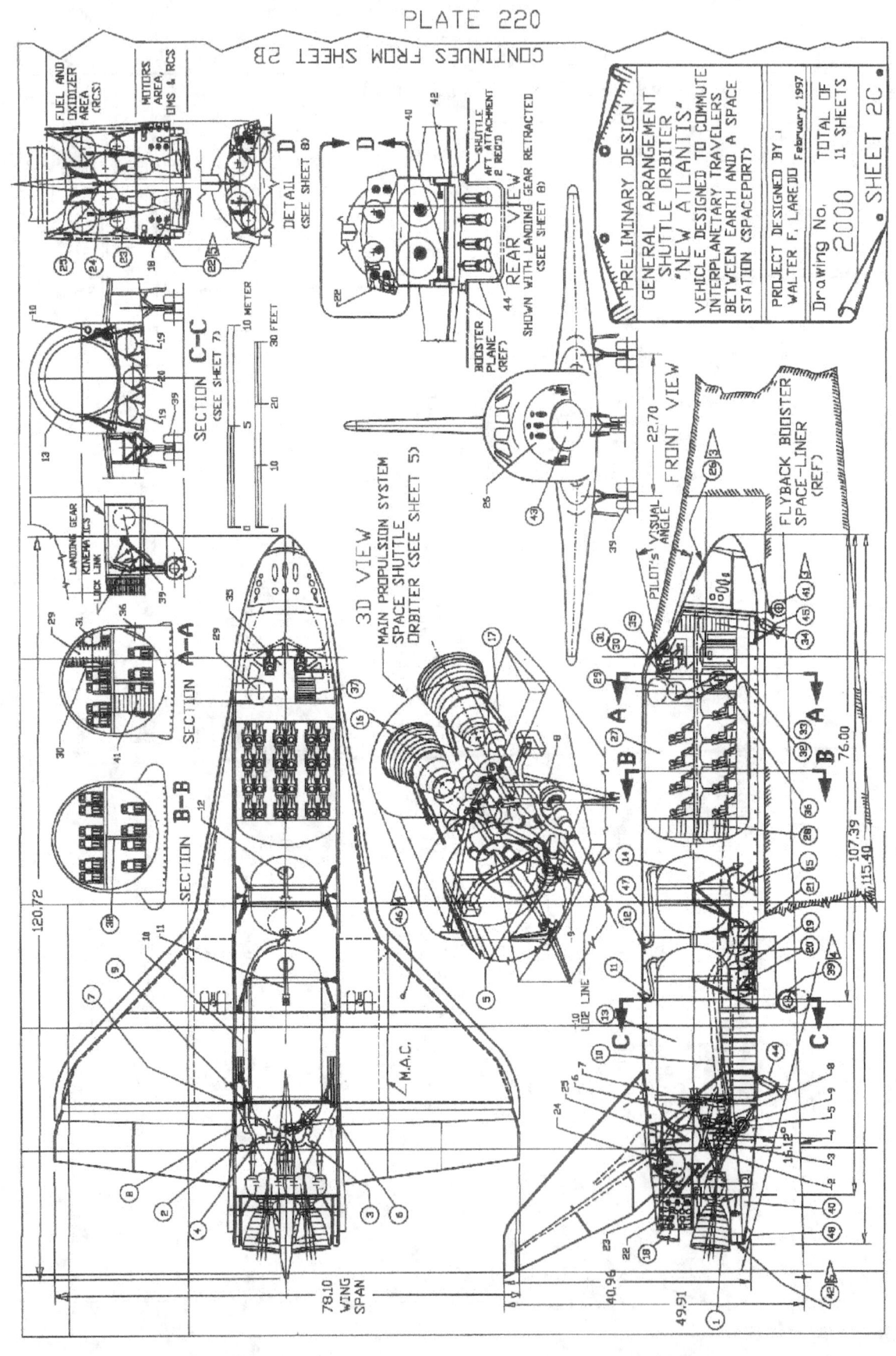

CONTINUES FROM SHEET 2B

FUEL AND OXIDIZER AREA (RCS)

MOTORS AREA, OMS & RCS

DETAIL D
(SEE SHEET 8)

REAR VIEW
(SEE SHEET 8)
SHUTTLE AFT ATTACHMENT 2 REQ'D
SHOWN WITH LANDING GEAR RETRACTED
BOOSTER PLANE (REF)

SECTION C-C
(SEE SHEET 7)

10 METER
30 FEET

SECTION A-A

SECTION B-B

LANDING GEAR KINEMATICS
LOCK LINK

3D VIEW
MAIN PROPULSION SYSTEM
SPACE SHUTTLE ORBITER (SEE SHEET 5)

FRONT VIEW
22.70
PILOT'S VISUAL ANGLE

FLYBACK BOOSTER SPACE-LINER (REF)

76.00
107.39
115.40

M.A.C.

W.L.O2 LINE

16.12°

120.72

78.10 WING SPAN

40.96

49.91

PRELIMINARY DESIGN
GENERAL ARRANGEMENT
SHUTTLE ORBITER
"NEW ATLANTIS"
VEHICLE DESIGNED TO COMMUTE
INTERPLANETARY TRAVELERS
BETWEEN EARTH AND A SPACE
STATION (SPACEPORT)

PROJECT DESIGNED BY :
WALTER F. LAREDO   February 1997

Drawing No.   TOTAL OF
2000   11 SHEETS

SHEET 2C

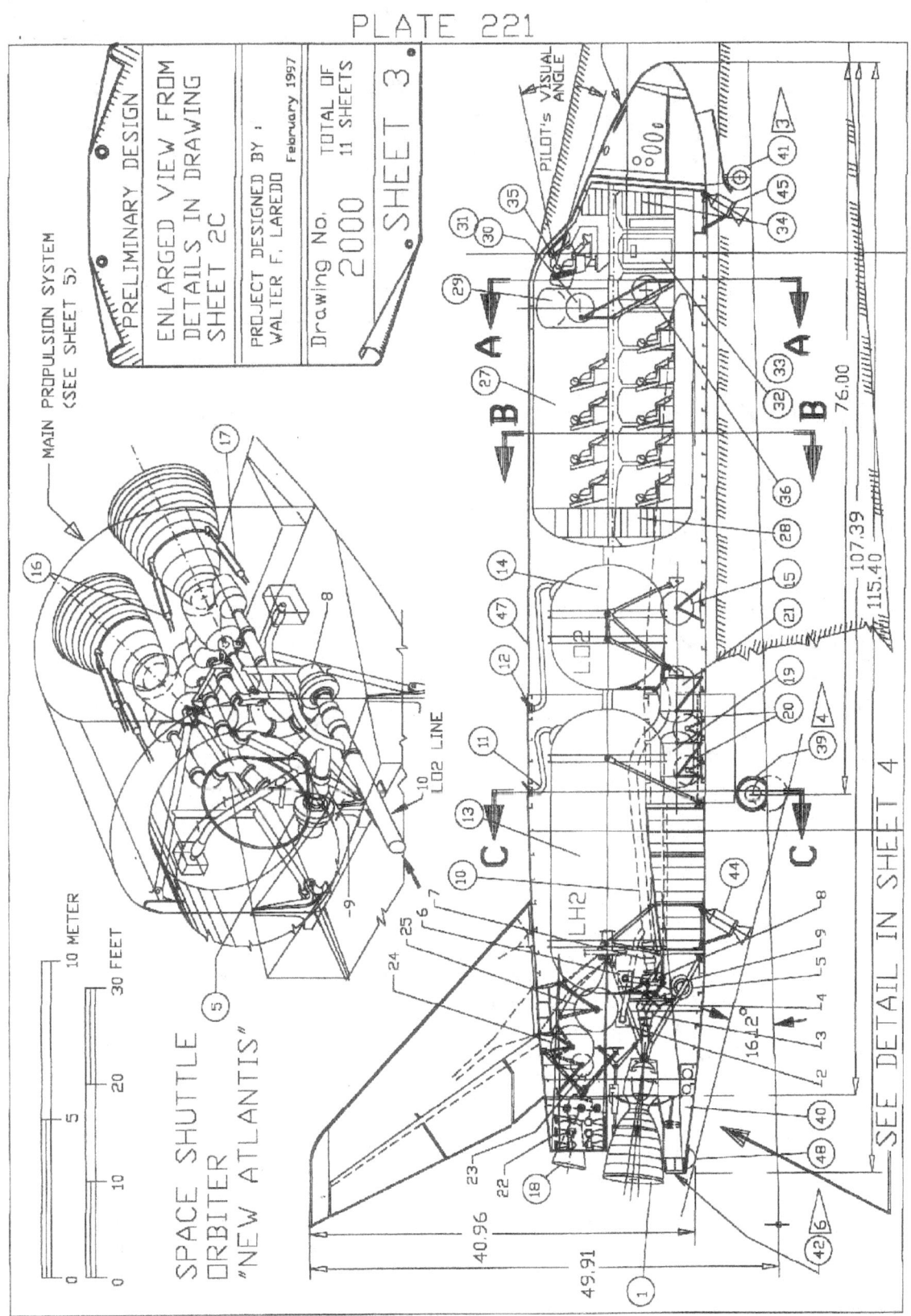

PLATE 221

PRELIMINARY DESIGN

ENLARGED VIEW FROM
DETAILS IN DRAWING
SHEET 2C

PROJECT DESIGNED BY :
WALTER F. LAREDO   February 1997

Drawing No.   TOTAL OF
2000   11 SHEETS

SHEET 3

MAIN PROPULSION SYSTEM
(SEE SHEET 5)

SPACE SHUTTLE
ORBITER
"NEW ATLANTIS"

10 METER
30 FEET

LO2 LINE

LO2

LH2

PILOT's VISUAL ANGLE

SEE DETAIL IN SHEET 4

76.00
107.39
115.40

40.96
49.91

16.2°

PLATE 222

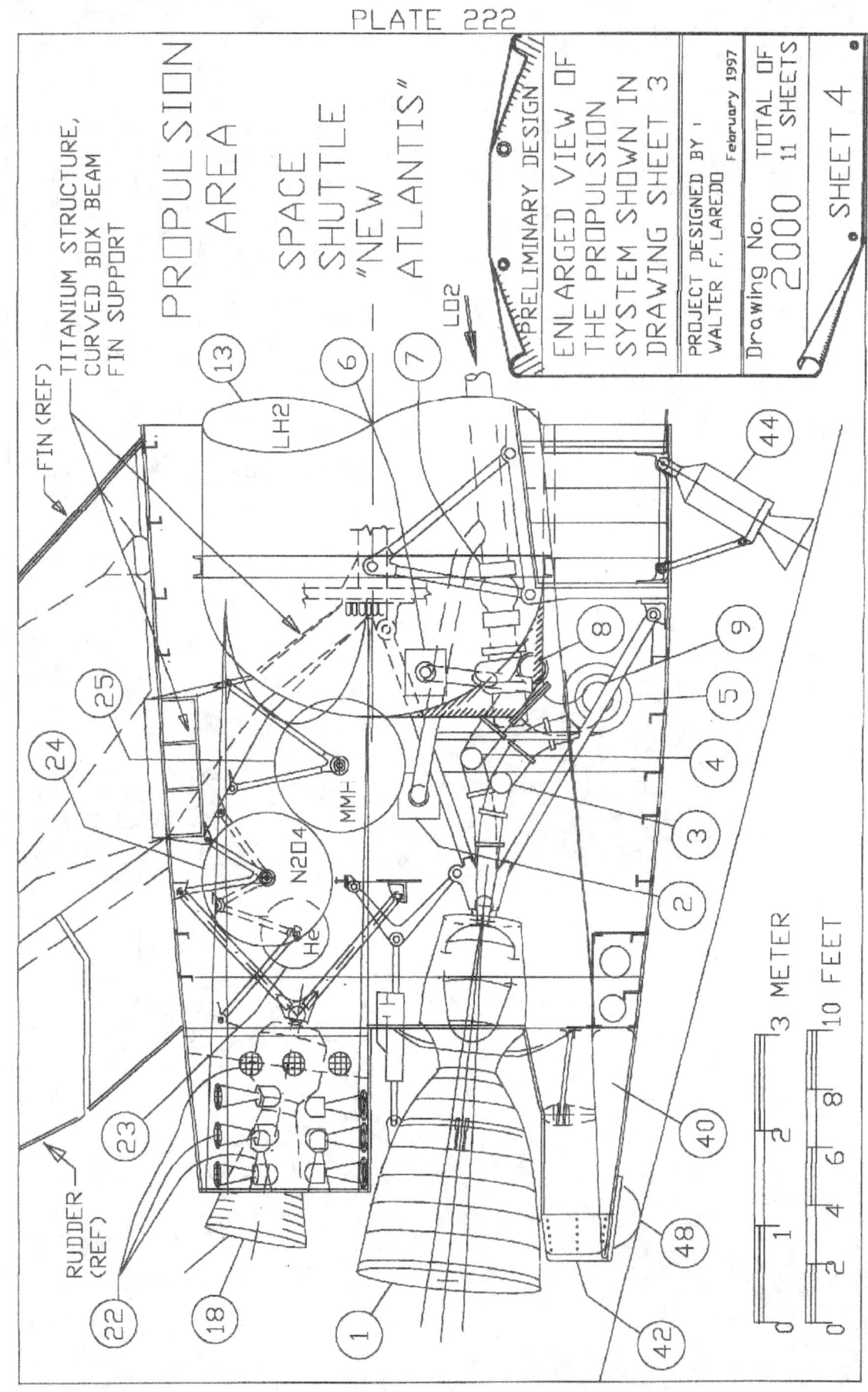

PROPULSION AREA

SPACE SHUTTLE "NEW ATLANTIS"

TITANIUM STRUCTURE, CURVED BOX BEAM FIN SUPPORT

FIN (REF)

RUDDER (REF)

LH2

MMH

N2O4

He

LO2

PRELIMINARY DESIGN

ENLARGED VIEW OF THE PROPULSION SYSTEM SHOWN IN DRAWING SHEET 3

PROJECT DESIGNED BY : WALTER F. LAREDO

February 1997

Drawing No. 2000   TOTAL OF 11 SHEETS

SHEET 4

METER

FEET

PLATE 223

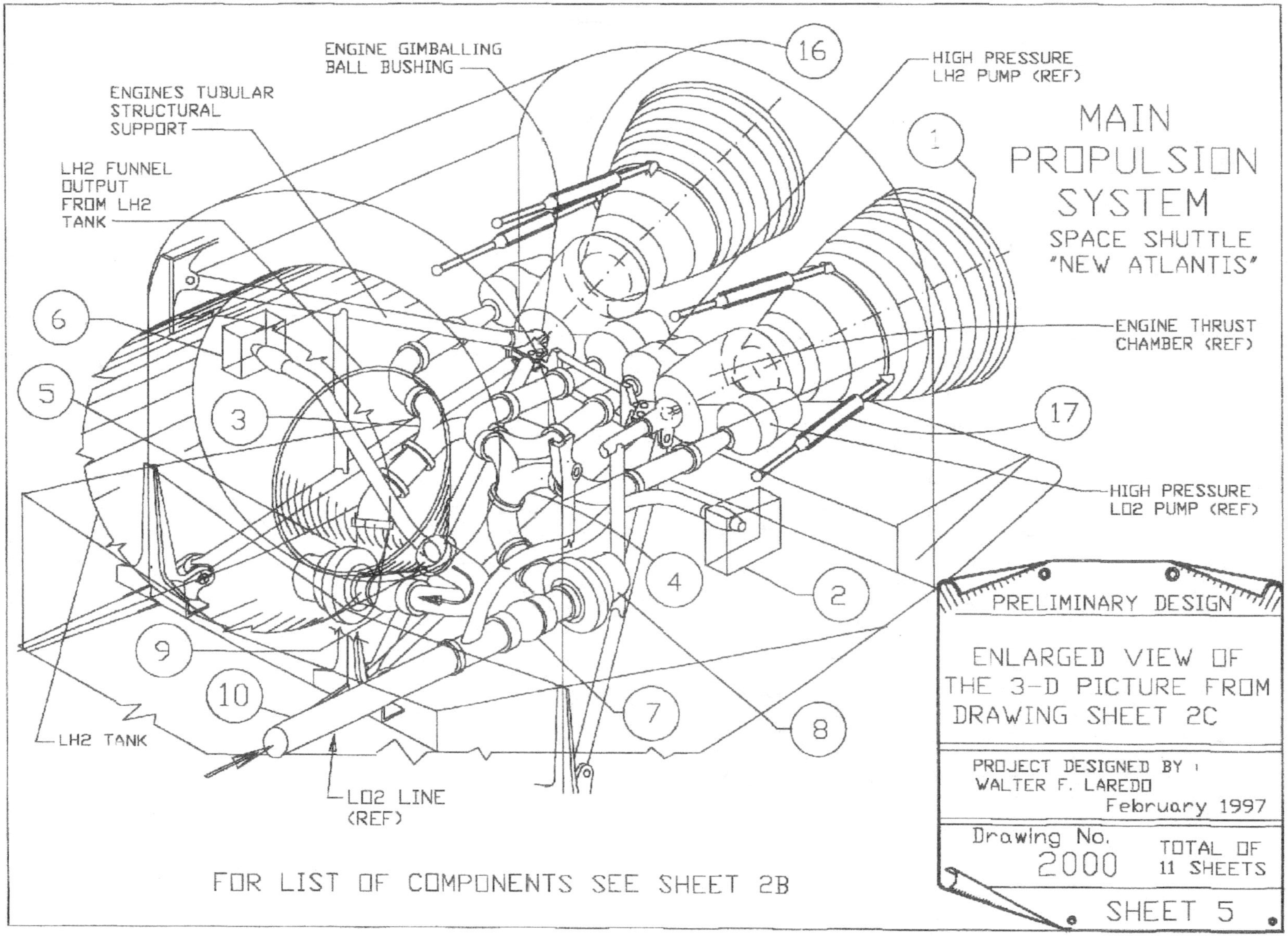

MAIN
PROPULSION
SYSTEM
SPACE SHUTTLE
"NEW ATLANTIS"
ENGINE GIMBALLING
BALL BUSHING
ENGINES TUBULAR
STRUCTURAL
SUPPORT
LH2 FUNNEL
OUTPUT
FROM LH2
TANK
HIGH PRESSURE
LH2 PUMP (REF)
ENGINE THRUST
CHAMBER (REF)
HIGH PRESSURE
LO2 PUMP (REF)
LH2 TANK
LO2 LINE
(REF)
16
1
6
5
3
17
4
2
9
10
7
8
PRELIMINARY DESIGN
ENLARGED VIEW OF
THE 3-D PICTURE FROM
DRAWING SHEET 2C
PROJECT DESIGNED BY :
WALTER F. LAREDO
February 1997
Drawing No.
2000
TOTAL OF
11 SHEETS
SHEET 5
FOR LIST OF COMPONENTS SEE SHEET 2B

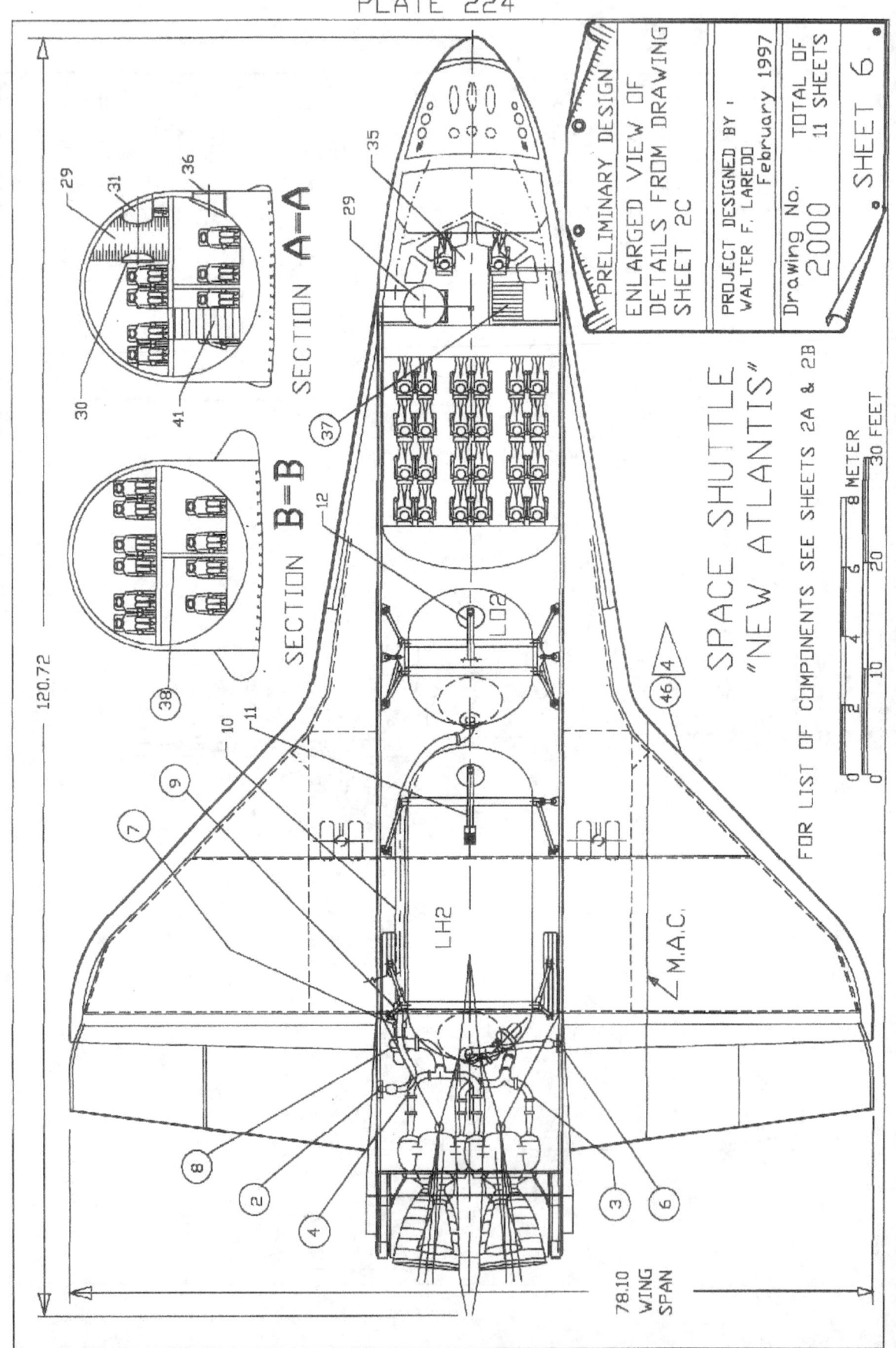

PLATE 224

SECTION A-A

SECTION B-B

SPACE SHUTTLE
"NEW ATLANTIS"

FOR LIST OF COMPONENTS SEE SHEETS 2A & 2B

PRELIMINARY DESIGN

ENLARGED VIEW OF
DETAILS FROM DRAWING
SHEET 2C

PROJECT DESIGNED BY :
WALTER F. LAREDO
February 1997

Drawing No.   TOTAL OF
2000   11 SHEETS

SHEET 6

120.72

78.10
WING
SPAN

M.A.C.

LH2

LO2

METER

FEET

PLATE 225

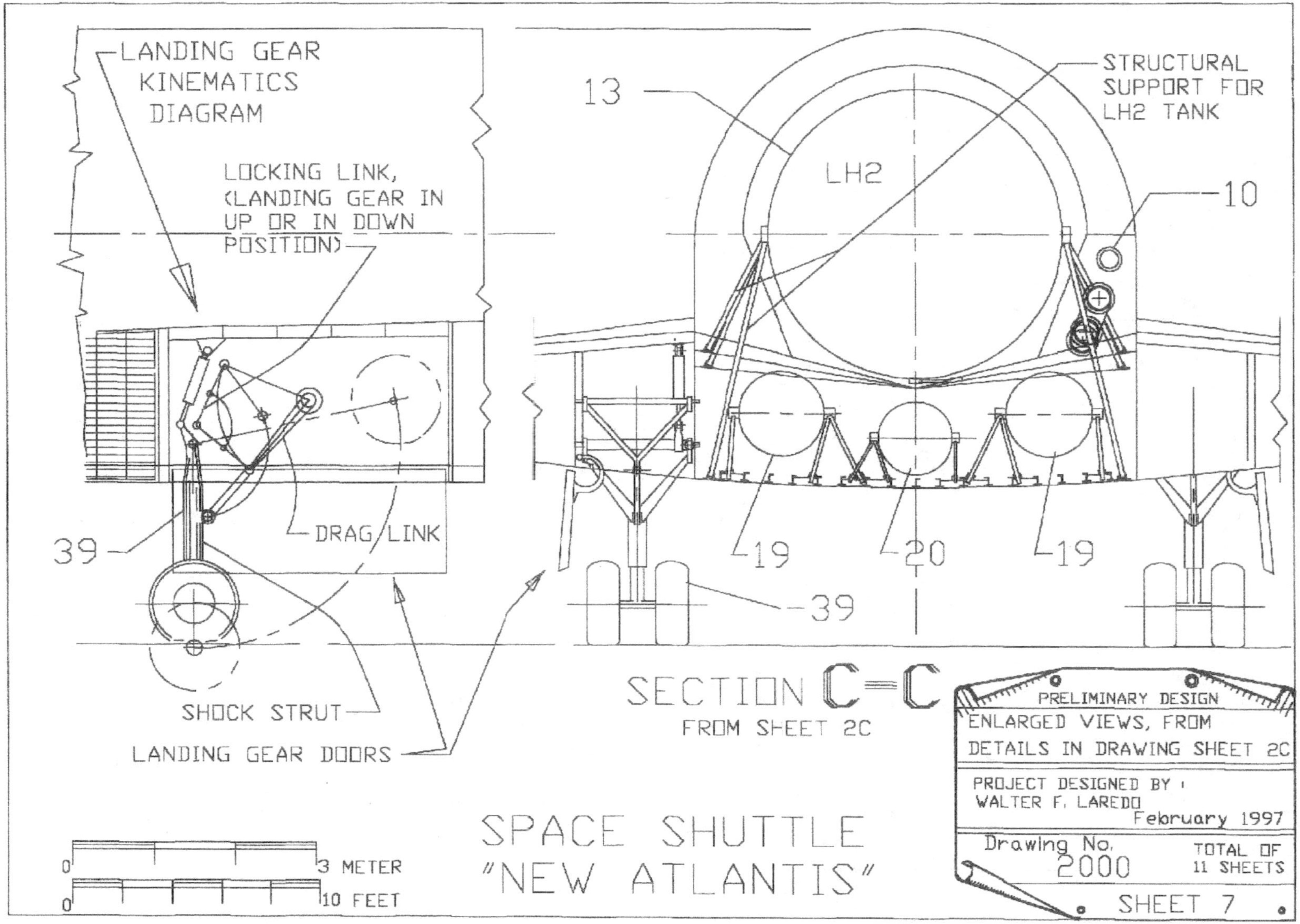

STRUCTURAL
SUPPORT FOR
LH2 TANK

—10

—19

LH2

—20

13 —

—39

—19

SECTION C—C
FROM SHEET 2C

PRELIMINARY DESIGN
ENLARGED VIEWS, FROM
DETAILS IN DRAWING SHEET 2C

PROJECT DESIGNED BY :
WALTER F. LAREDO
February 1997

Drawing No.   TOTAL OF
2000   11 SHEETS

SHEET 7

SPACE SHUTTLE
"NEW ATLANTIS"

LANDING GEAR
KINEMATICS
DIAGRAM

LOCKING LINK,
(LANDING GEAR IN
UP OR IN DOWN
POSITION)

DRAG LINK

SHOCK STRUT

LANDING GEAR DOORS

39

3 METER
10 FEET

0
0

PLATE 226

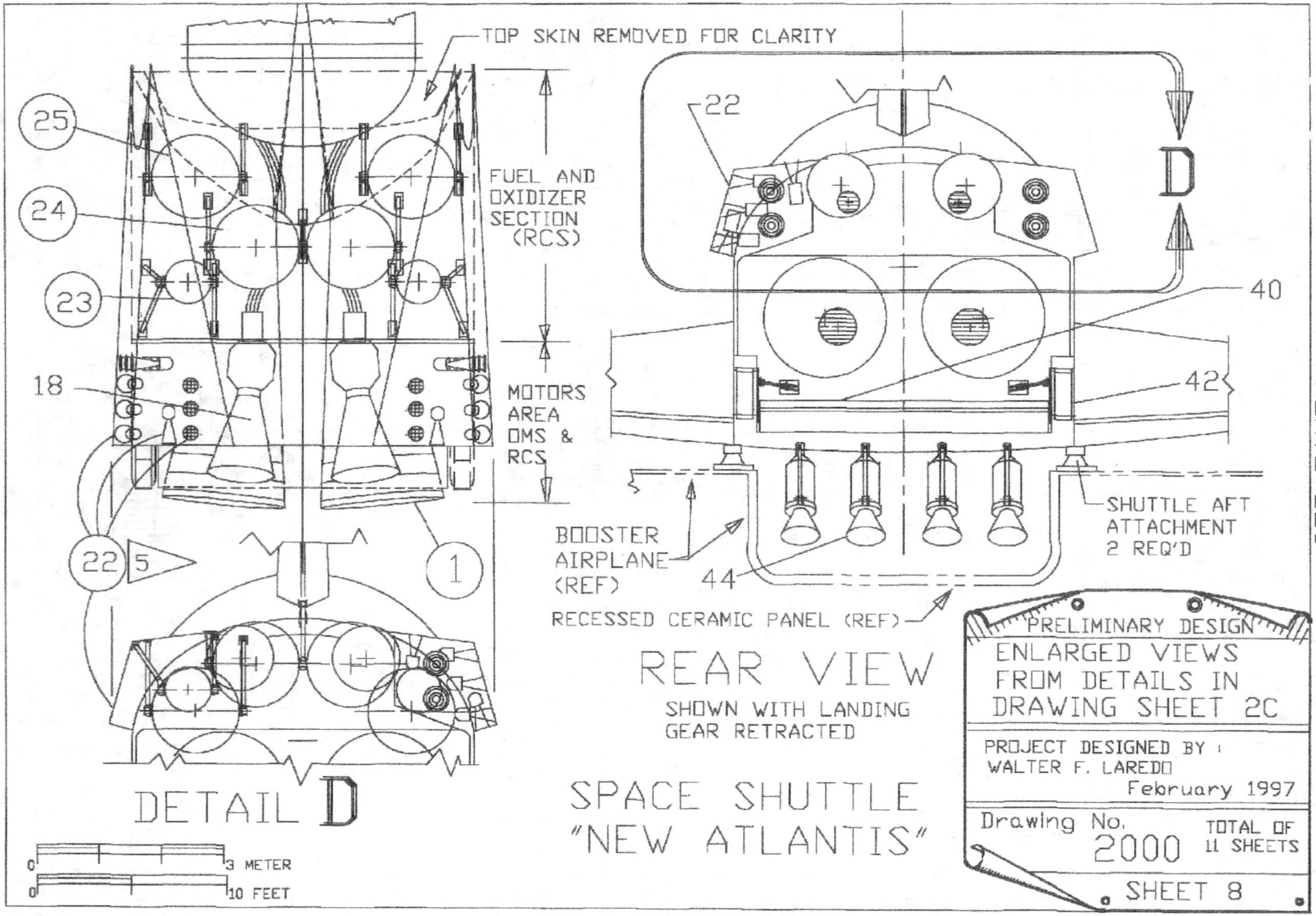

40

42

SHUTTLE AFT
ATTACHMENT
2 REQ'D

22

TOP SKIN REMOVED FOR CLARITY

44

RECESSED CERAMIC PANEL (REF)

BOOSTER
AIRPLANE
(REF)

REAR VIEW

SHOWN WITH LANDING
GEAR RETRACTED

FUEL AND
OXIDIZER
SECTION
(RCS)

MOTORS
AREA
OMS &
RCS

1

25

24

23

18

22 5

DETAIL D

3 METER

10 FEET

0

SPACE SHUTTLE
"NEW ATLANTIS"

PRELIMINARY DESIGN

ENLARGED VIEWS
FROM DETAILS IN
DRAWING SHEET 2C

PROJECT DESIGNED BY :
WALTER F. LAREDO
February 1997

Drawing No.
2000

TOTAL OF
11 SHEETS

SHEET 8

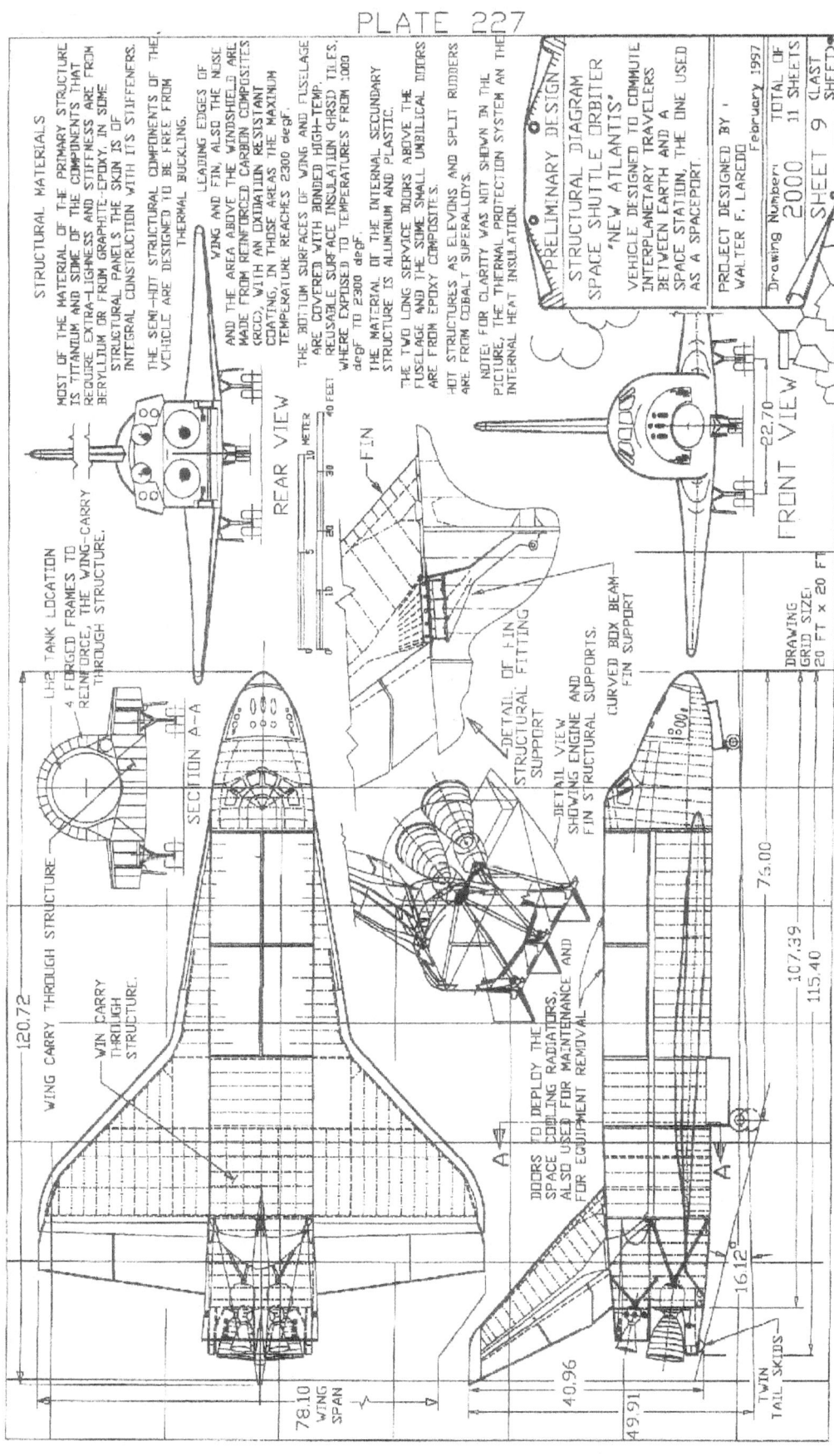

PLATE 227

STRUCTURAL MATERIALS

MOST OF THE MATERIAL OF THE PRIMARY STRUCTURE IS TITANIUM AND SOME OF THE COMPONENTS THAT REQUIRE EXTRA-HIGHNESS AND STIFFNESS ARE FROM BERYLLIUM OR FROM GRAPHITE-EPOXY. IN SOME STRUCTURAL PANELS THE SKIN IS OF INTEGRAL CONSTRUCTION WITH ITS STIFFENERS.

THE SEMI-HOT STRUCTURAL COMPONENTS OF THE VEHICLE ARE DESIGNED TO BE FREE FROM THERMAL BUCKLING.

LEADING EDGES OF WING AND FIN, ALSO THE NOSE AND THE AREA ABOVE THE WINDSHIELD ARE MADE FROM REINFORCED CARBON COMPOSITES (RCC), WITH AN OXIDATION RESISTANT COATING, IN THOSE AREAS THE MAXIMUM TEMPERATURE REACHES 2300 degF.

THE BOTTOM SURFACES OF WING AND FUSELAGE ARE COVERED WITH BONDED HIGH-TEMP. REUSABLE SURFACE INSULATION (HRSI) TILES, WHERE EXPOSED TO TEMPERATURES FROM 1000 degF TO 2300 degF.

THE MATERIAL OF THE INTERNAL SECONDARY STRUCTURE IS ALUMINUM AND PLASTIC.

THE TWO LONG SERVICE DOORS ABOVE THE FUSELAGE AND THE SOME SMALL UMBILICAL DOORS ARE FROM EPOXY COMPOSITES.

HOT STRUCTURES AS ELEVONS AND SPLIT RUDDERS ARE FROM COBALT SUPERALLOYS.

NOTE: FOR CLARITY WAS NOT SHOWN IN THE PICTURE, THE THERMAL PROTECTION SYSTEM AN THE INTERNAL HEAT INSULATION.

REAR VIEW

FIN

10 METER
40 FEET
30
20
10
5
0

PRELIMINARY DESIGN

STRUCTURAL DIAGRAM
SPACE SHUTTLE ORBITER
"NEW ATLANTIS"

VEHICLE DESIGNED TO COMMUTE INTERPLANETARY TRAVELERS BETWEEN EARTH AND A SPACE STATION, THE ONE USED AS A SPACEPORT.

February 1997

PROJECT DESIGNED BY :
WALTER F. LAREDO

Drawing Number: 2000

TOTAL OF 11 SHEETS

SHEET 9 (LAST SHEET)

FRONT VIEW

22.70

LH2 TANK LOCATION

4 FORGED FRAMES TO REINFORCE, THE WING-CARRY THROUGH STRUCTURE.

WING CARRY THROUGH STRUCTURE

WIN CARRY THROUGH STRUCTURE.

SECTION A-A

DETAIL OF FIN STRUCTURAL FITTING SUPPORT

DETAIL VIEW SHOWING ENGINE AND FIN STRUCTURAL SUPPORTS.

CURVED BOX BEAM FIN SUPPORT

DOORS TO DEPLOY THE SPACE COOLING RADIATORS, ALSO USED FOR MAINTENANCE AND FOR EQUIPMENT REMOVAL

DRAWING GRID SIZE: 20 FT x 20 FT

75.00

107.39

115.40

120.72

A

A

16.12

78.10 WING SPAN

40.96

49.91

TWIN TAIL SKIDS